ORIGINS

ORIGINS

THE LIVES
AND WORLDS
OF MODERN
COSMOLOGISTS

ALAN LIGHTMAN
AND
ROBERTA BRAWER

HARVARD UNIVERSITY PRESS

CAMBRIDGE, MASSACHUSETTS

Copyright © 1990 by Alan Lightman and Roberta Brawer
All rights reserved
Printed in the United States of America
10   9   8   7   6   5   4   3   2

First Harvard University Press paperback edition, 1992

Library of Congress Cataloging-in-Publication Data

Lightman, Alan P., 1948–
  Origins : the lives and worlds of modern cosmologists /
Alan Lightman and Roberta Brawer.
      p.   cm.
   Contains interviews with 27 persons.
   Includes bibliographical references and index.
   ISBN 0-674-64470-0 (alk. paper) (cloth)
   ISBN 0-674-64471-9 (paper)
   1. Cosmology—Miscellanea. 2. Astronomers—Biography.
I. Brawer, Roberta. II. Title.
QB981.L54 1990
523.1—dc20

90-4623
CIP

CONTENTS

Preface vii
Introduction to Modern Cosmology 1

51 FRED HOYLE
67 ALLAN SANDAGE
85 GÉRARD DE VAUCOULEURS
102 MAARTEN SCHMIDT
120 WALLACE SARGENT
136 DENNIS SCIAMA
154 MARTIN REES
170 ROBERT WAGONER
186 JOSEPH SILK
201 ROBERT DICKE
214 JAMES PEEBLES
232 CHARLES MISNER
250 JAMES GUNN
266 JEREMIAH OSTRIKER
285 VERA RUBIN
306 EDWIN TURNER
324 SANDRA FABER
341 MARC DAVIS
359 MARGARET GELLER
378 JOHN HUCHRA
395 STEPHEN HAWKING
399 DON PAGE
415 ROGER PENROSE
435 DAVID SCHRAMM
451 STEVEN WEINBERG
467 ALAN GUTH
483 ANDREI LINDE

Dates and Places of Interviews 497
Note on Citations and References 498
Notes 499
General Reading in Cosmology 512
References 513
Glossary 531
Illustration Credits 562

PREFACE

Origins explores the ways in which personal, philosophical, and social factors enter the scientific process. The role of such factors has been increasingly illuminated by recent scholarship in the history, philosophy, and sociology of science. While previous studies have often been retrospective or historical, our project uses direct interviews with active scientists, all cosmologists. Cosmology is a rich subject for such a study. Perhaps the grandest and most speculative of all sciences, cosmology sits at the boundary between what is considered science and what is not. Furthermore, cosmological thinking has recently been shaken by an explosion of new theories and observations. The interviews help document the history and impact of these new ideas, at a time when the dust hasn't settled.

This preface explains the goals of the project and the aim of the interview questions. The Introduction provides background to the subject of cosmology itself.

In the last 15 years, a revolution has occurred in cosmology—associated in part with the application of subatomic physics to theories of the beginning of the universe and with new observations of the locations and motions of galaxies—and cosmologists have reframed their subject and the questions they ask. In such a time of upheaval, the human side of science is more easily seen. Accordingly, we have restricted our interviews to cosmologists and have focused on a common set of questions, many about current attitudes. Thus, the interviews are more directed than traditional oral histories. Our own background as physicists has allowed us to engage the scientists on their own turf, so that nonscientific factors in the scientific process can be probed in their natural context. In these interviews, the scientists discuss themselves and their thinking in ways they do not in their papers and lectures. The strikingly different personalities brought out here clearly show that there is no single scientific personality or single approach to science.

Each interview consists of three parts. The first part focuses on the personal experiences and work of the individual scientist; the second deals with the interviewee's reactions to recent developments in cosmology.

The final two questions ask the interviewees to put aside their natural scientific caution and talk philosophically about their personal feelings toward their subject.

The questions begin with childhood experiences and then go on to explore education and early attitudes about cosmology. What attracted these people to science? What books were influential? What role did their parents or others play? Nearly all these individuals knew they wanted to become scientists at a young age, and many recall the same books. Furthermore, most felt that their parents were supportive of their scientific interests, despite a wide range of economic, social, and educational backgrounds. Further questions explore experiences in college and graduate school and the reasons for deciding to specialize in cosmology. Early preferences for different models of the universe are probed, as well as how those preferences have shifted in time.

How do scientists choose the problems they work on and the questions they ask? To explore these issues, the scientists were asked about some of their own research. For example, Robert Dicke was a driving force in the prediction and interpretation of cosmic radio waves, which provided a critical confirmation of the big bang theory. Alan Guth and Andrei Linde proposed an important theoretical modification of the big bang theory called the inflationary universe model. Vera Rubin's observations of the rotational speeds within galaxies have provided compelling evidence for the existence of dark (unseen) matter. What motivated this work? What was the role of serendipity versus premeditation? What was the role of philosophical preference?

The second half of each interview concerns the interviewee's reactions to recent developments in cosmology. Much of modern cosmology has been shaped by new ideas from subatomic particle physicists, who ironically once considered cosmology too speculative a subject for their attention. What caused the recent union between cosmology and particle physics, and how do the scientists feel about it? During the same recent period, new results in observational cosmology have also been dramatic. Maps of the locations and motions of galaxies have revealed clustering and voids not previously imagined in the simple big bang model. Such observed lumpiness has caused some cosmologists to rethink their notions about the basic structure of the universe.

One product of the union of subatomic physics and cosmology has been a major modification of the big bang model called the inflationary universe model, proposed in 1980. According to this model, the very early universe went through a brief and rapid expansion, after which it returned to the more leisurely rate of expansion of the standard big bang model.

The inflationary universe model has been extremely influential, yet there is little observational evidence to support it. We investigate the reactions of scientists to this new cosmological model. Belief in the inflationary universe model requires belief in about ten times as much mass in the universe as has been detected, the so-called missing mass. Why do some cosmologists believe in the missing mass? Why do scientists accept a theory in apparent contradiction to the observations?

The attraction of the inflationary universe model comes in large part from its resolution of two outstanding difficulties with the standard big bang model: the so-called horizon and flatness problems. The horizon problem asks why the universe appears to be homogeneous over a much larger region than could reasonably be expected—unless it began that way. The flatness problem raises the question of why the universe began with its gravitational energy and its kinetic energy of expansion so closely balanced, like a rock thrown upward with almost precisely the minimum speed needed to escape the earth's gravity. The horizon problem had troubled many scientists long before the inflationary universe model. The flatness problem, however, gained broad recognition only *after* its proposed solution by the inflationary universe model.

What decides whether a question is scientific or not, worth worrying about or not? Why do some questions gain legitimacy only after their solution? The interviews explore these issues in part by investigating the range in attitudes about the horizon and flatness problems. Both are related to the problem of "initial conditions" in cosmology. In other fields of physics, the initial conditions of one experiment follow from the final conditions of a previous one, in a chain of determinism. The exception may be cosmology, which goes back to the first link in the chain, the origin of the universe. Are cosmologists willing to accept that the universe has certain properties, such as the close balance between cosmic gravitational energy and kinetic energy of expansion, simply because it began that way, by chance, or do they demand a fully deterministic explanation in terms of equations they can derive? Indeed, do questions about the initial conditions of the universe lie within the domain of science? Do observers differ from theorists in their attitude toward initial conditions and, more generally, their attitude toward what constitutes a legitimate scientific question? (Roughly speaking, observers in astronomy collect and analyze data from telescopes while theorists confine themselves to mathematical calculations.)

How do scientists respond to new empirical results that challenge their previous thinking? One line of inquiry here involves the interviewees' reactions to recent discoveries of large-scale patterns in the distribution of galaxies. Until recently, most maps of the locations of galaxies have been

2-dimensional, as on a photograph. However, recent advances in technology have allowed detailed surveys of the 3-dimensional locations of galaxies, revealing large clusters of galaxies, long chains of galaxies, and huge voids in space with no galaxies at all. How have these discoveries been received by the community? A fundamental assumption of the big bang model, called the cosmological principle, states that the universe is homogeneous on very large scales. Are the newly observed structures causing cosmologists to question this basic principle?

We are also concerned with the role of individual style in the scientific process. For example, some scientists approach science as a process of model building and exploration of powerful principles, while others see science as discovery of the facts of nature. (This difference in approach cuts across the categories of theorist and observer.) The varied reactions of the interviewees to the flatness problem, the missing mass, and related issues help identify these two different approaches. Sometimes, the tension between the two approaches is apparent within a single scientist.

In ending, the scientists are asked: "If you could design the universe any way that you wanted, how would you do it?" Many cosmologists include life in their ideal universe. Traditionally, modern physicists have not attributed any special significance to the evolution of life in their pursuit of the laws of nature. However, some leading physicists have recently suggested that if our universe were slightly different, it would not permit the existence of life, and they have used this hypothesis as a basis for scientific arguments. How do scientists feel about such arguments? In a related question, the cosmologists are asked their opinion about whether the universe has any purpose.

Our questions, in general form, are not new to the study of how scientists think and work. We have not tried to analyze the interviews here. Rather, in this book, we wish to let the interviews speak for themselves.

Finally, a few words about methodology. For practical reasons, we could not include all cosmologists doing important work. The interviewees were chosen to cover a range of interests and ages, with a rough balance between observers and theorists. Many of the scientists responsible for the new cosmological discoveries and theories were included. A number of student-teacher combinations were also included. Margaret Geller was the student of James Peebles, who was the student of Robert Dicke. Dennis Sciama supervised the thesis work of Stephen Hawking and Martin Rees and had a strong influence on Roger Penrose while he was a graduate student. As a graduate student, Sandra Faber worked with Vera Rubin. John Huchra and Edwin Turner were the students of Wallace Sargent.

The interviews were conducted by A. L. in person, with a tape recorder. Each interview was about 90 minutes long. The tape of each interview was transcribed twice, with the second transcription done by one of us. The transcript was then lightly edited and footnoted and sent back to the interviewee for corrections and additions. Generally, the interviewees made only small changes; the few cases of more substantial revisions are noted in the text. Tapes and transcripts of the full interviews, which average about 11,000 words, have been placed on file at the American Institute of Physics in New York and are available for scholars. Additional and follow-up interviews are being conducted by R. B. For this book, we reduced each interview by about a third, removing sections that were especially technical or less relevant to our themes.

The cosmology interview project began in the summer of 1987. Prior to our work, extensive interviews of astronomers and astrophysicists were conducted by Spencer Weart and David DeVorkin during 1976–1979, as a program of the American Institute of Physics Center for History of Physics.

For their generous encouragement and support of the project, we are grateful to Irwin Shapiro and the Harvard-Smithsonian Center for Astrophysics; and to Kenneth Manning, the Writing Program, and the School of Humanities and Social Sciences at the Massachusetts Institute of Technology. We thank Gerald Holton and Spencer Weart for their early encouragement. We are grateful to Martin Rees for his guidance on the scientific issues throughout the project. Edmund Bertschinger, Margaret Geller, and Edwin Turner also helped with scientific matters. For critical comments on portions or all of the evolving manuscript, we thank Susan Carey, Owen Gingerich, Richard Goodwin, Evelyn Fox Keller, David Meltzer, Lisa Olson, James Peebles, Andrew Pickering, Michelle Preston, Martin Rees, Carol Rigolot, Paul Schechter, Diane Steiner, Edwin Turner, and Charles Weiner. We thank Mercedes Dahlquist and Diane Steiner for transcribing many of the interviews and Jason Puchalla for help with the references. Our editors at Harvard University Press, Angela von der Lippe and Susan Wallace, made many fine suggestions about presentation and editing of the manuscript; we are grateful to Linda Howe for early editorial suggestions. We also thank Jane Gelfman for her continuing advice and support.

Finally, we thank the interviewees, who uniformly supported the goals of the project and gave of their time and themselves.

AN INTRODUCTION
TO MODERN COSMOLOGY

Cosmic questions start inside. Sometime in childhood we become conscious of ourselves as separate from our surroundings; we become aware of our own bodies, our own thoughts. Then, we question. How far back can we remember our parents, or anything at all? What was it like before we were born? What would it be like to be dead? We see pictures of our great grandparents and try to imagine their talking and living. In a game, we try to imagine *their* parents, and so on, backward in time, back through the generations, until we don't quite believe that this all could have happened. Yet we are here.

In elementary school we are told that the earth is not flat as it seems, but bends on itself in a huge, mottled ball. We are told that the sun, that bright, small light that circles the sky, is actually far larger than the earth. We are told that the tiny points of light in the heavens at night are suns themselves. We close our eyes and mentally glide to a star, through the blackness, to stare back at the speck that is earth. In space, events happen slowly, if at all. The sun looks the same, day after day. The stars never vary. In the vast reaches of space, time seems to stretch forward and backward without end, engulfing us, our great grandparents, all human beings, the entire earth. Or perhaps, there is some limit, some enormous boundary that holds time and space.

In recent years a group of men and women have devoted their lives to exploring some of these questions through science. They are cosmologists. Cosmology is the branch of astronomy and physics concerned with the universe as a whole. Has the universe existed forever? If not, when and how did it begin? Will it end? How is the universe changing in time? Does space extend infinitely in all directions? How did the matter in the universe come into being, and why is it arranged as it is? At the turn of the century most of these questions were considered to lie outside science.

Even today, cosmology is a speculative science. The most widely held cosmological theory, the big bang model, rests on four observational facts: the outward motion of galaxies, discovered in 1929 and interpreted as evidence for the expansion and explosive beginning of the universe; the approximate agreement between the age of the universe, as gauged by the rate of motion of galaxies, and the age of the earth, as measured by the radioactive disintegration of terrestrial uranium ore; the bath of radio waves from space, predicted as a necessary remnant of a hot, younger universe and then discovered in 1965; and the overall chemical makeup of the universe—approximately 25% helium and 75% hydrogen—which can be explained in terms of atomic processes in the infant universe. Aside from these few critical observational tests, it is theory, assumption, and inference that support the big bang model. Cosmology, of all sciences, requires the most extreme extrapolations in space and in time.

The Birth of Modern Cosmology

Modern theories of cosmology date back to 1917, when Albert Einstein published a pioneering theoretical paper.[1] Using his new theory of gravity, called general relativity, Einstein proposed the first detailed model for the large-scale structure of the universe. Between its publication in 1915[2] and 1917, the theory of general relativity had been tested by only a single observation, the orbit of the planet Mercury. Einstein's new theory of gravity passed this test with flying colors, explaining a tiny effect in the orbit that could not be accounted for by Newton's older theory. The application of general relativity beyond the solar system, however, remained uncertain. Although Einstein understood that gravity was the dominant force for describing the cosmos at large, he had little working knowledge of astronomy. Not a single astronomical number appears in Einstein's paper on cosmology.

Einstein made two critical assumptions: the universe does not change in time, and the matter of the universe is evenly scattered through space. Given these two assumptions and his mathematical theory of gravity, Einstein was able to derive equations describing the overall structure of the universe.

There was no compelling evidence for either of Einstein's starting assumptions. Although astronomical observations were *consistent* with a static universe, many astronomers of the day were aware that the view seen through big telescopes was only a snapshot, revealing little about the long-term evolution of the cosmos.[3] The observations had nothing to say on this point. On the other hand, the notion of a static universe was deeply ingrained in Western thinking, dating back to Aristotle, and was one of the few astronomical beliefs not overthrown by the Copernican revolution. Einstein's second assumption, of homogeneity, greatly simplified the equations, but it too was made on faith. In fact, as far as astronomers could tell, it was clear that the universe was highly lumpy, with most visible stars gathered up in a disk called the Milky Way. Until 1918 astronomers had only a poor estimate for the size of the Milky Way; until 1924 astronomers were not sure whether other gatherings of stars existed in the space beyond the Milky Way. Einstein simply assumed that space would appear smooth when averaged over a sufficiently large volume, just as a beach appears smooth when looked at from a distance of a few feet or more, even though it appears grainy when looked at from close range.

The assumption of homogeneity may be required to manage the mathematics of cosmology. As of 1990, theorists have succeeded in solving the equations of cosmology only for homogeneous models, except for special and implausible circumstances. Of course, simple equations and reality are two different things. Nature may not have been so accommodating as to avoid inhomogeneities just because physicists cannot conquer the associated math.

A brief digression on models in science is warranted here. We will encounter a number of cosmological models: the big bang model, the steady state model, the inflationary universe model. A scientific model begins with a real physical object or system, replaces the original object with a simpler object, and then represents the simplified object with equations describing its behavior. Like a toy boat, a scientific model is a scaled-down version of a physical system, missing some parts of the original. Deciding what parts should be left out requires judgment and skill. The omission of essential features makes the model worthless. On the other hand, if nothing is left out, no simplification has been made. In making a model of a swinging pendulum, for example, we might at first try to include the detailed shape of the weight at the end, the density and pressure of the air in the room, and so on. Finding such a description too complex to manage, we could replace the weight by a round ball and neglect the air completely. This much simpler system, in fact, behaves almost exactly like the original. If, on the other hand, we left out gravity, the resulting theoretical pendulum would not swing back and forth. By solving the equations of a model, predictions can be made about the original physical system and then tested.

In 1922 a Russian mathematician and meteorologist, Alexander Friedmann, proposed cosmological models for a *changing* universe.[4] Friedmann adopted Einstein's theory of gravity and Einstein's assumption of homogeneity but rejected his assumption of stasis, pointing out that it was unverified and nonessential. Beginning with the equations of general relativity, as Einstein had, Friedmann found an alternative solution, corresponding to a universe that began in a state of extremely high density and then expanded in time, thinning out as it did so. Friedmann's model, later rediscovered by the Belgian priest and physicist Georges Lemaître in 1927, eventually came to be called the big bang model.[5] Einstein reluctantly acknowledged the mathematical validity of Friedmann's evolving cosmological model but initially doubted that it had any bearing on the real universe.[6] In any case, both Einstein's and Friedmann's models were all pencil and paper. Little was known from observations about the true structure or evolution of the universe.

Discovery of the Expansion of the Universe

A major stumbling block in all of astronomy, and particularly in cosmology, was the problem of measuring the distances to the stars. When we look at the sky at night, we can perceive width and length, but not depth. From our vantage, stars are just white dots on a black canvas. Some are certainly closer than others, but which ones? Because stars come in a range of intrinsic luminosities, a star of a given *observed* brightness could be either very close and intrinsically dim or very far away and intrinsically bright.

Measurements of astronomical distances were placed on much firmer ground around 1912, when Henrietta Leavitt of the Harvard College Observatory found that certain stars, called Cepheid variables, oscillate in brightness, growing dim, then bright, then dim again, in regular cycles.[7] Leavitt analyzed a group of Cepheids that were clustered about each other and thus known to be at the same distance. Within such a cluster, a star that *appeared* twice as bright as another was indeed twice as intrinsically luminous. Remarkably, Leavitt found that the time for a Cepheid to cycle depended on its intrinsic luminosity, or wattage. For example, Cepheids 1,000 times as luminous as our sun complete a light cycle every 3 days. Cepheids 10,000 times as luminous cycle every 30 days. Once this behavior has been calibrated for nearby Cepheid stars, it can be used to measure the distance to remote Cepheid stars. By measuring the cycle time of a particular Cepheid star, one can infer its intrinsic luminosity. By then comparing the star's intrinsic luminosity to its observed brightness, one can determine its distance, just as the distance to a light bulb may be inferred from its wattage and observed brightness.

Cepheid stars found at various locations allowed astronomers in 1918 to measure the size of the Milky Way.[8] In 1924 the American astronomer

Edwin Hubble found a Cepheid in the faint patch of stars known as the Andromeda nebula, allowing him to measure its distance.[9] He discovered that the Andromeda nebula was a congregation of stars far beyond our galaxy.

In the following years, Hubble and other astronomers measured the distances to many of the faint misty patches, called nebulae, that had been seen and puzzled over for centuries. Many were found to be separate galaxies of stars. With these discoveries, galaxies, not stars, became the basic unit of matter in the universe. A typical galaxy, like our Milky Way, contains about 100 billion stars, which orbit each other under their mutual gravity and extend about 100,000 light years in diameter. (A light year is the distance that light travels in a year, about 6 trillion miles. The star nearest the sun, Alpha Centauri, is about 4 light years away.) On average, galaxies are separated by about 10 million light years, or about 100 times the diameter of one galaxy. Galaxies, therefore, are isolated islands of stars, surrounded by mostly empty space. Einstein's assumption of homogeneity would have to be tested on volumes of space that encompassed many galaxies.

In 1929 Hubble made what was perhaps the most important discovery of modern cosmology: the universe is expanding.[10] Using data taken from a telescope at Mt. Wilson, California, Hubble concluded that the other galaxies are moving outward from us in all directions. Two kinds of measurements were needed in Hubble's analysis: the speed and distance of neighboring galaxies. It had been known since the early 1900s that many of the nebulae were in motion, speeding away from the earth.[11] This had been determined by a technique known as the Doppler shift. Galaxies, like all sources of light, emit light of particular colors (wavelengths), related to the chemical composition of the galaxy. When a source of light is in motion, its colors shift, analogously to the shift in pitch of a moving source of sound. For example, the pitch of a train's whistle drops when the train moves away and rises when the train moves closer. In light, the analogue of pitch is color. If a light source is moving closer, its colors are shifted up in frequency, toward the blue end of the spectrum; if the source is moving away, its colors are shifted down, toward the red. From the amount of the shift, one can infer the speed of the moving source of light. Although the effect in light is usually tiny, sensitive instruments can detect it.

The colors of many of the nebulae were shifted to the red, showing that they were speeding away from us. This change in color of cosmic objects came to be called the redshift. By using Cepheid stars to measure the

Leavitt

Friedmann

Hubble

Einstein

de Sitter

Lemaître

distances to about 18 nebulae, Hubble found that the nebulae were entire galaxies, lying beyond the Milky Way. More important, he discovered that the distance to each galaxy was proportional to its recessional speed: a galaxy twice as far from us as another galaxy was moving outward twice as fast.

This last quantitative result was just as had been predicted for a uniformly expanding and homogeneous universe. A simple example, with homemade equipment, shows why. Place equally spaced ink marks on a rubber band and declare one ink mark to be your reference point (for example, the Milky Way), measuring all distances and motions relative to it. Holding the reference mark fixed against a ruler, stretch the two ends of the rubber band. Upon stretching, you will find that each ink mark moves a distance proportional to its initial distance from the reference mark. For example, when the ink mark initially 1 inch away moves to 2 inches away, the ink mark initially 2 inches away moves to 4 inches away. Since these increased distances are accomplished in the same period of time, the second ink mark moves twice as fast as the first. Speed is proportional to distance. In fact, *any* uniformly stretching material produces the law that

Spiral galaxy M 104 (the Sombrero galaxy).

speed is proportional to distance. If the material is lumpy, so that some places stretch at a faster rate than others, then speed is no longer proportional to distance. Conversely, the proportionality of speed to distance means that the material is uniformly stretching. It is also easily seen that the expansion has no center or privileged position. *Any* ink mark can be chosen as the reference mark, and the result is the same: the other ink marks move away from it with speeds proportional to their distances. No ink mark is special. The view is the same for all.

Replace the ink marks by galaxies and the stretching rubber band by the stretching space of the universe, and you arrive at Hubble's result. Galaxies are moving away from us because space is stretching uniformly in all directions, carrying the galaxies along with it. Hubble's discovery in 1929 gave strong observational support for cosmological models in which the universe is uniformly expanding—but without a center to the expansion. The static universe of Einstein was ruled out.

If the galaxies are moving away from each other, then they would have been closer together in the past. At earlier times, the universe was denser. If we assume that this backward extrapolation can be continued, there was some definite moment in the past when all the matter of the universe was crammed together in a state of almost infinite density. From the rate of expansion, astronomers can estimate when this point in time occurred: about 10 to 15 billion years ago.[12] It is called the beginning of the universe, or the big bang . The original estimates of Hubble, in error due to various technical problems, gave an estimate of about 2 billion years for the age of the universe. For simplicity, we will assume 10 billion years in all subsequent discussions.

There is a completely independent method for determining the age of the universe. Radioactive dating of uranium ore, developed about two decades before Hubble's discovery, suggests that the earth is about 4 billion years old. What should this have to do with the age of the universe? Most theories of the formation of stars and planets indicate that our solar system could not be a lot younger than the universe. So its age should be less than but not a lot less than the age of the universe. In astronomy, where the ages of things span a huge range, 4 billion years—as opposed to a million years or 10 trillion years—is considered very close to 10 billion years. The match is good. Thus with two totally different methods, one using the outward motions of galaxies and one using the rocks underfoot, scientists have derived comparable ages for the universe. This success has been a powerful argument in favor of the big bang model.

Cosmology and geology share even more. Digging down to deeper layers of the earth is traveling back in time, back into our human past. Peering out to greater distances in space is also traveling back in time. When our telescopes detect a galaxy 10 million light years away, we see that galaxy as it was 10 million years ago; we see ancient light, which has been traveling 10 million years to get from there to here. When we detect a more distant galaxy, we gaze upon an even older image. Cosmological observation is a kind of excavation, a search for origins above the ground, a glimpse of not an earlier earth but an earlier universe.

The Big Bang Model

The big bang model logically follows from Einstein's theory of gravity and a small number of assumptions. This model provides a mathematical description for the evolution of the universe. According to the big bang model, the universe began in a sort of explosion, starting from infinite density and temperature, and then expanded, thinned out, and cooled. This was not like an ordinary explosion, in which a localized region of flying debris spreads out into a surrounding region of nonmoving space. Instead, the big bang explosion occurred everywhere. There was no surrounding space for the universe to move into, since any such space would be part of the universe. The concept boggles the imagination, but it is a little easier to visualize if one pictures individual particles in the universe. Since the big bang, all particles in the universe have been moving away from each other, carried along by the expansion of space, just as the ink marks move apart on a stretching rubber band.

Even though the universe expands, its parts tug on one another owing to gravitational attraction, and this slows down the expansion. The competition between the outward motion of expansion and the inward pull of gravity leads to three possibilities for the ultimate fate of the universe. The universe may expand forever, with its outward motion always overwhelming the inward pull of gravity, just as a rock thrown upward with sufficient speed will escape the gravity of the earth and keep traveling forever. Such a universe is called an open universe. A second possibility is that the inward force of gravity is sufficiently strong to halt and reverse the expansion, just as a rock thrown upward with insufficient speed will reach a maximum height and then fall back to earth. For a universe of this type, called a closed universe, the universe reaches a maximum size and then starts collapsing, toward a kind of reverse big bang. Such universes have both a beginning and an end in time. The final possibility, called a flat universe, is midway between a closed and open universe and is analogous to the rock thrown upward with precisely the minimum speed needed to escape from the pull of the earth. Flat universes, like open universes, keep expanding forever.

All three possibilities are allowed by the big bang model. Which one holds for our universe depends upon the manner in which the cosmic expansion began, in the same way that the path of the rock depends on the rock's initial speed relative to the strength of the earth's gravity. For the rock, the critical initial speed is 7 miles per second. Rocks thrown upward with less than this speed will fall back to earth; rocks with greater initial speed will never return. Likewise, the fate of the universe was determined by its initial rate of expansion relative to its gravity. Even without knowledge of these initial conditions, we can infer the fate of our universe by comparing its *current* rate of expansion to its *current* average density. If the density is greater than a critical value, which is determined by the current rate of expansion, then gravity dominates; the universe is closed, fated to collapse at some time in the future. If the density is less than the critical value, the universe is open. If it is precisely equal to the critical value, the universe is flat. The ratio of the actual density to the critical density is called omega. Thus, the universe is open, flat, or closed depending on whether omega is less than 1, equal to 1, or larger than 1, respectively.

Omega can be measured. The rate of expansion of the universe is estimated by measuring the recessional speed of a distant galaxy (found by its redshift) and dividing by the distance to the galaxy. In a uniformly expanding universe, as we have seen, the outward speed of any galaxy is proportional to its distance; thus, the ratio of velocity to distance is the same for any galaxy. The resulting number, called the Hubble constant, measures the current rate of expansion of the universe. According to the best measurements, the current rate of expansion of the universe is such that it will double its size in approximately 10 billion years. This corresponds to a critical density of about 10^{-29} grams per cubic centimeter, the density achieved by spreading the mass of a poppy seed through a volume the size of the earth. (The notation 10^{-29} is shorthand for a decimal point followed by 28 zeros and a one; 10^{15} stands for a one followed by 15 zeros; and so on.) The best measured value for the actual average density— obtained by telescopically examining a huge volume of space containing many galaxies, estimating the amount of mass in that volume by its gravitational effects, and then dividing by the size of the volume—is about 10^{-30} grams per cubic centimeter, or about one tenth the critical value. This result, as well as other observations, suggests that our universe is of the open variety.[13]

However, there are uncertainties in these estimates, mostly connected with uncertainties in cosmic distances. If the universe were precisely homogeneous and uniformly expanding, then the rate of expansion of the universe could be determined by measuring the recessional speed and distance of any galaxy, near or far. Conversely, the distance to any galaxy could be determined from its redshift and the application of Hubble's law. However, the universe is not completely homogeneous. Because of local inhomogeneities, the rate of expansion of the universe and the average density of matter should be measured over as large a region as possible (after which we must assume that such a region is typical of any region of the universe). Accurate distances to galaxies are needed for both of these measurements. The rate of cosmic expansion, for example, is obtained by dividing the recessional speed of a galaxy by its distance, if the latter is known. Individual Cepheid stars cannot be used to measure distances beyond about 10 million light years, because they become too dim. Instead, entire galaxies must be used as "standard candles," that is, objects of known intrinsic luminosity. Unfortunately, galaxies, like stars, come in a wide range of luminosities. There are no standard candles. The best that can be done is to search for some empirical relation between the intrinsic luminosity of a galaxy and another observed property, such as the orbital speed of its stars.[14] (Such relations are analogous to the relation between the intrinsic luminosity and light period of a Cepheid star.) After determining and calibrating such a relation for nearby galaxies, where distances can be measured by other means, the method can then be applied to much farther galaxies.

The pitfall is that looking to greater distances in space is equivalent to looking back in time. The light we see today from distant galaxies was emitted when they were much younger and has been traveling for hundreds of millions or billions of years to reach us. By contrast, the light from nearby galaxies has been traveling for a much shorter time and therefore shows them at a much later stage of their evolution. In other words, a distant galaxy, *now observed*, may be very different from more mature galaxies nearby that have been used to calibrate the luminosity–orbital speed relation, and the relation may not apply to the distant galaxy very well. Many modern astronomers have devoted their work to understanding the long-term evolution of galaxies. But at present, we have not found any astronomical objects as well understood and reliable as Cepheid stars for using as standard candles.

The difficulty in measuring large distances limits our ability to determine the average density of the universe. The average density of mass in the largest volumes of space we have measured, extending in size up to a 100 million light years, is estimated by how the motions of galaxies are affected by local concentrations of mass. If the mass of the universe were spread smoothly, then each galaxy would move directly outward from us, with a speed exactly proportional to distance. Indeed, this is approximately what we see. However, in "local" regions, of 10 or 100 million light years, the cosmic mass is clumped into galaxies and clusters of galaxies. The "lumpy" gravity of such mass clustering bends and alters the course of nearby galaxies—just as, for example, the ball in a pinball machine gets deflected this way and that as it runs into bumpers, even though falling down on the average. A comparison of the "peculiar" motions of the galaxies with the "normal" motions expected for a completely smooth universe, together with a knowledge of the "bumpers" producing the peculiar motions, determines the average density of matter in the region. The problem is that only the *total* motion of such galaxies can be measured; to know how much is peculiar and how much normal requires knowledge of the rate of expansion of the universe and the *distance* to the galaxies. (Recall that the normal outward speed is the rate of expansion multiplied by the distance.) If the distance is not known well, the normal speed of expansion will not be known well either.

A different method of determining omega involves gauging how the rate of expansion of the universe has been slowing down in time. This is closely related to measuring both the expansion rate and the average density of matter, since the gravity of the latter is the assumed cause of the slow down. In practice, the expansion rate is measured at greater and greater distances, which probe the universe at earlier and earlier times. Unfortunately, these measurements, which were begun by Edwin Hubble in the 1930s[15] and continued by Allan Sandage of the Mt. Wilson Observatories in the 1950s,[16] require either that accurate distances be determined for very distant objects or that a set of standard candles of known intrinsic luminosity be found. Thus, they too suffer from the difficulty of measuring large distances in the cosmos and the absence of standard candles.

Despite these uncertainties, cosmologists are fairly sure that the value of omega lies between 0.1 and 10.0—that is, within 10 times of 1. Even with the uncertainty in the rate of expansion, enough matter has been identified so that omega cannot be less than 0.1. On the upper end, an omega larger than 10, together with the current rate of expansion, would translate to an age of the universe much less than the age of the earth as determined by radioactive dating. In the realm of astronomy, where the sizes of objects range from 10^{11} centimeters for stars to 10^{26} centimeters for galaxy clusters, two numbers within 10 times of each other are considered almost equal. That our universe is apparently so near the boundary between open and closed is considered extraordinary by some cosmologists.

Einstein's theory of gravity, which underlies the big bang model, makes a theoretical connection between the evolution of the universe and its size. According to the theory, if the universe is closed, then it has a limited size. One might ask what lies outside the boundary of a universe that has only a limited size. The answer is that a closed universe has no boundary. It bends around on itself, in the same way that the surface of a sphere bends around on itself. Begin walking in a straight line and you come back to where you started. You travel around your entire world, covering a finite distance, but you never fall off an edge or meet a boundary. In three dimensions, this picture resists the imagination, but it can be expressed mathematically. Open and flat universes, by contrast, have unlimited size and extend infinitely in all directions. There is a further distinction between closed, flat, and open universes. Flat universes obey Euclidean geometry. For example, the three angles of a triangle formed by connecting three galaxies with straight lines add up to 180 degrees. In closed universes, the angles of such a triangle add up to *more* than 180 degrees; in open universes the sum is *less* than 180 degrees.

People are often confused about what it means to speak of the expansion of open or flat universes, which already extend infinitely in space. Expansion means that the distance between any two galaxies or other points is increasing. When we say that the universe is currently expanding at a rate such that it will double in size in 10 billion years, we mean that the distance between any two galaxies will double in 10 billion years. Such a definition has sense in open, flat, or closed universes.

The overall geometry or size of the universe has not been directly measured. The measured quantities are the rate of expansion and the average density. Only after combining these measured quantities with the big bang theory and its mathematics can we infer the geometry and the fate of the universe. Thus, much depends upon the theory and its key assumptions.

It is also important to realize that even if the universe is of infinite size, only a limited volume, called the observable universe, is *visible* to us at any moment. That is because we can see only as far as light can have traveled since the big bang. As we look farther into space, we are seeing light that has been traveling longer to reach us and therefore that was emitted earlier. When we look at the Andromeda galaxy, for example, we see light that was emitted 2 million years ago; when we look at the Virgo cluster of galaxies, we see light that was emitted 50 million years ago. Eventually, at some distance, the light just now reaching us was emitted at the moment of the big bang. That distance marks the edge of the currently observable universe. We cannot see farther because there hasn't been time for light to travel from there to here since the big bang. Today, the observable universe extends out about 10 billion light years, the distance light can travel in 10 billion years.[17] A billion years from now, when the universe is 11 billion years old, the observable universe will extend to 11 billion light years; we will be able to see to a distance of 11 billion light years. New regions of the universe, now beyond our horizon, will have come into view. We can never see further back in time than the big bang, but as time goes on we can see more and more of the universe at the big bang. Each day the observable universe grows a little larger. Each day, the light emitted from slightly more distant objects has had the needed time to reach our telescopes.

The big bang model does more than relate the evolution of the universe to its mass density and geometry; the model describes the broad history of the universe. Imagine a movie of cosmic evolution played backward in time, starting from the present. The universe contracts. The galaxies move closer and closer together until they touch and merge. As the universe grows denser and denser, galaxies and finally individual stars lose their identity, and the matter of the universe begins to resemble a gas. Like any gas being compressed, the cosmic gas becomes hotter and hotter. Eventually, the heat becomes so high that atoms cannot retain their electrons, and they disintegrate into atomic nuclei and freely roaming electrons. At a still earlier stage, as the big bang gets nearer, the atomic nuclei themselves disintegrate into protons and neutrons under the high heat. At an even earlier time each proton and neutron disintegrates into three elementary particles called quarks. The universe becomes a sea of careening subatomic particles.

The big bang model is quantitative. It specifies the average density, expansion rate, and temperature of the universe at each point in time, given the measured values of those quantities today. According to the theory, the temperature of the universe was about 10 billion (10^{10}) degrees centigrade and its density was about 100,000 (10^5) grams per cubic centimeter one second after the big bang.[18] At that moment, the universe consisted of a very hot gas of subatomic particles, uniformly filling space. By the time the universe was about 30 million years old, the epoch thought by some scientists to be roughly when the first galaxies began forming, its temperature and density had dropped to about zero degrees centigrade and 10^{-25} grams per cubic centimeter, respectively. (Absolute zero is -273 degrees centigrade. The measured cosmic temperature today is about -270 degrees centigrade, or 3 degrees above absolute zero, and is still dropping.)

In addition to providing an explanation for the observed expansion and age of the universe, the big bang model has successfully met two other major tests against observations. It explains why the universe is approximately 75% hydrogen and 25% helium, as previously observed. (The heavier chemical elements, such as oxygen and carbon, make up only a trace of the total mass in the universe.) The big bang model also predicted that space should be filled with a special kind of radio waves, created when the universe was much younger. Such cosmic radio waves, sometimes called the cosmic background radiation, were discovered in 1965, after their prediction. The big bang's successes with helium and with cosmic radio waves—the first a good explanation of a previously known fact and the second a prediction of a to-be-discovered fact—were crucial, not just for the science but for the attitudes of scientists. The agreement between theory and observation on these two phenomena convinced many scientists for the first time that cosmology had some contact with reality, that cosmology was a legitimate science.

According to the big bang model, the universe was once so hot that none of the chemical elements except hydrogen, the lightest element, could exist. Hydrogen is nothing more than a single subatomic particle, a proton. All other elements consist of a fusion of two or more subatomic particles, which could not hold together under the intense heat of the infant universe. As the universe expanded, it cooled. When the universe was a few minutes old, its temperature had dropped to a billion degrees, the critical temperature at which subatomic particles could begin sticking together via the attractive nuclear forces between them. According to theoretical calculations done by Fred Hoyle and Roger Tayler of Cambridge University in 1964 and by Yakov Zel'dovich of the Institute for Cosmic Research in Moscow at about the same time, and then refined by James Peebles at Princeton University in 1966 and by Robert Wagoner, William Fowler, and Fred Hoyle at the California Institute of Technology in 1967, nuclear fusion in the first few minutes after the big bang should have converted about 25% of the mass of the universe into helium, the next lightest element after hydrogen.[19] It is believed that all heavier elements were manufactured much later, in the nuclear reactions at the centers of stars. This accounting is in good agreement with the observed abundances of hydrogen and helium in space.

The other important experimental confirmation of the big bang theory, the cosmic background radiation, was first predicted by Ralph Alpher, George Gamow, and Robert Herman at George Washington University in 1948, and later independently predicted by Robert Dicke, James Peebles, P. G. Roll, and David Wilkinson of Princeton University in 1965.[20] Both of these groups pointed out that when the universe was a few seconds old and younger, a special kind of radiation, called blackbody radiation, would have been produced throughout space. Such radiation arises in any system of subatomic particles that collide with each other very rapidly, as would have been the case in the high heat of the infant universe. Small amounts of blackbody radiation are also produced today, in isolated regions such as in stars, but the universe now is much too cool as a whole to produce blackbody radiation filling all space. Blackbody radiation is easily identifiable by its universal spectrum of colors, that is, the amount of energy at each wavelength. Blackbody radiation can be uniquely characterized by a single parameter, which corresponds to the temperature of the radiation. According to theoretical calculations, blackbody radiation should have been produced uniformly through space in the early universe and would have continued bouncing off subatomic particles until the universe was about a million years old, when electrons and atomic nuclei combined to make atoms. After that, the radiation would have traveled freely through space, appearing today with a dominant wavelength corresponding to radio waves and a temperature of about 3 degrees above absolute zero. In 1965 Dicke's collaborators Roll and Wilkinson had just constructed an apparatus to search for their predicted cosmic radio waves when the radiation was discovered accidentally by Arno Penzias and Robert Wilson at Bell Laboratory.[21] Penzias and Wilson were awarded the Nobel Prize for their discovery in 1978. To date, the most precise measurements of the cosmic background radiation have come from the Cosmic Background Explorer, a satellite launched in November of 1989. This satellite has confirmed that the spectrum of the cosmic background radiation is extraordinarily close to that predicted by the big bang model.

The discovery of the cosmic background radiation in 1965 gave strong support to the idea that the universe was much hotter in the past. Just as importantly, the observed cosmic radiation seems to confirm the hypothesis of large-scale homogeneity of the universe. The radiation has the same intensity from all directions in space—that is, it is isotropic. If we assume that we do not sit in an unusual place in the universe, then we can infer that the cosmic radiation is isotropic at *any* location in the universe. This means that the universe was very homogeneous when the radiation last collided with matter, about a million years after the big bang. If the universe were lumpy or of uneven temperature at such an epoch, the cosmic radiation would have scattered off the lumps in uneven intensities and directions and would not appear so uniform today. Because it has been traveling since the universe was only a million years old, the cosmic radiation now detected has traveled much further than the distances to the visible galaxies, so it tells us about the smoothness of the universe on a much larger scale.

Observational confirmation of the large-scale homogeneity of the universe is vital for the standard big bang model and perhaps for all tractable cosmological models. As early as 1933, the British cosmologist Edwin Arthur Milne suggested that the assumption of large-scale homogeneity might be logically necessary for any cosmological model.[22] Milne named this assumption the cosmological principle, which has since become a starting point for most theoretical work in cosmology and has so far proven a necessary simplification for being able to solve the difficult equations of the subject. If future observations cast doubt on the assumption of large-scale homogeneity, the gross features of the big-bang model might still be correct, but the details would certainly be wrong.

Why didn't scientists immediately follow up on the original predictions of Alpher, Gamow, and Herman?[23] (Indeed, Dicke was unaware of the earlier predictions and came to his conclusions completely independently.) There could be several reasons. The predicted cosmic radio waves were thought to be undetectable by the instruments of the 1950s. In addition, as Alpher and Herman recall, "Some scientists had a philosophical predilection toward a steady-state universe."[24] In such a universe the temperature would always be what it is today and never hot enough to produce blackbody radiation. Finally, many scientists in the 1940s and 1950s considered cosmology too speculative a subject to be taken seriously. There was practically no contact between theory and experiment.

Other Early Cosmological Models

One variation of the big bang model, first discussed extensively by Richard Tolman of the California Institute of Technology in the early 1930s, is called the oscillating universe model. An oscillating universe is closed, but instead of ending after its collapse, it begins a new expansion, repeating the process of expansion and contraction through many cycles. If our universe were an oscillating universe, then it could be very much older than the estimated age of 10 billion years, which just measures the time since the last cycle of expansion began.

An apparent difficulty with this model results from the second law of thermodynamics, a basic law of physics that requires any isolated system to become more and more disordered until a state of maximum disorder is achieved. After many cycles, an oscillating universe would be expected to be much more chaotic than the universe we observe. Tolman was aware of this problem with oscillating models but argued that a state of maximum disorder might be impossible to define for the universe as a whole, thus rendering the objection uncertain. He concluded that "it would seem wise if we no longer assert that the principles of thermodynamics necessarily require a universe which was created at a finite time in the past and which is fated for stagnation and death in the future."[25] Today, physicists are still not sure whether an oscillating universe would be theoretically ruled out by the second law of thermodynamics or whether the second law applies to the universe as a whole.

The oscillating universe model was in vogue in the late 1950s and early 1960s. In fact, it was a preference for an oscillating universe that led Robert Dicke to predict the existence of the cosmic background radiation. Dicke and his collaborators began their classic paper in the *Astrophysical Journal* in 1965 by pointing out that an oscillating universe, having existed for all time, "relieves us of the necessity of understanding the origin of matter at any finite time in the past."[26] Taking such a model as a working hypothesis, Dicke then argued that if our universe had indeed gone through many cycles of expansion and contraction, its temperature would have to reach at least 10 billion degrees at each point of maximum contraction, in order to disintegrate all the heavy elements created in stars during the previous cycle and to restore the matter of the universe to pure hydrogen. Otherwise, the nuclear reactions in stars would by now have converted most of the matter of the universe into heavy elements, contradicting the observations. Dicke then pointed out that at a temperature of 10 billion degrees, the reactions of subatomic particles would be sufficiently rapid to produce blackbody radiation. (In actuality, the production of such radiation does not *require* that the universe oscillate, only that the cosmic temperature be high enough at the beginning.)

In 1948 a group of restless young theoretical astrophysicists at Cambridge University, not completely happy with the big bang model in any form and casting about for other possibilities, came up with the steady state model.[27] This cosmological model, conceived of by Thomas Gold,

Fred Hoyle, and Hermann Bondi, was not a variation of the big bang model. It proposed that the universe, on average, does not change in time. For example, the average density of matter does not change in time, and the temperature does not change in time. The steady state model reconciles itself with Hubble's observations of the outward motion of galaxies by postulating that new matter and galaxies are continuously created throughout space, compensating for the spreading apart of individual galaxies and allowing the average density of galaxies to remain constant. In this way, the universe maintains a steady state.

In their papers of 1948, Bondi, Gold, and Hoyle mention several reasons for proposing the steady state model. For one, they express dissatisfaction that the big bang model forces physicists to apply the laws of physics as observed today to a distant time in the past, when the conditions of the universe would have been far different. In the big bang model there is no way of knowing for sure whether the laws then were the same as now, yet no calculations can be made without assuming so. On the other hand, the scientists argue, a steady state universe is "compelling, for it is only in such a universe that there is any basis for the assumption that the laws of physics are constant."[28] A second motivation for the steady state model was more quantitative: The rate of expansion of the universe as measured by the relatively uncertain techniques available in the 1940s translated to an estimated age for the universe of only 2 billion years; this was less than the geologically determined age for the earth. Some people considered this a problem for the big bang model. The steady state model also appealed to many scientists because, like the oscillating universe model, it eliminated the necessity of confronting the birth of the universe and all the uncertainties and incalculables attendant with that beginning. In the steady state model, the universe always was and always will be as it appears now. It has no beginning and no end. Initial conditions do not have to be specified or accepted. Furthermore, some physicists and astronomers believed that the range of possibilities in such a universe would be much more limited than in the big bang model and thus easier to calculate. This was another attraction of the steady state model. Most physicists prefer theories that they can fully calculate. The steady state model was popular in the 1950s and early 1960s and considered the leading competitor to the big bang model.

Today, almost all cosmologists believe that the steady state model has been ruled out. Besides the lack of evidence for the continuous creation of mass out of nothing or an explanation of how such a process could occur, the steady state model has been refuted by the discovery of the cosmic radiation and other observations that suggest the universe was very different in the past. For example, the locations of certain astronomical objects called quasars (for "quasi-stellar radio sources") strongly suggest that the universe has changed over time. In 1963 Maarten Schmidt discovered these extremely luminous and distant pointlike sources of energy.[29] A typical quasar lies 2 to 10 billion light years away and has the luminosity of a hundred galaxies. In 1965 Martin Rees and Dennis Sciama analyzed the data for the quasars known at that time and found that the number of quasars per volume of space increased with distance from the Milky Way.[30] Since looking out to larger distances is equivalent to looking back in time, that meant that there were more quasars in the past. Rees and Sciama and others interpreted their result to contradict the steady state theory, which demands that the universe cannot change from one epoch to the next and hence cannot alter its population of quasars or galaxies or anything else.

Difficulties with the Big Bang Model

Despite its successes, the big bang model has suffered from a number of problems. One troubling concern, ironically, is why the universe appears so uniform on the large scale. In particular, the incoming cosmic background radiation is remarkably uniform in all directions, varying in intensity by less than 1 part in 10,000 from different regions of the sky. The observed uniformity of this radiation indicates that the material gas of the universe had a nearly uniform density and temperature when the radiation last collided with it, about a million years after the big bang. Although such uniformity and homogeneity is *assumed* in the big bang model, it still must be explained, or at least be made plausible.

There are two possible explanations. Either the universe *began* with a high degree of homogeneity, or else any initial inhomogeneities eventually smoothed themselves out, much as hot and cold water in a bath tub will come to the same temperature by exchanging heat. However, heat exchange takes time. The regions of space that produced the cosmic radiation, when the universe was 1 million years old, were about 100 million light years apart at that time—much too far apart to have had time since the big bang to exchange heat and homogenize. Thus, the second explanation simply doesn't work in the big bang model. The first explanation is considered unsatisfactory by some scientists because it seems to sweep the problem under the rug, relegating it to whatever unknown and currently uncalculable processes determined the initial conditions of the universe. Furthermore, it seems unlikely to many scientists that the universe would have been created so homogeneously. If nothing else, fluctuations in matter and energy arising from quantum processes, which will be discussed later, would naturally have produced lumpiness and irregularity in the very early universe. The problem of accounting for the large-scale uniformity of the universe has been called the horizon problem.

The term horizon denotes the largest volume of space at any moment that could have homogenized since the big bang. Since heat exchange, or any other homogenizing process, cannot travel faster than light, the size of the horizon at any moment can be no larger than the distance light can have traveled since the big bang. For example, the size of the horizon 1 million years after the big bang is about 1 million light years.[31] The size of the horizon today, the so-called observable universe, is about 10 billion light years. Each point in space has its own horizon, which can be pictured as a sphere centered on the point and encompassing the largest region that could have homogenized with the central point. As time goes on, the horizon of any point grows. The smoothness of the cosmic background radiation suggests that different regions of the universe separated by *more* than each other's horizon nevertheless appear to have exchanged heat— thus the horizon problem.

The horizon problem seems to have been first clearly stated in print in 1969 by Charles Misner of the University of Maryland.[32] Although Einstein simply assumed large-scale homogeneity, he would have had no trouble explaining how it came about. Since Einstein also assumed that the universe had existed forever, there would have been plenty of time for any two regions, arbitrarily far apart, to have exchanged heat and homogenized. However, such an explanation does not work in the big bang model.

A closely related, and more controversial, issue is the so-called flatness problem: Why should the universe today be so near the boundary between open and closed, that is, so nearly flat? In other words, why is the measured value of omega—the ratio of cosmic mass density to the critical density needed to close the universe—today so close to 1? It follows from the big bang model that as time goes on, omega should differ more and more from 1, unless it started out exactly 1, in which case it remains 1. In an open universe, omega begins smaller than 1 and gets smaller and smaller in time; in a closed universe, omega begins larger than 1 and gets larger and larger.

Omega is analogous to the ratio of gravitational energy to kinetic energy of motion of a rock thrown upward from the earth. If the rock is launched with precisely the critical speed, that ratio will start out 1 and remain 1. If the rock is thrown with less than the critical speed, the ratio will start out greater than 1 and continuously increase, becoming infinite just when the rock reaches maximum height and is about to fall back to earth. At this point, the rock has zero speed, its kinetic energy of motion is zero, and the ratio of gravitational energy to kinetic energy is therefore infinite. In contrast, if the rock is thrown with more than the critical speed, the ratio will start out less than 1 and continuously decrease, approaching zero as the rock escapes the earth's gravity altogether and journeys off into outer space. Finding the cosmic omega so close to 1 today, so long after the big bang, is as if we sighted the rock a long time after it was launched, very, very far from earth, and found its gravitational energy and kinetic energy of motion almost equal. Such an event is highly unlikely because it would require the two energies at launch to have been balanced to extraordinary precision. For example, if a rock is thrown upward with an initial ratio of energies of 0.75, that ratio will have dropped to 0.1 by the time the rock has reached a distance of 27 earth radii; for an initial ratio of 0.9, the ratio will have dropped to 0.1 at a distance of 81 earth radii. For the rock to reach a million (10^6) times its initial distance before the ratio falls to 0.1, the initial ratio had to be 0.999991. The numbers behave similarly for ratios larger than 1.

Physicists believe that the initial conditions of the universe were set when the universe was about 10^{-43} seconds old. In order for the value of omega to still remain between 0.1 and 10.0 today, after 10 billion years and after the universe has expanded to 10^{30} times its initial size at "launch," the initial value of omega had to lie between $1 - 10^{-59}$ and $1 + 10^{-59}$. Equivalently, the kinetic energy and the gravitational energy of the universe had to be initially equal to a part in 10^{-59}. What physical processes could have set so fine a balance? And there is another puzzle. If the gravitational and kinetic energies are not exactly equal, why are they becoming unbalanced at this particular moment, just when *Homo sapiens* happened to arrive?

The flatness problem seems to have been first clearly posed and put into print by Robert Dicke in 1969.[33] Although several British cosmologists, including Brandon Carter and Stephen Hawking, independently noted the problem shortly thereafter,[34] it was not widely known or understood until it was restated in an article by Dicke and Peebles in 1979.[35]

There is a broad range of attitudes about the flatness problem. Some scientists consider the initial value of omega to be an accidental property of our universe, a value that should be accepted as a given, and they see no validity in the flatness problem. Others agree with Dicke and Peebles that the required initial conditions seem too special to be accidental and that some deeper physical explanation is required. Among this latter group are scientists who say that for some reason the initial gravitational and kinetic energies must have been balanced *exactly*. Omega was and is exactly 1. This view requires the existence of a huge quantity of undetected mass. Since we have observed only enough mass per volume of space to make omega equal to 0.1, the belief that omega is really exactly 1 requires that there be, on average, about 10 times as much mass as has been observed in every cubic light year of space.

Before 1980 most cosmologists side-stepped or paid no attention at all to the flatness problem. After an influential modification of the big bang model called the inflationary universe model, which gave a natural explanation for the flatness problem, many scientists considered the flatness problem to be important. The controversial status of the flatness problem is evident in Alan Guth's seminal paper on the inflationary universe model, where an appendix is devoted to convincing skeptics that the problem should not be dismissed.[36] Even today there is still no consensus on the meaning or depth of this problem.

Another old cosmological problem has been the lack of a good explanation for the average number of radiation particles, called photons, relative to the number of baryons. (Examples of baryons are protons and neutrons, which make up the nuclei of atoms.) We do not know the total number of photons or baryons in the universe, but the ratio of these numbers can be estimated by counting photons and baryons in a large volume of space and then assuming that volume is typical of the universe as a whole. The measured ratio is about a billion photons for every baryon. What makes this number fundamental is that it should be constant in time, according to the theory. It is a fixed property of the universe. But what determined its value? As in the flatness problem, some scientists have invoked the accident of initial conditions to explain why the photon-to-baryon ratio has the value it does, saying, in effect, that the number is a billion now because it was a billion then. Other scientists believe that this number should be calculable from basic principles. The big bang model itself does not require the photon-to-baryon ratio to have any particular value, just as it does not require the initial value of omega to have been anything in particular.

Finally, a problem whose importance has been appreciated only recently concerns the entropy of the universe. In the nineteenth century scientists discovered the second law of thermodynamics, which states that any isolated physical system subjected to random disturbances will naturally become less ordered, that is, will naturally increase its entropy. Entropy is a quantitative measure of the orderliness of a physical system. For example, a deck of cards arranged with all the cards of each suit together is very organized. Such a well-ordered deck is said to have low entropy. A deck that has been shuffled many times, with its cards in random positions, is said to have high entropy. Intuitively, the second law of thermodynamics is easy to understand. If you start with a deck of cards arranged by suit and number and drop it on the floor, the odds are great that the regathered cards will not be arranged in good order. On the other hand, if you start with a randomly ordered deck and shuffle it 10 times, the chances are extremely small that the resulting cards will be arranged in ascending order. Similarly, eggs often break but never reform; skywriting fades but never comes back; unattended rooms gather dust but do not get clean.

In a series of papers beginning in 1974, Roger Penrose of Oxford University applied the second law of thermodynamics to the universe as a whole.[37] More specifically, Penrose estimated the entropy of the observable universe and found it to be fantastically small compared with what it theoretically might be, for example, if much of the cosmic mass were in the form of a huge black hole, rather than in galaxies. If one traces cosmic evolution backward in time, the second law of thermodynamics decrees that the universe began with an even greater degree of order. Penrose and others find it mysterious that the universe was created in such a highly ordered condition—as if we were dealt a royal flush—and believe that any successful theory of cosmology should ultimately explain this entropy problem. The big bang model, in its present form, does not. Indeed, the big bang model says nothing about the initial conditions of the universe.

Large-Scale Structure and Dark Matter

The basic assumption of homogeneity that underlies the big bang model is obviously not true nearby. The universe nearby is not evenly filled with a smooth and featureless fluid. Rather, it is lumpy. Matter clumps into galaxies and galaxies clump into clusters of galaxies and so on. Astronomers called such clustering of galaxies "structures." Structures of various sizes abound, and astronomers want to understand the nature of these structures and how they were formed. Until such understanding, it will be hard to decide whether the observed inhomogeneities are simply details in the standard view or hints of a radically different picture.

One of the first twentieth-century scientists to suggest a nonuniform distribution of mass in the universe was C. V. L. Charlier, who considered the possibility of a hierarchical cosmos, in which the structure and average density would change as one went to larger and larger scales.[38] In 1933 Harlow Shapley of the Harvard College Observatory commented that the observed irregularities in the positions of galaxies were too pronounced to be accidental groupings in a basically smooth background and suggested some "evolutionary tendency in the metagalactic system."[39] Fritz Zwicky of the California Institute of Technology suggested in 1938 that clusters of galaxies, of about 10 million light years in size, be considered the basic units of matter in the universe.[40] In 1953 Gérard de Vaucouleurs, then at the Australian National Observatory, discovered that the galaxies within about 200 million light years of the Virgo cluster of galaxies, which includes the Milky Way, were mostly confined to a giant disk.[41] He called this large congregation of galaxies a supercluster of galaxies. With this discovery and subsequent work, de Vaucouleurs challenged the assumption of homogeneity in the universe and, in fact, proposed a hierarchical universe, in which small structures are part of larger structures, which are part of even larger structures, continuing up indefinitely.[42] In this picture, not only is the universe inhomogeneous, but an average density of matter cannot be defined. The larger the volume one takes to measure the density, the smaller the density. If such a model were correct, few conclusions about the overall behavior of the universe could be drawn from measurements in our vicinity.

Further evidence for inhomogeneities has come in recent years. In studies of large groups of galaxies in 1975, G. Chincarini and H. J. Rood discovered lumpiness in the distribution of matter over distances of roughly 20 million light years.[43] At a symposium of the International Astronomical Union in 1977, W. G. Tifft and S. A. Gregory and, independently, M. Joeveer and J. Einasto of Estonia, reported the observation of clusters and chains of galaxies, and other "voids" with no galaxies at all, extending over distances of several hundred million light years.[44] In 1978 Gregory and L. A. Thompson found evidence for a large congregation of galaxies, called the Coma supercluster, with relatively empty space around it.[45] In 1981 Robert Kirshner, August Oemler, Jr., Paul Schechter, and Stephen Shectman found a huge void in space, in the direction of the constellation Boötes, with a diameter of about 100 million light years.[46] It appears that few galaxies inhabit this vast empty space. (For comparison, in our own cosmic neighborhood space, one can find a galaxy every few million light years or so.) About the same time, S. A. Gregory, L. A. Thompson, and

W. G. Tifft found a large number of galaxies apparently arranged in a long chain extending about 100 million light years, called the Perseus–Pisces chain, previously identified by Joeveer and Einasto.[47] Observations in 1986 by Margaret Geller, John Huchra, and Valerie de Lapparent of the Harvard-Smithsonian Center for Astrophysics revealed that the galaxies in a certain region of space appear to be located on the surfaces of bubble-like structures of about 100 million light years in size, with voids inside the bubbles.[48] These new observations were among the first to show the 3-dimensional locations of a large, connected sample of galaxies (1,100 in all). Extending their survey to several thousand galaxies, Geller and Huchra reported in 1989 evidence for a "wall" of galaxies stretching at least 500 million light years in length.[49] The recent 3-dimensional maps of

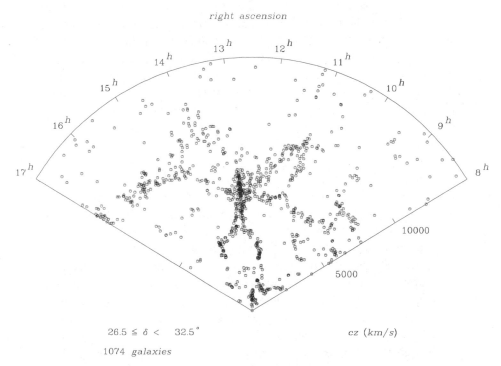

The Center for Astrophysics redshift survey (1986), showing the positions of galaxies in a thin pie-shaped slice through space. Each circle is a galaxy. Our position is at the bottom of the wedge. The radial direction directly measures recessional speed, which corresponds to radial distance in a homogenous universe. The farthest galaxy is approximately 500 million light years away.

large samples of galaxies have been made possible by advances in technology that allow the redshifts, and hence distances, of galaxies to be measured in fast and automated procedures.

It is not yet known whether the new cosmic structures found in a few selected regions of space are typical. What seems clear is that structures of *some* kind have usually been found at the largest possible scale in each survey of galaxies; that is, a survey that looks over a region of 100 million light years usually finds some chain or disk or *absence* of galaxies extending roughly a 100 million light years in size; a survey of a 200-million-light-year region finds structures of 200 million light years, and so on. Galaxy surveys are now in progress that extend out to a billion light years.[50] It remains to be seen whether these surveys will also show recognizable patterns and structures over such large regions.

Some cosmologists are worried that the observed inhomogeneities in regions of several hundred million light years might extend to larger sizes and threaten the foundations of the big bang model. Other scientists, including James Peebles,[51] are persuaded by surveys of very faint radio-emitting objects[52] and the nearly uniform cosmic X-ray emission[53] that the matter of the universe becomes smooth when averaged over several billion light years or more. That could be the distance over which the individual grains of sand on the beach would no longer be seen. Most cosmologists have confidence that the universe must be homogeneous when viewed on scales of 10 billion light years, since the cosmic background radiation is smooth and comes from such distances. If in the future we find filaments and bubbles and voids with sizes of a few billion light years, several times larger than those now mapped, then there would be a direct contradiction with the uniformity of matter implied by the cosmic background radiation. The big bang model could be thrown into crisis. At the present, many cosmologists feel that the observed inhomogeneities in the distribution of galaxies certainly have implications for the formation of galaxies but do not yet conflict with the big bang model and its assumption of homogeneity on very large scales. In any case, the large-scale structures must be reckoned with.

Related observations of cosmic structure involve the peculiar velocities of galaxies, that is, velocities that depart from the recessional velocity expected in a perfectly smooth and uniformly expanding universe. In 1987 David Burstein, Roger Davies, Alan Dressler, Sandra Faber, Donald Lynden-Bell, R. J. Terlevich, and Gary Wegner (dubbed the Seven Samari) suggested that a large group of galaxies within about 200 million light years of us are moving together, as if attracted by some large mass, called the Great Attractor.[54] The speed of this peculiar motion is about 10% of the expansion speed at that distance. Such large-scale peculiar motions were also present in the earlier data of Vera Rubin and co-workers.[55] Significant peculiar motions could complicate an accurate measurement of the overall rate of expansion of the universe.

An obstacle to understanding the distribution of mass in the universe and motions of galaxies is that about 90% of the detected mass in the universe is invisible. It emits no radiation of any kind—not optical light, nor radio waves, nor infrared, nor ultraviolet, nor X-rays. It is truly invisible. This detected but unseen matter is called dark matter. We know that dark matter exists, because we have observed its gravitational effects on the stars and galaxies that we see, but we have little idea what it is. The problem was first noticed in 1933 by the Swiss-American astronomer Fritz Zwicky.[56] Zwicky was able to estimate the mass of a cluster of galaxies in orbit about one another by measuring the amount of gravity needed to hold the cluster together. He discovered that the total mass thus inferred was about 20 times what could be accounted for by the visible stars in the cluster.

Zwicky's startling discovery was not taken seriously for many years. In 1973 Jeremiah Ostriker and James Peebles did theoretical analyses suggesting that the amount of visible matter in the disks of typical rotating galaxies was not sufficient to keep those galaxies from flying apart or drastically changing shape.[57] The observations of such galaxies quietly spinning around therefore suggested that their outer regions were filled with a halo of unseen mass, comparable to the visible disk mass, holding the galaxy together by its gravity. The next year Ostriker, Peebles, and Amos Yahil compiled observations of inferred mass and visible mass of various astronomical systems, from individual galaxies to large clusters of galaxies, and claimed that 90–95% of the mass of the universe is in an invisible form.[58] J. Einasto, A. Kaasik, and E. Saar independently arrived at the same conclusion.[59] Four years later Vera Rubin and colleagues at the Carnegie Institution of Washington and Albert Bosma of the University of Groningen estimated the mass of several galaxies by measuring the speed at which gas orbits around the center of each galaxy.[60] These researchers found direct and compelling evidence for about 5 times more mass in spiral galaxies than can be accounted for in visible stars. In similar measurements for groups of galaxies orbiting each other, 10 times as much dark matter as visible matter has been found.

It is important to distinguish between this dark matter, which has been *detected* through gravitational studies but not seen, and the unseen mass *hypothesized* by scientists who believe that omega equals 1. The latter we will call missing mass. To review, the light-emitting, visible mass provides enough density of matter to make omega equal to 0.01. Including the gravitationally detected but invisible mass—the dark matter—leads to an omega of about 0.1. A value of 1.0 for omega requires the existence of 10 times more mass, not only unseen but undetected altogether—the missing mass. We know that dark matter exists. The missing mass may or may not.

What is the nature of dark matter? Does it consist of numerous planets, or of collapsed stars called black holes, or of subatomic particles that interact with other matter only through gravity, or perhaps some new, as yet undiscovered type of subatomic particle? Depending on what it is, the dark matter could alter our theories of subatomic particles or of the formation of galaxies; it must be identified for any secure understanding of the cosmos. Astronomers have been deeply disturbed to realize that the luminous matter they have been staring at and pondering for centuries makes up a mere tenth of the inventory.

Putting aside for the moment the uncertain identity of the dark matter, a number of different theories have been proposed for the formation of galaxies and their distribution in space. In any such theory, two things

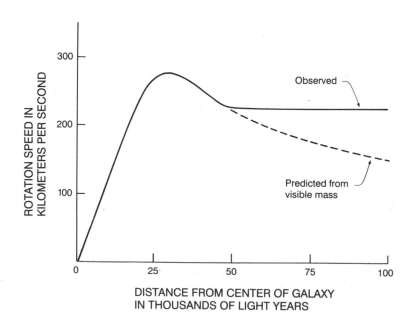

must be specified: the initial positions and motions of lumps in the distribution of cosmic mass, and the subsequent forces acting on the lumps. Scientists have traditionally assumed that it is mainly the force of gravity that acts on the initial lumps. One of the earliest pictures, called the gravitational hierarchy model, was first sketched out by Georges Lemaître in 1933.[61] Lemaître supposed that the mass in the early universe was nearly uniform but was bunched up very slightly here and there, like small ripples on the surface of a pond. The origin of these ripples he left for a later theory. In a place where the mass was bunched up, there would be slightly stronger gravity. This would cause nearby mass to bunch up more, attracting more surrounding gas. Gravity would get stronger, and the process would continue until a strong concentration of mass had formed. Small initial ripples and lumps would eventually produce single

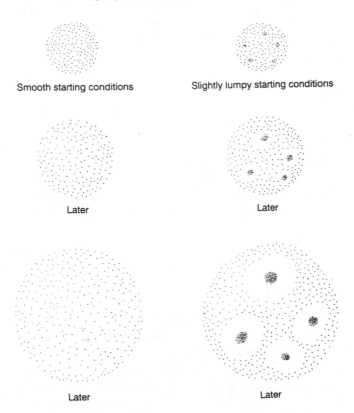

GRAVITATIONAL CONDENSATION
TO FORM GALAXIES

Smooth starting conditions

Slightly lumpy starting conditions

Later

Later

Later

Later

galaxies, larger ones would produce clusters of galaxies, and so on, with a hierarchy of structures. The locations of galaxies in space today would then be explained by the locations and sizes of the initial ripples, together with the subsequent action of gravity. In recent years, beginning with the work of James Peebles in the mid 1960s,[62] this theory has been made quantitative, but it does require specification of the sizes and strengths of the initial ripples. Modern theoretical investigations of the gravitational hierarchy model have used computers to simulate the growth of lumps in an expanding universe. In a typical such simulation, called an N-body simulation, 10,000 to 100,000 mass points, each representing a galaxy, are placed at initial positions, given an initial outward velocity corresponding to the expansion of the universe, and then allowed to interact via their mutual gravity. Dark matter and missing matter may also be added, making up some assumed fraction of the total mass and distributed in some assumed way. Despite the freedom in choice of initial conditions, the computer simulations have had difficulty in producing results in accord with the observed distributions and motions of galaxies.

In 1967 Joseph Silk, then at Harvard, calculated that lumps of matter initially smaller than about a thousand times the mass of a galaxy might not be able to hold together under the effects of radiation.[63] This result was incorporated into the "pancake" model, developed in the early 1970s by Y. B. Zel'dovich, A. G. Doroshkevich, and others in Moscow.[64] In this model, the first lumps of mass to begin growing were very large. There would have been many such lumps, of course. As each lump cooled, it would collapse under its own weight, with the collapse tending to be fastest in one direction. The result would be a thin "pancake" of gas, which would then fragment into many small pieces. Each piece would be an individual galaxy. In this picture, galaxies would tend to be distributed in sheets, following the shape of their parent gas cloud.

The gravitational hierarchy proposal is a bottom-up model for the formation of cosmic structures, with small lumps of matter condensing first and then growing larger and larger. On the other hand, the pancake model is a top-down model, with large aggregates of matter forming first and then splitting apart into smaller structures.

If galaxies are typically distributed on 2-dimensional surfaces, as suggested but not yet established by recent observations, then the pancake model might be favored over the gravitational hierarchy model. However, it is controversial whether even the pancake model can explain the thinness of the sheets of galaxies observed in some regions of space. (Here, thin means that the width is far less than the height or breadth, although that width might be a million light years across.) The big bang model assumes that gravity is the major force in determining the evolution and structure of the universe. And conventional wisdom holds that gravity acting by itself produces smoothly varying features in the locations of masses, with comparable widths, heights, and breadths of any groupings of galaxies. According to this view, additional physics or special initial conditions are needed to produce sharp features in the mass distribution, such as linear strings and thin sheets of galaxies. Conventional wisdom has been challenged by recent computer simulations, which show that sharp features can indeed occur if the initial inhomogeneities are sufficiently pronounced in small patches and short distances. These new calculations, by Changbom Park at Princeton University and by Edmund Bertschinger and James Gelb at M.I.T., use several million mass points in a version of the gravitational hierarchy model called the cold dark matter model.[65]

However, even the new computer simulations, which are the largest to date, may have problems. Any convincing theory of large-scale structure must explain the motions as well as the positions of galaxies. If the peculiar velocities of galaxies are caused by gravity, then they should be related to the inhomogeneities in the mass distribution. This relationship seems far from simple. Some large clusterings of galaxies seem to cause large peculiar velocities while others do not, unless the dark matter (and perhaps missing mass) is distributed in a special and complex way. Furthermore, the observed peculiar velocities on large scales seem too big relative to the peculiar velocities on small scales. Finally, there is the cosmic background radiation. Whether or not the matter of the universe is sprinkled about evenly on scales of billions of light years, it is certainly uneven and structured on smaller scales. If the observed structures have grown from small lumps in the distant past, then those initial lumps would have produced some uneveness in the cosmic background radiation. No such unevenness has yet been observed. If some unevenness is still not found in the next generation of measurements, with ten times more precision, then there will be a serious problem. In recent years, astronomers have become increasingly worried about reconciling the smoothness of the cosmic background radiation with the lumpiness of matter nearby.

Some scientists believe that the observed inhomogeneities, and especially the sharpness of those inhomogeneities, require other physical forces beside gravity or special initial conditions to explain what we see in our local region of the universe. For example, Jeremiah Ostriker of Princeton and Lennox Cowie of the University of Hawaii have proposed that gas pressure, generated by the explosions of stars, may have been the major force in forming galaxies and groups of galaxies.[66] A similar idea was proposed much earlier by Doroshkevich, Zel'dovich, and Novikov.[67] Such pressure waves might travel outward from various centers of explosions, expelling all the gas from a spherical cavity and then depositing it at the edge, where galaxies could then form. But even this nongravitational explanation for the observed distribution of galaxies apparently cannot explain inhomogeneities on scales as large as 100 million light years and larger. At present, many cosmologists feel that *no* current models are satisfactory. Both theoretically and observationally, an understanding of the large-scale structure of the universe is at the top of most people's list of the outstanding problems in cosmology.

Granted the bubbles and walls and strings of galaxies, the universe is still remarkably smooth compared to what it might be. The density of galaxies and the rate of expansion of the universe are approximately the same in every direction. And the intensity of the incoming cosmic background radiation varies by less than a part in 10,000 as our radio telescopes sweep the celestial sphere. Cosmologists certainly have to explain why galaxies cluster as they do, but cosmologists must also explain why the big picture is so smooth.

Instruments and Technology

We have said little about the importance of instruments and hardware in astronomy, but the dramatic progress of the 1980s in mapping the locations and motions of galaxies was fueled in large part by technology. New electronic light gathering and recording devices and computers have allowed images and colors of galaxies to be recorded faster, digitized, and processed in automated procedures.

To illustrate the electronic revolution in astronomy, we will consider optical astronomy. Most astronomical objects emit light at many different wavelengths, including wavelengths too long to be detected by the human eye, such as infrared or radio, and those too short, such as ultraviolet or X-ray. Optical light is light at wavelengths the eye can see.

There are two basic steps in detecting a light source: the light must be collected, and it must be recorded. In our own eye, the pupil collects the light, and the retina records it. Telescopes were invented to collect more light than the human eye can. One of the first big telescopes in the United States was the 100-inch telescope at Mt. Wilson, California, built around 1920 and used by Edwin Hubble and others. The "100-inches" refers to

Moonlight view of the 200-inch Hale Telescope on Mt. Palomar.

the diameter of its opening. With such a large pupil, a telescope can gather much more light and thus see much fainter objects than can a human eye. The largest telescope in the United States is the 200-inch at Mt. Palomar, California. It was completed in 1949. In the spring of 1990 the Hubble Space Telescope was placed in orbit about the earth, where it is immune to the distorting effects of the atmosphere. And in 1991 it is expected that the 396-inch Keck telescope, located in Hawaii, will be completed. Telescopes in the 50-inch range are referred to as moderate sized; telescopes larger than 100 inches are considered large.

Until recently, light from astronomical objects was recorded by a photographic plate placed at the "back end" of the telescope. The grains in a photographic emulsion, however, respond to only about 1% of the incoming light. To record the weak light from a distant galaxy, long exposure times were needed, making it difficult to undertake large surveys of many galaxies. In the early 1960s, telescopes were fitted with the first electronic imaging device, the image tube. The image tube sits above the photographic plate and greatly amplifies the incoming light. The photographic plate still records the light, but so much more light comes in that the needed exposure time is about 10 times less. The image tube was the first big advance in technology.

The next leap forward came in the period 1970–1972, when the photographic plate was replaced by a computer-driven digital detector. Such devices were developed independently by John B. Oke of the California Institute of Technology, Alexander Boksenberg of the Greenwich Royal Observatory in England, and Joseph Wampler and Lloyd Robinson at the Lick Observatory of the University of California at Santa Cruz. A digital detector translates incoming light into electrical signals, rather than darkened grains on a photographic plate. The electrical signals can be digitized and stored in a computer, and the stored data can be easily manipulated. For example, if it is known that the image of a distant galaxy was partly scrambled by the light from a foreground star, the computer can electronically subtract out the unwanted light and reconstruct a clear picture of the galaxy. Digital detectors were made possible, in part, by the revolution in computer microchip technology of the late 1960s. In the mid to late 1970s, Stephen Shectman of the Mt. Wilson and Las Campanas Observatories designed a cheap and simple digital detector that was reproduced in a number of observatories. Sometimes called the z-machine or Shectograph, Shectman's device has been used in the Center for Astrophysics redshift survey, for example.

The latest high-technology device to transform optical astronomy is the charge-coupled device, or CCD. The CCD is a digital detector with such high efficiency and sensitivity that an image tube is not needed. Indeed, charge-coupled devices represent an additional improvement of a factor of 5 or 10 over image tubes in exposure time needed. The CCD replaces both the old photographic place and the image tube. First developed in 1969 at Bell Laboratories, CCDs were introduced into astronomy largely through the work of James Gunn of Princeton and James Westphal of the California Institute of Technology.

An important consequence of the new high-technology instruments is that big observational programs once requiring large telescopes can be now carried out with moderate-sized ones. Given the heavy demand for large telescopes, the possibility of doing a lengthy project on a more available, smaller telescope can mean the difference between doing or not doing the project.

Initial Conditions and Quantum Cosmology

Initial conditions play a peculiar role in cosmology. The initial conditions and the laws of nature are the two essential parts of any physical calculation. The initial conditions tell how the forces and particles are arranged at the beginning of an experiment. The laws tell what happens next. For example, the motions of balls on a pool table depend both on the laws of mechanics and on the initial positions and speeds of the balls. Although such initial conditions must be specified at the beginning of an experiment, they can also be calculated from previous events. In the case of the pool balls, the initial arrangement was the result of a previous arrangement, which was ultimately a result of how the first ball was struck with the cue. Thus, the initial conditions of one experiment are the final conditions of a previous one. This notion fails for the initial conditions of the universe. By definition, nothing existed prior to the beginning of the universe, if the universe indeed had a beginning, so its initial conditions may have to be accepted as an incalculable starting point—the particular arrangement of balls in the rack before the first break. This distresses physicists, who want to know *why*.

The flatness and horizon problems are especially compelling if we believe that the universe could have begun with many possible initial conditions and physical processes, only a small fraction of which would have led to a universe as homogeneous and nearly flat as our own. It is certainly possible to imagine that the universe began with a uniform density and temperature and began with a near perfect balance between gravitational energy and kinetic energy of expansion. The question is whether such initial conditions are plausible. Are they probable or improbable? Probability arguments traditionally require that an experiment be carried out in a large number of identical systems or that an experiment be repeated many times on a single system. For example, you can sensibly speak about the probability of a car coming by your house between 8:00 a.m. and 8:01 a.m. on a Tuesday morning if you have looked out your window every Tuesday morning for a thousand Tuesdays and compiled some statistics. A thousand universes are not available.

How might the initial conditions of the universe have been determined? Did the universe suddenly appear at $t = 0$? Although the standard big bang model, based on Einstein's theory of gravity, requires that the universe exploded into existence from a state of infinite density, scientists agree that this model is not complete at extremely high densities of matter. In particular, Einstein's theory of gravity does not incorporate the physics of quantum mechanics. All other modern theories in physics do. In the 1920s physicists discovered that all phenomena of nature have a dual particle-like and wave-like behavior. In some cases, an electron acts like a particle, occupying only one position in space at a time, and in other cases it acts like a wave, occupying several places at the same time. The theory of this strange behavior is called quantum mechanics. The wave–particle duality of matter leads to an intrinsic uncertainty in nature, that is, an uncertainty not arising from our ignorance or our inability to measure but an absolute uncertainty. Nature must be described by probabilities, not by certainties.

Physicists have tried and so far failed to find a complete theory of gravity that includes quantum mechanics. However, it can be estimated that quantum mechanical effects would have been crucial during the first 10^{-43} seconds after the beginning of the universe, when the universe had a density of 10^{93} grams per cubic centimeter and larger. (Solid lead has a density of about 10 grams per cubic centimeter.) This period is called the quantum era or Planck era, and its study is called quantum cosmology. Since the entire universe would have been subject to large uncertainties and fluctuations during the quantum era, with matter and energy appearing and disappearing out of a vacuum, the concept of a beginning of the universe might not have a well-defined meaning. However, the density of the universe during this period was certainly huge beyond comprehension. For all practical purposes, the quantum era could be considered the initial state, or beginning, of the universe. Correspondingly, whatever quantum processes occurred during this period determined the initial conditions of the universe.

Using the general ideas of quantum mechanics, but without a detailed theory of quantum gravity, Stephen Hawking of Cambridge University, James Hartle of the University of California at Santa Barbara, and others have recently attempted to *calculate* the expected initial conditions for our universe.[68] This is very different from observing what the universe is today and working backward to figure out what the universe was like near its beginning. Hawking and Hartle are proposing to calculate how the universe *had* to be created—consistent with the general concepts in quantum theory and relativity theory—and then to work forward from there. The details of such a calculation must await a theory of quantum gravity; even then, the calculation may be too difficult to carry out in practice. Yet if such a calculation could be reliably done, the initial conditions would not have to be taken as a given. Initial conditions would be on the same footing as the laws of nature. All aspects of the universe could in principle be calculated and explained.

For some time, many scientists thought that the notion of an ultra-high-density beginning of the universe was an artifact of the idealized assumptions of the big bang model, such as the assumption of homogeneity. However, in the mid 1960s Roger Penrose and Stephen Hawking mathematically proved that even if the universe is not homogeneous at all, its current expansive behavior, together with the theory of general relativity, require that the universe had to have been enormously denser in the past—at least as far back in time as classical general relativity applies, that is, to the Planck era.[69] Thus, it seems that quantum cosmology must eventually be dealt with to understand the initial state of the universe.

Particle Physics and the New Cosmology

In the 1970s an important change occurred in theoretical cosmology. A group of physicists with expertise in the theory of subatomic particles joined astronomers to work on cosmology. They brought a fresh stock of ideas and a new set of intellectual tools to bear on the question of *why* the universe has the properties it does, not just *what* those properties are. In the "old cosmology," before the 1970s, most cosmologists concerned themselves with measuring the distances and motions of galaxies, the formation and composition of galaxies, the rate of expansion of the universe, and the average density of matter. In the "new cosmology," scien-

COSMIC TIME LINE

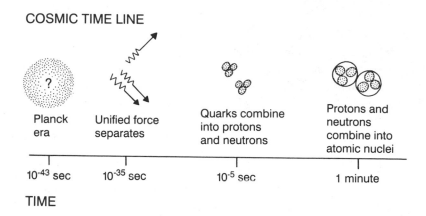

| Planck era | Unified force separates | Quarks combine into protons and neutrons | Protons and neutrons combine into atomic nuclei |

10^{-43} sec 10^{-35} sec 10^{-5} sec   1 minute

TIME

tists have seriously begun to ask why matter should exist at all. Where did it come from? Why is the gravitational energy of the universe so nearly equal to its kinetic energy of expansion (the flatness problem)? Why does the cosmic background radiation, arriving from billions of light years away, appear precisely the same regardless what direction the telescope is pointed in (the horizon problem)? Why is the ratio of photons to baryons in the universe a billion to one, rather than some much bigger or smaller number? Why did the universe begin in such a high degree of orderliness (the entropy problem)? The question of *why* was added to *what* and *how*. Some of these questions had been posed before, by a handful of scientists, but they had been largely dismissed or abandoned because no one had good ideas about solving them. Many scientists had regarded such questions as lying outside the purview of science. Particle physics expanded the science of cosmology.

A prelude to the future collaboration between particle physics and cosmology could be heard in the 1960s and concerned the number of different types of elementary particles. Theories of elementary particles and forces depend crucially on how many types of elementary particles there are, just as Aristotle's theory of the universe depended on his five elements: fire, water, air, earth, and aether. According to theoretical calculations first done in the 1960s, the amount of helium produced in the nuclear reactions of the early universe should have depended on the number of types of certain subatomic particles called leptons. (The electron, for example, is one type of lepton.) The more types of such particles, the more helium should have been produced. Thus, from the actual abundance of helium, which is measured to be about 25%, we can determine the number of types of leptons.

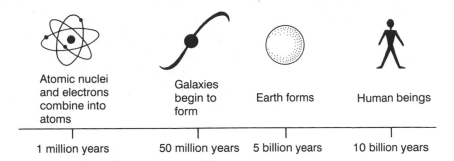

Atomic nuclei and electrons combine into atoms	Galaxies begin to form	Earth forms	Human beings
1 million years	50 million years	5 billion years	10 billion years

The expected relationship between helium abundance and lepton types was first suggested by Hoyle and Tayler in 1964 and later independently suggested by Robert Wagoner of Stanford in 1967.[70] V. F. Shvartsman of Moscow State University did the first quantitative calculation of the effect in 1969.[71] In 1977 Gary Steigman of the Bartol Research Foundation, David Schramm of the University of Chicago, and James Gunn of Princeton University, without knowledge of previous work, rediscovered the effect and did a more detailed calculation.[72] At the present time, three types of leptons are known—electrons, muons, and taus, and their associated antiparticles and neutrinos—but some theories of particle physics predict that there could be many more. Steigman and his collaborators claimed that *at most* two new types of leptons could exist, for a maximum total of five. Otherwise, the fraction of helium produced in the early universe, as calculated by the big bang theory, would disagree with the fraction observed. Recent experiments to probe the number of types of leptons, carried out in the giant CERN particle accelerator in Geneva[73] and at the Stanford Linear Accelerator in California,[74] indicate that there are *no* new types of leptons. The three so far observed are all there are. This confirmation of a result predicted from cosmology, using particle-physics technology, delights physicists and astronomers alike.

Most of the physicists involved in the above calculations had a background in astronomy and cosmology. However, in the mid 1970s particle physicists with barely any roots in cosmology began venturing into the field. A principle motive was to test the new grand unified theories.[75] Grand unified theories, often called GUTs, propose that three of the four fundamental forces of nature are actually different versions of a single, underlying force. (In a similar way, electricity and magnetism are not really two separate forces, because they can generate each other. A magnet moving through a coil of wire produces an electric current, and an electrical current flowing through a wire produces a magnetic field wrapping around the wire.) The three fundamental forces combined in grand unified theories are the electromagnetic force; the strong nuclear force, which holds together the subatomic particles in the atomic nucleus; and the weak nuclear force, which is responsible for certain kinds of radioactivity. Gravity, the fourth force, cannot yet be unified with the other three because a theory of gravity that includes quantum mechanics has not been formulated.

A grand unified theory has long been the holy grail of physicists. Since ancient times, physicists have sought minimalist explanations of nature. Theories with four basic particles are considered better than theories with ten. A single force that explains the fall of apples and the orbit of the moon is better than two. There is still little experimental evidence that grand unified theories are correct. One practical difficulty of testing them is that their effects become significant only at extremely high temperatures, much higher than can be created on earth or even at the centers of stars. Such extreme temperatures could be achieved at only one place, or rather, at only one time—in the infant universe, when all the matter of the universe was in the form of an ultra-high-temperature gas of subatomic particles. Thus, particle physicists became interested in cosmology.

In 1978 and 1979 several groups of particle physicists, including Steven Weinberg, then at Harvard and now at the University of Texas, pointed out that processes resulting from the grand unified theories, acting about 10^{-35} seconds after the big bang (well after the uncertain quantum era), could explain why the ratio of photons to baryons had the value it did.[76] Although there was a fair amount of uncertainty in these calculations, arising in part out of an uncertainty about which grand unified theory was most likely correct, cosmologists were enormously excited that such a previously mysterious cosmological quantity could be calculated at all. Weinberg was a highly respected theoretical physicist in the area of subatomic particles and was soon to win a Nobel Prize for his earlier work in particle physics. Encouraged by Weinberg's example, a new generation of young particle physicists decided that they too would work on cosmology.

The Inflationary Universe Model

One of the young physicists inspired by Weinberg was Alan Guth. In late 1978 Guth learned about the flatness problem from a lecture at Cornell given by Robert Dicke and was impressed by it. About a year later, while a postdoctoral fellow at the Stanford Linear Accelerator, Guth proposed a modification to the big bang model that provided a natural explanation of the horizon and flatness problems.[77] Guth's new cosmological model is called the inflationary universe model and has caused a major change in cosmological thinking. Some of the ingredients of the inflationary universe model had been discussed by others, including R. Brout, F. Englert, and P. Spindel of Belgium; Demosthenes Kazanas of the NASA Goddard Space Flight Center in Maryland; Martin Einhorn of the University of Michigan and Katsuhiko Sato of Japan; and A. A. Starobinsky of the Landau Institute in Moscow.[78] But it was Guth's clear statement of the model and its assets that galvanized the scientific community.

The essential feature of the inflationary universe model is that, shortly after the big bang, the infant universe went through a brief and extremely

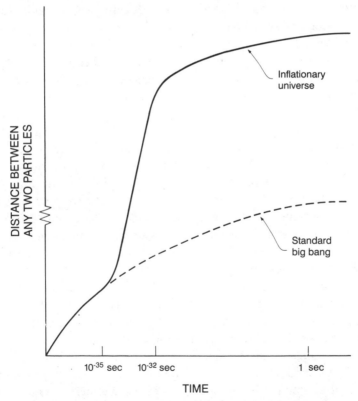

Expansion of the very early universe in the inflationary universe model versus the standard big bang model. The symbol ⌇ in the vertical axis means that the scale in the two graphs is different.

rapid expansion, after which it returned to the more leisurely rate of expansion of the standard big bang model. By the time the universe was a tiny fraction (perhaps 10^{-32}) of a second old, the period of rapid expansion, or inflation, was over. Under some conditions, the inflationary behavior of the infant universe is a natural consequence of grand unified theories. These theories predict that at the moment when the single unified force began acting as separate forces, the energy and mass of the universe existed in a peculiar state called the false vacuum, which behaves as if it had negative gravity. Negative gravity repels, so that instead of retarding the rate of expansion, it speeded up the rate of expansion. The inflationary period would have ended when the energy and mass of the universe changed from the peculiar state back to the normal state, with attractive gravity.

The epoch of rapid expansion could have taken a patch of space so tiny that it had already homogenized and quickly stretched it to a size larger than today's entire observable universe. Quantitative estimates can be made, although these estimates are uncertain due to ignorance about the details of the underlying grand unified theory and the resulting ignorance about exactly when the inflationary epoch began and ended. For purposes of illustration, we will assume that the inflationary epoch began when the universe was 10^{-35} seconds old and ended when it was 10^{-32} seconds old. At the beginning of the inflationary epoch, the largest region of space that could have homogenized would have been about 10^{-35} light seconds in size, or about 10^{-25} centimeters, much smaller than the nucleus of an atom. At the end of the inflationary epoch, this tiny homogenized region would have been stretched to something like 10^{400} light years. For comparison, *at that time* the region of space from which we now detect the cosmic background radiation was only about 1,000 centimeters in size, and today's entire observable universe was about the same size. Thus, the inflationary expansion would have homogenized the universe over an extremely vast region, far larger than any region from which we have data. The horizon problem is solved. Regions of space that appear to have never been close enough to have exchanged heat, according to extrapolations into the past based on the standard big bang model, were actually much closer, based on the inflationary universe model.

The inflationary universe model also solves the flatness problem. Regardless of the initial curvature of the universe—whether curved in the manner of an open universe or curved in the manner of a closed universe—any observable patch of the universe would be very nearly flat after the period of rapid expansion, just as a dime-size area of the surface of a beach ball would appear nearly flat after the ball has been inflated to a mile in diameter. The inflationary universe model firmly predicts that the universe today should be extremely close to flat. Equivalently, the average density of matter should be extremely close to its critical value—much closer than the one-tenth the critical value estimated from current observations. On this basis, the model can be either ruled out or supported from observational evidence, depending on whether the hypothesized missing mass is found to exist.

The inflationary universe model explains some of the outstanding problems of cosmology without resorting to the explanation-by-initial-conditions argument. In fact, it provides a physical mechanism by which the initial conditions of the universe may have been *irrelevant*—a notion that pleases and relieves many physicists. And the theory is elegant, combining subatomic and astronomical physics as it does. The inflationary universe model also has led to a new picture of the universe. Since the universe was enormously stretched out during the inflationary epoch, it may be vastly larger than we thought. Consequently, what we see from looking out of our largest telescopes—the small patch of space that we call the observable universe—may tell us very little about the universe as a whole. In this sense, the inflationary universe model has made us even smaller than we were before.

Andrei Linde of the Lebedev Physical Institute in Moscow has taken this idea further in what he calls chaotic inflationary universe models.[79] Linde speculates that under certain conditions an inflating universe can separate into different pieces, completely cut off from each other, in effect different universes. Each of the new pieces can repeat the process, in a random way, with each universe spawning many new universes. Some of the new universes might have very different properties from ours. In such a scenario, it would be impossible for us ever to learn about more than a tiny fraction of the possibilities and realities of nature.

The original inflationary universe model, as proposed by Guth in early 1980, had difficulties. In Guth's model, the special kind of matter and energy with negative gravity did not smoothly fill the universe but was scattered here and there. As a result, some parts of the universe expanded rapidly while others expanded slowly. The rapidly expanding pieces became more and more homogeneous, but the universe as a whole was left with a highly riddled structure, in conflict with the observations. This difficulty was solved in 1982 by a new version of the inflationary universe model developed by Paul Steinhardt and Andreas Albrecht at the University of Pennsylvania and independently by Andrei Linde.[80] In this revised model, all parts of the universe began inflating and stopped inflating at the same time. Overall homogeneity was maintained. Unfortunately, the new inflationary universe model of Linde, Albrecht, and Steinhardt also has serious problems. Theoretical calculations suggest that the model predicts large inhomogeneities in the early universe, spoiling the observed smoothness of the cosmic background radiation.[81] Moreover, a somewhat special kind of grand unified theory is needed to lead to a suitable phase of rapid expansion.

Furthermore, there is no direct observational evidence to confirm the inflationary universe model. In fact, one of the critical predictions of the model is inconsistent with the observations. Almost any version of the inflationary universe model predicts that the value of omega should be extremely close to 1, even 10 billion years after the big bang. As mentioned before, the actual measured value of omega is about 0.1, although this value is somewhat uncertain. In other words, we observe about one tenth as much cosmic mass as is required by the inflationary universe model. Scientists who believe on theoretical grounds that the model is right must therefore have faith that an enormous amount of mass is hiding from us, escaping detection, perhaps in a uniform and tenuous dark gas of particles between galaxies.

Despite its difficulties, the general features and results of the inflationary universe model are so appealing that most physicists believe some form of the idea is correct, even if it may differ in detail and mechanism from the original proposals of Guth, Linde, Albrecht, and Steinhardt. However, the inflationary model may not offer a good solution to Penrose's entropy problem. Some physicists, therefore, believe that the model will be replaced by a more fundamental understanding of the origin of the universe.

The Anthropic Principle

For the last couple of decades, a small but eminent group of physicists have addressed the problem of initial conditions in cosmology in terms of the conditions needed to create life. It seems plausible that not any conceivable physical conditions would allow life to form; the fact of our existence, therefore, might limit what possibilities should be considered. This notion is called the anthropic principle.

There are two forms of the anthropic principle, the weak and the strong. The weak form states that life can arise and exist only during a certain epoch of our universe. The strong form states that only for a special kind of universe could life arise at all, at any epoch. The weak anthropic principle restricts itself to the universe we live in; the strong implicitly refers to many possible universes. Some cosmologists have not only accepted the validity of the anthropic principle but also have used it as an explanation for various aspects of the universe.

Modern anthropic arguments in physics and cosmology began in 1961, with a short paper in the British journal *Nature* by Robert Dicke.[82] To understand Dicke's argument, we must go back to 1938, when the Nobel-prize-winning physicist Paul Dirac pointed out that a certain combination of fundamental constants of nature, when multiplied and divided the right way, happened to equal the current age of the universe, about 10 billion years.[83] (The fundamental constants of nature are such things as the speed of light, 186,000 miles per second, and the mass of an electron, 9.108×10^{-28} grams; such numbers are usually assumed to be the same everywhere and at all times. Indeed, every ray of light ever clocked in empty space travels at the same speed; every electron ever weighed has the same mass.) For Dirac, the coincidence between the two kinds of numbers seemed too unlikely to be chance, and he suggested that there must be some link between the fundamental constants and the evolution of the universe. Since the age of the universe clearly increases in time, the fundamental constants of nature would also have to change in time, maintaining the Dirac relationship.

Dicke explained Dirac's coincidence in a completely different way. Physicists, he reasoned, can exist only during a narrow window of time in the evolution of the universe. The carbon in physicists' bodies required a star to forge it, so a universe inhabited by physicists and other living beings must be old enough to have made stars. On the other hand, if the universe were too old, the stars would have burned out, eliminating the central source of heat and light that makes their orbiting planets habitable. Putting these limits together, physicists can exist only during an epoch when the age of the universe is approximately the lifetime of an average star. Dicke calculated this last quantity in terms of basic principles of physics and found it equal to the same combination of fundamental constants of nature that Dirac had noticed, numerically equal to about 10 billion years. Dicke declared the constants to be constant, as usually assumed. Much earlier or much later than the present time, Dirac's combination of fundamental constants would not equal the age of the universe, but physicists wouldn't be here to discuss the situation.

Dicke's argument exemplified the weak anthropic principle. As an interesting footnote, Dirac published a short reply to Dicke's paper, saying that Dicke's analysis was sound but that he, Dirac, "preferred" his own argument because it allowed for the possibility that planets "could exist indefinitely in the future and life need never end."[84] Aside from Dirac's objection, Dicke's weak anthropic principle is accepted by most cosmologists. Its statement and application concern only our actual universe.

In 1968 Brandon Carter, then at Cambridge University, stated the more controversial strong anthropic principle:[85] The values of many of the fundamental constants of nature must lie within a restricted range in order to allow life to emerge at all, even during Dicke's window of time. Carter argues, for example, that the emergence of life requires the formation of planets, which, in turn, may require the existence of stars that can shed spinning pieces of themselves. When Carter analyzes the conditions necessary to form such "convective" stars, he finds that the values of certain of the fundamental constants are restricted to a certain range. Such a range, of course, includes the values in *our* universe. Implicit in such an argument, and in all statements of the anthropic principle, is the assumption that galaxies and stars and other special conditions are necessary for life. Such an assumption is difficult to test, since our experience with life is limited to terrestrial biology.

In recent years, leading cosmologists have used the strong anthropic principle to explain certain properties of the universe.[86] According to this line of thinking, the particular universe we inhabit has some of the properties it does—including its values of certain fundamental constants and initial conditions—because only a universe with those specific properties could allow our existence. For example, the ratio of the mass of the proton to the electron, observed in the lab to be about 2,000, could not be 2 or 2,000,000 in *our* universe because such values would produce a physics and chemistry and biology incompatible with living substance. Values of 2 or 2 million might exist in other universes, perfectly satisfactory in every way except in their ability to allow our existence. Out of all these other possible universes, only a small fraction would have a proton-to-electron mass ratio suitable for life. As another example, Collins and Hawking used the strong anthropic principle in 1973 to discuss the flatness problem.[87] Starting with the assumption that galaxies and stars are necessary for life, they argued that a universe beginning with too much gravitational energy would collapse before it could form stars, and a universe beginning with too little would never allow the gravitational condensation of galaxies and stars. Thus, out of many possible universes, with many different initial values of omega, only in a universe where the initial value of omega was almost precisely 1 could we have existed. This anthropic solution to the flatness problem is accepted by some scientists and rejected by others.

Underlying these strong anthropic arguments is the notion that some properties of our universe are not fundamental and thus do not require fundamental explanations in terms of the laws of physics; instead, these properties are simply accidental outcomes drawn from a large hat. The numbers chosen by our particular universe are what they are because anything much different would exclude us, and our existence is a given. Other numbers were chosen by other universes, but we could not exist in most of them. The logic is similar to that in the explanation of why it rains on the days the parent drives his child to school and not on the days the child walks.

Scientists argue with each other about whether the anthropic principle constitutes a valid *explanation* of nature. For one thing, the explanations have force only if we invoke a large range of possible universes, with widely varying properties. Otherwise, we are back to explaining why nature is what it is. However, as mentioned earlier, some scientists are uncomfortable with postulating different universes. We inhabit just our one universe, and arguments that must go outside of that universe in order to explain it may have also gone outside science. For another thing, most scientists would rather explain nature in terms of basic laws that they can derive and prove, instead of conjectures that might be right but can never be proved. To many scientists, a fundamental theory showing that the ratio of the mass of the proton to the electron *must* be 1836 and nothing else—just as the ratio of a circle's circumference to its radius must be 3.1416 in Euclid's geometry—would be much preferred to an anthropic explanation. Some scientists consider that anthropic explanations of cosmic puzzles are proposed only when no better explanations can be found. Finally, the anthropic principle is sometimes tinged with teleological and religious suggestions. If life is indeed rare in all possible universes, was our particular universe designed with a purpose? Such questions, of course, run counter to the notion that our universe was simply an accident.

On the surface at least, anthropic arguments seem to involve life in a primary way. But there is a related question that sidesteps the issue of life. Could a universe exist with very different laws of physics, very different values for the speed of light and the mass of the electron and the initial value of omega? Or is ours the only universe possible, the only self-consistent set of laws and parameters? For some physicists, the search for the laws in our one universe has become a search after this much larger question.

FRED HOYLE

Fred Hoyle was born on June 24, 1915, in Bingley, Yorkshire, England. He received his A.B. in mathematics from Cambridge University in 1936 and his M.A. in physics from Cambridge in 1939. He was elected a Fellow of St. John's College at Cambridge in 1939 and in 1945 became a junior university lecturer in mathematics. In 1958 Hoyle became Plumian Professor of Astronomy and Experimental Philosophy at Cambridge. Hoyle was a staff member of the Mt. Wilson and Palomar Observatories from 1956 until 1965, when he founded and became the first director of the Institute of Theoretical Astronomy at Cambridge. He is a member of the Royal Society and a foreign member of the American Philosophical Society and of the National Academy of Sciences. Hoyle was knighted in 1972. In addition to his work in science, Hoyle has authored a number of works of fiction.

Hoyle has a wide range of interests in theoretical astrophysics, relativity, and cosmology, including gas accretion processes, stellar and big bang nucleosynthesis, and the steady state theory. In 1948 Hoyle, along with Hermann Bondi and Thomas Gold, proposed the steady state model of the universe, which invokes continuous matter creation to produce a universe that appears approximately the same at all times, with no beginning and no end. Hoyle has played a major role in understanding the origin of the chemical elements. In 1964 Hoyle and Roger Tayler did the first calculation of the cosmic abundance of helium expected on the basis of nuclear reactions in the first few minutes after the birth of the universe, according to the big bang theory. In this same paper Hoyle was the first to point out that the cosmic helium abundance depends on the number of types of subatomic particles called neutrinos, thus anticipating the recent link between subatomic particle physics and cosmology. A few years later, in 1967, Hoyle collaborated with William Fowler and Robert Wagoner to do much more extensive calculations of the elements and isotopes that should have been produced in the early universe.

The phrase "big bang" was coined by Hoyle during an influential series of radio talks he gave in England in the late 1940s. His popular books on astronomy, including *The Nature of the Universe* (1950) and *Frontiers in Astronomy* (1955), have stimulated a number of future astronomers. Since youth, Hoyle has challenged conventional wisdom. It is an irony of his career that many of his theoretical calculations have become the pillars of the big bang theory, yet Hoyle himself has continued to advocate the rival steady state model.

I wanted to start by asking you about your childhood. Did your parents have any influence on your deciding to become a scientist?

It's quite possible the answer is yes. But I must say that the whole of our valley was an engineering valley. It was connected with the wool trade. So I was brought up with the clatter of machinery from the factories, as I walked to school everyday. Now, that has to have some effect, as well as the home background. My father was always interested in scientific things. He had gotten a number of friends with similar interests, none of whom had a university education, but they tried to understand what was going on at the time. For example, from about 1922 onwards they built radio equipment. This was a great mystery in our village, and there were 20 or 30 people who were wiring up their own little radio receivers. There was a far greater feeling that it was possible for nontrained people to understand science than there is today. My father also had a chemistry set. I don't remember ever seeing him operate it, but it was lying there in a cupboard, so when I was about 10 years old I took it down and began to use it. I had very little use for formal education in those days. I didn't want to have anything to do with school.

How did your mother feel about that?

She knew that legally I had to go to school. So the struggle was between the legal requirement and my determination. That lasted about 6 months, before it was sorted out. During this time, I read the one book in the house on science, a chemistry book. It was after that that I took the chemistry set out. And I thought, "Well, if I have to go to school, I have to go," but at least I had the concept that I could read books and so forth. Then I became old enough to go to the town library. It was reasonably well stocked with books. It must have been about this time that popular books on relativity, like Arthur Eddington's, were beginning to arrive. So, with all these things put together, it wasn't an unlikely proposition that I would be interested in science. My mother was trained as a musician. My father played the violin. I could see that I didn't have the gifts that were necessary for music. That might otherwise have been an alternative, but I quickly rejected it. By age 12, I had pretty well decided that I was interested in science.

Did you get encouragement from your mother as well as from your father?

Yes. My mother was a school teacher, and by the time I was 3 she had taught me the multiplication tables. That's why I was bored in school. The English classes I regarded as absurd—for a reason that one of my colleagues at Cambridge many years later expressed very well when he said that every child of 8 has a perfect knowledge of syntax [laughs]. He said, "So why do we teach English?" I very much had that attitude toward the humanities from an early age. I would do them if I had to do them, but only if I had to. The tragedy of my life is that I never had a teacher who pushed me, from the age of 6 to the age of 18. Until I went to Cambridge, I was just allowed to drift, more or less.

You mentioned Eddington's book in relativity. Did you read that at the age of 10?

Not his technical book, but a more popular book. I was fascinated, but I didn't understand it. By the time I was around 12, I had understood the difference between being a layman who is interested and having a real understanding. Out of this grew my strong feeling that before one can be involved in science, one has to have a tremendous sense of craftsmanship. I still get mountains of letters from the lay public, proposing various ideas, and you can chuck it all out because you know that the brain isn't properly ordered to understand the problems they aim to solve.

Were there any other particularly influential books in science that you remember from this age?

Not of the Eddington type. But I would often grab hold of encyclopedias if I wanted information. We had nothing like the *Encyclopedia Britannica* in the house. My problem always, even before the teens, right to the time I went to Cambridge, was to find *anybody*—local community, school teachers, or otherwise—who could solve problems. I had reached the stage that there were technical things I couldn't solve, and there was nobody I could go to. It's *still* true. It's still the biggest bugbear in education generally. Whenever I go to Cambridge—my granddaughter is there in science— the first thing they say is, "How do we do this one?"

> *At this young age, did you think at all about the universe as a whole, about cosmology?*

Very much so, because of Eddington's books. He was always on about this. Cosmology was a very powerful theme in his writings. What he was really trying to do was to offset the importance of physics by the importance of the universe, trying to match the strength of the Cavendish Laboratory with astronomy in the internal struggles of Cambridge University. It was a political struggle that I inherited myself, years later.

> *Let me ask you a little about your education at Cambridge. Who had the biggest influence on you?*

Nobody had much influence on me in my first two years. I packed what is four years work into three, which was foolish for a person from a small, remote part of the country, not coming from one of the big, public schools. The first two years were formal, learning things like mathematical analysis. The third year was preparation for research. Then, our lecturers were world-class figures, and they had very considerable influence. People like Paul Dirac and Eddington would have had a huge influence. At the end of my second year, I made an effort to read Dirac's *Quantum Mechanics* during the summer, and Eddington's *Relativity*. At last I was tackling relativity properly. I had dropped the mathematical technique pretty well stone dead by then, which is a pity because there was a lot still to learn. I had completed the normal courses. Of those that were advanced, it wasn't really obvious what one should take. Mathematics is like a huge ocean. You feel like if you are thrown into it, there is an infinite distance to swim, and you still get nowhere. Whereas, if I had known what to do, it would have been fine.

> *When you read Eddington's books, and other books that presented cosmology, and you encountered the big bang theory, did you have any preference for open versus closed universes?*

No, I don't think so. If I go back to my student days, I simply learned relativity unemotionally. I didn't have any feelings. I just wanted to know what had been done. In those days, the Robertson–Walker metric was not available. It might have been available in the avant garde literature at that time, but I didn't encounter it. From 1936 I was doing research in nuclear physics, so cosmology was only a side issue to me. But had I been doing research in cosmology, I would have paid a lot of heed to H. P. Robertson's famous article on cosmology in *Reviews of Modern Physics*.[1] I finally did read it in a rather strange way. After the war, I began to learn technique again because I had to teach undergraduates—that's when you really have to understand, when you have to teach. I used to work quite a bit with Hermann Bondi in those days. Bondi was not sure what he should do. His expertise was in partial differential equations, fluid motions, things of that sort. So it was a big step when I suggested to him one day that maybe he should look at cosmology. I think he had been asked by the Royal Astronomical Society to write what was then called a Note, and we were discussing what he should write.

Why did you suggest cosmology?

Because, to my mind, the subject had been in abeyance for a long time. The Note was supposed to bring together a body of knowledge, if it seemed timely that that body of knowledge should be described. It was like a review article. Bondi thought this was a good idea so he began to read in relativity and cosmology, and then he came across Robertson's article. At that stage, I also went through it in some detail. It was only then that I began to have technical feelings about cosmology. If you remember Robertson's article, it's fairly encyclopedic in its structure. He doesn't express himself emotionally in favor of this or that. He just covers the possibilities. And so we naturally thought, "Has he really thrown his net wide enough? Are there any other possibilities?" That's when we began our own cosmological speculations—from the point of view of getting the range of possibilities as wide as we could.

And was that the seed of your later work on the steady state?

Yes, I think so. We had a strong hunch that there must be something else, because there was no creation of matter in it, and we felt that that had to be given expression at some point.

Why did you feel that creation of matter was an important element of a cosmological theory?

The actual historic sequence of events was this: It was Thomas Gold who had the notion that one might have a steady [state] universe. Bondi and I quickly pointed out that you couldn't have a steady universe unless you had creation of matter. Bondi, in particular, rejected it for a while. I accepted Hermann's rejection temporarily, but certain events in the next 8 months returned my ideas to it, and in early 1948 I discovered how to deal with creation mathematically. Then Bondi returned to it. That was the sequence of events. Gold had the idea. I was very interested that when I got the mathematical way of doing this—a crude way of coupling a scalar field to gravity—I found that the Gold steady [state] idea came as a consequence. Then Bondi and Gold returned to it and tried to use the steady idea as a philosophical principle, which to me has always been the cart before the horse. I've always felt that one must get the physical equations [first].

> *Going back to the late 1930s, when you were at Cambridge as a student and you became aware of the Robertson paper, what was the general attitude about the big bang theory? Was the interpretation of Hubble's discovery accepted?*

Oh yes. There was never any question at all about that. British astronomers were never plagued very much by the Shapley–Curtis controversy.

> *You mentioned that in developing your own version of the steady state theory you felt that the equations and the physics should come first, and the philosophy should come second. Is that something that is just basic in your approach to science?*

Yes. That might have been Dirac's influence. Dirac always said, "Find the mathematics first, and think what it means afterwards." That was probably a canonization of his own great experience with negative energy states. I was one of his few students. There was a great introduction to one of his papers—I think it was the paper in which he infers the magnetic monopole—in which he says that the nature of physics is that one must have the answer in mathematics first.[2] Using previous ideas, bits and pieces of previous physics, is not the way to make real advances in physics. The better way is to study abstract mathematics. It was almost impossible to be the student of a man who held that extreme view and for it not to rub off a little.

> *Was Dirac your thesis advisor?*

Yes. Well, actually, I had a strange business. I got an appointment [as a fellow at St. John's] before my thesis was due. So I never bothered very much about the thesis. But I did discover that when I began to get a salary, as long as I could claim to be a student, I was free of income tax. That is why Dirac agreed to be my research advisor. He thought it amusing. He didn't want students, but here was a student who didn't want a research director. That was the typical Dirac sense of humor. But he had a conscience, because he used to invite me to tea once a term.

I very quickly divined that the German style of discussing physics, hammering away at it and discussing it, was what Dirac hated. What he liked was to think quietly about a problem, work out the solutions, and then present the solutions. He didn't want a lot of talk at the blackboard. I realized this, and so I would talk to him about other things, like mowing the grass—mundane matters. And I discovered that he had a keen interest in practical problems. He's always given out as a person who was impractical. That's just wrong. The nearest I ever came to having a serious row with him was when we were building a house in Cambridge, and we didn't fit a heat pump. He said that was ridiculous. This was typical of Dirac. If he ever built a house, it would have to have a heat pump, but to my knowledge he never did.

Let me ask about some of your work on the synthesis of the chemical elements. When you realized that you couldn't make enough helium in ordinary stars, did that shake your faith at all in the steady state theory?

I never had any faith. I don't take much stock in faith. I'll explain that to you later.

Well, how did you fit that result into your cosmological views?

It had always been with us. Actually, Bondi, Gold, and I wrote a short paper called "Black Giant Stars" in 1955 in which we said that you had to have more helium than you could make in stars, in visible stars.[3] We had it made in clouds, which subsequently became the giant molecular clouds.

But I said before that I don't really work in terms of belief. I didn't go beyond saying that the steady state theory is a *possibility*. When I have defended it, I always defended it on the grounds that what were claimed as disproofs were things that a mathematician would not regard as disproofs. They were full of holes.

In both your paper with Tayler in 1964 and your paper with Wagoner and Fowler in 1967, you consider the big bang as well as massive stars for the production of elements.[4]

That's right. I asked Bob Wagoner to do a lot of calculations on that to see if we could get the [cosmic chemical] abundances right in what we called a little big bang.

There is another interesting point where we went wrong. You can take a local object, and you can ask what its mass would have to be in order to produce the standard ratio of photons to baryons—something like 3×10^9—that you get from the primordial synthesis of deuterium and lithium. If you do it, as we knew physics at that time, the mass is about 10^{20} suns. But we didn't do that case in our paper because it would have seemed almost too ridiculous to consider a mass of 10^{20} suns for a little big bang.[5] That's practically the whole [observable] universe. But the point that has emerged now, in the 1980s, is that there may be nonbaryonic matter—a hundred times more than baryonic matter, perhaps. Then you can calculate what is the *total* mass of such a little big bang, still sticking to the same ratio of 3×10^9 for photons to baryons. And the answer is 10^{16} [suns]. That's just a nice big cluster [of galaxies].

And this could all be done within a picture of a global steady state?

Yes. But I have had a problem with the steady state in the following sense. If you work from the mathematics to the universe, there's nothing I can see in the mathematics that would require anything more than average values to be maintained. That's to say, if I average over a certain time span, which should be at least a few Hubble periods, the situation is steady. Whereas, Bondi and Gold, with their "perfect" theory, require things to be steady on a timescale of, say, a hundredth of a Hubble period. Now that is a much more rigorous theory. To defend it against observational attack is much harder than to defend my version. If I had to give an opinion on the attack that was made against the Bondi–Gold restricted form of the [steady state] theory, I would have to say today that it is a lot weaker than most people think, but it is probably strong enough. But it isn't strong enough to rule out the version of the theory averaged over a whole Hubble period or longer.

In the last decade, there has been a lot made of the ability to limit the number of neutrino species from big bang nucleosynthesis calculations and the observed helium abundance. My understanding is that you and Tayler were the first to have pointed this out, in your paper of 1964.[6] Did you take that seriously at the time?

Well, we knew, of course, of the muon neutrino, and I suppose we realized that this would make a difference. I remember Steve Weinberg's asking me why didn't I use the abundances to argue back to the temperature.[7] It wasn't doable, unfortunately, because a slight variation in the temperature makes a big change in the abundances. And we didn't know the abundances accurately enough to do the calculation in reverse.

Would that have been the same calculation that would have allowed you to give a specific number of neutrino species [as constrained by the helium abundance]?

I think we could not have seen it as anything like the importance it is given today. If I were to claim that we saw its ultimate importance, that would surely be wrong. We weren't working with that level of accuracy.

Again in your paper with Tayler, after calculating that the ratio of helium to hydrogen should be greater than 0.14, you say that if it could be "established empirically that the ratio is less than this in any astronomical object, we can assert that the universe did not have a singular origin." Was a possible disproof of the big bang theory one of your motives in this work?[8]

No. My motive, as always in every piece of work I do, is to find out. The truth is more important than one's own predilections, a maxim which I feel is largely ignored at the present day.

Although you have been one of the advocates of the steady state theory, many of your theoretical calculations have been used by others as strong support for the big bang theory. Does that present any conflict to you?

Well, it does today. It wouldn't have done when Wagoner and Fowler and I were writing our paper in 1967. In those days, I didn't have any strong feelings about it. But today, I do have rather strong feelings that I don't think the big bang is right. I happen to get those views from something that hardly anybody else believes. I just don't think that the huge complexities of biology could have evolved in a mere 10^{18} grams of material on the earth. (The biosphere of the earth is 10^{18} grams.) I don't think that chemical evolution on the earth could possibly have produced the biological system. I think this has to be considered as a cosmological issue.

You think it requires more than the Hubble time to produce the biology we see? We need more time than the big bang gives us?

Yes. That's right.

When did you develop this idea?

Most of the last 10 years I've spent reading technical biology. It's come as a result of understanding the complexity of enzymes and all the rest of the biochemistry that is involved.

Your radio talks in the late 1940s and some of your popular books have been very influential. What motivated you in this kind of popularization of science?

It was really very simple. The radio talks started because Herbert Butterfield, I think, was asked by the BBC to give a course of five lectures. The idea had been started by a fellow of my college [St. John's], Peter Laslett, who had persuaded the BBC that there might be a high-level market for university-style lectures. Laslett was a historian, and Herbert Butterfield was a fellow historian. At a rather late moment, Butterfield contracted out, saying he just didn't have time to prepare the necessary material. So Peter came to me after dinner one evening at St. John's, and said, "Look are you in need of earning some money?" I replied, parrotlike, "My God Peter, am I in need of earning some money!" To which he said, "I can get you a contract to give five talks on astronomy, at something like 50 pounds a time. Are you interested?" So that's why I did it. I think my whole salary for the year was 400 pounds at the time. So 250 pounds was quite a lot. You know what Dr. Johnson said. "Sir!" he said, "there are few more innocent ways in which a man can be occupied than in the getting of money" [laughs]. That's the answer to your question.

I've heard that you originated the name "big bang." Is that correct?

Well, I don't know whether that's correct, but nobody has challenged it, and I would have thought that if it were incorrect somebody would have said so. I was constantly striving over the radio—where I had no visual aids, nothing except the spoken word—for visual images. And that seemed to be one way of distinguishing between the steady state and the explosive big bang. And so that was the language I used.

Are visual images important to you in your actual research, besides their pedagogical use?

No. I was never a very good geometer. I had to do all my geometry algebraically. I'm not very good at visual imagery.

What would you say are some of the critical events that led to cosmology's becoming a respectable subject of study?

I think it really dates from the particle physicists seeing cosmology as a vehicle, a convenient laboratory, for the execution of their ideas. I think if cosmology had had to depend on astronomers, it would be in a much weaker state. A great deal of sound and fury hasn't really led to any improved understanding of astrophysics that I can see. I think if you asked the young kids taking a course in cosmology today what are the reasons for believing in the correctness of the big bang, their first two reasons, probably their only good reasons, would be the reasons that we already had in 1965—the deuterium, helium, and lithium abundances and the microwave background radiation. The microwave background now is in awkward shape because of the recent Japanese submillimeter measurements. If those measurements hold up, there is potential trouble.

But you think it is the recent interest of the particle physicists, rather than those two observations you just mentioned, that gave the subject its respectability?

Yes. I sometimes ask myself, "If I had to attack this popular view, who are the hardest people I'm going to have to push against." And I feel it is the particle physicists, not the astronomers. The big guards on the front line of the scrimmage are the particle physicists.

Do you remember when you first heard about the horizon problem? I realize that in the steady state there is no horizon problem.

Actually, we did have a problem even in the steady state, because by the 1960s Narlikar and I began to see the steady state as an asymptotic solution. Basically, just as in any other expanding material system, random motions disappear with adiabatic expansion. But you aren't sure that random variations in the metric are going to adiabatically expand away. So we did a lot of work on that.[9] It's really a question of Mach's principle, which is another way of putting it. It took a long time, from the 1940s to the 1960s, before people were generally convinced that Mach's principle was outside general relativity. You have to add cosmological boundary conditions to general relativity.

Within the big bang model, did you think that the horizon problem was serious?

Well, I don't think it is solved, no.

Did you ever consider specialized initial conditions as a legitimate solution to the problem.

I never did, no. If you had asked me in the early 1960s what would be my objections to the big bang model, specialized initial conditions would have been one.

Would you feel the same way about the flatness problem? Or do you consider that to be a problem?

That's the closure problem, isn't it. I think the observations would vote for an open universe. Except if one makes an ad hoc addition of missing mass.

Do you put any stock in the flatness problem?

Well, I sort of feel that it's not really my problem. It's a problem of the people who want to argue for the big bang scenario. It's their problem. But I should tell you that in any of these arguments, I never overweight the arguments against my opponents. I think that [the flatness problem] is a strong argument against the big bang, but I always discount arguments against the other side because I feel you want to keep a balance. For instance, if a person has a theory that I am attacking, and the form in which they present the theory can be knocked down, but there is another form that evades my objections, I never just concentrate on knocking out the published version. I look at the other one.

I believe that [the flatness problem] is serious. If I were arguing for the big bang, it would worry me very much. But, as I said before, it's their problem, not mine.

Let me ask you a couple of questions about the inflationary universe model. When you first heard about the work of Alan Guth and other people, how did you react to it?

I was strongly caught up in biological studies at that time, so I didn't react very strongly. I've returned to cosmology only in the last two years, and so I've begun to construct what I think is the equivalent of the Guth inflationary model. In a sense, I wish I had studied it in 1980, when he produced it, because to me it is only a way of explaining why we should have a cosmological constant. And that doesn't seem to me to connect to the particle physics in a way that I would find interesting. I know that they go into the inflationary state for something like 65 orders of magnitude of the expansion factor, but they mysteriously come out with the radiation and the particles, and I don't see where the particles and the radiation have come from.

You said that you were working out the inflationary models in your own terms.

Well, when I found out that it led to the de Sitter metric, which is empty [of matter], then I wasn't very interested. The steady state equations lead to the de Sitter metric, but the model is not empty. That's a difference. So I feel it's probably a mathematical fiction, not the real physical thing.

So in order to get out the so-called benefits of inflation, you would prefer to stick to the steady state?

If I am going to get something that looks like inflation, but that really does create matter, then I find that the steady state is more attractive.

Why do you think that the inflationary universe model has been so influential?[10]

Because it obtains some of the advantages of the steady state theory (e.g., isotropy, flatness . . .) while still adhering, at least formally, to the big bang.

When you first heard about the work by de Lapparent, Geller, and Huchra on the distribution of galaxies on spherical-like surfaces, did that alter your picture of cosmology at all?

Yes, it altered my picture, in the sense that it suggested to me that if I am going to have a model in which there are many creation centers, what we used to call little big bangs, and not just one big bang, I had better make them related to that picture. That sets the scale. And that automatically told me it was going to be 10^{16} or 10^{17} solar masses.

In your recent paper in Comments on Astrophysics, *you wrote that the big bang theory, in its relation to observations, hasn't progressed one inch since 1965.*[11] *Do you consider that to be a problem with the theory or with the observations?*

No, the problem is not with the observations, surely. The observations have been an avalanche compared to what had gone before. I think it's a problem with the theory. It may be unique in the history of science that a correct theory doesn't lead to profitable correspondences to observations. It would be the first time in science . . .

Let me ask you a broader question along the same theme. In the last 20 years, how do you think theory and observation have worked together in cosmology?

I think you could say that there has been some development in the primordial nucleosynthesis—this group of three or four key nuclei. They have done some pretty hard work trying to relate the cosmological values to the observed values, as a function of the photon-to-baryon ratio. How good this is is another question, about which one might be optimistic or cynical according to one's point of view. I'll just present the cynical view. The helium value of 24% seems to me to be based on a trace sample of material, namely the dwarf galaxies. It would seem to me quite clear that if there are samples around in the universe running all the way from 20% to 30%—and there are quite a few observational astronomers who will show you evidence for that being true—then if you work hard enough and confine your attention to a particular kind of object, you can get the value you want [for the helium abundance]. Now the lithium 7 to hydrogen value. Is it 10^{-9} or is it 10^{-10}? Well, 10^{-10} is a pretty good value for carbon spallation by cosmic rays. So what does that mean? It looks to me that there is a case from the population I stars that 10^{-9} would be a primordial value—from the point of view of the disk of our galaxy, at any rate. So, I am not altogether sold that these are very good numbers. I have a feeling that the astronomers who claim these as good results wouldn't be so assertive if they didn't feel that they had a strong theory behind them. If the observers were *leading* and had to stand on a value, it wouldn't be anything like as precise as they claim. That's my feeling. That, I would have thought, was almost the strongest piece of observational correspondence [to theory]. What else is there? Yes, there is the microwave background. A lot has been done at shorter and shorter wavelengths.

Do you think that the extrapolations back into the early universe (in the big bang theory), done mainly by the particle physicists in the last decade, are valuable at this time?

I can only answer that by saying that if I was back in 1970 and was adjudicating on the award of money to research grants, I wouldn't root very strongly for giving a lot of money to that research. I have a strong feeling that progress comes out of relating mathematical analysis with few parameters to observations. For this [new work in particle physics applied to the early universe], the theoretical parameters are almost infinite, it seems to me. Of course, there are no observations. I find it very difficult to think it's worth spending any time at all on problems that relate to the early universe. It's quite different with your point about the distribution of galaxies. I think that's very interesting.

Did it surprise you when you heard about that discovery?

Not entirely. I remember Donald Shane counting galaxies out to about 18th magnitude—just a straight count on the sky. And he got a large-scale pattern. He discussed with me whether there were clusters of clusters. So that is rather what I expected.

So you knew about the idea.

Shane didn't have redshifts. He just had the counts [positions] against the background. But it was already clear that you could see a possible hierarchy. So my mind had been prepared for it. I wasn't shocked.

What do you think are the outstanding problems in cosmology today?

My attitude is unorthodox. I don't know whether it is useful or not.

I would very much like to hear it.

I take the view that the laws of physics are not what people think they are. What we count as the laws are a combination of the true laws together with a cosmological influence. There are long-range interactions. When you look at a book on particle physics and look at the masses of the fundamental particles, if you believe in the canonical view of physics, then all that is a part of basic physics. I don't believe it is. There *is* a basic physics. But in my way of looking at things, I don't have to assume that the various peculiar aspects of physics—particular masses, etc.—depend wholly on the basic laws. They are also a product of the way the universe actually is. What we actually see in the laboratory is a product of two things: long-range cosmological influence and the laws, which are very very much more elegant and symmetrical than particle physicists believe. None of this awkward left-handedness—it's just a cosmological influence that is producing that. To give an example, I feel that modern cosmologists are the way astronomers were for almost 2,000 years, in thinking that the elements of the planetary orbits had a basic physical significance. We know now that the basic laws don't have any particular relation to those numbers. The elements [e.g., orbital radii] just depend on where things are.

So that's what I am working on in cosmology. I've gotten quite a long way. I'm not entirely happy, though. I am reminded in the whole of this problem by a remark of Richard Feynman's. He thought that what we see as gravity really has to be a statistical theory. The details he thought must be the innermost physics of all. Unless we really see the bottom of it, we can't understand it. I have that feeling all the time.

Let me end with a couple of even more speculative questions. You may have to put some of your natural scientific caution aside. If you could design the universe any way that you wanted to, how would you do it?

If the physicists are right, I wouldn't do it their way. I sort of feel that if my way of looking at it turns out to be right, then probably one wouldn't want to do it any better.

With Mach's principle and the long-range influence?

Yes. After all, you say to yourself, Schrödinger's equation admits the existence of enzymes that will promote chemical reactions involving only the *minutest* energy differences, in such a way that physical systems— bacteria and so on—will find the bottom of the thermodynamic potential well to an *incredibly* high degree of accuracy. Now who is going to design a world that's better than that? I grant you that there will be a day when physics and astronomy and biology are sufficiently well understood that we will be able to begin to argue about what other games could be played. I think Feynman said it's like someone discovering the rules of chess. Until you've found the rules, you can't play the game. We don't know enough about the rules to play the game. There will be a stage when we do eventually know enough about the rules to ask whether we could play a better game.

There's a place in Steven Weinberg's popular book The First Three Minutes *where he says the more the universe seems comprehensible, the more it also seems pointless. Have you ever thought about this question?*

Well, I think Feynman was nearer the point when he said that you have to know the rules before you can understand the game that's being played. Feynman's example was a child who watches two grandmasters. First, the child has to figure out how the pieces move. But it's a long step from there to understanding the game, and a still vaster step to being able to play a better game.

ALLAN SANDAGE

Allan Sandage was born in Iowa City, Iowa, on June 18, 1926. He received his A.B. in physics from the University of Illinois in 1948 and his Ph.D. in astronomy from the California Institute of Technology in 1953. Since 1952 Sandage has been on the staff of the Mt. Wilson and Las Campanas Observatories. Among his awards are the Eddington Medal of the Royal Astronomical Society in 1963 and the Russell Prize of the American Astronomical Society in 1973. Sandage is an elected member of the Academia Nazionale dei Lincei (Rome). His research interests include stellar evolution, photoelectric photometry, stellar kinematics, galaxies, quasars, and observational methods to determine the rate of expansion and age of the universe.

As a young graduate student in the early 1950s, Sandage "was sent up to Santa Barbara Street" to become the assistant of Edwin Hubble, the great astronomer who discovered the expansion of the universe in 1929. When Hubble died, Sandage inherited the older man's mission of using the 200-inch telescope on Palomar Mountain to map out the distances and expansion rate of the universe. In the early 1950s Sandage helped with the first determination of the ages of globular clusters, which are believed to be among the oldest objects in the universe and thus of cosmological importance. He has undertaken a number of extensive programs to measure the distances to galaxies, the rate of expansion of the universe, and the rate of deceleration of that expansion. An influential paper by Sandage in 1961 recast theoretical cosmology into a form that could be confronted by observational results. Sandage minored in philosophy in college and has maintained an interest in that subject throughout his career.

I want to start with your childhood. Would you tell me a little about your parents?

My father was raised on a farm in Iowa, and my mother was raised in a small farming town. My mother was born in the Philippines, of American parents. She had traveled across the Pacific six times before she was six years old. My grandfather on my mother's side was the Commissioner of Education in the Philippines, under President Harding. He had spent eight years there and came back to the U.S. with three daughters to Lamoni, Iowa, to become president of Graceland College, which is the church school of the Reorganized Church of Latter Day Saints. My father was one of four children who were raised on a farm by his father, who had not finished the eighth grade. My father was the first in his family to go to what was called academy at that time, which is the equivalent of high school now, and then from academy to college. He went to Graceland College, and then on to the University of Iowa for a Ph.D. So he went from farming, on what was almost the western frontier in the 1910s, to a Ph.D. He married the daughter of the president of the college in Lamoni. He's still alive at the age of 87. I was raised in an academic atmosphere. My parents were connected with some university for all the time I was living at home.

Were either of your parents interested in science, either as a hobby or as part of their career?

Not in any serious way. They didn't know very much about science. My mother was trained in music and my father in economics. I became interested in science during a two-year period spent in Philadelphia when my father, on leave from Miami University in Oxford, Ohio, was connected with the Census Bureau. I became interested because of a childhood friend, who had a telescope. Upon looking through this telescope in the suburbs of Philadelphia, it became clear that I had to be an astronomer.

How old were you at this time?

I was in the fourth grade. Every weekend, I would go down to the Franklin Institute and to the planetarium. James Stokely, who was head of the Fels planetarium, was my early hero. That and the childhood friend were the beginning of my interest in science. I was always interested in numbers, but I didn't know much about mathematics at the time. I more or less learned science as a hobby after that, and it became a *very* strong passion. All of science became a great interest.

Did you have any idea at that early age that you might want to become an astronomer?

Yes, I knew right then that that was what I had to do.

You say had *to do? Why do you say* had*?*

Compelled. Out of a sense, not so much of duty, but there was just nothing else that seemed as worthwhile. I have been fortunate in my life in knowing, in general, what I have had to do. Now, when the neighborhood children come to our house, and I ask them what it is they want to do with their life, they have no compelling idea. Even through college, some of them have not yet found their goal, their vocation. I've never been in that state, yet it has not always been so comfortable to be so driven.

And you think you were driven at the age of 9?

Yes. It seemed required that I must learn everything about everything. Otherwise, I thought that I was not doing things properly. I felt guilty at not knowing. It developed into a sense of duty, yet I can't tell you where that sense of duty has come from. I still have it as a requirement, something like a permit, to do research.

Do you think it came from your parents?

Yes, because it's in the chemistry. It must have come from my parents, in the genes and therefore in my chemical makeup—not in the training, I think. There is this compulsion in other family members on my father's side. My father has it. Several of my cousins have it.

At this age, were there any books that you read about science that you can remember?

Yes, there were many popular books. I read a great deal. One was called *The Romance of Astronomy.*[1] These were books for children. One was called *The Stars for Sam.*[2] I expect that I still might have those two books. I was self-educated in astronomy. I never have taken undergraduate courses in astronomy. I read Baker's third edition of his textbook in astronomy when I was 11 or 12 and knew then what the harvest moon was like, why at some times during the year the ecliptic is much more steeply inclined to the equator than at others. Many of the celestial-sphere situations I had taught myself, from about the age of 9 or 10.

Before you went off to college, did you read any books that dealt with cosmology?

Yes, the books by Eddington, such as *The Nature of the Physical World* and *Space, Time, and Gravity,* but these were quite difficult at the time.

Did you read Edwin Hubble's Realm of the Nebulae?

No, I read that later, before I went to Caltech. I read it at a moderately early age, 15 or 16, probably. I remember reading James Jeans, and I do remember reading about Rutherford and the splitting of the atom. I was in the science club in junior high school and would attempt to explain the nucleus of the atom to the other members of the science club.

Did you have a preference for any particular cosmology at that time, or did you think about cosmology at all?

No. I simply wanted to be a practical astronomer, not a cosmologist. I wanted to do proper motions and parallaxes and all that good stuff. I got thrown into cosmology quite by accident. I became Edwin Hubble's assistant by the circumstances of events, and the compulsion forced me, out of some sense of duty, to do the extragalactic distance scale after I was on my own. I suppose, now, that I would have rather done parallaxes, radial velocities, and proper motions of stars [laughs]. It would have been a lot less controversial. It would have been a smoother road to travel.

It might not have been as interesting.

Oh, I don't know. Somehow, the setting of a problem and solving it, of and by itself, is enjoyable—to develop the methods of solution, to go and do it, and then to write the paper. It doesn't quite matter what the subject is as long as the problem requires detail. It's getting the thing done. The value of the Hubble constant [which measures the rate of expansion of the universe] not being done, in other people's minds, is a thorn, although I think it's done, and I think we know the answer for sure. I also think we know why other people have gotten the wrong value. But the fact that the problem is controversial and not solved is not a happy situation—psychologically. I don't mind being wrong. But what is disturbing is to listen to presentations where I truly believe there are fatal flaws in the opposing argument, such as neglect of observational bias in the samples.

Tell me a little bit about your undergraduate education at the University of Illinois.

I first went to Miami University, where my father was on the faculty of the business school. Knowing I must become an astronomer, I started in physics and had a tremendous grounding there by the hard taskmaster, Professor Ray Edwards. He was the one-man physics department at Miami. He was so inspirational that something like 80 eventual Ph.D's went through his hands on their way to first-class graduate programs throughout the country. Another teacher of great influence was Professor Anderson of the mathematics department at Miami. These two men were good to me, but they demanded much. I did not have a naturally analytical mind to which things come easy. So I had to work hard to understand the material. Then the war came, and the two years away in 1944 and 1945 were crucial to the maturation process.

You were in the Navy?

Yes. There I was in a situation that was difficult because it was practical— the repair of electronic gear. The training was to find something that was broken and to fix it under pressure. Many people in the program later became astronomers.

When you then went back to the University of Illinois, were there any particularly influential experiences or people there?

Yes. I knew that I had to major in physics as an undergraduate. The pressure and the competition to complete the courses and to do well was strong. Robert Baker, the only astronomer at Illinois, permitted me to do a junior and senior thesis. He taught me observational techniques. Had I not had these experiences in real research, as a junior and senior—and I was happy beyond words in this work—if I hadn't been taught these techniques by Baker, I would not have reached Caltech with the experience that came to be crucial in later being sent up to the Mount Wilson offices and becoming Hubble's assistant at the age of 24. So everything worked in a linear chain.

I applied to Caltech in physics because there was no astronomy then at Caltech. But that was the year the astronomy department began, with Jesse Greenstein as the one-man department. I eventually got a letter from the admissions office at Caltech saying that I had been admitted in the first class in the new department of astronomy. I had written a letter of application saying that I wanted to be an astronomer, and I was applying to Caltech to be associated with the new astronomy that was happening. I had read *Realm of the Nebula,* and it had gotten into the seed of my soul. Also, the 200-inch project had been common news since the 1930s, and I wanted desperately to be near that telescope.

> *But you say that you still wanted to do classical astronomy as opposed to cosmology?*

I think that the grandness of the cosmological dream was something completely beyond me. I didn't think I was capable of working in that ephemeral realm. It was something that you did only after you had seen if you could succeed in Peoria somehow. You had to do an apprenticeship before you could enter the gates. But classical astronomy had the precision (then and still) that holds its own immense satisfaction.

> *It sounds like you took cosmology seriously, though.*

Oh, very seriously. It was like going to a cathedral. I had the feeling that the world was magic. I went to Caltech still feeling as I had as a child that the world *was* magic, and that it had enchantment. Everything I had looked at from the time I was a child was enchanting, and I had to grab it somehow. Then I went to Caltech, and that was a very hard experience. The magic disappeared, in the understanding of what was required to do real research.

> *What do you mean when you say the "world was magic"? Can you tell me about that?*

I can't. It's gone. But I can remember something of what the child did feel. The world was spirit. I would go into the woods and see flowers and simply become overwhelmed. It was a continual experience of surprise, joy, and amazement that there is something rather than nothing. By magic, I suppose I mean mystery.

> *To see what there was.*

Yes, although the day-by-day world of people was harder than the world of nature. I couldn't wait for night to come and for the stars to come out. I would stand in the backyard and look at the appropriate time and identify the stars as they became visible out of the twilight. It was like being, I suppose, in a sort of heaven. I can't explain it in words even today. I had that internal feeling about everything—about physics, about the way the world works, and about why we are.

You are referring to the natural world?

Yes, the natural world. The world of people was a world to be avoided because it was more difficult than the world of things.

When I finally went to Caltech and realized that to become an astronomer you had to become an analytical machine, it was something of a crushing blow.

This didn't jive with your poetic sense of the natural world?

You couldn't be an amateur and accomplish anything. Although that's not quite true, because at Illinois I took plates, and I measured them. I counted a million stars. I learned to solve the fundamental equation of stellar statistics to derive stellar densities. It was all apprenticeship, and it was still magic at Illinois. The pressure was my own; the competition was not so fierce. Caltech was considerably more difficult than I had imagined. And the magic disappeared. It became a *requirement* to do the apprenticeship, so as to qualify for a vocation.

Has the magic come back?

No. It never has. It is difficult after all these years to explain what it was as a child that I felt, but I know that after the first year at Caltech, my life fundamentally changed. The magic of existence was replaced by the mysteries in the textbooks. The childlike awe was replaced by the awe of the enormous complication and order of the world of physics that was to be learned.

So you were learning these analytical techniques at Caltech, and somehow that made the poem disappear?

Yes and no. Something replaced the magic. By doing science in the Caltech way, in the problem-solving curriculum of the physics tradition, the world, in fact, became more mystical, in the sense that the interconnection of all of physics with mathematics became so beautiful but also so difficult. Throwing everything into those years at Caltech, I came to believe that reality *was* the equations. You get the connectivity at the deeper levels, and it's quite a mystical experience, one that takes enormous preparation. So this magic was replaced by mystic. It's not really *mysticism* in the way that word is generally used. You cannot do science in a mystical way, but only in a rationalistic, reductionist fashion. It is in the way that man has imposed a system on the world, called laws, and believes that he has some understanding. Yet there is no real understanding. Why do differential equations describe the world? No one understands how the world knows to work like that, but it does. What is action-at-a-distance in Newtonian gravity? Or what is the curvature of space in Einstein's alternate description of the gravitational force? One mystery has been traded for another, and yet we still do not understand. And take quantum mechanics. The modern way that the observer himself changes the situation is not our early intuitive notion of cause and effect. The philosophy of modern physics, in the sense I mean the word, is mystical. With its virtual states, nonempty vacuums, the universe created by a random fluctuation out of nothing, and so on, modern physics has become nonscientific in terms of what we would have considered scientific a hundred years ago.

> *Let me ask you about your work on the 200-inch telescope. When you were there in the early 1950s, you and a small number of people had a really unusual monopoly on the most powerful tool in astronomy. Did that make you feel a sense of responsibility about what projects you should work on?*

Absolutely. I was told what to work on in the first three years upon being sent up to Santa Barbara Street from Caltech to observe for Hubble. I had worked for Hubble in the summer of 1950 on bright, variable stars in [galaxy] M33, but then he fell ill with his first heart attack. Dr. Bowen then employed me to get the plates for Hubble on his large Palomar program, which had just begun. After Hubble came back to the Mt. Wilson offices in 1951, we would sit each week, and he would tell me what to do. So the first three years as a student, when I was the observer for Hubble, I was working to his direction at Palomar.

> *What about after that?*

After that, I felt a tremendous responsibility to carry on with the distance-scale work. He had started that, and I was the observer and knew every step of the process that he had laid out. It was clear that to exploit Walter Baade's discovery of the distance-scale error,[3] it was going to take 15 or 20 years, and I knew at the time that it was going to take that long. So, I said to myself, "This is what I have to do." If it wasn't me, it wasn't going to get done at that period of time. There was no other telescope; there were only 12 people using it, and none of them had been involved with this project. So I had to do it as a matter of responsibility. But I believe now that I was more interested in stellar evolution and following up the globular cluster work of my Ph.D. thesis.[4]

Before we go on, do you remember during this period of time, in the early 1950s, whether you had a preference for any particular cosmological model?

It was very clear to me from the beginning that steady state was wrong. There was never a question in my mind, because having been a student of Baade's and then understanding the age dating of the globular clusters and the connection between globular clusters and elliptical galaxies, it was obvious that all galaxies were the same age in their oldest stellar content. And that could not be a steady state universe.

What about within the big bang theory? Did you have a preference for, say, open or closed or flat?

No, that was a matter of direct observation. At the time, I didn't deeply understand the theory of the Friedmann models, nor did many others who were connected directly with the observations. Humason, Slipher, Mayall, and probably Hubble did not entirely appreciate the details of the Friedmann models, the collapse above a critical density, and the explicit (closed form) equations relating density to space curvature. It took me about 10 years to understand, and out of that self-education came the 1961 paper on the ability of the 200-inch to discriminate between cosmological models.[5] I had taught myself cosmology out of Heckmann's book, called *The Theory of Cosmology*; there was a book called *The Expansion of the Universe* by Couderc, translated by Sedgwick, that held the key to many of the central ideas; and there was the wonderful book by Gamow called *The Creation of the Universe*. Out of those—I probably read them five or six times—I finally understood what the papers in the earlier journals were saying, and I put it together in the way that I could understand. It turns out that that's the way other people of my generation could understand it also. All I did was make explicit what an observer should know so as to do practical cosmology.

When you started working on your program to determine q_0 [the deceleration parameter, which measures the rate at which the expansion of the universe is slowing down], do you remember what the motivation was?

There were two crucial papers in 1958 and 1959 by Mattig, in *Astronomische Nachrichten*.[6] Mattig was a student of Heckmann's, and Mattig later said that these were merely class exercises, but if you read those two papers, they are the first papers in the literature where the solution of the redshift-distance relation is given *in closed form*, without a series expansion. This was 1958. Hubble had died in 1953. Nobody—McVittie, Robertson, Heckmann, Tolman—had put those equations in closed form. They were always in series expansion. Once you had the Mattig equations between luminosity, redshift, and q_0 in closed form, then you saw the nature of what happened at the large redshifts. These two Mattig papers, I think, changed the subject of practical cosmology.

I didn't know that the equation for the space curvature was $H_0^2(2q_0 - 1)$ until Fred Hoyle came over from England and gave a course to the students at Caltech while I was still a graduate student. Hoyle's understanding of everything and teaching it to the students was the cornerstone from which I then began to work. So it was not true that I taught myself entirely from the books. I had understood parts of Hoyle's lectures, and out of that then came the 1961 paper, distilled into a language of the observers.

You think that is when you decided that you wanted to determine q_0?

Hubble had tried to find what he called the "deceleration," by the second-order term in a series expansion, as he mentions in his Rhodes lecture, "Observational Approach to Cosmology," in 1938.[7] However, if you read this in detail, it is evident that he has the sign wrong as to whether deceleration has occurred. But there the statement *is* made that by going to high enough redshifts, you're looking far enough back in time to permit a measurement of a deceleration.

Hubble never really emphasized what the cosmic significance of the redshift was concerning a creation event. When Milton Humason would come down from the mountain with an ever larger redshift, there would be a press conference, in which Hubble was asked for the significance of the observations. He would simply say that they were trying to carry the Hubble law to as far as they could. He did not emphasize the time scale, nor did he talk about the curvature of space from the *redshift* measurement. He talked about the curvature of space from the *galaxy counts*—that he well understood, because in 1936 he had done the experimental geometry of whether the volumes increased more or less rapidly than r^3. But he did not emphasize the dynamics that come directly from the Friedmann equation.

And you were turned on to all of that by Hoyle and Mattig?

Yes, and also by H. P. Robertson. He was my professor in mathematical physics. I couldn't have a better background for the problem that we later carried out. Baade was my thesis advisor; I worked for Hubble; H. P. Robertson was the leading theoretical cosmologist at the time.

Moving to more recent periods, do you remember when you first heard about the horizon problem?

Probably only much later than everybody else had heard about it. I probably didn't hear about the horizon problem until five or ten years ago.

Did you read about it, did someone talk to you about it?

I didn't go to meetings often. When Robert Dicke stated the problem that parts of the universe could not communicate with each other, yet the 3-degree radiation is so uniform, I had not appreciated its significance until some time later. The thrust for the distance-scale problem, which is a purely technical undertaking, was taking all my efforts and seemed far removed from the *philosophical* theory of flatness, etc. These subjects, which are now the modern cosmology, seemed at the time to be simply conjecture. I also thought that talking about such things was unusefully speculative. When the grand unification idea came, there were some five years when statements like "The universe can be made from nothing, from a vacuum fluctuation" appeared in papers in *Nature*.[8] I thought this work was crazy. It was not the nuts and bolts type of cosmology that I grew up with—the kind requiring the type of observations I could make that would be useful in measuring the Hubble constant, for example.

When you heard about the horizon problem, you say you didn't pay much attention to it for a while.

You mean by the horizon problem the communication problem and the question of delta T over T in the microwave background? And tell me what the significance of that is to an observer whose responsibility was to determine relative and absolute distances to galaxies? I didn't care at the time how galaxies were formed, because we knew that galaxies *did* form. And, as an astronomer, not a "new" cosmologist, it was crucial to study what galaxies were like, not how they were made. I thought, until about five years ago, that *cosmogony* [the study of the origin of the universe] was the worst subject you could go into.

What happened five years ago?

It dawned on me that ELS [Eggen, Lynden-Bell, and Sandage] had something to say about the formation of galaxies.[9] Then, when I was compelled in the 1970s and 1980s to add to the material on the velocities and chemical compositions of the stars in the galaxy to try to find out how the galaxy formed, I had to understand a bit about the collapse phases. I still think that there is a large separation between astronomers and astrophysicists who are trying to explain *how* it happened. I want to know, *after* it happened, what is it like now? So, in a sense, I was a geographer. Many *cosmogonists*, instead, are trying to understand *why* the continents did what they did. I was trying to find out what was *on* the continents. The difference between cosmology and cosmogony is that one *accepts* the universe; the other wants to know *why* and then *how*.

What is your view of the horizon problem today?

Now I think it's the most important problem in the field. I have become a devotee of modern cosmology. I am as worried as anybody that if you can't find these fluctuations [in the microwave background] out of which the galaxies are supposed to have been made, then maybe the whole thing is wrong.

The whole big bang model?

If one requires fluctuations to form galaxies out of gravity—and that seems to be what Laplace told us about the formation of the solar system—then we had better find with COBE [the cosmic background explorer satellite] the fluctuations, or one has a terrible problem.

I have become terribly interested now because of the quite beautiful grand unification theories. Formerly, I had no curiosity because I first thought it was too speculative. And only finally, very late, after reading Weinberg's book *The First Three Minutes*, did it seem to gell. Now particle physics and modern physics are so central to cosmology. This is the modern cosmology. Although I am still a nuts and bolts engineering cosmologist, I did begin to learn about that, and of course it's exciting and perhaps even convincing.

What is your opinion of the inflationary universe model?

I'm taken philosophically with Dicke's original problem of the flatness. That seems more important than the horizon problem, although they clearly are related. Why is q_0 so close to $\frac{1}{2}$ or omega so close to 1? I knew in 1961 that the universe comes out of the big bang with $q_0 = \frac{1}{2}$, regardless of what it develops into, and why is it still so close to $\frac{1}{2}$ this far away from the big bang?

Do you remember when you first heard about the flatness problem?

Perhaps about 10 years ago.

When you first heard about the flatness problem, did you take it seriously?

No, not at all. I thought, "Omega has the value it has and philosophy doesn't make one wit. Let's go out and try to measure it." It was philosophy, not physics, that omega has to be close to 1. It was the reasoning that then developed into the anthropic principle, and the fine-tuning problem. I didn't try to speculate back that way. Rather than speculate, it seemed to me that those with telescopes should try to see what is there, as a geographer, by direct measurement.

Do you still feel that way?

The beauty of omega equals 1 has been growing on me. I am trying to see whether I can reconcile the time scales as a test. Clearly the two time scales, one from the expansion age and the other from the age of the galaxy, [based on] stellar-age dating, have to be determined independently.

If the idea of omega equals 1 has been growing on you, how do you reconcile that with the required factor of 10 times more matter needed?

My thinking about dark matter has changed in the last three years. I thought that the evidence for dark matter was not at all proved, and certainly when I read that the rotation curves of galaxies are flat, showing dark matter, you are not going to find there the factor you need to close the universe. You need a factor of 100 [between the density of visible matter and the density required for omega equal to 1]. There is not even a factor of 10 from the flat rotation curves.

So you need another factor of 10. How do you feel about that other factor of 10?

I'm open on the problem now, and I guess I have been seduced in some way by going to the first CERN conference,[10] and to an Erice conference[11] of particle physicists and astronomers, where these things were married. That's where I began to take grand unification quite seriously. It's so beautiful that you think it has to be true.

It is remarkable that an observer would find a theoretical idea so beautiful that he is willing to accept it even when it conflicts with the observations.

I haven't accepted it. I haven't rejected it. It does not conflict with observations if we require 100 times more matter than can be directly seen, but nevertheless feel via cluster dynamics, as Zwicky proposed.[12] And we need not *reject* observations. The time-scale argument permits a direct measurement of omega. We should keep an open mind and try to sharpen the observations until there can be no doubt.

You see, physicists are the cleverest people in the world. Some people are going to spend their life trying to find this stuff by building detectors that can detect single high energy particles. To spend one's life in this way takes faith in the things unseen. I listen to the theoreticians talk, and they're in a magic world. I want to think they are crazy. But science has gone far beyond string and sealing in believing in the reality of things unseen, rather than in the assurance of things simply hoped for. They have become Bishop Berkeleyites.

Do you think they're right?

There is no way for me to tell. The crucial thing is experiments.

Well, if experiments are the crucial things, then why are you willing to at least entertain this other factor of 10 of mass?

It's clear that the jury is out. Until recently, I thought for sure there was no way that omega could equal 1 on the time-scale argument. I don't believe the Hubble constant can go any lower than 40 [kilometers per second per megaparsec], and I was entirely convinced that the age of the globular clusters was at least 18 billion years. Then this fantastic discovery came that the oxygen does not track iron as the iron goes down in the oldest stars. That's an amazing statement, but as an observer, that's a solid piece of evidence. The ages have to be decreased at least by a factor of 1.25. That takes the age of the globular clusters and the halos of our galaxy down to about 13 billion years.

So that's already pretty close to omega equals 1.

Yes, but only if the Hubble constant is low and if you're not willing to put the cosmological constant in. I heard George Smoot talk here yesterday, saying that five years ago at meetings such as this, you go to a cosmology session and there were not very many people, and everyone else thought that these were the way outs who weren't doing science.[13] But cosmology has taken over astronomy and become respectable. Physicists now call themselves astronomers, which clearly makes them respectable despite their claim that 99% of the universe is unseen!

And you were there at the modern beginning.

I was already old at the beginning, being a practical (i.e., an engineering) cosmologist. The subject has become respectable because of the detailed measurements of the cosmic background radiation. I should say something else. I didn't think that the discovery of the cosmic radiation was very important in proving the Friedmann model because we already *knew* that the universe had a beginning, based simply on the linear form of the Hubble law and the agreement of the time scales. What has struck me so clearly during the past 25 years is that three time scales agree within a factor of 2. The uranium half-life, the age of the globular clusters, and the value of the Hubble constant. Now, that argument is not convincing, and the 3-degree radiation is, of course, crucial. What has turned out to be so convincing is that the cosmic background radiation really is almost a blackbody, it's everywhere, and it shows the relative, large-scale motion of the galaxy, a motion that is the right order of magnitude for the random motions as we know them.

Let me ask you about another recent discovery—that's the discovery of some of the large-scale structure by de Lapparent, Huchra, and Geller.

They did not discover it. It started with Rood and Chincarini.[14] The central paper was then by Gregory and Thompson.[15] There has been a travesty of justice done in the reporting of the discovery of the voids and the bubbles and the sheets. The discovery was made by Gregory and Thompson, Rood, Chincarini, and William Tifft.[16] Gregory and Thompson feel as badly about the rewrite of the history as Alpher and Herman feel about the similar neglect of their prediction of the 3-degree radiation.

Has this sequence of discoveries, beginning around 1980 onward, surprised you?

I guess it finally made my mind capitulate to their existence. By that I mean that the Shane–Wirtanen counts had hinted at such large-scale voids and structures. Then when the Peebles's map came, showing the filaments, I tended not to believe that. Hubble's 1934 paper, which was one of the most elegant in science, showed that the galaxies were distributed homogeneously.[17] For that picture of a homogeneous universe to collapse then, having been brought up essentially by Hubble, took all of this stuff we have just talked about.

Has it shaken your faith in the standard model at all?

The standard model is a good approximation to reality. On the small scale you have what [Harlow] Shapley always said—nonhomogeneity. On the large scale, you have what Hubble said.

You asked earlier whether I ever thought there were two viable theories, such as the steady state in addition to the big bang theory. The steady state was out from the beginning on two grounds: the Hubble law and the fact that all galaxies were the same age. Of course, both of those positions are now known to be fuzzy. At any given moment, you think you know something, but 10 years later you know that foundation was less strong than assumed, and I'm sure that happens now. Science is not the thing I thought it should be as a child. It is *not* the discovery of absolute truth. Science is not the way to find *truth*, but only probable truth. You get only an approximation that always changes, and there are no absolutes. You don't read a physics textbook written 300 years ago to learn about physics, and I expect that 100 years from now you're not going to read the *Astrophysical Journal* to learn about astronomy. Science is the only self-correcting human institution, but it also is a process that progresses only by showing itself to be wrong.

How do you reconcile that philosophy with your feeling that you really think you understand what the value of the Hubble constant is?

That's like asking, "Do you know the value of the charge of an electron?" That's a measurement. There is no question that the charge of an electron is a number, with a measuring error affixed.

Let me ask you about your view of the outstanding problems in cosmology today. In 1970 you wrote an article in Physics Today *that had the title "Cosmology: the search for two numbers."[18] Do you think that things have changed since then?*

Now is the new cosmology. That was the old cosmology. That is when I didn't really care *how* the galaxies formed, and I didn't care *how* the chemical elements formed. I thought then that both were topics probably beyond the realm of science. Friedmann cosmology *is* still the search for two numbers, but the field has expanded to ask questions that I thought science was not permitted to ask. Of the three questions "Where? What? and Why?" I believed that science was permitted to ask only "Where?"

and "What?" The new cosmology is trying to ask, at the deepest levels, "Why?" "Why is there matter?" The modern cosmologists have been terribly successful. What really was a revelation to me was the discovery of the W and Z particles. I've asked people like A. De Rujula and John Ellis and others were they surprised, and they say, "Oh, no, it had to be there. It simply had to be there." I asked Abdus Salam that, and he said, "There is no question. The theory was just so beautiful, that if they hadn't found them, it was the experimenters who were wrong."

What do you think are the outstanding problems in modern cosmology now?

Omega and the dark matter. That's been forced upon us by the beauty of the theory. So you see I have kind of changed from a pure observer, hoping to see the absolute, to a mystic believer in beauty. It's called maturation, I suppose.

I want to end with a couple of philosophical questions. If you could design the universe any way that you wanted to, how would you do it?

If I were present at the creation, would I give the Creator better advice, and you are asking for that better advice. I wouldn't want to destroy all the mystery, as we do in reductionist science. The greatest mystery is why there is something instead of nothing, and the greatest something is this thing we call life. I am entirely baffled by you and me. We were both there near the beginning. The atoms in our bodies were made then, yet their sum now, in a living thing, is greater than the whole.

You ask how would I create the world differently. A better question would be, "What would you like to know in science?" I would like to understand the meaning of life. More to your point, perhaps the universe is the only way it can be for us to exist. If that's true, to ask to create a different universe is to ask to enter into genocide. Now that's the anthropic principle, but the more I think about how everything is so finely tuned, the more that principle makes sense. Everything that you and I need to live is given to us on this earth. Maybe the universe cannot be any different than it is now for us to exist. So my answer to your question would be that I wouldn't change a thing if I want to live and be the same as I am now.

There is a place in Steven Weinberg's book The First Three Minutes *where he says that the more the universe seems comprehensible the more it also seems pointless.*

I think that's a silly statement because the answer is not known. Pointless?

The universe is so mysterious in being tuned the way it is that I am willing to keep the option open and not ask for an absolute answer, which, I suspect, can never be gotten anyway. I would not have said it this way 15 or 20 years ago, but to end up like Nietzsche, sitting by a window for 7 years rocking, not talking to anybody because of his nihilism, is not the way—even if we don't know the path. Nihilism finally ends up in insanity, at least in Nietzsche's case in Basel. To avoid that, I'm quite willing to believe there is a purpose. But it *is* a belief. Weinberg, in his sentence, also states a belief, and why he's driven to that is probably as complex as why I am driven to the opposite pole. But I am not willing to be a Nietzsche nihilist, because I think that is much more pointless.

GÉRARD DE VAUCOULEURS

Gérard de Vaucouleurs was born April 25, 1918, in Paris. He received his L. ès. Sc. in 1939 from the University of Paris and his Ph.D. in physics in 1949. After working at the Australian National University (1951–1954), the Yale-Columbia Southern Station (1954–1957), the Lowell Observatory in Flagstaff, Arizona (1957–1958), and the Harvard College Observatory (1958–1960), de Vaucouleurs joined the department of astronomy of the University of Texas at Austin in 1960 and has been professor of astronomy there since 1964. Among his awards, Professor de Vaucouleurs received the Herschel Medal of the Royal Astronomical Society of London in 1980 and the Russell Prize of the American Astronomical Society in 1988. He is a member of the U.S. National Academy of Sciences and a corresponding member of the National Academy of Sciences of Argentina.

A classical astronomer, de Vaucouleurs' research interests have included photography, photographic photometry, variable stars, novae and supernovae, planetary studies, extragalactic astronomy, and cosmology. Among de Vaucouleurs' contributions to cosmology are the identification of the local supercluster of galaxies, the compilation of extensive reference catalogues of galaxies, the advocacy of an inhomogeneous and hierarchical distribution of matter in the universe, and a critical re-evaluation of the cosmic distance scale and the rate of expansion of the universe. De Vaucouleurs is known for taking controversial positions that often turn out to be true. On his blackboard, in neatly chalked letters, are past quotes from colleagues who once disputed de Vaucouleurs' claims. De Vaucouleurs married the late astronomer Antoinette de Vaucouleurs in 1944, and they shared their careers together.

I wanted to start by asking you some questions about your childhood and early background. Could you tell me a little bit about your parents?

I was born and educated in Paris, France. My parents were divorced, and I was raised by my mother, Raymonde, a talented painter, who gave me strong moral principles. She died in 1987 at age 94. I read my first book in astronomy when I was 12, in 1930. It was *L'Architecture de l'Univers* by Paul Couderc. That was my first contact with cosmology. It was an excellent book, but just a little above my head. Couderc was a teacher of mathematics and a superb popularizer of science. In fact, he was my professor in high school in 1932 and became an astronomer at the Paris Observatory in 1944. My mother's brother, Georges de Vaucouleurs, let me read the book. He had been a member of the French Astronomical Society since 1925, and he had bought the book in 1930. On summer vacation I read most of it, but it was only when I was 14, in 1932, that I decided to become first an amateur astronomer and then a professional astronomer.

What was the motivation for your decision?

I think there was something in the family. As I said, my uncle was a member of the French Astronomical Society. In the 1920s, my mother had bought a book called *Le Ciel*, which was one of those big Larousse books about astronomy, and I read it in the late 1920s.[1] The decisive influence was that I read four booklets by a French popularizer of science and astronomy, Abbé Moreux, who was a professor of mathematics at the Seminary in Bourges. The titles were something like: "Where Are We?" "Where Are We Going?" "Where Do We Come From?" "Who Are We?" The last one was more about biological evolution and the Catholic point of view. These books attracted my attention. My mother bought me my first "spy glass" in June 1932, and then I started observing. I became interested in galaxies almost immediately—through the observations of some nebulae in 1932 or 1933. And I tried to observe galaxies with the telescopes of the French Astronomical Society, which I joined in 1933. I was not quite 15. So, I've been a member of the French Astronomical Society for 55 years.

Did your father also encourage you?

No. My father had no influence whatsoever. I saw little of him and he did nothing to encourage my scientific interests. He could not see any future in my "scribblings."

At this early age, around 14 or 15, when you decided that you wanted to become an astronomer, did you think about the universe as a whole?

No, not too much. I was more interested in looking at the planets, the moon, the sun, and then the stars. Nevertheless, I began to fill five booklets of bibliographic references on nebulae and star clusters in 1933 or 1934, which you might say were the first seeds of the reference catalogues. By 1935 I had access to the 6-inch and 7-inch refractors at the observatory of the French Astronomical Society. I used them extensively, sometimes at the expense of my school studies. Trying to observe all night and attend class the next day is a strain. I offered to help the librarian of the French Astronomical Society to straighten out the mess and clean the dusty shelves of the library, so in this way I had access to things like the *Harvard Bulletin*, the *Mt. Wilson Contributions*, the *Astrophysical Journal*, *Popular Astronomy*, and all the publications that were not usually read by members of the Society, but which I devoured assiduously.

And you read those journals at the age of 15?

No, not 15. Seventeen to 20, in the last few years before the war. Essentially from 1935 to 1939. I was an active member, and I received some early prizes and medals from the French Astronomical Society. Then I became a lecturer at the Paris planetarium in 1937. I was part of a little team of students and amateur astronomers, headed by Reysa Bernson. She was the secretary of the Association Astronomique du Nord, that is, the North of France. She was Jewish, and she was deported and killed by the Germans during the war. We had a good group of enthusiastic amateur astronomers. We lectured at the planetarium to something like a half million people during that summer. For myself, I gave some 250 lectures, which was a good start.

Good practice for teaching at the university.

Yes, a full-scale planetarium is a wonderful teaching instrument. Then I began observing. I looked at practically all the Messier objects, in the years 1933 to 1936, first with a 3-inch refractor at home, then a 7-inch refractor at the French Astronomical Society. I began taking photographs.

I remember photographing two supernovae. The famous one in IC 4182 in 1937, and one in NGC 4636 in 1939.

This was after you had finished at the University of Paris?

No, I was then an undergraduate student. I had earned my bachelor of science degree—that's the final degree of high school in France, you know—in 1935 and 1936. And then I did one year of special mathematics at the St. Louis Lycée in Paris, just across the street from the Sorbonne. But I was fully aware that the Second World War was coming, and it was affecting my schedule of studies. I felt I had to obtain some university degree before the world collapsed. So instead of taking the long and chancy route through one of the "Grandes Ecoles," I crossed the street to the Sorbonne to do a quick "Licence" in mathematics, physics, and astronomy, which I finished in 1939, just before the war. In 1939 I was introduced to Julien Péridier. He was a wealthy amateur astronomer, and he had been a member of many astronomical societies, since the early part of the century. He was, when I met him, the general director of the Paris transport system—the whole thing: buses, trams, and subway. In the 1930s he had built a very nicely equipped observatory of professional quality in the southwest of France, at Le Houga. He had a superb twin 8-inch refractor and a 12-inch reflector, which are both here in Austin today. He was looking for a young astronomer to work at the observatory, and I was fortunate to be introduced to him by an engineer, Mr. Saget, who was the head of the astrophotography section of the French Astronomical Society. I was very actively interested in photography in those days, and later I wrote several books on photography.

And so I began to work for Mr. Péridier in July 1939. First, I observed the planet Mars, which was in opposition in 1939, and then I did some experiments in photographic photometry. I designed some out-of-focus cameras to do stellar photometry. But the war came, and I had to spend a year and a half in the French army, the inglorious French army of 1939–1940. I was lucky and came out unscathed. I returned to the Péridier Observatory in July 1941.

During your time at the University of Paris, were there any people who were particularly influential besides Péridier?

Yes, my professor of physics, Jean Cabannes, was the leading French physicist in molecular optics. He was the director of the Physics Research Laboratory at the Sorbonne. As soon as I graduated in 1939, I contacted him and asked if I could become a grad student in his lab, and he said yes.

This was postponed because the war came. In addition to the credit classes, I took classes from Georges Bruhat, a physicist who was interested in astronomy; he had written an excellent text, a high-level popular book on the sun. In 1939 he gave a one-semester course of lectures on the physics of the sun. I remember taking that. There were very few students taking this class because it was not for credit. Bruhat was also killed by the Germans during the war, for helping students to escape deportation.

By the time I graduated from the Sorbonne with my undergraduate degree, I was living for astronomy 24 hours a day. I had access to the Paris Observatory library. I read books there and translated many research papers from English and German into French. So I learned a lot of astronomy on my own.

> *In your studies of astronomy at this time, in the late 1930s or so, had you studied the big bang model?*

I was aware of it because I read books and the review articles. There were excellent review articles in those days, high-level review articles—in *L'Astronomie,* the French bulletin—which we don't often find today in popular magazines. We had excellent articles about stellar astronomy, including articles on extragalactic astronomy by Henri Mineur, who was a young and very active promoter of the new astronomy in France.

> *Can you possibly remember what your early impressions were of the big bang model?*

Well, there was great excitement with the discovery of the redshift and the Hubble law. When Couderc wrote his book in 1930, the Hubble law was one year old, and it was the first book that brought the subject to the French reading public. I must say that I did not spend too much time speculating about beginnings and philosophical implications of the big bang. The expansion of the universe was an observed fact. It was not called the big bang, then, of course. It was the primeval atom of Georges Lemaître.[2] I didn't have a strong reaction one way or the other. I was interested in what is today inside the universe, not in speculating about origins, although I was aware of some of the implications.

> *Did you accept the interpretation of Hubble's measurements, as indicating an expansion?*

Yes. Yes. I had no question at that time. I accepted it. Why? It was what the authorities were saying. I started like an innocent, believing what the authorities said. Only later you begin to question the dogmas. Yes, I accepted it, that it was valid, that it was reasonable. But I was more interested in the planets, especially in Mars, and the stars. I did a lot of stellar photometry, observed variable stars. It was not for a lack of interest in galaxies, because I was continuing my bibliography [catalogue of galaxies]. But I had no telescopes suited to such studies. I did photograph a few nebulae, a few supernovae. But I realized I did not have the tools.

Let me go back to the early years in my career. After I had spent about a year and a half at Péridier Observatory, I received a fellowship to come as a graduate student to the Physics Research Laboratory in Paris. I went back to Paris in the spring of 1943. Then I started the research work for my dissertation, which was in molecular physics. The decisive step occurred as a result of a mishap. After the liberation of Paris, the following winter, we were still at war, and it was very cold. There was no heating at the Sorbonne, so the lab was closed for a couple of weeks at one time. As usual, someone forgets to turn off the water when this happens. So when the thaw came, there were burst pipes. One day I arrived at the lab, and the assistant said, "Oh, Monsieur de Vaucouleurs, it's terrible. Your office is flooded." And there was several inches of water on the floor. Unfortunately, some of my plate collections had been ruined. I turned around and immediately went to the telephone, and I called Daniel Chalonge, a leading astrophysicist in those days. He was the acting director of the Institute of Astrophysics. I said "Monsieur Chalonge, I'm fed up with the old Sorbonne. Can you give me a lab at the brand new Institute of Astrophysics?" I had hardly finished, and he said, "Why, of course. When do you want to come?" That was my admission test, to enter the Institute of Astrophysics as a grad student [laughs].

I finished my dissertation and graduated in the spring of 1949, which I think is the year when I published more papers than at any other time. I published something like 25 papers, including popular articles, in 1949. My wife, Antoinette (whom I had met in 1944), and I were busy with various projects. Then we moved to London, because with the background of having been an amateur astronomer interested in Mars, and a physicist interested in galaxies, there was little future in French astronomy. As Chalonge sadly remarked to me one day, "Il faut faire ce qui est à la mode" (one must do what is in fashion). You had to have a powerful patron, or otherwise you were lost.

You needed a powerful patron to get time on a telescope?

No, but to be promoted. To be protected. Also, the salary scale was pitiful in those days. It was just after the war. And then the instruments available were very limited. There was a 32-inch reflector in Haute-Provence, and a 50-inch, which had a glass so old and full of stresses that one day it exploded. We used them after the war. I had met Harlow Shapley in 1949 also, and I wanted to come to this country because that's where galaxies were being well studied. But at the time he was looking for solar astronomers.

I wanted to ask you some questions that possibly related to your meeting with Shapley.

Oh, it was very simple. I had read his books. I had written to him.

Which of his books?

All of his books, especially *Galaxies.* I had long ago read Hubble's *The Realm of the Nebulae.* I still have them. I was very familiar with all that had been published in the *Astrophysical Journal, Astronomical Journal, Astronomische Nachrichten, Harvard Bulletin, Harvard Annals.* I still have a good collection of *Harvard Annals* myself; I bought it before the war. I started building a sizable astronomy library as soon as I could afford it.

I wanted to know your motivation in the early 1950s when you began working on the local supercluster.

That was later. My first work on galaxies was done in France, at the Haute Provence Observatory. I did photographic photometry, and that's where I introduced the $r^{1/4}$ law that expresses the brightness distribution in elliptical galaxies. I published this in 1948.[3] My concern at the time was this: I realized fully that there was no point in competing with Hubble when you had only small-sized telescopes. Studying Hubble's work in detail, I discovered that he had made some technical mistakes in photographic photometry. Also, having been trained as a physicist, I was not impressed by this eyeballing of photographs to decide the Hubble type of a galaxy. I thought, "To be scientific, we must be quantitative. What are we doing when we look at a photograph?" Well, we look at the distribution of surface brightness, or specific intensity. Therefore we must do photographic surface photometry to be able to classify galaxies and measure things quantitatively." So, my training in photography and photometry was very helpful, and I started a program of photographic photometry of galaxies in 1945–46 at the Haute Provence Observatory. I defined a number of quantitative parameters to measure the size, brightness, and so on, of galaxies, which are still used today. And I stumbled onto the $r^{1/4}$ law, which for 20 years was denied as being impossible. Now everybody is making hay out of it.

Wasn't it around then that you were doing your work that ultimately questioned the assumption of homogeneity in cosmology?

In 1949 I had already started a revision of the Shapley–Ames catalogue of galaxies.[4] I had measured the diameters of galaxies on as many plates as I could use, including those of Isaac Roberts, a wealthy British amateur at the turn of the century. He had made the first large collection of photographs of galaxies, using a 20-inch reflector. It was in the course of this work on revising the Shapley-Ames catalogue—providing better galaxy types, sizes, and so on—that I began realizing that there was a stream of galaxies across the northern galactic hemisphere. It was either a very rare chance alignment, or, more likely, it was a physical association of galaxies, on a scale much larger than clusters or clouds of galaxies. Of course, I was aware of the work of Shapley on very large-scale density gradients [inhomogeneities] across the universe.[5] I didn't have much to say about that, but I was aware of it. I was also aware of the work of Fritz Zwicky, claiming that clusters of galaxies, not galaxies, are the building blocks of the universe.[6]

It occurred to me that this belt of galaxies had to be a real physical association. This I reported in the *Astronomical Journal* in 1953.[7] I was given the incentive, or the courage, to publish this because Vera Rubin had just published an abstract of her master's thesis, where she tried to detect a "universal rotation" of the world.[8] She had used a hundred redshifts that were available from Humason in 1936. When, later, she sent me her full master's thesis, I realized that her "universal equator" was nothing more than the local supercluster equator. As you know, probably, she was so poorly received when she presented her paper at a meeting of the American Astronomical Society in 1951 that she never touched the subject again for 25 years. They really scared her away, and for about 10 years I was the only astronomer who was corresponding with her on this subject.

How did people receive your work when you reported it in 1953?

Very badly, very badly. But I am more pugnacious, probably, and if it's there, damn it, I'm going to say it's there. Now, trying to be very conservative, I said this line of galaxies must be the edge-on view of a flat system, analogous to our Milky Way. So I realized two things. If it is a disk with a center near the Virgo cluster of galaxies, because it's flat it must be rotating. That was the kind of reasoning I was making; of course it's absurd today.

Were you aware that your work was challenging some of the assumptions of the big bang model?

Yes. Not the big bang model in general. What I was challenging was the assumption of uniformity. That assumption is obviously incorrect, and what puzzles me is that it took 20 years before my view became accepted. Because it's absolutely obvious.

Why do you think people resisted so much in 1953?

Longer than that. First, for several years, there was complete silence. Then there were denials. I still love those denials so much I've written them on my blackboard here. You can read them. For example, Walter Baade, in 1956, said to someone who came to interview him: "We have no evidence for the existence of a local supergalaxy." And then in 1959 Zwicky said: "Superclustering is nonexistent." It's surprising because Zwicky was open to new ideas. These are good quotations.[9] In fact, Gart Westerhout, who is now the scientific director at the U.S. Naval Observatory, told me that when he was a student in the 1950s, at Leiden, the students were interested in this concept of superclusters and wanted to make some studies. But the great professor there, Jan Oort, told his students, "It's complete nonsense. You shouldn't pay any attention." But in 1983, at the Trieste meeting, Oort was one of the defenders of the supercluster. Well, people learn.

Why do you think that people resisted?

Number one, because it did not come from a member of the establishment. As one of them told me years later, "If it doesn't come from us, I don't believe it." There is only one true church.

Do you think that if Oort had been saying that there was a supercluster people would have believed it?

Yes, of course. They would have acclaimed it as something great. The greatest discovery of the great man. That was very clear for many years. I think it took a new generation and just the overwhelming accumulation of evidence. Also, I must say, the inhomogeneous structure complicates life to those who try to determine H_0 [the expansion rate of the universe, or Hubble constant] and q_0 [the rate of deceleration of the universe]. The homogeneous model is necessary to do calculations. No one knows how to handle the mathematics in an inhomogeneous universe, except by numerical simulations. So the existence of superclusters and large-scale inhomogeneities made life difficult. I remember a discussion I had with Allan Sandage about this in 1957. He was very upset because he could see this would complicate his life. He said to me, "If what you say is true, what would you do to measure H_0?" I said, "I would try to find a rich Coma-type cluster of galaxies near the south galactic pole, at about the same distance as the Coma cluster in the north, and then measure the relative redshift and their distances. Then you would have an approximation of the Hubble constant." Of course measuring distances was the catch, but you could not do such a measurement from nearby galaxies, as was done at the time, because, I insisted, excess density in groups of galaxies and obviously in clusters would locally reduce the expansion rate. There was nothing revolutionary about it. I even double-checked with some theoretical cosmologists. My statement was perfectly Newtonian and Einsteinian. There was nothing wrong with saying that an excess density must slow down the expansion rate. Why this was resisted has always been a puzzle to me. I think there is a combination of reasons. It complicates life for those who want to [determine] H_0 and q_0 in their own lifetimes. And it was not from the establishment.

> *In some of your later papers, you also pointed out the problem that there may not be any average density in the universe.* [10]

One of the early supports for the concept came from George Abell's catalogue of clusters, where he found some weak correlation indicating that clusters tend to be associated. But here I have repeatedly stressed, in vain, that superclusters are not made up of clusters. Superclusters are populated by groups and clouds of galaxies. They are not clusters of clusters. That's what I think Zwicky had in mind when he said that superclustering was nonexistent. One should not think in terms of big globular clusters of galaxies of the Coma type. What I was speaking about was plane groupings, populated mainly by poor groups of galaxies, rich groups, and what I call clouds of galaxies—not by rich clusters. Of course, rich clusters do show associativity, and participate in superclustering, but superclusters are not made of rich clusters. This has confused some of the discussion.

Another thing was purely a question of nomenclature. I remember having Zwicky for lunch in 1957 or 1958, when we were at Lowell Observatory, in Flagstaff, Arizona, where we had a house on "Mars Hill." Although we had this disagreement, Zwicky was always extremely generous towards me. One could discuss freely with him. I said, "How big can a cluster be in your nomenclature?" He said, "Oh, 30 or 50 megaparsecs." I said, "but that's the size of a supercluster." And I pointed out that we could say that New York City and Johnson City, Texas, are both "cities," that is, agglomerations of houses, but surely there is a structural difference, a very deep difference between a megapolis and a little village.

These inhomogeneous structures disturb some people, shake their confidence in the big bang model, in the assumption of uniformity.

This is something I've discussed repeatedly, beginning in 1960. In 1960 I made studies of the virial theorem in nearby groups of various sizes, and then clusters.[11] I found that the mean density of the groups of galaxies decreases as the size increases. Then I realized that this was the same thing that Ed Carpenter, the former director of the Steward Observatory in Tucson, had said all along in the 1930s.[12] Nobody understood what he was talking about because he gave a funny title to his paper, which was "A density restriction in the metagalaxy."

A constant density is certainly one of the assumptions of the big bang model.

That's right. At first, I said that maybe we have a hierarchical universe of the type discussed by Charlier, in which the density of increasing volumes decreases indefinitely, and at the limit of very large volumes it could be zero.[13] The fact that the same mean density for the universe was found repeatedly from the 1930s until now—except for distance scale corrections—proves simply that we are just probing always roughly a hundred million light years, give or take a factor of two or three. Over that small range, within the errors of distance scales and of mass estimates, you always get about the same numbers, 10^{-30} to 10^{-31} grams per cubic centimeter. I believe today that the deep radio source counts show that on scales of thousands of megaparsecs, we do have homogeneity, with the proviso that these deep radio source counts do not cover the sky but are just narrow strips that have been explored at great depth.

There is another battle that still has to be won, but it's making progress rapidly. In the 1960s and the 1970s, Kiang[14] in Ireland, Saslaw[15] working with him in England, and Kalinkov[16] in Bulgaria—nobody ever tells you about the work of Kalinkov in Bulgaria, but he's done good work—all found that on scales of hundreds of megaparsecs, there is still positive correlation [inhomogeneities]. Recently, Neta Bahcall has found the same thing, not knowing the previous work.[17] Karachentsev and Ozernoy, about 15 or 20 years ago, showed that the density contrast between one order of the hierarchy and the surroundings decreases as you increase the size.[18] In other words, there is a varying contrast between stars and galaxies, between a galaxy and a group, and so on. By the time you reach a supercluster and the surrounding volume, the mean density ratio is down to about three. On the hypercluster, or "super-duper-cluster," scale— that is, on scales of hundreds of millions of light years—there is still a density contrast.

I think the work of Kiang and Saslaw, based on the Abell clusters, the work of Kalinkov on general galaxy counts, and the work of Neta Bahcall, again on the Abell clusters, proves there is still a weak positive correlation, an association, on scales of hundreds of megaparsecs. So it means that we have to go out to a radius of about a thousand megaparsecs before we define a fair sample of the universe.

And that's within a factor of a few of the scale of the cosmic microwave background radiation.

That's what worried me for a while. I thought that perhaps the only fair sample of the universe is the universe itself. In which case, what does the concept of homogeneity mean? Homogeneity means that if you take equal disjoint volumes, they have the same mean density, within statistical fluctuations, as molecules in a glass of water. But if the only fair sample of the universe is the whole universe, where can you take another sample? I believe now—fortunately for theoretical cosmologists, and if the faint radio source counts are correct—that if we take a volume of a thousand or two thousand megaparsecs across, the mean density of that should be the same as in another equal and disjoint volume.

I can see that you have your fingers crossed.

Yes. Because it's at the limit of the data. I would like to see uniform radio surveys of faint sources across the whole sky, so we can be sure that there are no big gradients across the sky.

What were your reactions to the recent work by Haynes and Giovanelli on the Perseus-Pisces supercluster, and the work of Geller, de Lapparent, and Huchra? Is this all consistent with your earlier thinking?

Oh yes. Absolutely. I make the point that what we are discovering now at intermediate and even great distances—on the Lick maps, for example— is entirely consistent with the structure we see nearby. But because of the instrumental limitations, I've never attempted to study great distances.

I've never been given access to very large telescopes. I've always had small or medium telescopes.

Do you remember when you first heard about the horizon problem?

I don't remember. Perhaps in the late 1960s.

Did you regard it as a serious problem for the big bang model?

No. This did not interest me. I'm an observer of galaxies. My own interest in cosmology has been the concept of heterogeneity—the hierarchical structure of the universe—and, of course, the work on the Hubble constant, in the past fourteen years now.

Did your view of the horizon problem change at all after the inflationary universe model came out?

You are making me talk about a subject I don't know. My reaction is this: The big bang model is the dogma. It's got to fit and be saved at all costs. And you will notice that cosmological models always have more free parameters than facts. When something doesn't fit in the standard model, like the homogeneity of the 3-degree background radiation, someone comes up with a bright idea to save the standard model. And this makes some people, like Alfven, very mad. Basically, I think that the big bang model has been so imbedded now in our thinking, it has become the ultimate truth.

May I ask you a question about the inflationary universe model, even though you said that it's not your interest?

Yes. Inflation was invented in order to fix a very serious defect of the standard model. If you are allowed one or two more assumptions each time there is one difficulty, where is the check in the theory? That's what worries me. There is no check.

Well, the inflationary universe model does predict that omega is equal to one. If we can show that omega is not equal to 1 . . .

Yes. But that's beyond our power. We are still arguing about the dark matter. I know it's the fashion today. There is dark matter everywhere, but time and again people say that in this or that particular system they don't see any dark matter. I'm wide open to new evidence here. I will accept anything that's based on observation, that can be repeated and checked.

Do you think that the inflationary universe model is speculative in this sense?

I am tempted to believe that the concentration of efforts in making the big bang fit, no matter what, is . . . Maybe it's the right direction if we know that we have hit the truth. But if we have hit the truth now, if we can calculate the universe back to the Planck time, what will the cosmologists do in the twenty-first century? Is it plausible that we have all the answers now? I keep asking that question. I think it is very dangerous when we don't want to consider other possibilities. I am upset that the establishment will not spend the time necessary to show where the nonorthodox cosmologies go wrong. There have been some discussions, but many of the nonorthodox cosmologies are just ignored.

I have just a few more questions. Do you remember when you first heard about the so-called flatness problem?

I'm not sure I understand what your question is.

This is another one of these problems with the big bang model. Some physicists feel that it is unlikely that omega would be so close to 1, 10^{60} Planck times after the beginning of the universe, that such a result seems to require a very fine balance. Do you put any stock in this problem?

No. I believe this is not the type of topic I am interested in because I don't know what it means. I don't know how a whole universe can be created out of nothing. When we were in Rome in 1979 for a cosmological meeting, we lecturers had a private audience with the Pope. He gave us a fine speech in Italian, about astronomy and space and about the big bang. He made it very clear that he was all for the big bang, because, essentially, "We told you so." Here, I am just a layman, but I cannot really believe all the intricate physics that is being done on scales that are far beyond anything we are dealing with. What the standard model, or the modified big bang, claims is that an enormous mass of matter and energy occurred out of nothing at the beginning of the universe. I have a question that I ask my students to illustrate my position here. "What was God doing during the time before?" I know what the answer of physicists is: "This is not a question you should ask because this is a mystery. There was no time and there was no space." The theologians used to make it even simpler. "You should not ask those questions, brother, because otherwise the Inquisition will start being interested in you." So, I really believe it is premature to deal with such large, enormous extrapolations. I'm not saying I favor some other cosmology. I say I don't know.

Do you think that in the last 10 years theoretical cosmology has gotten too far away from the observations?

I think so. The observations are struggling to reach a z [redshift] of 4. There was a paper some years ago at the American Astronomical Society meeting that was entitled "$z = 5$ or bust." The author was looking for large redshifts, you see.

But you would probably say that's very far from talking about the first microsecond after the big bang?

Look, I'm totally ignorant of detailed nuclear physics. People will say, "This is out of his depth." But I think the nuclear physicists don't realize we just haven't got enough data. Even with nearby galaxies, we keep discovering new properties, new things. I have spent my life trying to study the nearby galaxies, where you have some resolution, where you can get spectra. You know this excess of blue galaxies in distant clusters, and these galaxies at large distance that don't quite fit this and that? How do we know that the universe is chemically homogeneous on very large scales? How do we know that the types of stars in distant galaxies are the same that we find in our galaxy and in nearby galaxies? Repeatedly, we have looked at our neighborhood and made extrapolations that can be false. When we measure distances—you have not asked me any questions about H_0; perhaps you ought to avoid it—we assume that the fellows beyond the river are the same size as us. We measure their apparent diameter and from this infer their distance. But, if they are very distant, we just see these creatures moving on two legs. Suppose they are penguins? We misidentify. So there are all sorts of problems that come up when you extrapolate to very large distances.

One point I wanted to make about the ad hoc modifications of theories, to force fit the standard big bang, is the question of the space density of quasars and evolution of quasars. We find in the standard model more and more quasars per unit volume as we go to larger z. This is fixed by postulating a time evolution of quasars. But as Souriau and his students have pointed out, if you had a cosmology with a nonzero cosmological constant, you could have a uniform density of quasars.[19] And, of course, you can also have this result in nonorthodox cosmologies, like Segal's.[20]

I have two more questions that are more speculative than the ones that I have asked you. You'll have to put aside some of your natural scientific caution. If you could design the universe any way that you wanted to, how would you do it?

That's a question that never crossed my mind. There is only one universe. It's here and we'd better accommodate our thinking to it.

O.K. Let me ask you one final question. There's a place in Steven Weinberg's book The First Three Minutes *where he says that the more the universe seems comprehensible, the more it also seems pointless. Have you ever thought about this question of whether the universe has a point?*

Oh, you mean finality? The question of final causes?

However you would interpret the statement. Weinberg doesn't elaborate. You can interpret it as finality if you wish.[21]

I tend to agree with Weinberg's opinion, but I fail to see why the universe should have any purpose (understandable by us). The remark sounds almost nostalgic. The concept that the universe should have a goal strikes me as a remnant of anthropocentric thinking (cf., the recent emergence of the so-called "anthropic principle"). Surely we do not believe, four centuries after Copernicus, that we are the center and wonder of the world and, a century after Darwin, that we were created in his own image by an anthropomorphic God (how insolent to the Deity!)

From what we know of the incredibly small mean density of matter in the universe, of the frequency of solar type stars, of the fiercely hostile environment of most of the universe, of the improbability of producing self-propagating forms of life spontaneously by cosmic processes, and of primitive biota evolving to rational intelligence capable of creating science and religion and philosophy, I wouldn't be surprised at all if it turned out that we were the only ones in the universe. A sheer accident. Now, I know all the arguments in favor of the plurality of inhabited worlds. This has been very popular with people through the ages because that's what they want to believe—that "we are not alone." Why do you think we still have astrology and religion and superstition, permeating practically all mankind? That's what people want. They want the world to be concerned with their well being, their future. Scientists may even wish it "made sense" in some way. But as I point out to my students sometimes, if the universe was created to produce mankind—and especially rational man and his modern civilization and everything that man has developed in the past ten thousand years (a split second on the time scale of the cosmos)—it's an incredibly inefficient mechanism. The idea that there is a "purpose" to the universe (usually implying some specially significant role for mankind, since who else is there to decide on "purpose" except us?) just doesn't make sense to me. So, while I personally would concur with Weinberg's view, we must also at all times recognize the limits of our knowledge and keep open minds. We don't know what new facts and concepts will emerge in the future.

Let me finish with two quotations. One is a witty remark by the great French catholic thinker and scientist Teillard de Chardin, who once wrote that he was always amazed that there are people "whose goal in life is to prove that life has no goal." The other is from Shakespeare, who has Hamlet admonishing his philosopher friend: "There are more things in Heavens and Earth, Horatio, than are dreamt of in your Philosophy" (*Hamlet*, Act 1, Scene 5). This neatly sums up what has been the guiding principle of my life as an astronomer.

MAARTEN SCHMIDT

Maarten Schmidt was born December 28, 1929, in Groningen, the Netherlands. He received a B.S. degree at the University of Groningen in 1949 and a Ph.D. in astronomy from the University of Leiden in 1956. After working at the Leiden Observatory, he moved to the United States in 1959 to join the astronomy department of the California Institute of Technology. In 1964 Schmidt was awarded the Warner Prize of the American Astronomical Society and in 1968 the Rumford Award of the American Academy of Arts and Science. He is an associate of the Royal Astronomical Society of England and a foreign associate of the National Academy of Sciences.

Schmidt's early research involved the structure, dynamics, and evolution of the galaxy; his current work concentrates on the cosmic distribution of quasars and on statistical properties of extragalactic X-ray sources. Schmidt is best known for his 1963 discovery of quasars, which are highly energetic objects at great distances from the Milky Way. Because of their great distances, quasars can reveal much about the structure and evolution of the universe, and Schmidt has continued to be a leader in this work. Schmidt's scientific style is to concentrate on a small number of topics and to understand them well. As part of his personal style, he is rarely seen without a bow tie.

Do you remember any early influences—either people, your parents, or things that you read as a child—that got you interested in science?

I know that after I got through the stage where I wanted to have the same profession as my father, who was an accountant—probably at the age of 9 or 10 or perhaps a little later—I became interested in chemistry. I had a chemistry lab at home and did the usual things. It was really my uncle, a brother of my father's, an amateur astronomer himself, who got me onto astronomy. As far as I can remember, it was in the summer of 1942 that I visited him and my aunt in central Holland. He showed me his telescope and things he was doing with it. Then I went to my grandfather's, my father's father, who lived in the country. He was a house painter, but retired by that time. He was sort of a tinkerer himself. So I rummaged through his things, and I found a big lens, probably about 3½ inches across and thick in the center. I asked him whether I could have it, and he said all right. My uncle had told me that if I took a small eye-piece like those used for biology and matched up the focal points, you could make a telescope. I did that at home back in Groningen where I lived, with a toilet roll in between. It gave me about the right distance. It had to be focused of course, so I had another tube in it. That's how I first got an image. I then wrote to my uncle and I said, "I got the thing together. What do I do with it?" He wrote back and said, "See whether you can split this and this double star." I had not the faintest idea where to find the thing, so I had to go to the library to find a book on astronomy, and that's what started it. In looking up the double star, I became interested in what was in the book, and the rest is history. So it was really my uncle. He's alive and, until a few years ago, still active as an amateur astronomer.

Did your parents encourage your interest in astronomy when you were this age?

Yes, they were certainly ready to encourage anything I wanted to do. They were always supportive. By the end of high school, probably in the year 1946, when it was clear that I wanted to have astronomy as a profession, my father became somewhat concerned as to whether there was a future in that. He talked with Adriaan Blaauw, the Dutch astronomer, who at that time himself was in the beginning of his career at Groningen. The way Blaauw talked to my father was obviously sufficiently positive that there were no objections when I then made clear that I wanted to go on in astronomy. The situation at that time was, by the way, not very favorable. I seem to remember from a book that Marcel Minnaert wrote about astronomy that there were at that time 14 permanent positions in astronomy in Holland. When you looked around and you saw vital people like Minnaert and Jan Hendrik Oort, younger people like Jan van de Hulst and Blaauw, and a number of others who [were] just starting, you just wondered how you would ever get in there. So although my father's objections perhaps were alleviated to quite a degree, the situation was not all that positive.

But you were thinking about a career in high school?

No, I don't think so. I think that book probably appeared later, and I'm probably now thinking back about how it must have been at that time. But at that time I couldn't have cared less about 14 positions. You're just optimistic, I think.

When you looked at stars with your telescope at this age, 10 to 12, did you ever think about the universe as a whole—how far space went, or questions like that?

Well, not so much while working with that telescope, but reading about galaxies and so on, slowly your interest would go that way. Initially I think I didn't know at all what I wanted to do in astronomy. It was just something of major interest. Since it was the war when I started, 1942–43, there was a blackout in Holland, which made observing even from the city of Groningen possible and quite good. Youngsters these days have no chance of really doing it from the cities where most of them live. At that time, it was all right.

When you were reading your astronomy books, did you read anything about cosmology?

Not that I remember, although it must have been in the books that I looked at. I would copy whole paragraphs or tables out of books onto my own notebook. At that time, in the war, books often would not be available anymore. You would get a copy out of the library. I copied whole tables about the solar system, the masses and sizes of planets. In hindsight, it was a strange way of getting into a field. But that's just what happened. I don't remember cosmology particularly well. I think the bigger thing that I had in mind was our galaxy, the Milky Way. But you start small. You start nearby.

Do you remember when you first got interested in cosmology? Was that in college?

It was certainly not very early. I think that happened in a very gradual fashion. My interests really blossomed out slowly, first towards the galaxy. Even before I had my Ph.D., I did work on the 21-centimeter spiral structure in the northern hemisphere. That ended in about 1955, while I was doing my thesis work, a mass model of the galaxy, which was finished in 1956.[1] So I did those two jobs almost in parallel during my graduate years, and that clearly set my interest in the galaxy. I think that there never came a time when I had a major interest in galaxies as a whole. I remember that when I became a Carnegie Fellow in 1956 through 1958 in Pasadena, I was still interested in mass models. When material came out on M31, I think I published a mass model for M31 then.[2] But it sort of was an extension of my work and my interest in the galaxy itself. When radio galaxies and quasars finally became my area of interest in the early 1960s, it happened that way. There was never a stage in which I was really interested in extragalactic astronomy, apart from these [individual objects] and quasars.

Let me back up just a second. When you went to university, did you know from the very beginning that you wanted to go into astronomy?

Yes, that's what I wanted to do. I entered university in 1946. There were certain options that you had in the university, and I concentrated on the one that was aimed at a professional career in astronomy.

Were there any particular people in your university or graduate career that had a strong influence on you?

Yes. Adriaan Blaauw, whom I mentioned before. He was sort of my role model. Then Lucas Plaut, who was also an assistant at Groningen, was also an example for me. He essentially did all the teaching of astronomy at Groningen, because Professor van Rijn had tuberculosis during the war and was laid up. Those were the two at the time, except distant people like Minnaert and Oort, whom you would see during an annual summer conference for all of Dutch astronomy.

Did you study cosmology in graduate school?

No. I never was formally taught in the field of cosmology. I must say that when I started on galaxies—and that was in the form of radio galaxies— there was no such lofty purpose as to try and decide about how the universe is built. I wish it were true, but it just came together in a fairly practical way.

While I was a Carnegie Fellow, it seemed natural to follow up the radio work I'd done in Holland which established parts of spiral arms—whether one could get the equivalent optical structure, say from open clusters. So I started to work on a few open clusters that seemed of interest because I had hoped that they were very distant. I was, after all, at Mt. Wilson with opportunities for observing that were quite remarkable. That work never crystalized into a publishable paper. It's a tribute to the freedom that one has as a Fellow that even when something doesn't pan out as well as it should, you still can undertake it and don't feel too guilty about not finishing it. While I was at Mt. Wilson, I became interested in the consequences of the change of gas density in our galaxy—perhaps in galaxies in general—as a consequence of star formation. It had just been established around that time that star formation was really active, that it took a certain amount of gas out of the interstellar medium. So I started on that, and that kept me very busy in the year 1957.[3] After having gone back to Holland, I came back to Caltech in 1959 to take up a permanent position. I was interested in the helium abundance, and as soon as I could, I did some observational work with the 200-inch telescope. I did some very difficult observations of the helium abundance in HII regions in the center of M31, to try and see whether the helium abundance, as people thought at that time, did go just as we thought metal abundance should go in galaxies. And the discovery, although it was never properly published, was that it didn't at all. Just at that time it became clear that, in fact, it shouldn't. But it all was part of the whole evolving picture of the evolution of stars and of the process of chemical enrichment by stars.

Did you think at that time that helium might have been produced primordially in the big bang?

Well, I was told that. In 1957, when this idea had just started, people thought that the abundance distribution of helium would just go as that of the metals [made in stars]. Then when I did these observations in 1960 and found that it didn't, and I talked with William Fowler and Fred Hoyle when Fred was in Pasadena, they hardly gave any attention to what I had to say, and they asserted that indeed the helium was all primordial.

Then, in the summer of 1960, Minkowski retired, just after having discovered [the extragalactic radio source] 3C295's redshift of 46%.[4] He found it in May and he retired on the 30th of June. Then it became a sort of unofficially shared responsibility between a few people like Guido Münch, Jesse Greenstein, myself, and perhaps somebody else, to take care of the radio galaxy identification that Tom Matthews was doing from radio data taken mostly at Owens Valley. But what happened in practice was that nobody else did, and I did it all. I became gradually interested, and Tom was feeding me identifications of small-angle radio sources because he was interested in going to high redshifts. So that's essentially how I got into the radio galaxy business. I think that in 1963 I published a paper with 30 or 35 redshifts or so, which was the production that I had achieved at that time.[5]

By the early 1960s, were you aware of different cosmological models?

Yes, I think by that time I was. I think that came mostly just by proximity, because Allan Sandage, of course, was very much working on his q_0 test [measurement of the deceleration parameter], the redshift versus the magnitude of the brightest cluster of galaxies. I couldn't help but hear and read about it. It really happened not in a formal course but essentially just by Allan's great activity in that field. So one couldn't help but be interested at that time in the question, particularly of what q_0 was. Somehow, at that time the question of what H_0 [the expansion rate of the universe] was was slightly less important. The darn thing changed all the time.

Do you remember having a preference for any particular model—such as open versus closed or homogeneous versus inhomogeneous?

No, no. Inhomogeneous was essentially never mentioned at that time, and open versus closed was something that I had no views about. I am totally different now, because my prejudices are definitely in the direction that q_0 is a half—probably the effect of physicists on us.

Did you think that you might measure q_0 yourself some day? Were you that interested?

No. In fact, if I think about what I hoped for at that time, it was incredibly modest. I remember that during one of my runs I collected data from a particularly difficult radio galaxy. I used the fastest camera at the front focus of the 200-inch, and I'd taken a spectrum that had a quite long exposure time. I don't remember whether it was four hours or nine hours, and the result was quite disappointing. I remember that Dr. Bowen, the director, came up the next day after lunch. He came to the 200-inch and asked me whether he could see what I'd done that night. I perhaps had only two plates or so, two spectra, which was not unusual, but the longer one really was very poor. And I well remember how positive he was about it because he looked with great interest at that spectrum. I explained what I'd been trying to do and why I did it and he said, yes, it was very difficult but he appreciated that the 200-inch was being used for things that were exceedingly difficult. I think that he didn't necessarily approve of people who used the 200-inch to do lots of things very fast, in order to accelerate the work, but for these borderline problems, he really felt that was the way to go. I think he was right in a sense. That was very reassuring. I remember at that time, my greatest expectation was that, in my life, I would get to a redshift of two thirds, with radio galaxies. Tremendously modest, as you see [laughs]. That doesn't sound like a lofty purpose, does it? I mean I was not going to solve world problems.

Why did you want to get to a large redshift?

Ah, well, that must have been just instinctive. Clearly things became very much more difficult in that work when you got to a redshift of the order of 0.3, and it was in a sense the challenge. But it was a fairly narrow challenge at that time. I was not thinking of solving the cosmological problem. I was not thinking of solving *the* problem of where the radio galaxies came from with all their energy. I just wanted to contribute in a particular direction. It sounds very narrow minded, but that's the way it was at the time. I may be underestimating myself. It may be that I had loftier things in mind.

I think your career shows that one can get good results from that approach.

Yes, but, for instance, in the chemical evolution of galaxies, which was the field that I mentioned somewhat earlier, I think my purposes were much deeper, and I had a program in mind, namely, to truly understand how star formation affected everything, the formation of the elements, the light of the galaxy. I hoped the change in light could contribute to Allan's q_0 problem. There I had a real program. But I got disenchanted by it when it soon turned out that the number of free parameters that you had seemed to increase so much. For instance, I worked with a closed system, but it soon became clear that infall of gas from outside the galaxy could affect things, and I became disenchanted with the fact that there were so many possibilities. I think that in terms of the style or the type of things that I'm interested in, as soon as either there are too many facts, but *especially* if there are too many degrees of freedom, I start to lose interest somehow. I seem to want to work in an environment where there is *not* too much freedom. Perhaps it's working at the edge where you don't know yet what it's going to be, but where a good set of observations and a conservative interpretation of the results are of value. So it was a good program in chemical evolution, but not in what I then started, which was radio galaxies.

Do you remember when you first heard about the horizon problem—the fact that we see regions of the universe very far apart that seem to be at the same temperature and yet haven't had time to exchange heat since the big bang?

I cannot remember when I first heard about it. I'll have to guess. I think it was 10 or 15 years ago. I don't know who posed it first.

When you first heard about it, did you regard it as a serious problem in cosmology?

Yes, very. It seemed indeed like a very fundamental problem to me, as soon as I heard about it. It could have been explained to me by Charles Misner of Maryland.[6] Yes, I saw it immediately as a crucial problem that had to be resolved, so that you need not wonder anymore about what we were doing.

Did you have any ideas about how it might be resolved?

Nope. Absolutely not.

But you did think that it was a serious problem that struck at the foundations of the whole big bang model?

Yes, absolutely, yes.

Has your view of the horizon problem changed as a result of the inflationary universe model? Do you feel like there might be a solution of it now, or do you still think that it's a serious problem for cosmology?

That I don't dare say much about. The fact that the horizon problem is addressed so squarely is very good. The inflationary universe model has gotten a lot of attention. As to whether one is now on the right road, I have not the faintest idea. I'm impressed by what inflation does all in one fell swoop. But I must also say that I'm very impressed, and sometimes alarmed, by the *cleverness* of theoretical physicists and astronomers. When I say alarmed, what I have in mind—and I'm not meaning to be facetious— is that one wonders at times whether if one *invented* some observations— which none of us ever thinks of, by the way—and presented them in the right way, one wonders whether or not quite soon there would be a number of reasonably authoritative explanations about them. This will not be tried, of course. But somehow one has a feeling that the fact that something can be possibly understood in certain terms, sometimes very complicated terms, need not give one the confidence that it's the right road at all.

Sometimes you almost wish that the things that can just *barely* be understood with great effort happen to be understood by just the right one [the right theory]. But there is no such matching, I think, of nature and imagination. But, as you know, I'm not one who looks at the theories and can make up my mind independently. I'm an observer of the scene. Hopefully a perceptive one, but no more than an observer.

Do you feel comfortable or uncomfortable with invoking initial conditions as solutions to cosmological problems?

I'm not unhappy with that. I can't explain that. It's just the view you have about the universe as to whether, in any possible case, it had to come about, or whether the one in which we live happened to be slightly special—dependent on those initial conditions. And since we have only *one* universe, that seems just about the most extraordinarily difficult question to answer, and I see no reason to be unhappy if initial conditions have to be evoked.

How do you regard the inflationary universe model? Do you feel it's speculative, or on the right track?

Well, totally without any authority, it looks to me as if it may be all right. But to me it becomes increasingly complicated. There were so many different inflationary scenarios that one hears about. It seems a fascinating possibility, and it might well be right—one of the variations of the model. It sounds as if it's a possibly good solution, but . . .

But you wouldn't bet money on it.

Oh no! Absolutely not.

Why do you think the inflationary universe model has been so widely accepted, or gotten so much attention?

Well, I mentioned that not only did it solve the communication problem [horizon problem], but also the prejudice that many physicists have that omega should be one. And there is yet a third problem that is quoted that I can't remember. The inflationary universe model seemed to be something that addresses more than one of the problems in cosmology. Perhaps the reason it's gotten so much attention partly has to do with its name. I'm now slightly facetious, because I think inflation is a marvelous name for it [laughs].

We just mentioned another cosmological problem, the fact that omega is so remarkably close to one, which many people call the flatness problem. Do you remember when you first learned about the flatness problem?

Perhaps slightly later than the other one, but it's also about ten years ago, I think.

Do you remember how you reacted to that problem when you heard it? Did you think it was a serious problem, at the same level as the horizon problem?

Yes and no, because there's immediately a solution. You make omega 1.

At the beginning. But that requires certain initial conditions, right?

That's right.

But you could have also given that solution to the horizon problem. You could have said the universe began completely uniform.

Well, here you may well point out that I'm not the deep thinker that you wish I were. On the communication problem I had no such thoughts. In the flatness problem, I felt immediately that omega equal to 1 was obviously the preferred solution to the problem, because otherwise omega should have been by now either very large or very small, but anyhow not near 1.

You saw omega equal to 1 as somehow a special value that might have been singled out by something?

Well, the way I understand it is that if omega at the moment is slightly smaller than 1, for instance, the value of 0.2 or 0.1 that you get from the abundances in the early universe, then if you look backwards in time, omega has been an ever-increasing function where asymptotically it was 1 at very early times. After on the order of 10^{60} Planck times, it's *extraordinary* that it then only went down to 0.1 or 0.2. One would expect it to be 10^{-80} or 10^{-30} or whatever. I see that as a very powerful argument. It appeals to me very much.

And you don't worry about why it was exactly equal to one at the beginning. That doesn't particularly bother you?

That's correct. And it may even be consistent with what I said earlier— that I'm willing to accept initial conditions. No, that doesn't bother me.

It sounds like you were more troubled by the horizon problem than by the flatness problem.

Indeed, yes. For the horizon problem, somebody had to come up with a solution that I had not the faintest feeling for. I'm just an observer, and I reacted to it that way. With the flatness problem, I immediately felt an affinity for the value of 1, once the problem was explained to me.

Let me ask you about the results of de Lapparent, Geller, and Huchra and of Haynes and Giovanelli in the last few years on the large-scale structure. Do you remember how you reacted to those observations when you first heard them? Did it change your thinking a lot, or is it something that you already had an opinion about?

It perhaps started, I think, with the void, this big void that . . .

The void found by Kirshner, Oemler, Schechter, and Shectman?[7]

Yes, that may have slightly preceded the Geller et al. work. I think that there, in the beginning, I was worried. In the beginning, I wasn't just willing to accept it. I'm still not sure that I'm convinced that the effects have to be as large as we interpret them to be because we really think of these bubbles and these voids as being entirely empty. But in the mean-time a few objects of a certain nature have been found in them, so they *are* not entirely empty. But nonetheless, I do think that they have a powerful effect on one's thinking about the structure of the universe as a whole. It does thoroughly change one's idea about the overall structure.

How does it change your ideas?

I was brought up with strong statements by authorities like Oort and many others who in the 1950s really would insist that the universe *has* to be homogeneous on the larger scale. So as soon as inhomogeneities were found at that time, they were said to be *absent* on the next larger scale. I think, in fact, those inhomogeneities were the evidence of Gerard de Vaucouleurs for the local supercluster. You now only have to look at a Shapley–Ames catalogue projection on the sky, where you immediately see these big bands of galaxies. Oort himself shows it these days. But at that time it was not something that was preferentially shown.

You think even though the data were there . . .

The data were there. People didn't give any attention to it. I think the tendency in the 1950s was very much to believe that things *had* to be homogeneous on the larger scale, and so one took it [homogeneity] at a scale that was just above the one at which you observed that it [the distribution of matter] wasn't homogeneous. And the fact that this darn business [of inhomogeneity] can just go on to these very large scales, I think, is a major change in our concept of the universe. Initially, I was somewhat critical of these things. I wondered if the people who were interpreting the data realized that at most of these larger distances you look only at the very bright end of the luminosity function, so if there were minor changes from position to position, perhaps you could get this [the observed very inhomogeneous structure]. Nonetheless, I think it's a major change, a major improvement in our concept of the universe. And at the moment, it seems to me one is not willing to trust *any* scale that things are homogeneous over, except for the microwave background, which comes in with this fantastic homogeneity. If we didn't have the microwave background, I wouldn't be surprised if some people, perhaps myself too, would be willing to believe . . . Wasn't it de Vaucouleurs who talked about a hierarchical picture, inhomogeneous on all scales?[8] I think I would have believed that by now. I would be a happy believer of that by now if only that microwave background didn't put a lid on things. So I think the microwave background has had an enormous effect on our thinking. Anyhow, yes, I'm impressed by it [the recent observed structure]. It is very remarkable. But I think it's also very good. I see it as positive, in an optimistic view, because scenarios of galaxy formation will have to explain that. I imagine there is much more information in all these structures than if, at more than 50 megaparsecs, it was all beautifully homogeneous, without structures.

So you see these structures as having more explanatory power than no structures—even though it shakes up the paradigm.

Oh, but that doesn't matter. In astronomy, so often it happens that when you meet people who have found a problem, they are very down. They say, "Ah, I can't understand this. This is going so badly." And you say, "What's wrong?" And they say, "Ah, this doesn't fit, and I can't understand that." And your tendency is to commiserate with them. You say, "Yah, that's too bad. No, that's not good," etc. But of course, when people are in that mode, that's very good. There is more to be gained, because this and this don't fit in the existing picture. So out of all this moodiness, there is a reason for great happiness, I think. There is more information to be gained. Of course, people only will feel happy for the first time—it's a strange attitude of scientists, I think—once they've come up with a reasonable explanation. They say, "Ah, I had a great day! I explained this and that." But just before it, when they are really in the data-gathering and the difficulty-gathering phase, they are really down. It's really interesting. They should be up.

I'm interested in the extent to which mental imagery plays a role in the work of different scientists. When you work, is it important to you to visualize things you're working on?

If you mean whether I have to have an idea in my mind how things came about, or that I must have a ready explanation for everything—like with quasars or in cosmology—even if things aren't worked out yet, I can do very well without that.

I don't mean an explanation, but I mean an actual visual picture. When you study quasars of greater and greater redshift, do you have some picture of space, or of these objects, or of the early universe?

No. I don't think I do. Although if you probe it further, perhaps you would find that actually I do, but I'm not aware of it.

Do you think that theory and observations have worked well together in the last 10 years? Or do you think they're going in opposite directions, or not paying enough attention to each other?

I think that they are well-attuned to each other. There is a reasonably good interchange and equilibrium. As a practical fact, I feel that often theory contributes strongly to the explanation of observations once they have been made. I think it is rather rare that theory helps the observers a lot in telling them what to do. Now there are good counter-examples to that, too. But I must say that when I'm thinking of the observational programs done over the last 20 or 30 years, it was rarely so that I found myself very strongly influenced by what theorists were telling me I should do. There is no great pressure on me or other observers to direct the way the observations should go. Only at times, such as when we thought, in a perhaps somewhat unguarded moment, that we had a gravitational lens with a separation of two and a half minutes of arc. *Rapidly* from the theoretical side came a number of very good tests, which we reacted to within a month and found ourselves firmly deeply believing that we didn't have a case of a gravitational lens at all. But that was the exception rather than the rule. In my main programs, I *think* I'm not terribly affected by what theorists tell me, but more by the opportunities as they present themselves to me—technologically, in terms of observing.

Do you think theorists have been sufficiently responsive to the data?

I believe so. Yes. Of course, you have all sorts of theorists. There are theorists who practically never give any attention to what's going on, and they are among those that I mentioned earlier. I have great admiration for their imagination and what they can achieve, but they are not the best of barometers as to whether something is feasible as a practical explanation of the universe. So when I give an answer, I naturally give an answer about those theoreticians that have some contact with the observations. I think they pay proper attention and are receptive and responsive. As I just mentioned, it is rare that the observers heed the theoreticians too much, except in rare cases.

But you think that's a good thing?

No, I think it is something that is typical of astronomy. I'm sure that in physics it's quite different. Most of the experiments that are done come out of theoretical developments on the basis of previous experiments. It turns out that you should go for the *XY* particle, so there you go. That follows out of theory that comes out of the previous experiment. But I think the difference has to do with the fact that astronomy is a passive science and physics is active. So often in physics . . .

You can turn up the knob.

That's right, or you can kick it differently so that it *will* do what would yield the *XY* particle. But it is rarely so in astronomy. In extragalactic astronomy, you can never be active. It is rare that when a theoretician has a good idea about what you should try, practical circumstances in the universe— telescopes, sensitivities, and energy ranges and emissions available— allow you to do that at the right level.

What do you think are the major problems in cosmology today, from your perspective?

A major problem that I see at the moment is the time scale and the distance scale of the universe. At the moment, the better *distance* scales that we have in the universe lead to a Hubble constant of the order of 100 or slightly less. The best *time* scales that I'm aware of lead to a Hubble constant of the order of 50. I find it very interesting that this is the third time in history that those two have been out of whack. The first time that happened was in the 1950s, and we got the steady state theory, and it was a major theoretical perturbation. And then 15 or 20 years later there was a mismatch again, simply because the one went [was modified by updated observations] faster than the other. They were again out of whack for a while and the Brans–Dicke theory appeared. To my mind, this time they are out of whack in by far the most serious fashion that we've seen yet. It's very interesting that now there's not even a theory that addresses it. People say that the problem might be resolved by the cosmological constant, but again that cosmological constant is also subject to one of these arguments that is somewhat similar to the flatness problem, namely, that the value that we play around with in astronomy is just *totally* different from the expected value. So we're in a very strange situation. Perhaps very new methods or the space telescope will help solve this problem. But I see this as a major problem.

> *The second time scale, the one that is not related to distances, comes from the cosmic clocks and ages of globular clusters and things like that?*

Correct. I find it interesting that whereas in earlier epochs, this was considered such an important problem that whole new theories were developed that were based on it, this time when it may be more serious, there is not much attention spent on it. Everybody says happily, "Of course, we know things only to within a factor of two." But suppose that persists. We'd have one of these long periods in which we are very down, until there comes a solution of this. But who knows? It may be fundamental. I've been impressed by the fact that I like both arguments. I like both the distance scale and the time scale arguments, but they lead to different results.

> *Do you have any opinion as to why this discrepancy is not drawing a lot of attention right now?*

I think partly because the debate between mostly de Vaucouleurs—but then later Marc Aaronson, Jeremy Mould, and the others—on the one side, and Sandage and Gustav Tammann on the other side, got too boring. People are just fed up with it, I think. They both come up with solid statements of what's so bad about the other method. As a good Pasadenan, until recently, I had tended to believe that Sandage was probably right. But one or two of the fundamental pillars of Tammann and Sandage's argument have not done well lately. And that was the one that gave 50 and would be in agreement with the long time scales.

What are other major problems? Well, strangely enough, I wouldn't say that the flatness problem is a problem. I would just say, well, omega is 1.

The communication [horizon] problem I would still consider a problem. I apparently don't feel that it is really resolved, although it's a fantastic attempt that has been made. Then, finally, it is my feeling—although I'm not sure that others share this—that all these big motions and these big mass concentrations that one needs are just an incredible mess. I'm not necessarily inclined to believe that whole story. I don't know where it's going.

The Seven Samurai work?[9]

That's right. And then Alan Dressler with his great attractor.[10] They are many close and dear colleagues of mine. I'm not absolutely sure that I believe everything that's going on there. But it's fascinating. And it's certainly true that the Rubin–Ford thing[11] for ScI galaxies gave this funny effect around the sky. They really started the thing, and apparently what they found is still around. So something is going on, but I consider what's happening in that field—although the people who are in it are probably deliriously happy—as a potential problem, where I wonder whether most of the statements that are made really will hold up very long. But it's just a feeling.

Let me ask you to take a big step back. You need to put some of your natural scientific caution aside, perhaps. If you could have designed the universe any way that you wanted to, how would you do it?

Well, that is surely the ultimate question that doesn't allow one to be cautious. God! What a wild question. This is really the worst question I've ever heard. The first thing I would say is that I would never have constructed a universe in which it was even allowed to ask such questions [laughs]. That's a question to which I'm not sure I can give an answer. It's like asking me to take off my clothes and start walking around here. One naturally has a certain shyness about things, and I think that it is absolutely in conflict with my philosophy about doing science.

I volunteered earlier that I don't have strong ideas about how things in the universe work if I have had no input on which to base those ideas. So in the very early days of quasars, for instance, even before black hole scenarios with accretion disks, people would often say to me, "Well, you have been so active in quasars. What are they and what do they do?" I say, "I've not the faintest idea. This is possible, that's possible. This is what's been brought up, mostly not by me." And they would say, "Well, what do you think?" I say, "I don't know. I have no feeling for it." And here you can see how unreasonable your question is to ask me to construct a universe and to indicate a preference. It's something that is totally out of my mode of thinking and my mode of working.

If I can be allowed to say something else, since I didn't give you a good answer to this question. Briefly, as one of the perpetrators of looking at the distant universe, I find it extraordinary that it is possible with human means, with pieces of glass that are no larger than this room, to see things that are interestingly far out in the universe. Sometimes it strikes me that the universe is much smaller than . . . All right, here we go. I would have constructed a bigger universe. I think the universe is *small*. There we go. If I'd had my rathers, I would do that. I find the universe too confined. I find it amazing that it is so small.

Can you explain what you mean by small?

All right. We have a certain concept of what is a reasonable distance that we have some feeling for. In astronomy, you probably think that many astronomers would have a feeling for the distance to Andromeda [a galaxy about 2 million light years from the Milky Way]. You talk about it so often and it's fairly big in the sky, and so on. It's at one half megaparsecs. Now we know that the observable universe is of the order of—what is it—6,000 megaparsecs. It is extraordinary that we can get out to a distance where the light travel time is a substantial fraction of the age of the universe, in only a few thousand times the distance to Andromeda. It could have been any arbitrary number of course. It could have been 44 million times the distance to Andromeda. So, I find the universe small. I'm surprised. I would have made it much bigger, I think. But I got to that because I was trying to explain to you why I find that with a fairly small piece of glass you can catch enough light to study fairly distant parts of the universe and have some chance to observe these things. But that's really the same question. It's the size of the universe.

So you would rather the universe be much bigger . . .

Of course, I wouldn't rather, but I think that if I, in my innocence, had constructed the universe, I would probably accidentally have made it much bigger. I wouldn't have thought of making such a small universe. Anecdotally—and I'm not sure whether it was right—I heard that when it was clear that the turnover of the counts of radio sources really had been understood well by people in radio astronomy, Martin Ryle had expressed that that result was almost disappointing. Because here was the end of the universe. And I share that. I mean you'd almost feel claustrophobic in a universe that is so small that when you just look at one of the first results in radio astronomy, you already see the effect of the end of the universe. That's amazing.

There's a place in Steven Weinberg's book where he says that the more the universe seems comprehensible, the more it also seems pointless. Do you ever wonder about the point of the universe or whether the universe has a point?

No, in fact, I would probably say that in terms of my *personal* experience, my thinking would be the other way around. To me, the universe becomes more and more incomprehensible, and this probably more reflects a personal view. Weinberg understands more and more of the universe. I understand less and less. When you think you understand things more, I think it's really just an extension of your awareness of the problems.

WALLACE SARGENT

Wallace Sargent was born on February 15, 1935, in Elsham, Lincolnshire, in Great Britain. He attended Manchester University, where he received a B.Sc in physics in 1956, an M.Sc in astrophysics in 1957, and a Ph.D. in astrophysics in 1959. From 1959 to 1962 he was a research fellow in astronomy at the California Institute of Technology. After two years back in England at the Royal Greenwich Observatory, and another two years at the University of California at San Diego, Sargent returned to Caltech, where he is now the Ira S. Bowen Professor of Astronomy. He is a Fellow of the Royal Society and of the American Academy of Arts and Sciences. In 1969 Sargent won the Warner prize of the American Astronomical Society.

Sargent has worked mainly in observational astronomy. His current research interests include quasars, Seyfert galaxies, Lyman alpha clouds, and the cosmic microwave background radiation. Trained as an observer of individual stars, Sargent first took an interest in "peculiar" stars and then, under the influence of the eccentric but brilliant Caltech astronomer Fritz Zwicky, shifted his interest to peculiar galaxies. In the early 1970s Sargent began a program to identify "young" galaxies in the process of formation. Sargent has also led the search to investigate the still mysterious extragalactic clouds of gas between us and distant quasars. Raised in poor circumstances and the first from his high school to go to university,

Sargent vividly recalls seeing a picture of the Mt. Wilson 100-inch telescope in an encyclopedia at the age of 10. He is married to the astronomer Anneila Sargent.

I wanted to start with your childhood and ask you a little bit about your parents.

Neither of them were well educated. In fact, none of my relations went to high school. I was the first person in my high school to go to university. When I was small, my father was a gardener at a large country house in England. My mother had worked as a shop assistant before she married him, and then she became a full-time housewife except for one small period.

Were either of them interested in science?

No. But my mother was the one who pushed me into trying to become educated. During the period she worked as a shop assistant, the son of the owner of the shop went to a decent grammar school in the village where we lived, and then went to Oxford. Following that time, she had the view that her children would go to a university, preferably Oxford.

My parents couldn't give me any practical help about going through college, but they did let me stay on at school. When I was 11, I took the exam that kids in England took at that time, which determined whether you would go to the grammar school or the secondary modern school. Roughly 20% of the population went to the grammar school, which was to give you an academic education. The remaining 80% would leave school at 14. I was the first person in my family to make the step of going on after 14. I was in a peculiar situation, because my father had been in the Air Force during the war. After the war, he didn't go back to gardening. He went to work on the steelworks in a town near the village where I was brought up. Because this town had a steelworks, there was a thing called a technical high school, which was intermediate between a grammar school, which was purely academic, and a secondary modern school, which was for the dim, the ones who were going to be drawers of water, etc. The technical school was supposed to prepare you for being a higher craftsman on the steelworks—a draftsman, for example.

Is that where you got interested in science?

No. I would say I started being interested in science before I even went to that school. For a brief period during and slightly after the war, my mother worked for a woman in the village who was crippled and couldn't do her own housework. They were somewhat better off than we were, so they could afford to pay somebody else to come in once a week. They had a set of encyclopedias for their children called *The Children's Encyclopedia,* which I read when I was around 10 or 11. It had several volumes, and there were articles about science and, in particular, about astronomy. I remember vividly a full-page photograph of the Milky Way, with all these thousands of stars. On another page, there was a picture of the 100-inch telescope, which I was later to use. That turned me on to science. And then, when I passed this exam at the age of 11, which meant I could go on to a better school, a traveling salesman came around to the houses of the few people who had passed the exam. He sold my parents another set of encyclopedias, this time more advanced. There were courses of approximately early university level on physics, mathematics, biology, art, history—lots of interesting things. I read these when I was around 12 and realized that astronomy and physics were the things I was interested in. But I didn't think I would be able to go on professionally. I thought the best I could do would be to become a lab assistant someplace, and that was my intention.

Was there anything about cosmology in those encyclopedias?

Yes. There was some discussion of the immensity of space. I remember being aware around that age of the possibility that space wasn't as simple as it looked, that it wasn't Euclidean. I don't know what the terminology was, but they used the word "curved space."

Did you try to visualize that?

Yes. And, of course, I thought I could, as a kid. Later, I discovered I couldn't. I did well at this technical school, and by the time I got into the upper form—the school was streamed into four streams according to ability—I was the top kid in most subjects in the top stream. I realized at that point that I should have gone to the grammar school where you could study Latin and Greek, for example, for the university. But the school I went to did not prepare people for the university. It was aimed toward trade. You were supposed to leave at 16 rather than 14. Gradually I became aware that I wanted to stay on longer. When you're 16, you take some exams in England called ordinary level to prepare you for more specialized courses in more restricted subjects, which you then take for the last two years from 16 to 18. For this ordinary level, most of the kids took only a few subjects, but I took nine subjects and did well in all of them, including English and things like that. The teachers realized that I should go to the university, and there was a possibility of transferring to another school in the town. But, for some reason, I was kept on at the original technical high school, and they made special arrangements for me to study for another two years along with two or three other kids who wanted to become school teachers. For the last two years, I took physics in a class of one or two. The teacher had been a coal miner in Nottingham until he was 28 and then had gone late to university. He didn't know all that much about physics, but he tried. I guess the bottom line was that I taught myself a lot in the last two years. I specialized in physics, mathematics, and chemistry, but I also took French and English as well, because I was really interested in humanities.

Did your parents support you in this change, this altered direction?

My mother was always keen for me to go on to the next step. My father was sometimes doubtful, particularly because he was ill. He suffered from various illnesses from about the time that I was 12 until the rest of his life. He was often home from work, and we would have to draw some sort of welfare. Particularly when I was 16, it was pretty bad, because basically we had only potatoes to eat. I felt very guilty about staying on at school when I could have been working. But nevertheless, I was so stuck on being a scientist of some kind that I ignored this and continued.

When I was 15 or 16, Fred Hoyle gave some lectures on the radio which became the subject of the book *The Nature of the Universe*. I heard those lectures, and they were absolutely brilliantly done. I learned, for example, that you could know about the temperature in the interior of the sun by measurements of the outside. It came as a surprise, although to some extent it had been revealed in the encyclopedias I had read earlier. Then Fred had a quite strong Yorkshire accent, and I realized that people without standard English accents—BBC accents—could actually do that kind of work. That was a tremendous liberation. The last talk he devoted, in some sense, to religion. I had been brought up as a Protestant. My parents were never very religious but made me and my brother go to church when we were young. I dropped out of going to church. But Fred made me violently antireligious, and I got into trouble with the school because of it. You were supposed to read prayers in the morning, in English schools. It was a state religion, so there was a short religious service at the beginning of each day. Fred got me into serious trouble. I had fights with some of the teachers with whom I was on otherwise good terms.

What is it that he said that made you violently antireligious?

I would need to go back to the book again to be sure, but he was propagating the view that a sort of personal God that we had been brought up to believe in was really not very likely. He might be willing to believe in an all-embracing creator who took an eye on things, but he produced arguments that it was unlikely that God interfered in our lives—interfered is probably the wrong word—influenced by prayer, for example, to move blocks of stuff from one place to another, or anything like that. I think that probably I had this idea all along, but didn't recognize it. That was also liberating. I think now that Fred has gone back to some sort of cosmic intelligence, which I would have said he was against at that time. Anyway, Fred was a great influence, but I didn't meet him until many years later.

Tell me a bit about your work at Manchester.

As an undergraduate, I took physics, and there I discovered that I'd been badly prepared, which was hardly surprising. I did well in physics, but I wasn't among the very top students. Part of the reason was I didn't bother too much as to whether something I was studying was for exams or not. I hadn't been brought up with this approach. Nick Woolf, who's now at the University of Arizona, was in the same class as me. He and I used to spend a lot of time talking about various scientific things that had absolutely nothing to do with the classes we were taking.

Did you build anything at that time?

Only in labs, which I was not good at. Although I'd supposedly been trained at the technical school to be good at that sort of thing, in fact, my weakest accomplishments in the technical skill were building things or drawing.

Were there any professors you had that were particularly influential?

Yes. Maybe I should tell you a bit about Manchester. When I got there in the early 1950s, it had a very distinguished history in physics. Ernest Rutherford was there, followed by William Bragg, followed by P. M. S. Blackett. All three got Nobel prizes. It was a wonderful period, which lasted maybe 40 years. Just as I got there, Blackett was leaving to go to Imperial College, and they didn't really replace him for awhile. The place was drifting a bit. But, in the meantime, he'd helped to bring in radio astronomy and other kinds of astrophysical work. Wolfendale, who worked on cosmic rays, was there. The first few days I was at Manchester, we had to be told which classes we were taking. I remember going along and the man in charge said, "Dr. Wolfendale cannot lecture until next Thursday because he's in Sardinia flying a balloon." This immediately changed my perspective, because I realized now that not only did you get to do science, but you got to visit Sardinia. That was pretty good. His lectures on cosmic rays were excellent, too. Manchester had been very strong in cosmic rays, but it was a dying field.

At the end of the first three years, I got an upper second-class degree, which was okay considering the field. It wasn't brilliant. But they had a scheme whereby the better students from the second year could apply during the third year for a grant—like an NSF grant—to do research. So I applied for one, and I was 1 of 10 out of a class of 55 who were selected to go on for a Ph.D. In those days, most people in England, and certainly at Manchester, stayed at the same place to do their Ph.D. Once you were given this grant from the government, you could do any sort of physics. I looked around, and I decided that Jodrell Bank, where the radio astronomy was, which was about 20 miles south of Manchester, was too isolated. And also I felt that a lot of the work was boring. Although I had built radios as a kid, I didn't think that was my forte. So I signed on to do theoretical astrophysics with Franz Kahn, who at that time had just done a very important piece of work on the expansion of HII regions.[1] Also Lighthill was a professor of applied mathematics at Manchester. It was a very strong school in fluid mechanics, and the bottom line was I decided to study fluid mechanics applied to astrophysics. I did one piece of work first on the radiative viscosity figures, which was published in *Astrophysical Journal* in 1959.[2]

Why was the publication so delayed?

Writing it up took some time, but it wasn't too long. We also had to pay for it. I wrote it with another student, and we didn't have any money; eventually, the editor of the *Astrophysical Journal* published the paper for nothing. So I worked on radiative viscosity, partly because it looked like it might be important in some of the fluid mechanics problems I would want to do in astrophysics. Then I worked on the expansion of supernova remnants into the interstellar medium. While I was doing the second part, Kahn came to Princeton for a year and spent part of the time in Pasadena. He recommended me to Jesse Greenstein, who was then running a thing called the abundance project in Pasadena, financed by the U.S. Air Force. He had 10 postdocs. Observations would be carried out at Palomar and Mt. Wilson by this flock of postdocs. In those days, it was possible to join a thing like that without any previous experience. I had never seen a telescope.

So this was a big shift for you, since you had done a thesis in theory.

Yes. In fact, the intention was that I would go on doing some sort of theoretical work with Greenstein, maybe on the abundances of elements. I knew no nuclear astrophysics, but I had taken a course in nuclear physics in postgraduate school, and a course in quantum electrodynamics. After I'd been at Pasadena for about two or three months, I went up to Mt. Wilson to see the telescope that I had seen a picture of all those years before. They allowed me to guide for a few minutes, and I immediately decided that this was the thing to do. I quickly started to learn stellar spectroscopy, which I found amazingly easy. You know, different people have different natural abilities. I have a very good memory. I think I'm very good at pattern recognition. So when somebody presented me with a spectrum—you may have seen these old spectra with all these lines across it—I found it very easy to remember that this one must be strongly ionized silicon and this one must be europium. In fact, I can still remember the wavelengths. I last worked on them 25 years ago.

During your time at Manchester, say up to the age of 24, were you familiar with the big bang model?

Yes. I'd learned about it in Fred Hoyle's radio lectures—although, of course, the way he described it there was rather derogatory. In fact, I think he coined the term big bang. I was very familiar from around the age of 16 with the steady state–big bang controversy. Then when I went to Manchester, the first radio counts were being talked about. I knew about that, yes.

Do you remember, during that period, having any particular preference for one type of cosmological model versus another?

Yes. I was definitely in favor of the steady state.

Do you remember why?

It's hard to remember, but I think one of the reasons was Fred's claim, which I now find to be specious, that having things created all at an instant put the act of creation beyond the scope of science. Whereas, if you're creating matter all the time, you could then study it. It seems to me that while that might be possibly true, it was nevertheless no reason to believe in one hypothesis rather than another.

Any other reasons why the steady state appealed to you?

I think also because it did away with God, in what I would now regard as a very superficial fashion. You could regard creation, if it was going on all the time, as more of a natural process. I tended to think of God as a being, perhaps the creator of natural processes, but also the person who interfered in them from time to time. And doing away with that possibly appealed to me, but I wouldn't have that attitude anymore. I don't think I treated very seriously the arguments having to do with the Hubble constant, the age of the universe, and the age of the earth as evidence for the steady-state theory. Those didn't impinge on me as being very important arguments. I can't reconstruct why. A sensible thing to say now is that one's knowledge of the relevant numbers was so poor that it wasn't the clinching factor. I think it was more aesthetic—the same thing that drives people now to think that omega is one.

Let me ask you about some of your early research work. You started off working mainly with stars, and then in the mid-1960s you began doing extragalactic work. Can you tell me why there was that shift in direction?

Pure accident. In Pasadena, at that time, there was a division of labor, and interest, between dark-time astronomers and bright-time astronomers. The bright-time people were people like Paul Merrill, Jesse Greenstein to a large extent although not completely, and Olin Wilson, who worked on the details of stars. The dark-time [when there was no interfering moonlight] people were people like Edwin Hubble, Allan Sandage, and Minkowski, who worked on galaxies. You were not allowed to switch between the two. In fact, if you were hired at Mt. Wilson at Santa Barbara street, you were hired into either the nebulae department or the stellar department, and you were not supposed to move from one to the other. If you were hired into Greenstein's abundance project, you were supposed to slave away on stars. I tried to get out of that. For example, I was interested in Seyfert galaxies. Now the reason I was interested in Seyfert galaxies—I became interested before they became fashionable—was that they had peculiar spectra. My whole career has often revolved around peculiarities in things, recognizing peculiarities. I worked on peculiar A stars, for example, at least in part because that's what Jesse happened to be doing when I got there.[3] But I was really attracted to the field by the peculiar nature of the spectra, rather than by the intrinsic scientific interests. Other people were working on the spectra of the oldest stars of the galaxy, which I think are more interesting from a fundamental point of view, because you want to know what matter was like early on. But I was absolutely interested in the peculiar A stars because they were peculiar. I read about Seyfert galaxies—I probably read Seyfert's paper—and became aware that there were these galaxies with very unusual spectra.[4] That seemed reason enough to investigate their properties. I guess the course of my career was that gradually I became interested in more and more fundamental problems, while at the same time having an interest in the peculiar. I gradually moved from bright stars, peculiar A stars, to stars in the galactic halo that I studied with Leonard Searle in the late 1960s and stars in globular clusters.

By the end of the 1960s, a little less than 10 years after I started as an observer, I switched completely to working on extragalactic things. The first major piece of work I did was on the Zwicky compact galaxies,[5] and again, the reason was that they were a class of things that were outside the normal accepted body of things that people were investigating. It seemed to me that there was a good opportunity to find out something, although it was not entirely clear what it was going to be. I vividly remember my interest in that started in 1964 in the Hamburg IAU meeting, where Fritz Zwicky gave a talk about his compact galaxies. I had known Zwicky while I was a postdoc earlier. I had talked to him in the basement of Robinson while we were both measuring things. He was measuring his pictures, and I was measuring my bright stars. I'd realized he had lots of interesting views. Then I left Pasadena to go back to England. As I recall, Zwicky gave the first talk that I'd heard on the compact galaxies. He showed some pictures, and I remember Margaret Burbidge asking the question, "Are the spectra emission line or absorption line or continuous?" And Zwicky said, "Everything." That indicated it was a rather broad class of objects. At the end of his talks, Fritz left out on the desk some lists of the positions of his compact galaxies. In fact, the first list is in his book.[6] Not very many people took copies, but I took two and took them home. By this time, I'd been at the Royal Greenwich Observatory for two years. After three years as a postdoc with Greenstein, I had to find a job. I wanted desperately to go back to England. I went to the RGO [Royal Greenwich Observatory], which I disliked intensely. The only good thing I did there was to get married. But I left, as soon as I'd spent the two years that you had to spend outside the United States in order to get back in again. After I'd been at the RGO for a year, I met Geoffrey and Margaret Burbidge at a conference on peculiar A stars in Germany, and I said I was miserable and needed to be rescued. They appointed me to San Diego, where I went as an assistant professor of physics in 1964. So in 1964, I came back to the

United States with Zwicky's first list of compact galaxies. Then I was able to observe at Lick. To begin with, I worked on the stars because I was still supposedly a stellar person. I'd been at San Diego for only a year and two months when I was offered a job to go back to Caltech again, which I immediately jumped at, particularly since, when I told Geoff, he said, "Well, you'd better take it then." All this business about place A maneuvering with place B to keep somebody did not work out. Then at Caltech I was still supposed to work on bright stars, but after a year I was asking for dark time.

You were transgressing the order.

Yes. By that time, it was beginning to fold up. Various sociological processes led to its demise. I was very lucky, since with Zwicky's compact galaxies I wasn't trespassing on anybody's territory, because most of the nebular department people said they didn't exist. So I could hardly be accused of treading on their territory by studying the components of the universe that didn't exist. That got me into two subjects of interest. One was very high star formation in small galaxies and the other was Seyfert galaxies that were very much more luminous than the classical Seyfert galaxies. In fact, they were intermediate between quasars and Seyfert galaxies. There was a small paper somewhere in the *Astronomical Journal* in which I think I used the term "missing link."[7]

Do you think that the training of observers in astronomy has changed since your own student days, or your early days?

Well, I guess I got no training at all. I was merely shown.

Well, the experience you had.

I think there are two factors that are working in opposite directions right now. One is that the equipment is becoming sufficiently standardized that people trained as radio astronomers can use optical telescopes and vice versa—as long as it's routine measurements. In that sense, it's more and more possible to do [multi-wavelength] synoptic studies. On the other hand, people don't now get the hands-on experience that you really require to get the very best out of the observations, in many cases.

Why is it that people don't get the hands-on experience that they used to?

Because there isn't the opportunity. If you're a graduate student, you have a choice of what you do, but if you're a postdoc as I was, and you want to work in some other subject, it's very difficult. The advertisement says,

"Wanting somebody with expertise in this, this, and this." And that's what you do, and you go there. I think that's bad. I think that one of the best things that could be done for science nowadays would be to have postdoc positions that are sufficiently broad in their scope that people can do anything as long as the equipment is available.

Do you think that would allow them to get more hands-on experience?

Yes, *and* they would discover what they were best at, because often the first thing you do is more or less by chance. I think people's abilities are quite specialized. I know in my case that I'm terrible working with pictures, and I think probably better than most people working with spectra. And I'm quite sure there are people who are sort of hum-drum theoreticians like I was and would be much better as observers. Not necessarily, because I don't subscribe to the view that observing is an easy activity for anybody who is good at theory. There are more factors involved than mathematical intelligence. The ability to see what are the problems you can do doesn't seem to me highly correlated with mathematical ability.

Do you think that your view of astronomy has been affected by the particular fact that you're an observer as opposed to a theorist?

Well, remember I started out doing theory. I think I still look at questions from a more theoretical standpoint than most observers do. I would have said I was a schizophrenic person, in that I'm driven to choose topics that are peculiar—often not because they have any obvious potential theoretical impact. But, on the other hand, I think the way I tackle questions is more that of a theoretician. I think I know more about mathematical physics than a lot of observers do.

Well, suppose I generalize the question and ask you whether you think observers as a group have a different view.

Yes, I guess in selecting a problem, an observer's viewpoint is fundamentally different. That is, you often select problems because they're in some sense doable or observationally attractive to do rather than because they have a deep significance.

Do you remember when you first heard about the horizon problem?

Not clearly, but it was almost certainly hearing Kip Thorne talking about such matters at Caltech when I was an assistant professor. I recall Kip giving colloquia and Journal Club talks on, for example, the "Mixmaster Universe," which was an early, unsuccessful, attempt to show that an initially anisotropic universe would become isotropic.

When you first heard about it, did you regard it as a fundamental problem in cosmology?

No. I didn't really think much about cosmology until quite late in my career, because of this business where I stepped from bright stars to more distant stars to nearby galaxies to further away galaxies.

So you didn't worry about it too much?

No. Also, I had a view, until quite late, that we didn't know enough to extrapolate the expansion of the universe back very far. It wasn't until probably a year or two after the microwave background radiation was discovered that I finally concluded what everybody else had concluded long before—that the background radiation gave you a very good excuse to extrapolate back the expansion of the universe by at least a factor of 1,000. Therefore, one ought to start taking seriously, shall we say, deeper questions.

Do you regard the horizon problem as a fundamental problem today?

Yes. I've worked on the microwave background angular fluctuations for the last few years.[8] So therefore, yes, I do regard it as a big problem.

Let me ask you about another problem, the so-called flatness problem. Can you remember when you first heard about that?

About why omega's so close to 1? Probably late. It's very hard to reconstruct this. My tendency is to work in subjects observationally and get to know what things are out there before I bother to find out what people think about them. I'm a sort of bottom-up, not top-down, person.

Do you remember approximately when you heard about the flatness problem?

I know it was before Guth, because when Guth made the first attempts to solve it, the problem didn't come as a surprise.

Did you know about it long before then?

I would have said so—maybe 10 years before then. But that's only a guess.

When you first heard about it, did you regard it as a serious problem in cosmology?

No, I didn't. I cannot give you a rational reason why. I was much more taken with the apparent homogeneity of the universe than I was with the actual value of omega.

So the horizon problem impressed you more than the flatness problem?

Yes.

Did your view of the flatness problem change any after Guth—after the inflationary universe model?

I regard it as a very important, serious problem after Guth, but my view as to whether omega is 1 has not changed. I belong to the class of people who say, "I don't know. It is a problem to be determined by observation."

Do you have any idea why you gave the flatness problem more credibility after inflation?

I suppose the fact that there was a potential explanation gave it more credibility. That's the only reason I can think of.

How much stock do you put in the inflationary universe model itself?

Well, this is perhaps tangential, but the thing that's come out of it that I didn't know before was the possibility of having a physical explanation or a physical accounting for the cosmological constant. I'd read for years about the fact that Einstein didn't think it was a very good idea to have introduced the cosmological constant. Other people are claiming that it came naturally out of general relativity. But I was very impressed with the idea that you could get negative pressures out of physical situations and, therefore, measure the effects of the cosmological constant. Now, of course, examining what the value of the cosmological constant is is another question. I think that's the main intellectual thing I got out of the inflationary universe model. For the rest of it, I would prefer to see what nature's actually like.

Do you remember how you first reacted to the results of de Lapparent, Geller, and Huchra on the large-scale structures that they found?

I would have to dig a bit deeper. I was already familiar with the idea that there appear to be voids and narrow structures. So that confirmed views, or pictures, of the universe that I had for awhile. But I have to say that that whole field has been the biggest observational surprise to me. When I started on extragalactic astronomy, the absolute word was that there were field galaxies that were spread more or less uniformly and then clusters, and there was practically nobody who entertained the idea that there were voids. I think that's been one of the biggest surprises in our view of how the universe is arranged. The work that you mentioned has pushed that to the extreme, but it was already on the scene. Gregory and Thompson had done it, and others.[9]

Has that work, beginning in the early 1980s, shaken your faith in the big bang model at all, or not?

No. Because I don't see how it can contradict the idea that the microwave background radiation is coming from large redshifts. Therefore, I think the universe has expanded by three orders of magnitude or so. It doesn't strike me as being all that surprising that it shouldn't have been expanded by some more. I think the topology is very surprising.

Besides these examples that I've just mentioned, have any other of your major perceptions of cosmology changed in the last decade or so as a result of other developments?

I guess the thing that I thought a lot about is the Lyman alpha clouds, which I worked on for some time.[10] I can't say they've changed my perceptions of things, but they've sure as hell made me worry about what they could be and how they would fit into the kinds of schemes that are now entertained. Most likely, they're objects which—at least at the time when we observe them—fill the voids. That seems to be the canonical idea. But what happened to them and why there were things like that in the voids is an intriguing and completely open question.

Do you think that theory and observation have worked well together in modern cosmology?

Yes, in the following sense. It's very rare in astronomy that you have, as you have in physics, clear predictions about what to do. In that sense, I regard present-day astrophysics as being an ideal subject for somebody like me, because I would be bored and uninterested in just going out and checking that some number that somebody had predicted was correct. It's much more interesting than that. My business is to find out things and have some theoretician say, "Well, yes, this is what we expected or could have expected on the basis of such an idea. Now maybe this will be true if we do this." It's more of an interactive process. Occasionally, of course, it happens in physics too, where somebody almost accidentally discovers high-temperature superconductors. Then there is the theoretical interplay in trying to understand why the mechanism works, and that inspires other observations. But, in high-energy physics, in particular, it seems to me that the interplay is all in one direction. Astrophysics, until recently, has almost been too much in the other direction, whereby observers accumulated fairly unrelated facts; and then a picture was put together by theoreticians. But the observers could often do damn little about verifying the picture. Maybe by the time the picture was produced, all the evidence that you could have accumulated had already been gotten. I really like subjects in which there's interplay. It seems to me there is a good deal in cosmology now.

Do you think that the theorists are justified in extrapolating backwards in time in the recent early universe work?

You mean back to the Planck time?

Or 100 times the Planck time—back that far.

I'm sure that 10 years ago I would've said no, and now I would say yes. Because just as the history of actual discovery has been far more surprising than expected, it seems to me one ought to treat wild theories with more seriousness than one would have done. Everything we've learned indicates that we can actually know more about the universe than one would have suspected at the time when a particular point of view was formed. For example, when the expanding universe idea was put forward, I think there was no clear view that you could, in fact, look back to the microwave background radiation. Of course, as we know, the calculations were done that would have enabled you to have said clearly, "If you point a radio telescope, you should detect this radiation." But nevertheless, that didn't happen until 30 years after. And there are things like the primordial abundances, which I also worked on, where it took years for it to be realized that there was an actual connection between the observed world and something that one would have said was beyond our knowledge. Therefore, I am fairly sure that there will be other examples of that. Whether that allows us to go back to 100 times the Planck time, who the hell knows? But at least it's worth trying.

What do you think are the outstanding problems in cosmology today?

I occasionally think, being a keen reader of Watson's *The Double Helix*, if I were Watson and Crick working in astrophysics, what would I regard as *the* problem to try and solve. I think the dark matter problem is unquestionably the most interesting, and the one that is most likely to be solved, and the one about which I have no idea at all. I would be interested to know what other people think. For example, I am not particularly interested in knowing what the value of the Hubble constant is. I'm not all that interested in knowing whether omega is 0.2 or .99. But I would really like to know what the major constituents of the universe are.

One of the things that we're interested in is the extent to which scientists use metaphors or visualization in their work. Do visualization and images play any role in your scientific work?

I would say no, but they may be playing such a fundamental role that I just don't recognize it.

Have you ever tried to visualize the big bang? Has that ever been important to you?

Only when I'm trying to explain to a class how you are inside the big bang, and how the big bang can be in all directions. In that sense, yes. But not for my own research.

With this next question, you might have to throw aside your natural scientific caution. If you could design the universe anyway that you wanted to, how would you do it?

Well, that question could be dealt with on lots of levels. I don't think I'd want anything different to what we have here. It's sufficiently complicated to be interesting. Are you interested in answers like whether one would like a completely chaotic universe or close to the edge of a closed universe? I wouldn't want omega to be one for some aesthetic reason, for example.

If it turned out to be one, you would be satisfied?

I would be more satisfied at having understood why, than in the actual value.

There's a place in Steve Weinberg's book The First Three Minutes, *where he says that the more the universe seems comprehensible, the more it also seems pointless. Have you ever thought about the question of whether the universe has a point?*

I've thought about it occasionally. I have essentially zero religious impulses. Therefore, I don't see any reason why I should expect the universe to have a point. One wasn't asked at the beginning, "Do you want it to have a point or not want it to have a point?" And so you should just take things as they are. I certainly wouldn't belong to a class of scientists who have an endless striving to find a point that would lead them to believe in God or anything like that.

DENNIS SCIAMA

Dennis Sciama was born in Manchester, England, on November 18, 1926. He received his B.A. degree from Trinity College, Cambridge University, in 1947 and his Ph.D. from Cambridge in 1952. He was a Fellow of Trinity from 1952 to 1956 and a lecturer in the department of applied math and theoretical physics from 1961 to 1970. Sciama has had visiting positions at the Institute for Advanced Study in Princeton and at Harvard, Cornell, and the University of Texas. In 1970 Sciama became a Senior Research Fellow at All Souls College, Oxford, and in 1983 he joined Scuola Internazionale Superiure di Studi Avanzati in Trieste. Sciama is a Fellow of the Royal Society and of the American Academy of Arts and Sciences and a foreign member of the American Philosophical Society and the Accademia Nazionale dei Lincei.

Sciama's research has involved theoretical studies in general relativity and cosmology, including influential work on Mach's principle (the notion that the properties of nearby matter and space are affected by very distant matter), early studies of the steady state model, black hole thermodynamics, and the role of particle physics in cosmology. In the 1950s and 1960s, Sciama was the mentor to a generation of brilliant students in relativity and cosmology at Cambridge University, including Stephen Hawking, Roger Penrose, and Martin Rees. A strong and early supporter of the steady state theory, Sciama withdrew his support when he and Rees concluded that the theory was incompatible with new observational evidence on quasars. Sciama's enthusiasm and personal warmth, as well as such books as *The Unity of the Universe* (1959), have inspired many.

I wanted to start with your childhood. Can you tell me a little about what your parents were like, what they did?

My father was a businessman in Manchester. I grew up in Manchester. I then went to what we in England call a public school—that means a private school—from which I got a very good mathematical training. Those schools could afford to pay for the better teachers. In fact, my main teacher was a man who these days wouldn't go into school teaching. He got first-class honors in all three parts of the mathematical tripos in Cambridge, and he went into school teaching, and he helped me to get a scholarship to Cambridge.

Were either of your parents interested in science?

No, not at all. The atmosphere was entirely a business one. It rather surprised my father when I had this interest in science, which was outside his orbit. He was a very clever man, but he had left school at the age of 12 because his father had died, and he therefore wasn't used to higher education. Although he had a fine brain, it hadn't been trained. He was trained in the world but not trained in institutions. He didn't particularly know about higher education until I told him. I told him Cambridge was great and Trinity was great, and he accepted that. But it wouldn't have been anything in his world.

When he knew that you had an interest in science, when he became aware of that, did he discourage you or encourage you?

He tried to discourage me because he thought that I ought to go into his business.

What about your mother?

She helped me a little bit, but he was much the stronger personality. It was just that I was so motivated to do science and mathematics. I suppose at that age I didn't even distinguish them. I originally thought of myself as a mathematician, and only later did I move first toward physics and then to cosmology.

Do you remember any particular books that had a strong influence on you?

Yes, I can't remember how old I was when I read them, but I think it must have been in school—books by Eddington, in particular. Although I did read Jeans a bit, I found Eddington more challenging. He had several popular books. Perhaps now they've faded out a bit. At that time, they were very well known and considered the leading books of that kind. G. H. Hardy, the pure mathematician, wrote a lovely little book called *A Mathematician's Apology.* In it, he says from an early age his one ambition was to become a Fellow of Trinity. Again, this reads a bit old-fashioned now, and some people would even say it is no longer impressive, but at the time it thrilled me. I also read some Bertrand Russell, who again was associated with Trinity.

So you were interested in philosophy?

I've always had a mild interest in philosophy. In fact, I'm giving a talk on the philosophical aspects of the anthropic principle in a week or two. When I went up to Trinity in 1944, I attended a whole course of lectures by Wittgenstein, who was then still a professor and giving lectures. That was a very good experience. So, while I was basically doing mathematics, I had this interest in philosophical things, and it just so happened that many of the leading people at the time were or had been Fellows of Trinity. Trinity was the most prominent college. That was all part of the image of what a youngster would be attracted to—to strive, as it were, because there was this goal.

At this age, before you went up to Cambridge, did you have an intention to go into science or mathematics?

Yes, from about the age of 15 or 16, I suppose. Before that, I was very young, and I naturally said I would go into my father's business because that was the obvious thing to say.

Do you have any idea what caused you to be so taken with science and mathematics?

I can answer that question, but it's a bit wisdom after the event. In fact, I came to cosmology and astronomy relatively late. When I was doing my Ph.D., I started out in statistical mechanics. Only in the middle, partly under the influence of people here like Fred Hoyle and Hermann Bondi and Tommy Gold, did I start getting interested in cosmology and Mach's principle and so forth. Rather unusually, in the middle of my Ph.D., I switched to relativity and Mach's principle. They had to give me a new supervisor as a result. They gave me no less than Paul Dirac, in order to try and cope with this rather alarming change of subject. So, something inside of me must have burst out at that point. Although the statistical mechanics problem is very attractive theoretical physics, it doesn't, of course, have the connotations of understanding the origin of the universe. Once I started doing things beginning with Mach's principle, I then realized my real passion was for understanding the fundamental nature of the universe. Some people, perhaps the majority, do that by particle physics, and a few of us do it by cosmology. Of course, now the two things are linked together. So, I said, "Ah, hah, it's clear to me what it's all about, and I want to understand the way the world is made, where it comes from, and what it means in the scientific sense." That's my passion. Therefore, I've always tended not so much to work on very technical, detailed problems—although some of my students have—but rather on problems that in some way help to understand the great questions. But at 15, I didn't say all that. It expressed itself then as an interest in, say, mathematics. I remember enjoying projective geometry at school. I thought it was very beautiful and well ordered, and so on. Cosmology came much later.

Did you like well-ordered things?

Yes. Because, you see, if you do understand the universe . . . I mean, if Mach's principle had been true and sensible and worked well, or if superstrings or something are right, you are imposing order on the universe. And no doubt a psychoanalyst would have his own views as to why one wants to do that. If you impose order on the universe, then you help to achieve it yourself. Roughly speaking, what I like to say is that the universe is enormous—it is much stronger than you are—and your only way of hitting back at it is to understand it.

Can you tell me a little about your undergraduate and graduate work at Cambridge, just some of the high points?

The high point is that I was a disastrously bad student. No, that's putting it too strongly. I did get a minor scholarship in mathematics at Trinity, which was a great achievement. A large part of that was due to very good coaching by R. H. Cobb, the school teacher I mentioned earlier. It's a bit like training for a race or something, learning how to solve these problems. It's all book work, and you do many examples. You learn how to prove these things. I was good enough to be trainable to get a minor scholarship to Trinity, which was the great place in maths at Cambridge. But then I did so very badly in exams that when I finished I had to go into the army. This was just after the war, but there was still conscription, and I couldn't remain as a research student. I got a lower second in finals, and two thirds in my earlier exams. So I was in disgrace. However, during the two years that I had to be in the army, for 18 months of it I managed to get sent to a government research lab, which was called TRE in those days. They were studying photoconductors, or semiconductors. I was with a team, and I wrote internal reports. Hartree was one of the professors here at the time. I had seen him as I was leaving as a student, and I told him I wanted to get back into research. He helped me to get transferred to this government lab and then accepted me back as a research student when he had seen these internal reports. That is how I got back into the system.

So they thought you might have been dismissed out of hand from Trinity?

Well, I wouldn't take a student on with my exam records. It's all rather embarrassing when I now have to take students on. But Hartree took me back without a grant, and that's where my father being a businessman came in. I was able to live through the help of my father, despite his early discouragement.

How did he feel about supporting you in this intellectual pursuit?

Well, he was still terribly upset that I had rejected business, but he saw that I was so determined that he let me do it. Later, I agreed with him that I would stay in scientific research only if I got the research fellowship at Trinity—the thing Hardy had written all about. That would be a sign that it was worth the sacrifices. That was a crazy agreement, because even if I were very good—which I didn't know really at that time—it's very chancy whether you get a fellowship. You're competing with a whole group of people in a whole range of all subjects.

So you made him a business proposition.

I made him a business proposition. Exactly. But a very bad one. By sheer luck I did get the damn thing, so I was able to remain in an academic career.

When you decided to do cosmology, you said that you came under the influence of Hoyle and Bondi?

And Gold. They were all at Cambridge. They were senior to me, but I got a bit friendly, particularly with Tommy Gold, and to some extent with

Hermann Bondi. Hoyle was still older than that. They were all playing a strong part here. They were considered sort of rebels at that time. Hoyle was not Sir Fred Hoyle, Plumian Professor. He probably had a lectureship then, and I think Bondi did. Also, Bondi wasn't Sir Hermann Bondi, etc.

This was in the early fifties?

Yes. I got my fellowship in 1952, and I actually got the degree of Ph.D. in 1953. I started being a research student in 1949. The steady state theory, which was one of the dominating ideas in cosmology at that time, was published in 1948. The steady state was a very attractive idea to some of us. Hoyle, Bondi, and Gold were also concerned with astronomical questions. But in a lot of their work, they were introducing rather new points of view, which tended to be the kinds of points of view that got resistance from the establishment. They were the young rebels, and they were an exciting influence for a younger person like myself. Even when I was doing the statistical mechanics, I must have gone to their lectures, and their personalities were robust and exciting. I suppose that played a part.

You mentioned steady state. Obviously that was extremely important during this period. Can you tell me a little bit about why you were so attracted to the steady state theory?

I suppose because of its simplicity and predictive power. Even now, we're struggling to understand the big bang. I accept the big bang, although Fred Hoyle still doesn't. I now accept that basically the big bang picture is correct. But there is a naturally very complicated physics that goes on near the bang. [Previously], there were even questions like: can you be sure the laws of physics are the same in a changing universe. There might be philosophical reasons for worrying about that. This was all part of the original discussion. Whereas it's reasonable to say that if the universe always has the same large-scale appearance, it's less of an assumption that the laws are unchanged. There were various arguments of that kind. The whole picture you got of the universe was a rather simple, appealing one. And the steady state theory did have predictive power, which was good.

All those things didn't mean I believed it, as it were, but just that it was so attractive that I felt in a small way that I should try and make it work. When hostile evidence started to appear, you weren't sure what to make of it. I remember writing various papers at the time and having arguments with Martin Ryle about whether the evidence against the steady state was good or not. It was worth trying to save it, but as the evidence mounted, there came a point where one couldn't. But the reason for supporting it was not, as I say, that it *had* to be right, but just that it was very attractive, and the penalty of having creation of matter didn't seem to be such a terrific penalty. As they used to say at that time, continuous creation of matter is even less of a thing to introduce than the creation of a whole universe at one go. But, I never felt then and I don't now feel so alarmed about outrageous proposals in physics, unless they're easily disposed of by experimental evidence.

> *You said that you felt that steady state had predictive power, and that appealed to you. Did you feel that it had more predictive power than the big bang model?*

It did in some respects, because by denying the possibility of evolution of the average properties of galaxies, which would [require] many [unknown] parameters to describe, you could make much more specific predictions about, for instance, the number of sources as a function of redshift. Whereas, indeed as we all know now, the big bang model *requires* evolution.

> *At this time, during the 1950s, did you have any preference for a particular model in the big bang, say open versus closed or that kind of thing?*

I did, and that was linked to my interest in Mach's principle, although this was never fully worked out.[1] But, as did other people perhaps for similar reasons, I preferred the Einstein–de Sitter model, the $k = 0$, flat model. In the Newtonian analogue of these models, k is the energy—kinetic plus gravitational. If the energy is due to gravitation, à la Mach, rather than having a kind of spontaneous existence, then at least it might seem as though it would be rather natural to have one energy balance the other. Therefore, that would be the attractive model. But that turned out not to work later, because I had a student, Derek Raine, now a lecturer at Leicester University, who worked later on Mach's principle, producing a much better theoretical statement of the principle. When he made the most Machian statement he could—a statement that I approved of—he then found that all the Robertson-Walker models (open, closed, and flat) except the empty one would count as Machian. I had to accept that, but it was disappointing. Until that was done, I would have preferred the Einstein–de Sitter model.

> *Why were you thinking about Mach's principle at all? I didn't know that was on people's minds at the time.*

I probably picked up the idea from Bondi. If you look at the Bondi–Gold paper on steady state[2] and you look at Bondi's very lovely book on cosmology that came out in 1952, there was a lot about Mach's principle in both of them. I found the idea *extremely* attractive, and this has something to do with my psychology. I like simple ideas with very great power in physics—the idea that centrifugal forces and Newton's rotating bucket are mainly due to galaxies. As I have pointed out in my books, the main contribution came from galaxies beyond what you can see with telescopes—suggesting that the whole universe acts as one unit in this way.

Let me ask you about another project that you worked on somewhat later. Do you remember what motivated you to work with Martin Rees on plotting the distribution of quasar redshifts versus intensities?[3]

Yes! That was very funny. That was typical of a lot of my work, where the student really does it much better. At that time, the hostile evidence against the steady state theory was accumulating, but it was in the early days, and you could still try to save the steady state theory. The microwave background had just been discovered. But at that stage you couldn't be sure it wasn't due to things other than the big bang. In fact, I wrote a paper saying that there might be a type of radio source whose integrated radiation would mimic a black body spectrum over at least a limited range of wavelengths—which was all that could be measured at that time.[4]

So you were defending the steady state.

The idea was to defend the steady state, and also I learned astrophysics in the process. I knew from the great battle between Martin Ryle and Hoyle about the radio source counts that questions of counts would be crucial. Quasar data was beginning to come in during that period. Of course, quasars were just three years old or so. In fact, the great discovery by Maarten Schmidt of a quasar with a redshift of 2 came in 1965. So, I started plotting out the number of quasars as a function of redshift.

Why did you do that?

To see whether it agreed with the steady state. This relation between number and redshift is a unique prediction of steady state. You don't have to worry about whether the quasars evolve at different redshifts. So there was a specific formula. The question was: is there enough data accumulated to test this? You see, today there are far more people in the field, and this sort of thing would be done instantly. But at that time there were fewer of us. So I plotted out the number–redshift relation. The way I do these things, it was sloppy. And lo and behold, it fit the steady state prediction. I remember going to Martin Rees and saying, "Martin, I have plotted out N [number] as a function of z [redshift] and the steady state is supported." Martin was then a research student of mine, with whom I discussed the more astrophysical types of questions involving cosmology. He was always a bit skeptical about my enthusiasm for steady state. He is a very well balanced chap. He said, "Well, I'll have a look at it," and he went away to have a look at it, and he did it better. Two days later—I forget how long it took him—he came back and said, "I've done it properly, and it's very bad for steady state. The observed relation is quite different from that predicted by steady state." I looked at what he'd done, and I agreed that he'd done it properly. That was the thing that for me made me give up steady state. There was a conceivable let-out from people like Hoyle and Geoffrey Burbidge, who were then saying that quasars are local, but I didn't like that. It was piling one thing on top of another. It really wasn't reasonable. I said, "Okay, the quasars are cosmological, and therefore this decides it." So, for me at least—though not for most people—it was this study that was decisive, and I had a bad month giving up steady state. Then, of course, Maarten Schmidt did a much better job, getting this evolution, with more data, and it's now always attributed to him, and I think quite rightly.[5] But we were the first to actually point out that quasars evolve, so I'm quite proud of that. But it was Martin, not me.

That brings up an interesting question: You have been the advisor of a number of students who have gone on to brilliant careers. Can you tell me a little bit about your approach to advising students?

Let me first say that I always feel that I've been in a false position, particularly by being at Cambridge, and to some extent also in Oxford. We've had the best students in England. And so, if you have a very good student, you just sit back and let him go, and he does wonderful things. That's what's happened in quite a number of cases. My only role was enabling them to do relativity and cosmology. That required a certain structure and someone who was willing to take them on, but then they did their own thing.

Did you talk to them on a regular basis?

Oh yes. Well, let's say I'm the kind of person who suggests problems to people. A good example, actually, is Brandon Carter, who did some very important work on the uniqueness of the Kerr solution and other such things.[6] I remember saying to him one day early on when he was my student—and he still remembers this and he says he's grateful for it—I said to him, "Brandon, why don't you do axisymmetric collapse. I think there is a lot of richness and interesting things there." And he went away and did axisymmetric collapse [laughs].[7] So, therefore, I provoked them a little bit in some cases. In Steve Hawking's case—as I think Steve himself has recorded now in his book and elsewhere—for the first year or two he was struggling for a good problem. At that time, in the more relativistic side of cosmology, as distinct from astrophysical, there wasn't too much to do that was high-class. Then in 1965 Roger Penrose produced the singularity paper, which was a bombshell—but for a star, a collapsing star.[8] I know there are articles that credit me with saying one ought to look at the singularity theorems more generally. I can't honestly remember doing that. My memory is that Steve came to me one day and said, "I can adapt Roger's arguments for the whole universe and get the singularity of the big bang." I said, "Yes. Good. Do that." The last chapter of his thesis is his first singularity theorem.[9] So, I regard it as a matter of sheer luck that I've been associated with all these students.

Let me go back to the 1950s again, when you were here with Bondi, Gold, and Hoyle, and the steady state was in the air. Can you tell me a little bit about the general attitude in the larger community towards cosmology—cosmology in general, not steady state in particular.

Physicists regarded it very badly, I think. Physicists generally, and in particular particle physicists, would have said that cosmology is highly speculative. Everything is uncertain. They were very scornful. I remember Murray Gell-Mann was once a visitor at Cambridge, and he came to dinner—it must have been in the mid 1960s—and he said to me, "There has been no progress in cosmology since Friedmann in 1922" [laughs].[10] Generally, I think, it was then regarded as just speculation—not because of its intrinsic nature, but because of the lack of good observational evidence. Cosmology was not quite respected.

How would a general astronomer have regarded cosmology at that time?

I think an astronomer would not have had those particular feelings that the particle theorists did. Someone like Hubble was regarded as a great man. Astronomers would have been even more aware of the uncertainties of the data, but they would recognize it as a worthy enterprise, I suppose. The intellectual scorn was more characteristic of the particle-theorist type of person. Astronomers might have had a few smiles at the passions with which cosmologists argued. But there wouldn't have been the contempt. I don't think contempt is too strong a word in those early days, among physicists.

That changed, bit by bit, as the new era came in and particle physics ideas became important. Things changed when, for example, the physicists realized that cosmologists could do much better than the particle physicists at restricting the number of neutrino types.[11] All that came in later. Then the physicists had to admit that maybe the cosmologists have got something.

I wanted to switch gears a little bit and ask you about your reactions to some recent theoretical and observational discoveries. Do you remember when you first heard about the horizon problem?

I am vaguely aware that Robert Dicke had raised that point, but it was not in the forefront of, certainly, *my* consciousness until Alan Guth's paper. Although the history of inflation is complicated. There were people before Guth, who now never get mentioned, and that, I think, is not fair.[12] But then we are not discussing that. As far as I'm concerned, it was in practice Guth's paper that emphasized that the horizon problem had to be taken very seriously. And the business about the flatness. In fact, it was the flatness, perhaps, that Dicke had referred to even more than the horizon problem. Maybe I'm getting them slightly confused.[13]

When you did become vaguely aware of the flatness problem, did it worry you as a serious problem?

No, I don't think so. This was probably my concern with other matters or my not being smart enough to spot that it really was rather important.

You mentioned that you became much more aware of these problems after Guth's paper. When you read that paper, did you take these problems seriously in the sense that they were important problems that demanded solutions?

I do remember that I was a bit slow to appreciate the significance of what Guth had done—perhaps because I had other things to attend to. When his paper came out, I glanced at it, and I didn't say to myself, "Ah, hah, here is a great breakthrough. Whether true or not we must attend to this thing." I didn't quite even know fully what it was all about. It was only a few months later, I suppose, when other people started talking a lot about it, that I said, "Ah, hah, I'm getting left behind, I better find out what this is all about." Then I either read his paper again or read something by Mike Turner or heard a talk. I did my book work, and I learned the stuff.

Then, it all fell into place and I saw how potentially important it was. In fact, Guth came to the Royal Society in London for some meeting. He spoke, and at lunch I remember saying to him, "Do you realize that your inflationary epoch is just the steady state theory?" And he said, "What is the steady state theory?" He hadn't even heard of it. So that is just one of many reminders about culture gaps, or time gaps and culture gaps.

Once you understood the horizon and flatness problems, or thought about them more deeply, did they seem to you to be serious, fundamental problems?

Yes, I think they are genuine problems, and the reason we weren't all worrying about them is partly because until recently there were so few people in the field. What was worked on or worried about at that time was a very sensitive function of who happened to be in the field and what their interests happened to be. It's the same when you look at the history of cosmology and black holes, where rather strange views were peddled by top people like Eddington. They only got away with that because there weren't an army of technically equipped people to say the correct thing and push them aside. It's interesting when a subject depends for its development on so few people that it depends on their individual attitudes and interests. Whereas when hundreds of people do it, you very rapidly get a kind of streamlined view. Now, there is a whole army of researchers. For any new idea about particle physics, there are hundreds of people ready to apply it to the early universe. In those days there was only a handful of us, and if this handful hadn't paid attention to these problems, then they weren't in the literature or currently debated. I think that's the reason. I suppose once they are thoroughly pointed out to you and your nose is rubbed into it, then yes, they are very important problems. Whether inflation has solved them or not is a separate, technical question. But clearly they are important problems.

Putting aside inflation, do you have any view as to how the flatness and horizon problems might be solved?

There's a third problem that's also very important—and I agree with Roger Penrose that inflation doesn't solve it—and that's the smoothness. It's related to the horizon problem. One argument is that the early wrinkles get pulled out by inflation. But that is not a correct argument. What inflation does, if it works well, is that it provides a possibility for a transport process being slower than light to equilibrate different regions and remove temperature gradients. And that was all that was claimed originally. Then there was a kind of shift of view that came in almost surreptitiously, which said that, in addition, inflation already does the smoothing out for you automatically, because of pulling out the smaller scales to larger scales. But if the small scales are very rough and they're pulled out to larger scales, the larger scales are rough.

What is your view now about the inflationary model, either in the original form or one of the derivatives of it?

Well, in the end I think it's turned out a bit disappointing. It was a marvelous idea. It had various difficulties. It's now in what I call a Baroque state. There are so many variations, and there is no formalism, there is no reasonable grand unified theory and a cosmological formalism that gives a scheme that really does all that is required of it. Half a dozen people in the field have produced their own variations. Perhaps this is the nature of scientific research. I'm not saying therefore the idea is wrong, but it's a mess at the moment. I do think that it is oversold by some of the pundits. But it's still potentially a marvelous idea—we just need more particle physics first, to get a grand unified theory that we might have faith in.

Why do you think that the inflationary idea has caught on so widely?

Two reasons, I suppose. One is the very elegant link with the most advanced questions of particle physics. Cosmologists like me are happy that particle physics plays a key role, but also the particle physicists enter the arena. And partly that inflation delivered what it advertised. It solves some great problems. Those are two perfectly adequate reasons. Plus, it's not everyday that there is a great new idea in cosmology. There is the fighting for recognition. So therefore people jump at it.

Let me ask you about an observational discovery. Do you remember when you first heard about the work of Geller, de Lapparent, and Huchra on the bubble-like structure of the distribution of galaxies? How did you react to that work?

I was very excited. That seems to me extremely important. I've talked to Margaret Geller about it. She visited Trieste where I work mainly now, and she spoke to the summer school I was organizing. She was saying quite rightly that the irregularities continue to the largest length scale that she has observed, and therefore why shouldn't they go on forever, and maybe the whole idea of a homogeneous universe is lousy.

How do you feel about that?

I said to her afterwards, over a meal, "Look, there is one constraint that you have got to recognize, and that is the isotropy of the microwave background. If you put too much irregularity on too large a scale, you conflict with that, and that is therefore an overall constraint, although it doesn't come in at 100 megaparsecs." She said, "What would you do if we go on making the studies, and we keep finding this effect, let's just say out to 1,000 megaparsecs?" I said, "Well, that would be the most devastating thing in physics and astrophysics. I don't know what I would do." There is no obvious, easy way out. To say we've totally misinterpreted the microwave background . . . We considered that in the early days. There were jokes that if it's so isotropic, that's because your box that is measuring the thing is isotropic. But by now, it would be very, very difficult to reconcile a bumpy universe on a scale of 1,000 megaparsecs with the isotropy of the microwave background.

Does that worry you?

No. I feel confident that the universe has to smooth itself out on that scale. Obviously you can ask me a hypothetical question: "What would you do if it didn't?" But that would just be a crisis in physics. It's silly to speculate.

I would rather ask you about what your attitude is right now.

Well, my attitude is that it's an extremely important discovery, because, of course, galaxy formation has to be understood. And it's related to the nature of the dark matter that we haven't talked about—how galaxies form and so forth. It was totally unexpected from a theoretical point of view. Therefore, it's a very, very important scientific discovery.

I gather from what you have just said, though, that it doesn't shake your belief in the large-scale homogeneity.

Well, fortunately, up to the scale that's now been found, it wouldn't conflict with the isotropy, although it's interestingly coming close to it. There are plans afoot to improve the measurements of the isotropy another factor of 10. If they don't find anything then, that would also be worrying, even from other points of view. Just structures you can see in the sky would then work [show anisotropy] at the one in a million level. Therefore, I'm confident they will find something. I think that's reasonable confidence. But if not, then we will have this crisis. I would suppose that you wouldn't find the same effect [inhomogeneities in the distribution of galaxies] at a much larger scale. Perhaps a bit larger, but not 10 times larger. So, I'm not worried about this. I'm very much excited because it's got to be understood.

Since you're not necessarily a strong proponent of inflation, I assume you are not convinced of the missing matter.

With an omega of 1? I take inflation very seriously. I was only saying—it's an objective fact, I think—that the theory is in a bit of a mess. But some form of inflation may very well be correct. It's a marvelous idea. Whether it requires omega as 1, I'm not completely sure. There is an argument going on at the moment between two of my old students, George Ellis and Martin Rees, as to whether inflation does require that. Of course, there might be other reasons we don't yet understand why omega equals 1. It's a nice thing from the point of view of theoretical physics. So I would be very happy with an omega of 1 on these vague grounds of fundamental theoretical physics. It's great fun looking for a form of the dark matter, although equally you have to worry about galactic haloes anyway.

Do you think that theory and observations have worked together in modern cosmology?

I think extremely well. One example, which I mentioned, is this business about the number of neutrino types. It fits almost too well. If you take the present abundances of the helium-4 and the other light elements and do the theory of it and worry about the neutron half-life, which isn't quite as well in line, you still find that you are only allowed three or four neutrino types. Now amazingly, as I am sure you know, the recent supernova, from the same kind of argument about how much energy is emitted, limits the number of neutrino types to perhaps five or six. So all this involves observations of all different kinds—both particle physics and astronomical. It all fits together. I think that's very remarkable.

What do you consider to be the outstanding problems in cosmology right now?

I suppose it depends a bit if you are more interested in astrophysics or fundamental physics. For fundamental physics—and I'm only saying what everybody says—it is the essential vanishing of the cosmological constant, because the grand unified theory type of discussion will rather naturally throw out a cosmological constant of 10^{120} times bigger than any value you have astronomically. If you think more astronomically, there is a clutch of problems. Some of them are quite old, like is the universe going to expand forever or collapse or what? That is clearly still not settled. The nature of the dark matter is not settled. The way galaxies form is not settled. We don't even know, observationally, the ultimate scale of the universe. I would have said all of those are important problems. Plus the problems that inflation aims to solve.

> *Let me end with a couple of philosophical problems. Here you might have to put some of your scientific caution aside a little bit. If you could design the universe any way that you wanted to, how would you do it?*

Can I first answer evasively? I have a view that by-passes that question. The problem, of course, is that the universe has to be very fine-tuned to bring about the possibility of intelligent life and human beings, or if you like, myself. That is probably not controversial at all. The controversy is: what is the significance of that statement. There seem to me three possibilities. The one I favor relates to your question. The first is just chance, which I think is really unpalatable. You can't disprove it. The second is purposiveness, or God, or something: God exists and regards us as the highest point of creation; he wants us to come about, so he fine-tuned the universe to make jolly sure that we came about. And I find that unpalatable, although many people accept that. And then there is the third proposal, which I didn't invent, but I favor very much. According to this third proposal, there are many disjoint universes, where the laws and constants of nature are different from one to another. In fact, I would put it even stronger: any logically possible universe exists, not just for anthropic reasons. Of course, the anthropic theory clearly leads just to the type of universe we're in.

People might say to me, "What about Occam's razor? You're crazy." But I believe that this third proposal in a sense satisfies Occam's razor, because you want to minimize the arbitrary constraints that you place on the universe. If you imagine all these logically possible universes, then you've got to think there is a committee, or maybe just a chairperson, who looks at this list and says, "Well, we're not going to have that one, and we won't have that one. We'll have *that* one, only that one." Now, that could have happened, but it seems to me a remarkable thing that that happened. It's much more satisfying to say that there is *no* constraint on the universe. All logically possible cases are realized, and we're in one of the few that allow us. So, that's not quite answering your question, but I prefer to say it that way.

Could I add something, in case you or anyone would think that this is an untestable proposal. Let's consider all the universes that do lead to me. Now we would not expect that we're in a very special one of those. All I know is that I exist, and I'm happy enough with that. If the universe is unique, however, you might expect a very special initial condition, and

Roger Penrose and Steven Hawking have both made proposals for the special initial conditions.[14] My prediction is that Penrose is wrong and Hawking is wrong, because if there are these other universes, and ones very close to ours, then we should be in a generic universe of the set that could lead to me. Therefore, I would not expect a beautiful, elegant, mathematical ansatz, like the Penrose one or the Hawking one, to apply to the initial universe. The initial conditions would be messy, but not too messy, or life wouldn't emerge. Therefore, when you do a measurement, in principle, of the initial conditions, I would predict them to be messy, and not describable by a simple, mathematical, elegant statement.

> *There is a place in Steve Weinberg's book* The First Three Minutes *where he says that the more the universe seems comprehensible, the more it also seems pointless.*

I remember.

> *Have you ever thought about this question of whether the universe has a point?*

I have thought about it, and I can't think of any point it has. It's the old question about why there is something rather than nothing. If you're going to have some logically possible cases, even one, you ought to have a whole lot. But why have any? I find that quite inscrutable. Of course, the very concept of a meaning is perhaps too anthropomorphic. I don't know. Obviously I have thought about it, but I have nothing to contribute.

> *Your explanation number two for the anthropic idea was not unrelated to this.*

But it doesn't really explain. I'm allowing that when I talked in Venice, I permitted that as a conceivable explanation. In fact, it was a Jesuit astronomer who spoke after me, and he said, "I am prepared to have all Sciama's universes. I don't mind that these days. But there is God in all of them." But as far as I'm concerned, I'm afraid—and I'm not a professional here—the word "God" is just a word. When this Jesuit spoke after me, he knew so much about God. It was amazing. God was a person, he said. So we have to say "he," "she," or "it," because those are the only personal pronouns in English—not just that God was some force that made the world, it was a person. How can he possibly know such things? It's ridiculous. If you had a concept of something that made the world, and it was needed in order that the world be made, then who made that person or thing or whatever it was, and so on? These are old, standard arguments, but they still have force as far as I'm concerned. It's true that people have, internally, a religious feeling, which they use the word God to express, but how can a feeling inside of you tell you that a thing made the whole universe? There is no relation between the two matters of concern. Therefore, while I'm prepared for and I can't rule out that there is another order of structure than ordinary matter, I know nothing about that order. There could be many orders. The word God just doesn't denote any structure.

MARTIN REES

Martin Rees was born June 23, 1942. He received his M.A. in mathematics and his Ph.D. in astrophysics in 1967 from Cambridge University, where he was the student of Dennis Sciama. Since 1973 he has been Plumian Professor of Astronomy and Experimental Philosophy at Cambridge and is also director of the Institute of Astronomy there. Rees has been a visiting lecturer and professor at Princeton, Harvard, and the California Institute of Technology, among other institutions. He is a fellow of the British Royal Society and a foreign associate of the U.S. National Academy of Sciences. Among his awards are the American Institute of Physics' Heineman Prize for astronomy in 1984 and the Gold Medal of the Royal Astronomical Society in 1987.

Few areas of astronomy and cosmology have not benefitted from the clever and original ideas of Martin Rees. In the mid 1960s he was one of the first to point out that the distribution of quasars in space was inconsistent with the steady state model of cosmology, thus lending support to the big bang model. He has contributed to the theory of galaxy formation, galaxy clustering, and the origin of the cosmic background radiation, among other topics. Rees brims with new ideas, which he freely offers to students and others. He is known for his ability to quickly plunge to the heart of a problem, and he is also known for being able to keep several mutually inconsistent theories in his head at the same time.

Can you tell me any influential experiences that you had as a child, perhaps through your parents, that might have gotten you interested in science?

I have no scientific background in the family, really none at all. My parents were both school teachers, and I went to ordinary kindergarten. I don't recall any particular experiences that made me keen on science at that stage.

Do you remember when it was that you got interested in science?

I think I was always interested in natural phenomena and puzzled about everyday things. But I don't think at any stage in my early school days I had any definite commitment that I was going to end up being a scientist.

Do you remember any books about science that you read?

I was rather addicted to encyclopedias; illustrated encyclopedias of how things work were among my favorite books. They had nice cut-away illustrations of machines and things like that.

Were you more interested in those kinds of things than other sections of the encyclopedia, such as geography?

No, I don't think I was. I was equally enthusiastic about maps and statistics and figures of all kinds. I remember having a good memory for figures, like the heights of mountains, the wheel-bases of cars, and things of that kind, but no particular influence or pressure to go towards science.

When you went to secondary school, did you get particularly interested in science?

I was quite interested, but in the English system, by the age of 15 you have to specialize in science-oriented subjects or nonscience. Since I had to make a choice, I chose mathematics, physics, and related subjects. But I think honestly, even at *that* stage, it was not a firm commitment. It was simply realizing I was, for instance, rather bad at languages. On the other hand, I was always good at English and at history, so it was somewhat arbitrary that I specialized in science subjects.

When you were in secondary school, did you learn anything about astronomy?

Nothing in school. I read some popular science books and one or two popular astronomy books, but we weren't taught any astronomy at all in school. I remember one of the first I read was by Ray Lyttleton. I also remember an encyclopedia article on astronomy.

Did you know anything about cosmology at that age?

In some of the popular science I'd read, I'd learned about the steady state and the big bang. I'd read things by Hoyle, who was somewhat notorious for some radio talks he'd given in the early 1950s. So I knew that there was a debate about whether the universe was in a steady state or not.

Was it in your undergraduate work that you headed in the direction of astronomy, or was it later?

It was really later. When I got to Cambridge, I specialized in mathematics. Mathematics at Cambridge includes pure mathematics and applied mathematics, which is actually mainly theoretical physics. At that stage, I liked pure mathematics. I was better at that in the exams. I enjoyed the puzzle-solving aspect, but I knew already that I wasn't the kind of person who was going to be an academic mathematician. In my final undergraduate year, realizing that, I concentrated on applied mathematics and theoretical physics and also went to lectures on statistics and things. Of the areas of applied mathematics and theoretical physics, I found particle physics and field theory rather difficult, and fluid mechanics very dull. I was oscillating between either astrophysics or statistics and mathematical economics. I did think quite seriously about going in the latter direction, and it was a somewhat arbitrary choice to try graduate work in astrophysics.

When you went into graduate school, were there any particularly influential people you encountered?

Yes. I started my graduate work in the end of 1964, and for my first year I was very diffident and undecided. I didn't know whether I was doing the right thing, and I wasn't particularly happy in it. But, after a year I became enthusiastic and realized that it was an interesting subject that I should persevere with. There were two reasons for this. First, I was in a lively group. Dennis Sciama, my advisor, was a very stimulating and encouraging person. And, of course, those years were particularly interesting years because the quasars had just been discovered and the microwave background as well. One of the first seminars I went to as a graduate student was by Roger Penrose. I didn't understand much of it at the time, but it was actually one of the first occasions where he explained the singularity theorem, which he'd just discovered.[1] So, in retrospect, it was a very stimulating period in astrophysics and relativity because, on both the observational and the theoretical side, lots of new developments were just beginning.

Did you find cosmology to be particularly interesting as an area of astrophysics?

I would never say that I'd been primarily a cosmologist rather than an astrophysicist. The topics I learned about when writing my thesis were the background radiation, intergalactic gas, radio sources, radiation mechanisms, and things like that. It was already clear that high redshift quasars were of importance for cosmology, so I was involved right from the start in the astrophysics of these. My first student papers were, in fact, on astrophysical models for 3C273 and for radio sources and for the intergalactic medium.[2] Also, I think I helped to talk Dennis Sciama out of the steady state theory.

Tell me about that.

The steady state theory had originated in Cambridge and was perhaps never taken all that seriously outside Britain. But in Britain, because its advocates were a very loud and articulate trio—Fred Hoyle, Thomas Gold, and Hermann Bondi—one heard a lot about it, and it *was* taken very seriously. Dennis Sciama, although not an inventor of the theory, had always been a great advocate of it. When I started research, he was defending the steady state theory against the evidence of the radio source counts. The radio source counts had had a somewhat checkered history in the 1950s, but by the early 1960s they were thought by Martin Ryle to be fairly reliable. Dennis devised somewhat contrived models to reconcile the counts with the steady state theory, postulating that many of the radio sources were local galactic objects, and we were a "local hole" in their distribution. He believed that the quasar redshifts were cosmological, and therefore quasars were a population that one could not say was local. When the first few redshifts had been discovered, we plotted out the distribution of these objects in successive "shells" of equal volumes around us and showed that there were more at large redshifts.[3] This was important in changing Dennis's mind.

Did he struggle with you initially on this or did he accept your arguments?

They were our joint arguments. But Dennis always had a rather different attitude to science, I think, in that he always had to believe *something* strongly and defend it as an advocate. So his views underwent a fairly sudden transition. My attitude towards science has never been quite like that. There were, in the mid-1960s, active debates about the microwave background and about the interpretation of source counts and quasar redshifts. That was one reason why I became interested in cosmology.

At this time when you were still in graduate school, do you remember whether you had any preference for an open versus a closed model within the big bang context?

I don't think so. I think at that stage the main new thing was that the microwave background offered the first evidence for a big bang.

Did you know about the oscillating universe model?

I suppose so, yes, because in 1969—about two years or so after I got my Ph.D.—I wrote a paper on what would happen if the universe collapsed, showing that the collapse would be more irregular than the expansion and that stars would get destroyed not by hitting each other but by the night sky getting bright and hot.[4] I was certainly aware of the different consequences of these two eschatologies, but I can't remember having either "claustrophobic" or "agoraphobic" prejudices at that stage. I think I inclined towards a closed model because I liked Wheeler's idea of an ensemble of universes, which made more sense if each member of the ensemble was finite. There certainly wasn't then the present strong prejudice in favor of omega being *exactly* one. The Cambridge radio astronomy group under Martin Ryle was then leading the world in observational cosmology and the source counts. Historically, those counts were very important because they were the first evidence against the steady state, before the microwave background came along. I accepted that as important evidence. I was never part of Ryle's group, which had a very closed "fortress-type" mentality against outsiders. But I attended their seminars and followed the subject quite closely. I was very puzzled by the tremendous hostility shown by the steady state advocates to Martin's results, which seemed to me very convincing. I acquired a better understanding of the psychology of this only when, at a much later stage, I read more about the history in the 1950s. Ryle had then been equally dogmatic about controversial and, in retrospect, wrong data. Back in 1954 he had been unwilling to accept that the radio sources were extragalactic at all, although Tommy Gold was urging this, and the counts based on the earliest radio surveys were indeed erroneous. So when I read more about this history, I became better able to understand the reluctance of people like Hoyle, Gold, and Sciama to accept evidence that seemed to me to be quite compelling.

Let me ask you about how you've reacted to some of the recent results in the last ten years. Do you remember when you first heard about the horizon problem?

I certainly remember, in lecture courses I gave as early as 1970, emphasizing that it's a big puzzle in the big bang why everything started off in a synchronized way, because causal contact was worse in the past. That was certainly something that we were familiar with. Indeed, in 1972 I wrote a paper suggesting that different parts of the universe were initially unsynchronized, and dissipation of the random motions released energy that was thermalized to give the microwave background.[5] The development that had certainly made everyone in Cambridge well aware of this problem was the so-called Misner program.[6] Charles Misner spent a whole academic year—1967–68, I think it was—in Cambridge and lectured about this. His motivation for discussing anisotropic models was that they might have a different horizon structure, which allowed causal contact. So the horizon problem. Although the horizon problem may have been news to the people who came into cosmology from the particle physics side, it was very familiar to relativists and others, being the main motive of people like Misner.

You mentioned an attempt to resolve the horizon problem by having homogenization occur by dissipation. Did you have any other ideas about how the horizon problem might be solved?

Not really. I think most of us in 1970 felt it was the kind of thing that quantum cosmology would have to solve. You can extrapolate back to the Planck time, but no earlier. That's why the idea of the infinite number of oscillations in the mixmaster universe, discussed by Misner and the Moscow group, was never relevant because you could only legitimately go back 60 decades of logarithmic time.

What did you think about the possibility of having the horizon problem explained simply by proper initial conditions?

Well, there is the "anthropic" type of argument, which has a certain appeal as a last desperate attempt—the idea that we're here worrying about the problem only *because* of those special conditions. I think that people felt that there may be some answer, but it was premature to seek it at that time. It was obvious that back at the Planck time, quantum gravity would come in and would transcend the concept of a classical horizon anyway.

Do you remember when you first heard about the flatness problem?

I've never quite understood the distinction between the horizon problem and the flatness problem. The paradox I'm sure I was familiar with in the late 1960s was that the big bang must have been synchronized in a rather mysterious way. Different parts of the universe that were not in causal contact with each other at $t = 1$ second had to have "known" to start off expanding with the same local curvature, the same entropy per baryon, etc.

I see. When you phrase it that way, then it's the same problem.

Yes. There was no natural way to homogenize or synchronize things. The motive of the Misner program was to homogenize an initially anisotropic universe by having neutrinos or something able to communicate over large scales.

Yes, but would that communication also have made the potential energy come very close to the energy of expansion, or omega equal to 1?

No. I remember worrying about this limitation at that time. There was a tremendous amount of activity in the late 1960s and early 1970s on anisotropic cosmologies by George Ellis and many collaborators.[7] This work was pursued in the United States as well, and by the Russian group of course—Khalatnikov and Lifshitz.[8] One of the motives was to see if anisotropies would damp out naturally. If so, this would have meant that the *isotropy* of the universe could easily be understood if we assumed it was *homogeneous*. But I remember, at least informally, arguing with people back then that that did *not* answer the homogeneity problem, unless one could show that damping out the anisotropy everywhere, when only a small part of the universe lay within each horizon volume, would make the universe homogeneous.

But you wouldn't make it flat.

No, and I suspect that even the specialists were a bit confused about this. They seemed to believe that if they could show that an anisotropic homogeneous universe would isotropize, that was equivalent to saying it would flatten. I remember talking to lots of people about this problem. It became clear in the 1970s that the damping out of anisotropy was not going to explain homogeneity.

Did your feeling about either of these problems change after Guth's inflationary universe model?

The idea of inflation clearly offered an important new insight, suggesting a possible explanation. It meant that one would not necessarily have to go right back to quantum gravity to get the solution. Prior to that time, I had supposed—in an entirely ill-informed way—that it was the kind of question that only quantum gravity could answer.

Did you find a mechanism that solved the problem after the quantum gravity era more acceptable? Or would you be just as happy finding a solution to the problem in the quantum gravity era, at some future date?

Guth's version of inflation offered an answer to this question without having a full theory of quantum gravity. But if we *did* have a theory of quantum gravity that gave us an answer, then that would be equally satisfactory. As to the way things are going, I get the impression that the distinction between those two concepts is getting rather blurred, in that the detailed features of the early version of inflation—based on a particular phase transition—have not survived. In a sense, things are reverting to the pre-Guth version of inflation, Starobinsky 1979, which was based on a sort of sudden transition, some sort of quantum transition, from a de Sitter universe to a Friedmann universe.[9]

Had you been aware before Guth's work of inflationary universe ideas or ideas based on de Sitter expansion as a way to solve some of the outstanding cosmological problems?

I had heard seminars by Starobinsky. I'd heard, but not understood, a lecture by Englert, who had done rather similar things.[10] After the concept of the inflationary universe was introduced, I at least understood better what they were about.

Did you think the inflationary universe model was likely to be correct?

It didn't make a tremendous impact on me when I first heard about it. At a Texas conference in Baltimore, I remember people telling me about this and that it was very exciting, but it didn't really sink in. Over the years since then, I've come to think that there's almost certainly something in the general idea of a de Sitter phase, but not necessarily in anything closer to Guth's first formalism than to Starobinsky's first formalism.

Do you have any opinion as to why the inflationary universe model has caught on so widely?

Oh, quite cynically, I think it helped that it was boosted in America, so there was more ballyhoo about it. And, of course, Guth wrote a much clearer paper than anyone else about it. But I think it was really an evolution rather than a revolution, when one looks at the earlier work of Starobinsky, Sato, and others.[11] I think there's another reason, actually. Guth's version of the delayed phase transition was tied to GUT theories [Grand Unified Theories], and these same theories introduced the idea of baryon nonconservation. The weakness in the early models was that if baryon number is fixed, then it's no good exponentially inflating a region and diluting all the baryons in it. You've still got to explain how we ended up with 10^{80} baryons in the observable universe. So the most important change around 1980, perhaps, was that the GUT models gave people reason to doubt baryon conservation. Had that other consequence of GUT not come along—suppose that Guth had just talked about delayed phase transition as a way to get exponential growth—then his model wouldn't have carried any more weight than Starobinsky's.

On an observational discovery now, do you remember how you reacted to the results of de Lapparant, Geller, and Huchra on large-scale structure?

There again, I regard it as an evolution rather than a revolution. I think the first detailed discussions in print about large-scale structure, and particularly about voids, were at the IAU symposium in 1977, the one in Estonia, with the proceedings edited by Einasto and Longair.[12]

With this early work, were you beginning to accept the idea of a very inhomogeneous universe?

Everyone knew that galaxies were clustered and things were generally smoother on still larger scales. A popular scenario at that time was, of course, the adiabatic pancake model, which naturally gave these structures.[13] So in the late 1970s, "sheets" and "voids" were taken as support for the pancake model, and therefore the adiabatic perturbations. It was clear that the actual statistics of clustering were going to be important as a discriminant. When people talked about voids and sheets, which certainly came in 1977–78, then theorists related them to those ideas.

What about the question of the sharpness of the features? Did you agree that the sharpness of the features might indicate that gravity was not the dominant force in forming these structures?

I think it's suggestive. I first saw the CfA [Center for Astrophysics] maps in *Time* magazine, but I'd seen things like this before. In this plot, the radial coordinate is velocity, or redshift, not distance. When one subtracts off the obvious "fingers" that the Coma cluster gives in such a plot, and also when one realizes that systems that are in the process of turning around will naturally give you apparent sheet-like features perpendicular to the plane, the evidence becomes a bit ambiguous. When I first looked at the picture, I realized that the key thing was whether the sheets were or were not preferentially perpendicular. If they were preferentially perpendicular to the line of sight, then the natural thing would be to say that they are the effects of the region turning around, with a small velocity gradient. Clearly, when you're looking in velocity space, it's not the same as physical space. As you know, since then there have been all kinds of different [computer] simulations of clustering. I'm still not convinced that we have firm evidence for nongravitational effects. The pancake picture, which goes back to the 1970s, predicted collapsed sheets where there would be dissipation, galaxies being formed only in those sheets.

If we average over a time scale of say 10 or 15 years, in what way do you think cosmological thinking has changed?

Perhaps in two ways. First, the realization, triggered by developments in physics, that features of the universe that previously we had to regard as initial conditions, or premature to study, can now be seriously discussed. The baryon-to-photon ratio clearly can't be discussed at all until one had the idea of baryon nonconservation. The same goes for specific versions of inflation, and possibly the origin of the density fluctuations. These are fundamental questions that everyone accepts can now be discussed seriously. So there have clearly been very important developments over the last 10 or 15 years in the *scope* of theoretical cosmology. On the observational side, there have been developments in the quantitative analysis of clustering, starting with the correlation function. Also, there is the realization that the microwave-background-isotropy limits are a serious constraint on scenarios for galaxy formation: a baryon-dominated adiabatic pancake model can already probably be ruled out on the basis of the present microwave limits. So one can now confront different theories with observational tests—studies of clusters, high z galaxies and quasars, and the better microwave background limits. And, as I mentioned, one can now talk seriously about what caused the baryon-to-photon ratio and the isotropy and the fluctuations. At the same time, there have been new ideas such as strings which may be relevant to the fluctuations.

Do you think that any completely new cosmological questions have emerged in the last 10 years—questions that we weren't even asking before the last decade?

I suppose many of the questions we didn't ask seriously—or not in our working hours, as it were—because we thought they were premature. But I can't really think of any of the questions that we think about now that one didn't *hope* one would be thinking about.

> *Do you think our overall view of the universe has changed in the last 10 years, on the largest scale?*

I would say the most significant change is the evidence for dark matter. There again, this goes back 50 years. But the firm quantitative evidence goes back to the mid-1970s—the rotation curves and the more detailed studies of clusters. The general realization that the universe is dominated by dark matter, which may or may not be enough to give critical density, is an important change over the last 10 years. Again, there's been an evolution in the seriousness with which the idea has been taken. I can't think of anything that's been a really sudden transition—such as we'd have if strings turned out to be dominating the universe or we discovered some entirely new piece of physics that created fluctuations from nothing. But I don't think that the observations have pointed to anything that's really qualitatively new. There are some things that we *haven't* found. We haven't found any objects with a z of 20, for instance, though one wouldn't have been amazed if we had. It's surprising, actually, how well the standard model has stood up. The standard big bang model is a success. There are things that could have happened over the last 15 years that would have discredited the big bang model. We might have found zero helium abundance somewhere. We might have found some objects of an age vastly greater that 15 billion years. We might have found the Hubble constant to be at least 100, in which case there'd be a big age problem. The fact that none of those things has happened is gratifying for the extrapolation back to t equals one second. With the advances in physics, one has got a new set of questions to address what happened *before* the first second. The status of those discussions is rather like the status of the hot big bang at the time when Georges Lemaître and George Gamow were talking about it. They didn't have quite the right physics for the neutron-proton ratio and things like that. But over the subsequent decade or two, the physics of the first few minutes got firmed up. Likewise, we can now hope that ideas on the earlier stage will firm up. In one respect, we should be *less* hopeful, since the relevant physics is going to be more exotic and harder to confront with laboratory tests. But, on the other hand, in the 1950s only a few "eccentric" physicists took much interest in cosmology, whereas the subject now clearly engages the interests of a substantial fraction of the best mainstream physicists, and that ought to accelerate progress.

You've touched on this already, but how well do you think theory and observations have worked together in the last 10 or 15 years?

I think to some extent they've evolved independently. Obviously, there always will be theoretical investigations of things that are not yet ripe for comparison to observations, but I think there *are* now areas where there is a fruitful confrontation. One rather good example is that the detailed calculations of microwave background isotropy predicted in different models can be compared with the fluctuation amplitudes needed to get galaxies in clusters now. One needs quite sophisticated calculations in order to get a precision better than a factor of two, and that level of precision is merited by the observations now. And observers can now measure helium abundances to almost one percent accuracy, which allows important constraints to be set on the conditions at the epoch of primordial nucleosynthesis. N-body simulations of galaxies and galaxy clustering have been important in guiding the observers. And the studies of galactic evolution, which started with Beatrice Tinsley, have clearly been important.[14] Observers know what they can really expect to get out of their data and what is premature to try and get. The fact that in 1970 astronomers were optimistic about getting q_0 [the deceleration parameter] showed that they didn't then have an adequate appreciation of the complexities of galactic evolution. We now realize that the astrophysical theory and the studies of galaxies and the cosmology have to advance on a broad front, in the hope that all parts will come into focus. One won't, for instance, make any progress on geometrical cosmology until we have a better understanding of galactic evolution. Many of the observations are made regardless of theory, but I think we're getting to the stage where it's often prudent for observers to seek theoretical input in formulating systematic programs. Theorists can advise on how important it is to get an extra factor of two in the microwave background, how surprising it is to see gravitational lenses, just how puzzled we should be by the distribution of galaxies, and how to get a better handle on the large-scale velocities of galaxies.

Do you think that the theorists are justified in extrapolating backwards to $t = 1$ second?

Well, I would certainly bet more than 50% odds on that extrapolation and on the standard big bang nucleosynthesis story being OK. I think one should still be slightly open-minded about other possibilities, however.

What about earlier than 1 second?

Earlier than that, then I would have less confidence of any particular picture, simply because the physics is uncertain at energies about 100 MeV, the quark-hadron transition, when t is about 10^{-4} seconds. Going further back, it becomes still more uncertain. It will be a very long time before we can have the same confidence in any detailed calculations of, say, the photon-to-baryon ratio, in the way we now have confidence in the predictions of helium abundances in the standard big bang. Similarly, any really detailed theory for incorporating inflation in a precise cosmological model or understanding the origin of fluctuations will have to wait quite a long time. What's encouraging is that these topics can be addressed. I suppose there are really three ways in which one could advance. Developments in particle physics may allow us to understand the early universe more definitely and may, for instance, predict some species of stable particle that survives in sufficient quantities to provide all the missing mass now. The other line of attack is to explore the *cosmogenic* consequences of different postulates about the very early universe: to see if galaxy formation and clustering accords best with what we observe in models dominated by cold dark matter or hot dark matter, or if fluctuations were triggered by strings or random Gaussian noise. I suppose a third line is the more speculative discussion of the very early universe, which might lead to some new insight.

> *Could you give a list of what you consider to be the most outstanding problems in cosmology today.*

Clearly, there are the fundamental issues, whose solution will have to await a better understanding of ultra-high-energy physics. Those are the problems of initial conditions, fluctuations, etc., which depend on as yet uncertain physics, such as quantum gravity, phase transitions, and so forth. The questions of isotropy and flatness, the baryon-to-photon ratio and fluctuations all depend on the ultra-early universe in some sense. So that's one set of problems. And there is the other set, where the basic physics is known, but where things are complicated simply because they're nonlinear. Here I think we can hope for progress via a combination of theoretical modeling, numerical simulations, and observations. We'd like to understand the evolution from amorphous beginnings to a universe structured in galaxies and clusters of galaxies. Did galaxies have their disks at a z of 1? Are there quasars beyond a z of 5? If not, are there sources of energy beyond a z of 5? Clearly, the different models—the pancake model, the string model, the cold dark matter model, and so on—make very different predictions, and I think we can hope, by better observations of high z objects and background radiation, to gradually firm up on an evolutionary scenario. The kind of cosmology that I'm mainly concerned with—the evolution of galaxies and why the universe was different at high z—is not really different in conceptual status from what's done by a geologist or a paleontologist. The nature of the evidence we have may be slightly more tenuous, but in principle it's no different. We're trying to place our solar system and our galaxy in a grander evolutionary context and to trace the chain of cause and effect back further. But it's not different from the methodology of many other sciences, although the first set of problems I mentioned—the more "fundamental" ones—are perhaps different in that they do confront, for instance, the question of whether the universe is inherently unique or not.

Do you think about questions like that, also?

I wonder about those questions a lot, yes. And when ruminating on the anthropic principle, the universe, and the place of planet earth in it, one's attitude depends very much on whether one thinks that there is something unique in the laws of nature, whether there is going to be a unified theory, which is going to tell us that the constants could not be otherwise. Or, alternatively, could there have been a universe or different parts of our universe beyond our present horizon where things were different? That question, whose answer will have to await further developments in physics, is going to make a difference in one's attitude to the anthropic principle.

Let me end with a couple of questions that are of a more philosophical nature. If you could design the universe the way that you wanted to, how would you do it?

You mean so that I would enjoy contemplating it more?

If that would be a criterion that you would use. You can use any criteria that you want to.

I saw a title of a paper by someone—"Is our universe the simplest possible interesting universe?"—by which it meant did we have to have all the laws of physics we have, and four different forces, in order to get any kind of "interesting universe." I think the answer is probably yes in the sense that we do need to have some quantum mechanics and some microphysical laws, and we also need a force like gravity that can organize things on a large scale, generate thermodynamic disequilibrium, and allow large structures to evolve in the initial featureless universe. But, paradoxically, the weaker gravity is, the better. The weaker it is the more powers of ten there are between the microphysical and astronomical scales, and the more powers of ten difference between the time scales for global evolution and the time scales for the microphysical processes. We have—and our very existence requires—a universe that allows complexity and structure on many scales to evolve within it, starting off from simple beginnings. That universe has to have the property of flatness and also a force like gravity, which can produce large-scale organization without being so strong as to crush out all small-scale structure. As to whether one wants the universe to be infinite, to go on forever, I suppose I would *like* a complex and varied universe where different laws of nature might prevail in regions beyond our present observational horizon. If the universe went on expanding forever, other domains with different laws of physics and different properties could conceivably come within our observable range in the very distant future. Some people, of course, prefer the idea of a collapsing universe. Then you could put the idea of an ensemble of universes on a more serious footing, but you could do the same, I suppose, for this one universe, if you allow different parts of it to be disconnected and have some oases with laws of nature propitious for complexity, and others not.

> *Do you like the idea of having different parts of the universe with different laws, or would you prefer all the universe to have the same laws?*

I rather hope that there are unending complexities in the physical laws, rather than what many physicists hope for, which is that there will be a unique theory that is attainable. On the other hand, it's probably best that many physicists should believe in the final theory, because that's the strongest motivation that keeps them working hard at it. It's good that they should be looking for such a thing, even if one hopes they may never find it.

> *There's a place in Weinberg's* The First Three Minutes *where he says that the more the universe is comprehensible, the more it also seems pointless. What is your attitude about that idea?*

I don't understand the remark at all. First of all, I'm not quite sure what's meant by saying the universe has a "point" or not. Some people react that way because life and intelligence seem a small, unimportant part of the universe. I think that's an invalid reaction for two reasons. First, of course, there *may* be intelligence all over the universe—that's something we'd like to find out. But if there's no life elsewhere—if life is such a rare accident that it got started only on this one Earth (which is an entirely tenable point of view)—then we shouldn't think of ourselves as being the culmination of evolution. Even if the universe is going to recollapse, it's run less than half its course, and it may have an infinite expansion ahead of it. It's quite conceivable that, even if life now exists only here on Earth, it will eventually spread through the galaxy and beyond. So life may not forever be an unimportant trace contaminant of the universe, even though it now is. In fact, I find that a rather appealing view, and I think it could be salutary if it became widely shared. Then one could properly regard the preservation of our biosphere as a matter of cosmic importance. Despite our own species' unprepossessing characteristics, it may have potentialities. If you had clobbered the first fish that crawled onto dry land, you'd have destroyed the potentialities of all land-based life. Likewise, if we snuffed ourselves out, we'd be destroying genuine cosmic potentialities. So even if one believes that life is unique to the Earth now, then that doesn't mean that life is forever going to be a trivial piece of the universe. And if life exists elsewhere, then there's still less reason to believe that, in any sense, the universe is pointless. So I've never really felt the pessimistic connotations of Weinberg's remark.

ROBERT WAGONER

Robert Wagoner was born on August 16, 1938, in Teaneck, New Jersey. He received a B.M.E. degree (engineering) from Cornell University in 1961 and a Ph.D. in physics from Stanford University in 1965. After postdoctoral work at the California Institute of Technology, Wagoner joined the faculty of Cornell, from 1968 to 1973. Since 1973 he has been at Stanford, where he is now professor of physics. His scientific interests have focused on theories of gravitation and their application to astrophysical objects, such as sources of gravitational radiation. He has also worked on probes of the universe, such as element production in the early universe and the use of supernovae to determine the cosmological distance scale. Wagoner is a Fellow of the American Physical Society.

In 1967 Wagoner, William Fowler, and Fred Hoyle published the classic work on the synthesis of light elements in the big bang—a theoretical calculation that has become one of the cornerstones of the big bang theory. Wagoner has also proposed that the properties of exploding stars, called supernovae, might allow a measurement of their distance and thus help determine the rate of expansion of the universe. On the wall of Wagoner's small and cluttered office at Stanford are photographs of Fowler, Hoyle, and himself from the 1960s.

Let me start with your childhood. Can you tell me a little bit about your parents?

My mother's mother came from County Cork, Ireland. So I'm one-quarter Irish. The rest I'm not sure about. My father's parents grew up in Ohio. They had been there for many generations. I was lucky to grow up in Teaneck, New Jersey. I spent my first 18 years there, and we had a very good school system. I had a great math teacher, and I was on the math team. I had very good science teachers, except for physics. I had a terrible physics teacher in high school, and that's why I went into mechanical engineering at Cornell—because I didn't know any better.

What occupations did your parents have?

My father was a real estate manager for Western Union. My mother was a homemaker.

Did they encourage you in science?

No, not in particular. I just discovered it myself. I remember in ninth grade we took a test of interest or aptitude. I forget what it was called. I decided at that time, based on that test, that I wanted to be a forest ranger. I really wasn't exposed too much to science, although I did read a lot of books on rockets during one year. The next year I read a lot of books on golf, because I was really into golf at that time. I tended to investigate different areas intensely for a certain period of time. I won the science prize in my high school. I was interested in science, but I didn't know in what area.

Do you remember any books on science you read, besides the books on rockets?

Willy Ley's books on rockets were the ones that influenced me the most. He was one of the early German pioneers, along with Werner von Braun. He was a good writer.

At this age, did you ever think of any questions that we might call cosmological—like where did the universe come from?

No. The furthest out I thought about was the solar system—rockets to the planets and maybe nearby stars.

So you went into engineering because you were turned off of physics in high school?

And the climate at that time was such that there seemed to be a need for engineers, and there were good job opportunities. That was the reason.

You were already thinking about a job at that age?

My father encouraged me to think about what I wanted to do in a practical way. He didn't know anything about science, although my grandmother was a math teacher. She started teaching me calculus at the end of high school. She lived in Florida, so I didn't see her that often. She did motivate me towards math.

Did you build rockets yourself, or did you just read about rockets?

I tried, but that was my first indication that I was not an experimentalist. I had a chemistry set, which was another disaster. But I did build and fly model airplanes. I didn't do too bad at that.

When you went to Cornell, did you make a conscious decision between engineering and math?

When I got to Cornell, I was very happy in engineering, although, at that time, it was a five-year experimental program, which was soon abandoned. I was inundated in engineering, although I remember I had an excellent calculus teacher, Paul Olum. Until recently he was the President of the University of Oregon. That was freshman calculus. I remember taking freshman physics and it was somewhat daunting. I stuck it out.

Do you remember any other things that did make an impression on you at Cornell?

Yes, I had another good teacher in applied math. Another person who made an impression was Max Black, the philosopher. I took a logic course from him, which I really enjoyed. I started reading, on my own, a series of books by this crazy guy Alfred Korzybski. He had some wild ideas in philosophy. I really got interested in philosophy.

But the big event that turned me on was Fred Hoyle's Messenger Lectures in 1960 at Cornell. I think these series are still going on. Richard Feynman gave a series somewhat later. Fred talked about cosmology. That was my fourth year of the five-year program. I entered Cornell in 1956. Dennis Sciama was visiting Tommy Gold, who had just taken over the astronomy department. I think I met Sciama. But I just saw the light then and was really turned on by it, and I started reading Fred's books on cosmology that summer. I had a summer job at Cornell. I read Sciama's book, *The Unity of the Universe,* and Fred's book called *Frontiers of Astron-*

omy—quasi-popular paperbacks. That was just great. The next year I took an electricity and magnetism course in physics by a terrible lecturer. But, because I was motivated through cosmology, I got turned on to physics.

Do you remember any personal viewpoint on the steady state model?

No. I didn't know enough to really get much out of Hoyle's lectures except a feeling of excitement. And, of course, Hoyle is a fantastic lecturer, so that helped motivate me too. It was for a general audience, so it was a broad view of astronomy and cosmology.

So you decided you would go into physics at that point?

When I got my degree, I had an NSF fellowship, but in engineering. And I thought that I wanted to go into physics, although I was still interested in aeronautical engineering—going back to the rocket thing. So I looked around for a place where I could make my decision. Stanford had this great program, Masters in Engineering Science, something like that. You could take essentially whatever courses you wanted for a year and get a Masters. So I had my engineering fellowship, I signed up for that program, and I essentially took all physics, making up my physics background. That was 1961. I applied to the physics department during that year and managed to get in and switch my fellowship to physics. I was very lucky because my timing couldn't have been better. That was the optimal time to be a graduate student as far as support goes. There was plenty of money for everybody. Money had become available because of Sputnik.

When you went to Stanford and started doing physics, did you intend to make cosmology the area you were going to go into?

Talking to my fellow graduate students was one of the best aspects of education at Stanford and still is. My roommates were Heinz Pagels and Ron Adler, who's now working at Lockheed. They were both interested in general relativity. Heinz was more interested in particle physics. We used to have great discussions on philosophy, religion, physics. I got turned on to gravitation at that time, talking to them. But I still wasn't sure. I started my thesis work with Peter Sturrock, who was still in the physics department at that time, before he and some other faculty formed Applied Physics. I wasn't especially turned on, but I thought I wanted to work in astronomy. We couldn't find a good problem. So I started working with Leonard Schiff, who was getting interested in gravitation at about that time. In some ways it was good and in some ways it was bad, because he was department chairman and had been for a long time. He had to spend a lot of time on departmental matters. In those days, a chairman decided everything. It was a great system. Few committees. They just found someone and they hired them. So I saw Schiff maybe once a week or once every few weeks. I had to find a problem on my own. And that was another piece of luck. Quasars were identified in 1963, a year after I entered the department. And strong radio sources were known, so the motivation was to try to understand the mechanism of producing these strong radio sources and maybe quasars. My thesis was on the gravitational collapse of rotating massive objects—asking what would happen to a rotating collapsing body in general relativity, as a possible model of these objects.[1] It even involved some computer calculations.

It was really exciting in 1963, going to the first Texas Symposium in Dallas in December. Especially because of the Kennedy assassination, it was very emotional. At the meeting there were the Burbidges, William Fowler, Hoyle again, Maarten Schmidt, speaking about quasars. One thing I remember, which you'll appreciate. Roy Kerr stood up at that meeting—I was sitting near Chandrasekhar—and announced his solution to Einstein's equations for a spinning mass.[2] There were some questions asked, and it was obvious that the general feeling was that this solution was very special. Only Chandrasekhar stood up and said that this could well be a very important result. That meeting was so motivating for me as a graduate student, as you can imagine.

Did you present any of your own work?

No, I was in the very beginning phases of my thesis. I finished my thesis in 1965, another important year. When I went to Caltech, the microwave background had just been discovered.

Did you study cosmology while you were at Stanford?

Yes. I read every book I could get my hands on in cosmology.

The steady state theory was still in good shape then?

Well, there was a big debate about the radio source counts. And the quasars were starting to become a problem.

Do you remember having any personal viewpoint as to steady state versus hot big bang?

No. I was interested in cosmology, but at that time more in a philosophical way. I was more interested in doing calculations in general relativity with respect to the strong radio sources and quasars—understanding that. I wasn't interested in doing calculations in cosmology. I was very interested in aspects of cosmology such as Mach's principle—trying to formulate some realization of Mach's principle, understanding inertia, and things like that. Sciama's work on Mach's principle influenced me the most.[3]

Tell me a little about your work as a postdoc at Caltech.

The first thing I remember about Caltech was meeting John Bahcall, who was then an assistant professor there with Willie Fowler. The first time I met him, I introduced myself, and he said, "What area do you work in?" I said, "I work in relativity." He immediately got very upset and said, "You don't want to work in relativity. Switch to some field related to the real world." John was very frightening to me at that time. But the real reason I switched was the microwave background. I thought it was interesting to see what the implications of that were, although I did work on other things at Caltech.

When you say switched, you mean from something in relativity to something else.

To nuclear astrophysics.

Yes. You really hadn't had much background in nuclear astrophysics, had you?

None. I had taken the standard courses in nuclear physics.

What was your personal motivation in working on this problem of cosmological nucleosynthesis—as opposed to just doing what Fowler or Hoyle said you should do?[4]

Willie, as you know, is a great motivator. Willie made me realize that using the data that existed on nuclear reactions, you could really say something definite in cosmology—which was exciting. I had a well-defined model and could produce a specific answer.

By "well-defined model," do you mean the big bang?

The hot big bang. I began to feel that it was established at that time.

After the discovery of the microwave background.

Right. I wanted to explore the consequences of that. This was a physics problem that you could do in cosmology. Previously, there weren't many solid physics problems in cosmology where you could get definite answers.

Couldn't you have done this calculation some years earlier?

In fact, this calculation could have been done and should have been done in 1953. The history is that soon after Alpher, Gamow, and Herman began their investigations, Enrico Fermi and A. Turkevich developed a nuclear reaction code.[5] They had about 30 nuclear reactions. And within the Gamow model, they calculated the abundance of the elements.[6] Alpher, Follin, and Herman in 1953 then did the entire calculation systematically and formulated the hot big bang model in its full mathematical structure.[7] They incorporated all the correct physical processes. They produced the correct thermal history of the universe in their paper. In fact, they even commented on the helium abundance. Why didn't anybody then apply the Fermi and Turkevich reaction network within that model? It just lay dormant.

What's your opinion about that?

My opinion is that the initial motivation was to make all the elements in the big bang. It was beginning to be realized that stars were going to do most of it. Burbidge, Burbidge, Fowler, and Hoyle was on the horizon—the 1957 classic paper that showed how the stars produced the heavy elements.[8] That was in the wind, I believe. Fred Hoyle was pushing these views before 1957. He was the initiator. I think that's why the synthesis of the light elements in the big bang wasn't pushed. It's a very interesting question. You're absolutely right. In 1953 everything was there, although the cross sections weren't known as well. There are two big questions I'm trying to understand. How did Gamow and Fermi miss on the neutron-proton ratio, and why, when the mistake was realized, didn't they redo the calculations? I think it's because they wanted to make the heavy elements [in the big bang]. They realized the mass gap could not allow them to do that, at mass 5 and 8. So they said, "If we can't make everything, we're not interested." That's my guess.

It had to be all or nothing.

Well, Gamow was that way. He wanted to make everything in a simple way. He felt it had to be beautiful and simple. I don't know what Fermi wanted.

Getting back to 1966, was Fowler more influential than Hoyle in getting you to work on element production in the big bang?

Hoyle motivated Fowler. I don't know the exact sequence, but Willie motivated me directly. Another interesting question is Fred's motivation. Why does a steady state man want to get involved in a big bang calculation? In fact, I talked to Fred a little bit about this in Bologna last May. And I think he agrees with this statement, but I couldn't really get him pinned down. If you look at our 1967 paper, you'll see that more than half that paper is not on the early universe but on what he calls bouncing massive stars.[9] He wanted those guys, within the steady state theory, to make the helium, as well as the microwave background radiation. Fred pointed out—what people must have realized—that if you just thermalize the nuclear energy in the universe (at a moderate redshift), it gives you a temperature of 3 degrees. I was struck by that coincidence. I still am. That's the only natural way to get the microwave background without essentially any adjustable constant. But, of course, the big sticking point is thermalizing the radiation. But I was struck by that. That partially motivated me towards massive objects as well as the big bang. I try to maintain a balanced view.

When you started that calculation of nuclear networks in the big bang context, were you expecting to find the agreement with the observed helium abundance?

Oh yes, because Roger Tayler and Fred Hoyle in 1964 had gotten roughly the right number based on the correct physics.[10] Interestingly enough, I discovered, going back to that paper, Hoyle and Tayler say explicitly that if there are more types of neutrinos, you'll make more helium because of the increased expansion rate. Their priority of discovery of this result is contrary to what's claimed by everyone else. So Hoyle and Tayler were the first, not Shvartsman in 1969[11] or myself in 1967 in my *Science* paper.[12] Even Fred had forgotten. He told me that Roger had forgotten that, too. It seems everybody had forgotten. But they did explicitly predict that.

Do you remember how people reacted to your results? Were most people accepting of your results?

Willie is a very fatherly guy, and he took me to the National Academy of Sciences meeting in April 1966. The abstracts were published in *Science*, and I presented our results for the first time in April of 1966 in front of the National Academy. I would say it got a good reception. We had done the first careful calculation. In fact, it was great working with Willie. Almost everyday, I would get a note from Willie to look at a given nuclear reaction. We would look at it either theoretically or experimentally and try to track it down. We tried to understand every reaction—80 or 100 reactions. So it was a day to day interaction with Willie on the nuclear reactions, while I developed a computer program.

> *What do you think about the situation today? Do you still think that the nucleosynthesis calculations and the agreement between those predictions and the observations are one of the strongest supports of the big bang model?*

A few years ago I wouldn't have answered this way, but a couple of things have happened. In 1973 the interstellar deuterium was observed. The year before, I wrote a paper pointing out how important this would be.[13] When it was discovered, I felt much more confident, for the additional reason that people had been looking at other ways to make deuterium, such as in supernovae explosions and things like that. In most scenarios where you try to make deuterium, other than in the big bang, you make too much boron and beryllium. So when the interstellar value was announced, it was very exciting that it agreed for one value of the present density with the abundances of the other elements. That's when I first started to feel more confident. Because we didn't only have helium 4, we had deuterium, which is the only nucleus that couldn't be made in stars. You can make the helium, in principle, in an early generation of stars. But I feel you can't make the deuterium any other way. That's a strong point. The other thing that happened around that time was that Reeves, Audouze, and company did a careful calculation of the galactic cosmic rays interacting with the interstellar gas.[14] They calculated the lithium 6, the beryllium 9, the boron 10, and probably the boron 11, although it wasn't so sure at that time, and everything agreed beautifully with the observed abundances. So there was a beautiful, natural way to produce 6, 9, 10, and 11. Precisely those guys [elements and isotopes] that you cannot produce in the big bang. Everything else could be produced by stars, everything heavy. So it was a beautiful match. The big bang, the galactic cosmic rays, then stars. That was the next thing that gave you confidence.

The last thing happened a few years ago with the lithium 7. In our best-fitting model, the lithium 7 was always a little out of whack. That was the one that tended to agree worst with the observed abundances. Two things happened at about the same time. This was done by Schramm, Steigman, and their group, because I quit computer calculations in 1973 when I moved to Cornell from Caltech.[15] Essentially I have been out of the big bang game—in detail—since 1973. With the new cross sections, the predicted lithium 7 abundance that we calculated came up a little bit, and

Spite and Spite discovered these halo stars, which seemed to be better samples of primordial material.[16] And this abundance was lower by a significant factor, of order 5 to 10, than the standard lithium abundance. Both those came in the right way to get better agreement. Calculated abundance rose, and observed abundance went down.

Did this increase your faith in the big bang model?

Well, it was a confirmation of a prediction, in a sense, based on the other agreements. So that gave me more faith. But frankly, I'm still bothered by the Hoyle argument that I mentioned before—that if you release all the nuclear energy in the universe and thermalize it, you get the microwave background. It's so beautiful and natural.

Let me ask you about your reactions to some of the recent developments in theory and observation. To get a little bit of background on that, do you remember when you first heard about the horizon problem?

Alan Guth was at SLAC [Stanford Linear Accelerator], and he came to my office one day with this inflation idea, to try it out. That's the first time I thought about the horizon problem. I never worried about it much before.

Do you think you might have heard of it before?

Yes, but I can't document it.

But you knew you weren't worried about it before?

Right, I wasn't worried about an initial condition that seemed special. My feeling was then, and to some extent still is, that since we don't have a theory of quantum gravity, we don't know that there's a horizon problem. Because you cannot extrapolate back past the Planck time, you don't know whether you might solve the horizon problem before that time.

Did your view of the horizon problem change any as a result of Guth's work?

Frankly, not to a great extent. We were at a conference together the next year or so in San Francisco. A lot of famous particle physicists were there, like Murray Gell-Mann. Guth presented his views on inflation, and he was virtually ignored by the particle physicists. They did not think it was important at that time. His views were not accepted until a few years later. It was very interesting. I'm still not a believer in inflation.

Why?

I don't think it's necessary. As I said, one reason is that until we have a theory of quantum gravity, we don't know that there's a horizon problem. I'm perfectly willing to live with inflation. I'm just not an apostle of inflation.

Why do you think that inflation has caught on so widely?

Because it's the most natural solution to the horizon problem. It's certainly a possible solution. It's well founded physically.

But you, yourself, are not a believer of that?

I'm not motivated to adopt it. My position, which I've stated in some recent papers, is that one of my goals in my research has been trying to probe back into the universe. What I'm doing now with supernovae is a probe of the recent past.[17] Then there is my work on gravitational lensing[18] and the nuclear stuff. I'd rather probe with a *probe* whose physics we know, like supernovae or nuclear reactions, than with a probe whose physics we don't know, like galaxies or some speculation about particle physics. I think it's very dangerous to try to learn about physics from cosmology. I'd personally rather go the other way: learn about cosmology from physics. I know, in practice, it's hard at high energy. It's extremely exciting and interesting but dangerous to probe cosmology using probes whose physics we haven't verified. How do we test inflation other than looking to see if omega is one? That should be a test, but not a very selective test among other possibilities. In principle, there is large-scale structure, which might be an imprint of this early epoch, although to completely unravel that story could be difficult. I just am not motivated myself to worry about those problems right now. I'd rather understand problems that we can really get a handle on.

Let me ask you about a closely related question, and that is the flatness problem. Did you also first hear about the flatness problem in your discussion with Guth, or did you know of the flatness problem earlier than that?

Yes, I knew of it. I had worried about it and thought about it.

Do you remember when you first started thinking about the flatness problem?

Yes, when I worked on nucleosynthesis, we worried about whether we could close the universe, and the implications from our results that the universe was open. We wondered how natural that was. Omega being 1 was so natural.

So you were thinking about that in the late 1960s then.

But not to the extent that people did later—not deeply.

Did you ever think deeply about the problem?

I used to have this counter-argument. You just take the time-reversed argument with a star. Drop a star from any possible initial condition, with different energies, bound or unbound. Start a star collapsing and it's going to look like omega equals 1 when it reaches high density. So in the time-reversed problem, it's not unnatural that omega approaches 1.

As long as the initial energy of the star was small compared to the binding energy of maximum collapse.

Right. So I didn't really worry about the flatness problem too much, and that was one reason. To put it another way, I don't feel that our nucleosynthesis calculation necessarily requires that the universe be closed with some nonbaryonic manner.

When you say that the expansion of the universe today is like the time-reversed problem of the star, do you mean that the initial conditions . . .

Well, think of an oscillating universe, even though we can't calculate at the "bounce." At maximum, it can have arbitrary energy. So it looks flat at high density. So what. That's not a mystery. It's a consequence of dynamics, from an arbitrary initial condition.

Can you really argue by looking at the time-reversed problem?

I don't think any of these arguments are relevant because I think they are philosophical. Let observation decide what omega is. It's fine if the flatness argument motivates models like inflation that can be tested. Great. I'm interested in finding out what the universe really is, by observation. That's my own motivation. I'm trying to measure q_0 with supernovae. I don't want to get hung up with any prejudices. They are very dangerous in science, in my opinion. Although prejudice is a motivator, it can also be a dangerous motivator, if you're doing the observations or interpreting the observations.

Do you think it would be dangerous for you to have a prejudice, for example, about the value of q_0 in the work that you're doing right now?

It could be subconsciously dangerous even though I could convince myself I'm treating the data fairly. I think there have been cases where people have been misled subconsciously, interpreting their data by prejudices that motivated them initially to do the problem. Most people are motivated by prejudices, or a lot of people are. It's natural.

Would you say the same of theorists?

Yes. There have been a lot of ideas created to get omega equal to 1. People are motivated to have omega be 1, such as this quark-hadron phase transition work. That's fine. I'm motivated to *measure* omega—by looking at distant supernovae, or whatever. I think we need some probes that we understand, that we can use to measure omega. That's my approach, so I've become a less pure theorist. I've evolved in that direction. I used to be more doctrinaire, a long time ago.

What changed your thinking on that?

Experience. I worked on some theoretical problems based on observations that later turned out to be wrong. I've become wary. These observations were wrong in part because of prejudice, subconscious prejudices.

Let me shift to another topic. When you first heard the results of de Lapparent, Geller, and Huchra on the large-scale structure—and perhaps similar work by Haynes and Giovanelli—were you surprised?

A better word would be worry. They worried me for the following reason. As we have been looking further out in space, we're discovering inhomogeneities on larger and larger scales. At the same time, the microwave background, as we look with better sensitivity, remains smooth. Something smells very interesting. So it's mostly the *scale* that worried me, not the topology. We're seeing larger and larger structures. What's happening to the cosmological principle? It's working great in the microwave background; it's not working so well for ordinary matter. Another factor of 10 in the isotropy measurements of the microwave background could cause us to reach a real theoretical impasse. That would be exciting. In a way, I hope that happens. I'm not involved directly in that area, but that's my feeling.

When you say that you were worried by these observations, do you think that your faith in the big bang has been slightly shaken?

Yes. When I see the cosmological principle breaking down as it is continually tested at increasing scales, I get a little worried. The breakdown's not drastic, but I do get worried. More and more models are not consistent with those two facts, the microwave background and large-scale structures. So maybe we're in for a surprise. I hope so. It's unlikely.

Why do you say you hope so?

Because I think it would be great. We'd have to go back to the drawing board.

You would like that?

Well, we'd learn something new about the universe, something fundamental.

Even though you took great pleasure in finding agreement between your nucleosynthesis calculations and the observations, you would still look forward to the possibility of having to go back to the drawing board?

I've always been schizophrenic this way. I've always been wary about being misled, so I've always tried to think of alternatives. I've thought about cold universes and things like that. Again, I was initially motivated by this Hoyle thing and the fact that the theories of baryosynthesis are untestable, I believe. I'm a little uneasy. That's the best way to put it. If it turned out that the early universe was not the model we used in our calculations, I would not be unhappy.

Do you think that theory and observation have worked well together in cosmology in the last 10 or 15 years?

Let me start off with one area that's of particular concern to me, the distance scale, the Hubble program. I think that was an unfortunate program, basically because it was not based on a physical understanding of our probes of the universe. It was based on an empirical understanding of galaxies or stars or HII regions, believing they could be standard candles, without a physically well-defined model we could independently test. I think that was a sad part of the history of science, that Hubble program. There was all this folklore about these standard candles, which just kept disappearing with time, because there weren't independent checks of a well-defined physical model. We're going to get in trouble in cosmology if we don't have well-understood probes. That's been one aspect I've been unhappy about. The other aspect is the early universe. I think it's too speculative. It's exciting. It's fine for people to work on it, but I think it's a little over emphasized as compared to other areas in cosmology where we need more work, such as good distance measures, understanding quasars, etc. There are a lot of fundamental problems to be worked on in astrophysics compared to the number of people. For example, why do we have jets on galactic as well as extragalactic scales? Is there something fundamental going on there? There's been a lot of good work on jets, but it's still a great problem.

What do you think the outstanding problems are in cosmology right now?

The nature and distribution of the dark matter, I think, is the most immediate outstanding problem.

When you say dark matter, do you mean dark matter that we have evidence for that doesn't emit physical light?

Right. The dark matter we know is there, the clumped dark matter. There may be additional dark matter, but let's start with the clumped dark matter. I think that's the most outstanding immediate problem, in addition to the usual problem of H_0 and q_0. Of course, q_0 is another aspect of the dark matter problem. That's the only way to measure the total density of the universe.

What do you think is the most outstanding theoretical problem in cosmology?

The most outstanding theoretical problem is the evolution of structure—coming back to this problem of the connection of the microwave background and the present observations, and hopefully eventually relating back to the early universe.

Do you think the N-body simulations contribute to the understanding of that?

To some extent, but I'm sure they're misleading as far as their lack of hydrodynamics. But that's being fixed up. I think we'll gain much more understanding given the correct hydrodynamics, thermodynamics, radiative transfer, and so on. I think we should understand how to relate the present structure to the microwave background structure before we start making speculations about the early universe. Let's first pin down the nature of the recent universe.

Let me ask you to take a big step backwards. If you could have designed the universe any way that you wanted to, how would you have done it?

I've never thought about that question. If you want an off-the-cuff answer, I hope it would have been designed so that there are other civilizations that we can communicate with. It's getting a little worrisome because we haven't heard anything yet. It would mean an awful lot. That might be the most important observation made, communicating with other civilizations. I hope intelligent life is not unique. That's the first thought that pops to my mind. It would be an interesting universe, since, being populated, civilizations can learn something from each other. I'd hate to think that we're alone. It would seem very strange. I wouldn't understand it.

> *There's a place in Steven Weinberg's book* The First Three Minutes *where he says that the more the universe is comprehensible, the more the universe also seems pointless. Have you ever thought about whether the universe has a point or not?*

I used to think that it was very presumptuous of us to think that we were anything special in the universe. I really got worried about people being too concerned with their everyday life and not being aware of their cosmic environment, to put things in perspective. In that sense, I think it's good for people to be aware that we really are, in a sense, insignificant, as far as our position in and influence on the universe. I think it's good for people to be humbled a little bit in that way. Maybe we are insignificant. In certain senses we are, and in certain senses we're not. It's fantastic that we can contemplate the universe, that bits of matter can try to understand their complete surroundings. It's lucky. And maybe there are even luckier civilizations that we could communicate with, who have learned so much more. It would be fantastically exciting to learn what they might have learned.

JOSEPH SILK

Joseph Silk was born on December 3, 1942, in London, England. He received his B.A. in mathematics from Cambridge University in 1963 and his Ph.D. in astronomy from Harvard University in 1968. After temporary positions at Cambridge and Princeton, Silk went to the University of California, Berkeley, in 1970, and has remained there since, apart from sabbatical years at the Institute for Advanced Study in Princeton, the Institut d'Astrophysique in Paris, and the Mt. Stromlo Observatory in Canberra. He has been professor of astronomy since 1978 and is coordinator of the theory group of the NSF Center for Particle Astrophysics at Berkeley. Silk is a Fellow of the Royal Astronomical Society of England and of the American Association for the Advancement of Science.

Silk's research interests include theoretical studies of galaxy formation, star formation, galaxy clustering, the interstellar medium, and dark matter. His contributions to cosmology include one of the first theoretical calculations of how radiation affects inhomogeneities in the early universe, with implications for the size of galaxy clusters. Silk is also the author of a book on cosmology for the general public, *The Big Bang* (1980). The foreword of this book was written by Dennis Sciama, whom Silk remembers as the person who inspired him, while at Cambridge, to go into astronomy. He recalls that some of his most formative experiences growing up were on camping trips, when he would gaze at the sky at night.

I wanted to start with your childhood. Do you remember any particularly influential experiences that you had as a child?

I would say that one of my most formative experiences was spending most of my youth in the Boy Scouts. We did a lot of camping and spent a certain amount of time outdoors, at night, looking at the sky. That, somehow, must have instilled in me some curiosity about what was up in the heavens. It took a while for that to really come to fruition, because when I went on to university in England, I studied mathematics.

Tell me a little about what your parents did.

I was the first person in my family to ever aspire to go to college, so it was a real breakaway from tradition. My grandparents were immigrants from Eastern Europe. My parents grew up in poverty in the London slums. Basically, my father left school at the age of 14, and that was typical of the education level of the family. It was quite a breakthrough for me to make it through school, high school, and eventually go to Cambridge as I did.

Did your parents encourage you in academics, or in science in particular?

They were very supportive. They were very proud to have a son who was able to do something that they couldn't understand but admired. I suppose I was top in my class, or very close to the top, with sufficient motivation to keep on going. As long as I performed well, they encouraged that. It must have been a sacrifice to them, to some extent, because in my childhood it was assumed that I would leave school at 15 and go to work to try to support the rest of the family—one had younger sisters or brothers or whatever, or aging grandparents. That was what society expected, where we lived in London. Anyway, they supported me, and eventually I got a scholarship to Cambridge University.

Before you went to Cambridge, do you remember whether you read any popular books about science?

No, I don't think so. The English school system was very narrow. One specialized at a rather young age. From 15 years onward, I was doing only physics and mathematics. Everything else was dropped. There was no time, really, for any wider horizons.

What about before 15 years of age?

Before 15, everything was being crammed into us. I was taught four different languages apart from English, plus assorted subjects like geography, history, biology, and chemistry.

This was a public school?

It was a state school, a grammar school [corresponding to a public school in the U.S.].

Was science an interest that stood out for you?

It was mathematics. I had a mathematics teacher who inspired me and spent many hours with me after school. In those days, the apex of mathematics was at Cambridge University, and each college at Cambridge held special examinations for which the ordinary state school really had no expertise or track record in preparing its students. As a result, the public schools in England would dominate these examinations. This math teacher who took a special interest in me was able to coach me and direct me towards trying for this examination. In fact, I was the first person from my school to get into Cambridge University. I read mathematics at Cambridge. Then, in my third year, I did run into someone who was a very good popularizer of science, Dennis Sciama. He gave a course that I attended, on general relativity and cosmology. I think that was what really inspired me to go into astronomy, in addition to whatever thoughts I might have had earlier.

Out of all fields of astronomy, was cosmology of any particular interest to you at this time?

I was really just sitting in as an auditor on Sciama's lectures, but hearing Sciama talk in grand terms about Mach's principle and Olbers' paradox and all these things really opened my eyes to new ways of thinking. I would say that was the main influence on me. I spent the year after Cambridge at Manchester University, more or less looking around for some project to get involved in. It was during that year that I realized that astronomy was going to be the most interesting thing that I could see. I managed to get a fellowship to go to Harvard.

Tell me a little about your graduate experience at Harvard.

I was part of a class at Harvard that had a number of well-known astronomers. In my year, there were people like Ben Zuckerman, Frank Shu, Pat Palmer, and Jay Pasachoff. Harvard in those days was a rather smaller, more intimate environment than it is now. I was on close terms with a number of the faculty. David Layzer was my supervisor, and I worked with him and under him on cosmology. I have very good memories of Harvard. I went through my Harvard experience managing to avoid taking a lot of courses. I think I got through my entire graduate degree program by formally taking one course, on plasma physics.

When you were first exposed to cosmology, do you remember having any preferences for particular cosmological models—for example, open versus closed, or homogeneous versus inhomogeneous?

The first paper I wrote was actually on a very unconventional cosmological model called Gödel's universe, which is a rotating universe.[1] I suppose you could say that I started off in a very unorthodox way.

Did you believe in Gödel's model?

No, but Layzer was a great believer—and still is a great believer—in a very unconventional version of the big bang model, in which the universe is cold in the beginning.[2] I also worked on his approach to cosmology, on his theory. It took a certain amount of effort on my part, but I eventually broke away from that. I will say this for David, that he supported what I was doing even though I began working on the hot big bang, which he was philosophically and mathematically and experimentally opposed to. The breakthrough actually came when I spent one summer at Woods Hole, where the Woods Hole Oceanographic Institute organizes an annual summer school on fluid dynamics. In alternate years they tend to choose an astrophysical topic. This was in the summer of 1967, I think. They decided to look into the question of galaxy formation as a fluid dynamical problem in an expanding medium—idealized that way. I remember that there were a number of people there, but George Field, in particular, was one of the lecturers there. And he pointed the way to the standard view of the big bang, which one had really not been exposed to at Harvard at that time. I never really looked back from then on. In fact, the summer project I did at Woods Hole eventually turned into my thesis.

Was that the summer where you got inspired to work on the opacity of radiation and matter in the early universe?[3]

That's right. The people at the Woods Hole Institute included Ed Spiegel and a number of others. And so things really gelled together in pointing the way to an application of radiative transfer and fluid dynamics to cosmology. It was just a new area that no one had looked at. Another person there at that time who was a source of great inspiration and who died very shortly afterwards was Richard Michie.

When you began breaking away from Layzer's picture and started working in the standard hot big bang model, do you remember any early preferences for particular kinds of models—open versus closed, for example?

I was really dealing with the very early universe, and it doesn't really matter if it's open or closed, so the answer is no.

But you didn't have any additional personal prejudices besides?

I had no philosophical reason, nor any scientific reason, to believe that the universe was opened or closed. When one is dealing with what happened in the first million years of the big bang, it makes no difference.

When you began working in 1967 on the fluctuations and the primordial fireball radiation, do you remember what was the motivation for that work?

I had been hearing lectures on the coupling of radiative transfer to an expanding medium. The lecturers were people like Field and Spiegel and one or two other people who worked in fluid dynamics, like William Malkus and George Veronis. All of these were very inspiring lecturers, and I just started looking for a problem that would combine something that I was hearing about in the lectures with something a little bit different. It was clear that the early universe was an area that no respectable fluid dynamicist at that time would have worked on. It was really virgin territory.

In your Nature *paper of 1967, you start off by saying that the fluctuations are usually taken as a given.[4] Did you worry about what caused the fluctuations to begin with?*

No. I think at that time one just took the fluctuations as a given. One had no idea of what happened in the first second of the universe. The particle physicists hadn't discovered the big bang cosmology. Understanding of the early universe really began only with nucleosynthesis and the predictions of the light element abundances in the first 3 minutes. What happened before then was very obscure, I think. It was clear that the fluctuations must have been produced at the very beginning of the universe, before 1 second. And since one didn't have a theory, one just took it as a given at that time.

Why was it clear that the fluctuations had to have been produced so early?

Because it was very hard to imagine any causal process that could have gathered matter together on these very large scales. To make a galaxy, you need to collect together 10^{68} atoms. If you go back to the first instants of the universe—the first few minutes or even the first year of the universe— you find that the volume that you have to gather that matter up out of is so large that light cannot travel across it, which means that you need some noncausal process, something that is beyond the domain of any physics that we understood at the time.

Doesn't that problem just get even worse as you go back to an earlier time?

That's absolutely correct. So, the only resolution is that it's a property of the initial conditions of the big bang. You say, "I don't understand this; let me just suppose that it was there from the beginning, that the fluctuations were laid down and galaxies began from them." It's only been very recently that we have developed a theory about the origin of those fluctuations, and that theory is by no means universally accepted. It may even be that we still have to have some acausal process, some very special initial conditions.

One of the common threads through most of your work, coming up to the present time, has been the influence of various physical processes on the cosmic microwave background radiation. How confident do you think we are in our interpretation of the cosmic microwave radiation background?

When it was discovered in 1965, one really had no idea it was blackbody radiation, with measurements at one or two different wavelengths, and very limited knowledge of its isotropy distribution around the sky. It was on shaky ground. But in the more than two decades since then, we have learned that it is, to very high precision, blackbody radiation and also, to very high precision, isotropic. Those two properties are very strong arguments for an origin in the very early universe. The isotropy means that it has nothing to do with the galaxy, nothing to do with the supercluster of galaxies, nothing to do with the sun, or the local interstellar medium. And the blackbody nature is exactly what you expect in the very dense fireball in which the universe began. So, it fits in beautifully. That is not direct proof, of course, but it's by far the simplest explanation for the origin of the cosmic microwave background radiation. That's the philosophy that we take in physics. You start off with the simplest theory first. If someone can shoot the big bang down, then they'll have to think of a different origin for the microwave background.

I gather from reading your papers that you take the interpretation of the microwave radiation background so seriously you feel that you can rule out theories of galaxy formation based on it.

Yes. The reason is that the smoothness of the background is now measured to better than .01% on all angular scales, and that statement alone makes many theories of the origin of galaxies untenable because they would predict larger fluctuations. It's fair to say that until the fluctuations of the background are actually measured, we will not have any positive evidence for any theory. Hopefully, that measurement will come about some day. At the moment, it's sort of a negative argument. We're ruling out theories rather than finding the right theory.

I wanted to ask you about your reactions to some of the recent developments in cosmology, both theoretical and observational. To begin, do you remember when you first heard about the horizon problem?

I think it goes back a long way, to this question of the beginning of the origin of fluctuations. Cosmologists have been aware of the horizon problem for 20 years.[5]

Do you remember when you personally first heard about it?

One aspect of the horizon problem is that the fluctuations from which everything began would have to be acausal. You have to think of some process. That goes back to reading papers by Ted Harrison and Charles Misner in the late 1960s and early 1970s.[6] As far as the question of the horizon of the present universe—why is the present universe so smooth— all of this was focused, of course, very well by Alan Guth around 1981. But he was just gathering together a problem that was well-known.

When you first encountered the horizon problem in reading Harrison, were you persuaded that it was a serious problem?

Yes. I guess at that time the solution was simply one of initial conditions. You simply said the universe began smooth, or began with certain fluctuations that were almost smooth. Those fluctuations gave rise to the structure and the smoothness that we have today.

Was that a solution to the problem that you accepted?

It was not a completely satisfactory solution. But remember that we only have one universe, so we don't exactly have a statistical ensemble on which to start making comparisons and calculating the most likely possibility. So, yes, with reservations, I accepted it.

Did your view of the horizon problem change after the inflationary universe model?

It did change in the sense that I found Guth's arguments and the subsequent developments in inflation theories very persuasive, and I still do. I think if one really has some sort of "no-hair theorem" for cosmology, then you can start off with more or less anything and end up with a very smooth universe. That is really a marvelous development. I don't think we've achieved that today, but it is still a possibility.

When you say that you don't think we've achieved that today, do you mean that the inflationary universe model is not worked out to a satisfactory degree, or that it still has problems?

I think that you need to specify particular conditions in order for inflation to work. If the universe is too inhomogeneous initially, then inflation may not work. That's one of the worries, and there are other related problems, too. Now, it may be that we don't yet have a satisfactory theory of inflation. I think one is still waiting. If you really need to push inflation back to the Planck epoch, as some people believe, then you need to know the theory of quantum gravity at the same time.

Do you think that some version of the inflationary universe model is likely to be true?

I think it's sufficiently powerful as, if you like, a scientific philosophy. I find it very compelling, and it would seem to me that the ultimate theory of cosmology has got to contain something like inflation.

Why do you think it's been so popular?

It's been popular because its advocates, at least, have made it seem compelling, natural, and able to explain many things—not just the smoothness of the universe but the fact that the universe must have had fluctuations. If the fluctuations in the beginning were the wrong amplitude [strength], either too small or too big, we wouldn't end up with the universe that we have today. The inflationary advocates make persuasive arguments.

They're still having some trouble in getting the right amplitude of the fluctuations, but they get the spectrum, the distribution, the sizes to come out remarkably well, close to what we think we see.

The Zel'dovich spectrum.

Right. The Zel'dovich-Harrison-Peebles spectrum, to give full credit where it is due.

Let me ask you a related question. Do you remember when you first heard of the flatness problem?

I think the flatness problem, stated that way, was really concocted by Alan Guth. I first heard of it when I read his paper. Certainly, long before his theory of inflation came out, there were advocates of a flat universe, who regarded that as being the only natural model.

That's different from the flatness problem.

That's right. But the flatness problem is not so different from the question of why is the universe so old or why is the universe so isotropic. They're somewhat related aspects of the flatness issue.

So when you heard about the flatness problem, you heard the problem and received its solution at the same time?

Yes, that's right.

Suppose that you had not been presented with the solution to the flatness problem at the same time that you heard it. Suppose that you had heard about the flatness problem before Guth's paper. Would you have found it compelling?

Not at all, because in my view, it's all initial conditions—given the fact that we have only one universe. I see nothing whatsoever wrong with starting things off within an amount epsilon of being flat at the beginning so as to arrive today flat or almost flat. You may well have to make assumptions like that to get inflation itself to work. That is fine tuning. The flatness problem is just another version of fine tuning.

So you wouldn't have been particularly bothered by it?

No. I don't lose sleep at night worrying about the flatness problem. In fact, I personally think it's unlikely the universe is flat. One solution to the flatness problem, of course, is that the universe was born flat and always is flat. It may or may not be inflation that does that for us. That argument means nothing to me. In fact, I suspect that it could well be that the universe is open today. All the observations point to that.

If you are willing to make the argument that we have only one universe, so there is no reason a priori to expect one set of initial conditions to be more probable than another, and if therefore you are willing to take whatever initial conditions are needed in order to produce a homogeneous and nearly flat universe, then what does inflation really buy you?

Very little. In fact, initial conditions are an alternative to inflation. Inflation is really a way of trying to erase arbitrary initial conditions. And it hasn't succeeded. That's the worry about inflation. It erases a large subset of possible initial conditions, but it doesn't erase them all. So, I think one really has a choice. One can go with inflation and hope that someday someone will come along with a better inflationary theory, or else just take the simple view and say that the universe began in a certain way and here we are today.

Do you personally have any preference for one of those two approaches?

I would be delighted if inflation turned out to be right. It's such a beautiful mixture of particle physics and cosmology. So, philosophically, I sort of lean toward inflation. On the other hand, as an astrophysicist, I am perfectly willing to contemplate another universe [in which omega is not equal to one], just because that's what astronomers tell us is out there.

Does it worry you any that we have to reconcile the observations of omega equal 0.1 with the inflationary requirement that omega is equal to 1?

Presumably, if there is omega equal 1 worth of matter out there somewhere, we're going to detect it someday. It may be that we haven't been clever enough yet. But there are possible ways to look for this extra dark matter. Eventually, there's going to be an experimental test of this question. I would simply defer to that point. It's not really a worry because our experiments just are not good enough yet. We haven't gone deep enough.

The observations are only reasonably complete out to 10 to 30 megaparsecs. Within that range, it's clear there is not enough matter to close the universe if I use my local ratio of dark matter to light matter as a guide. However, it's completely an open question as to what might be on larger scales. We need better observations. You probably need thousands or even tens of thousands of redshifts. You have to go out to distances of 200 or

300 megaparsecs. A project like that will, I think, eventually directly measure the curvature of space. That's probably a few years off. I see the answer coming.

When you first heard about the results of de Lapparent and Geller and Huchra on the large-scale structure, and the related results of Haynes and Giovanelli, did those results surprise you?

I think it's fair to say they did. The idea of large voids had been around before the CfA [Center for Astrophysics] and the Arecibo surveys were made. You have to go back a ways, but the Russians, Einasto in particular, have been pushing this sort of thing for a long time, for several years before the U.S. papers appeared.[7] However, Einasto's work was very incomplete. He used very biased samples. Nevertheless, the suggestion was there. The beautiful images that Geller and co-workers produced— those shells and spherical-seeming voids—were convincing and surprising. No one had really expected structures like that.

Did those structures change your view about the homogeneity of the universe, or shake your faith in the big bang model?

No, because the structures that we are talking about are really on scales that are no larger than the superclusters of galaxies, 10 to 30 megaparsecs at the outside. Inhomogeneity on that scale is perfectly consistent with the big bang theory. If one measured a gradient or large void that extended over a thousand megaparsecs, then I think he or she would have to seriously question the big bang theory. But, we're a long way from anyone ever claiming that sort of structure. The structures of Geller and company are simply a measure of the large-scale inhomogeneities in the universe on intermediate scales really. It's just a property of the fluctuation spectrum at the beginning.

Do you still believe that on the large scales, let's say 100 megaparsecs up to a thousand, the universe is well approximated as a homogeneous medium?

The only evidence we have is that on scales of several thousand megaparsecs the universe today is extremely homogeneous and smooth. Those scales are probed by the microwave background radiation.

And you think that the standard big bang model has not really been threatened by these observations on scales of tens of megaparsecs?

What is at issue at the moment is not the standard big bang model but the fluctuations in the standard big bang model. Now if one takes a Zel'dovich spectrum of fluctuations, that is being challenged by some of the observations of large-scale structures, such as the large-scale velocity fields measured by several groups, although I have my reservations about those. That's one issue. And, there is another observation—the first report of anisotropy in the microwave background on large angular scales, of about eight degrees. If either of those observations turns out to be right, then the standard fluctuation models, including the Zel'dovich spectrum, would have to be rejected. But I don't think that will mean rejecting the standard big bang model. It means rather that one has to rethink the inflation theory. But I think the big bang is going to survive.

Have there been any other discoveries, either theoretical or observational, that have changed your thinking in the last 10 years or so?

The most interesting discovery has been the possible distortions in the microwave background.[8] The new experiment shows evidence for distortions near the peak of the black body spectrum. The distortions amount to about 10% in energy of the cosmic black body spectrum, and it's extremely hard to understand what could have produced that large a distortion in the standard big bang model. I would say that observation comes closest to shaking one's faith a bit in the standard big bang model. Let me give you my speculation as to what could be going on. Either you say there ought to be some sort of astrophysical sources, stars basically, giving rise to this energy. If you make that claim, then you need to put an enormous amount of stars in the very early universe, when the universe was dense enough and opaque enough to make the radiation nearly blackbody. It is stretching one's credulity quite a bit to do that, but it's not impossible. The other possibility is that there might be some new particle field, or scalar field, or a type of particle that is not in the standard picture of the big bang but is coupled somehow to radiation. Perhaps it's a particle that decays or a field that decays and gives you a uniform source of energy everywhere in the universe, starting in the very early universe. But as the matter density falls off as the universe expands, this particle source would become more and more important and could result in a heating up of the matter by the present time. That heating up is the sort of thing one needs to explain the distortion, because the microwave photons pass through the hot matter and get shifted slightly in energy, and this can give you the distortion. So, that's one way I can imagine the big bang changing. But again, it's rather a minor variation in the big bang to add some new particle or field.

What would you say are some of the major outstanding problems in cosmology?

That's one of them. Probably the greatest challenge at the moment is to find a theory of galaxy formation that simultaneously fits in with the observations of large-scale structures. Large-scale structures formed by gravitational processes alone, those are rather easy to understand. You can in principle simulate points on a computer under the force of gravity. You think you understand well what's going on. But galaxy formation is something quite different. There you have many other complex things going on, including star formation, and we just have the vaguest glimmerings of a successful theory. Our hope is that if you start off with some basic spectrum of fluctuations in the very early universe, on the large-scale end you make the galaxy clusters and the superclusters, and on the small-scale end the same spectrum produces the galaxies. So you have a unified picture. Our progress in developing this has not been overwhelming so far. One doesn't know that one has a unique set of initial conditions. One has the Zel'dovich spectrum, which can explain the large-scale structure. However, if one takes the new observations of the large-scale velocity field and the possible anisotropies in the background radiation, the Zel'dovich spectrum is in serious trouble. At the same time, we haven't really got very far in going from the Zel'dovich spectrum, or any other spectrum of initial fluctuations, to the observed galaxies. We see spiral galaxies and elliptical galaxies and many other complex things out there. Theory leaves a lot to be desired. One of the key things that one can hope to look for is a galaxy that is actually forming. If we can find forming galaxies, that would at least give us some yardsticks on the theory of galaxy formation.

Any other problems you would add to the list of important, outstanding problems?

I suppose that there is, of course, getting a satisfactory theory of inflation and, tied to that, what happened in the very beginning. The other key question is not just what is the Hubble constant, which we would obviously like to know, but whether the universe is open or closed. What does that mean? Where is the answer? That will tie down the theories a little better.

One of the things I'm interested in is whether or not scientists use visual images in their thinking. Are they important to you at all in your scientific work?

I find metaphors rather inspiring. That is, verbal descriptions can make a huge difference. One can take a fairly mundane theory, and if you dress it up in suitably vivid language, that can make enough of a difference to make one read the theory. One is bombarded with so many papers these days. It used to be that one had time to read the abstract, and now one barely has time to read the titles. Vivid language and the title of the abstract are very important. I personally don't use computers in my work at all so I don't dabble with visual images. I tend to work in terms of mathematical or physical concepts rather than visual concepts.

Changing gears a little, how well do you think theory and observations have worked together in cosmology, say in the last 10 or 20 years?

Well, it's been a bit of a rollercoaster, I would say. Sometimes the theories have been ahead of the observations, and sometimes the observations have been ahead. Theory was ahead of the observations for a while with the big bang. There was an awful lot of theoretical work done in the 1940s and 1950s, and one had to wait for the 1960s before the observations came along. Today, it seems clear that observations are well ahead of the theory. We're grasping for straws in theory at this point, to try to understand some of the observations. The only note of caution is that the observations, which are perturbing to theoreticians, are all still rather tentative. They could solidify or they could disappear in a year or two. It's not inconceivable that all the problems could go away. If one had real faith in one's theory, it may be that one might just persist with the theory, as some of my colleagues seem to do, and ignore the observations. That's one approach, to just keep on going. Some of the cold dark matter advocates do that. They may turn out to be justified in the end.

Let me ask you to take a step backwards. If you could have designed the universe any way that you wanted to, how would you have done it?

I wouldn't have bothered to make it expand. I think that's a complication. It makes it hard to calculate things. I would have made a simple static universe—steady state or static, just like the galaxy is. You can imagine that life proceeds very nicely in the solar system or the galaxy, and why bother with the complication of the big bang. We have a fairly good comprehension of evolution in the solar system. We have a fairly good comprehension of how the galaxies evolved over the past few billion years. The puzzles come in trying to understand the first few minutes of the big bang. One could well do without those rather horrifyingly high densities and high temperatures that make one's ideas so hard to work out. It's really just a throw back to the cosmology that prevailed at the time of Einstein— a natural view of the universe.

> *Let me end with one other philosophical question. There is a place in* The First Three Minutes, *Steve Weinberg's book, where he writes that the more the universe seems comprehensible, the more it also seems pointless. Have you ever thought about whether the universe has a point or not?*

I have thought about that, and I think that's an overly dramatic statement on Weinberg's part. Evolution proceeds regardless of whether our puny brains can comprehend it or not, and there seems to be no question that structure develops in a very organized way, whether we are talking about human scales or galactic scales. So I don't find what I see around me as pointless at all. I'm sure that we're a very long way from comprehending where it all began or where it's all going. I think it's very dangerous to try to compress everything into a phrase like that.

ROBERT DICKE

Robert Dicke was born May 6, 1916, in St. Louis, Missouri. He received his A.B. degree from Princeton University in 1939 and his Ph.D. in physics from the University of Rochester in 1941, in addition to several honorary degrees. He has taught at Princeton University since 1946 and is currently the Albert Einstein University Professor of Science Emeritus at Princeton. A member of the National Academy of Sciences, Dicke received the Academy's Comstock Prize in 1973.

Bob Dicke is one of a small fraction of physicists who are outstanding in both theoretical and experimental work. In the 1940s Dicke was a leader in developing new devices, called microwave radiometers, for detecting radio waves. Theoretical work by Dicke in 1954 laid the foundation for the laser. Dicke's contributions to relativity and cosmology include the first use, in 1961, of the anthropic principle; an extremely precise experimental confirmation in 1964 of the equivalence principle (all bodies experience the same acceleration in a gravitational field); a theoretical prediction of the cosmic microwave background radiation with his former student James Peebles and others in 1965; and the first statement of the flatness problem, in 1969. Dicke, a soft-spoken man of few words, has been the mentor to several generations of students in relativity and cosmology at Princeton.

Could you tell me briefly how you got interested in science as a child?

That was a long time ago. It's a little hard to say.

Did you build things when you were young?

Yes. I had chemistry sets, and I was interested in insects. I had insect collections. I got interested in astronomy from reading.

Do you remember any particular ideas or books that impressed you?

Millikan's book on the electron plus and minus.

You read that as a youngster?

As a young teenager. And I badgered my father into buying *The Source Book in Physics.*[1] I just did a lot of reading at the library I guess.

Were you studying physics at that age?

I had a high school physics course, but it didn't amount to much. My high school physics teacher wasn't really trained for the job.

Were you studying physics on your own, outside of school?

I was doing quite a bit of reading on my own, yes. I was reading calculus, too, in high school. In those days, there was virtually no high school calculus course. If you wanted to learn calculus, you did it on your own.

Were your parents interested in science?

My father was a patent attorney, trained originally as an electrical engineer.

Do you remember in Millikan's book, or in any other books that you read at this age, any particular scientific ideas that appealed to you?

I remember being interested in the cosmological question in high school. I actually did an experiment. It was absolutely crazy.

Can you tell me about that?

Yes. From the standpoint of action-at-a-distance, if a light bulb emits light, the light has to be absorbed somewhere. So I put a flashlight in a Wheatstone bridge and then pointed it up at the sky and pointed it at the floor. The idea was that if space was empty, the light wouldn't have anywhere to go.

And if not, it might be like the closed loop of a conductor?

Yes. If space were empty, then pointing the light up at space would unbalance the bridge. And if space were full, the light would be absorbed.

And there would be no difference in the resistance measured by the Wheatstone bridge.

No difference. So I guess from the experiment, you'd have to conclude that space was full [laughs].

That's a very clever experiment for a high school student. And you took that seriously?

Yes. It didn't have a very sound basis. I thought it was a serious question. I don't know what significance I put in the result.

As a high school student, were you aware of any cosmological theories?

Not really, no. I had no background in relativity, or anything like that. I remember I did some other scientific things in high school. In chemistry class I made a Wilson cloud chamber, to see alpha particle tracks. I was interested in physics generally. But I must say our high school course was pretty poor.

In the experiment you did with a flashlight and the Wheatstone bridge, when you thought about the possibility of the universe being filled with matter or being empty, do you remember trying to visualize a universe that kept going or a universe that had a finite size?

I'm sure I thought of it in terms of a finite size. I certainly had no idea of an expanding space.

Tell me a little bit about your undergraduate education. Did you immediately know that you wanted to go into physics?

I got into physics really quite by accident. I was admitted to the University of Rochester in electrical engineering. I was going to study that because my father had a background in that. And then, when I looked at the catalogue more carefully, the physics department first-year course looked just great compared with the nuts and bolts of the engineering curriculum, so I decided I would major in physics for one year and then transfer to engineering after that. I never transferred. One of the reasons was that I had such excellent teachers. Lee DuBridge put on demonstration lectures every week, and I learned a lot from him. By the end of the first year, I think I was pretty much seduced to taking physics.

Do you remember when you first got introduced to cosmology as a discipline of physics? Was that as an undergraduate?

No. Except for a few places—and Princeton was one of the exceptions, with Bob Robertson there, and Einstein—relativity and cosmology were not regarded as decent parts of physics at all.

Why was that?

I don't know. I asked Victor Weisskopf one time—he was at Rochester when I was a graduate student—shouldn't a graduate student pay some attention to relativity? And he explained to me that it really had nothing to do with physics. Relativity was kind of a mathematical discipline.

Was it in graduate school that you began studying relativity theory and cosmology?

I can't recall as a graduate student getting involved at all with those things. I got interested in cosmology as a tool for getting at the fundamental gravitational questions, which I had become interested in. It looked like cosmology was a handle at getting at these things.

Was your experiment[2] with the weak equivalence principle along the lines of the kinds of fundamental questions that you were asking?

This is the way I got into it. I was on a sabbatical leave at Harvard that year, and I wanted to do something quite different from what I had been doing, so I did some reading and got interested in the question of how Mach's principle was related to relativity. As an experimentalist, it became clear to me that the Eötvös experiment was really a very fundamental experiment, and it hadn't been done with modern techniques at all. I wanted to take a crack at that. I'm not sure I would have ever done it if I had known how long it was going to take and how much trouble it was going to be. From the Eötvös experiment, and my interest in the scalar-tensor theory,[3] it became clear that there were some implications for geology and astrophysics. So I sort of got interested in astrophysics through the back door. First I was interested in gravitation, and then this provided a tool for real observations.

You said that you got interested in cosmology as a handle on your general interest in gravitation. Do you remember when you first began getting interested in cosmology, and what were the problems that most interested you?

When I got interested in relativity theory, at that time, the only thing available were just model calculations—de Sitter theory, Einstein–de Sitter solutions, and so on. Tolman's book was my main source of information. I'm trying to think what the first step was. I think, on cosmology, the necessity for having a hot beginning in the connection with trying to understand the composition of stars was one of the leading motivations. In retrospect, I think our reasons for considering to look for the background radiation were a bit misguided.

I have some questions about that I'd like to return to later. When you first started getting interested in cosmology, do you remember having a preference for any particular cosmological models? The steady state model was around, and the standard open and closed Friedmann cosmologies. Did you have a preference?

No. And as I said, in those days, one didn't know how much physical significance to put in these things. In the mid-1950s, there was a lot of interest in the steady state model and in evolving cosmologies. I can't really recall just what my interests were at that time. I remember one thing, though. I remember our own water vapor measurements at one centimeter wavelength that set an upper limit to radiation coming from space, corresponding to a maximum temperature of some 20 degrees or so.[4] And I also remember reading about some experiments at Bell Labs in which they tried to set noise levels. I don't think they interpreted this in terms of radiation coming in, but they had gotten some residual temperatures of 3–5 degrees.

Was that in the late 1950s?

That was about in the mid 1950s.

Let me ask you a little about your motivations for considering and predicting the microwave background.[5] Do you remember your chain of reasoning?

Yes, I think I can construct that pretty well. First of all, I wasn't impressed with the thought that you could suddenly make all that matter that we see around us in 10^{-20} seconds or so. It seemed to me that a more reasonable thing was that the universe would be oscillating—collapsing and expanding again. At that time, I had heard some lectures from one of the astronomers at Caltech, in which he was talking about the Population I and Population II stars. From these lectures it was quite clear that the old stars in the galaxy were relatively free of metals, and later on they were getting dirtied up. In an oscillating universe, it was clear that you would have to clean up these dirty stars [that is, transform their heavy elements back into hydrogen in each cycle]. The only way to do that would be to get the universe hot enough to decompose the heavy elements. It would have to be pretty darn hot. So this led me to the view that the universe would have to be expanding out of a high-temperature state. I remember making a crude estimate that the temperature of the blackbody radiation today would be about 45 degrees or something like that. If it was much more than that, the energy density would be too high. I remember talking to the boys in the group about this.

This would have been in the early 1960s?

It was probably a year and half to a year before that 1965 paper. And we certainly talked about it at the evening seminars, too.

So, by having an oscillating universe, you wouldn't have to explain the creation of matter.

Well, you wouldn't have to explain the creation of all of it, because in every oscillation you'd create a little more matter and have it exponentially growing. This also gets around the entropy problem. You could add entropy every time [oscillation] but you add more matter to go with it. So the entropy per particle of mass stays more or less constant.

Wouldn't you have to explain the initial amount of matter whenever the oscillations started?

Hopefully, you could explain this as a single quantum fluctuation, with just a few particles created initially in a tiny universe that expands a little and then collapses again.

And those few particles, on subsequent oscillations, would increase in number to what we have now.

Yes. At that time, it wasn't so crazy to think about. We didn't know about the singularity theorems. Lifshitz was calculating models that would bounce—highly disordered models.[6] There was something wrong with them, but at the time a lot of people thought that these universes bounced. Now you can still make them bounce, but it's not so easy.

Could you calculate how many new particles would be produced in each oscillation?

I had no theory.

It seemed from your 1965 paper, which of course had only a piece of your full thinking, that somehow you were trying to postpone consideration of the initial conditions.

Well, we knew nothing about it [laughs]. The main idea at that time was just that we see the dirty stars, we see that the stars start off clean. Then, even a single bounce demanded that it [the early universe] be hot.

At the time, did you have a preference for the oscillating model?

Very definitely. It was the only way you could generate all that matter, from my viewpoint.

Let me ask you a little about the so-called flatness problem, which appeared in print in your article with Peebles in 1979.

It actually appeared in print before then. There's a little book I wrote for the American Philosophical Society, the Jayne Lectures of 1969. There are a few sentences in there.[7]

Was it around 1969 that you first began thinking about the flatness problem, or was it even earlier than that?

I can't remember whether it was any earlier than that. I think probably it was in connection with preparing those lectures.

It seems that in the astronomical community—for some reason or another—the flatness problem first caught on after your article with Peebles in 1979.

Actually, there was a period of time when I was quite interested in these paradoxes, and I was going around giving a set of colloquium talks on them. I gave one at Cornell, and I hadn't known this, but Alan Guth actually put in print that he was there. He was stimulated by this, which makes me very happy.

When you were going around giving colloquia and discussing this puzzle, do you remember what kind of reaction you got? Did people take it seriously?

I can't say really. At least one person did [laughs]. I can't even remember all the puzzles. I think I had some four puzzles. One was flatness, one was the causality one [the horizon problem], one was the Dirac coincidences.

Some astronomers take the view that we have only one universe, so any arguments that are based on the possibility of there having been an ensemble of universes with all different values of omega today and all different kinds of initial conditions don't make any sense. Other people take the flatness problem as something that demands a physical explanation.

Well, not necessarily physical. An anthropic explanation is possible, too, I think. With an ensemble of universes, this is the only kind we could live in.

When you thought about this puzzle, did the anthropic explanation occur to you?

No. At the time I wrote these notes [the Jayne Lectures of 1969], I doubt it. The anthropic explanation had been used earlier, in connection with the Dirac argument about the gravitational constant. But I don't think I was thinking of it in this relation. I had the feeling that this [the close balance of gravitational and kinetic energy] implied the universe was very nearly flat for a good physical reason.

And they were just reasons that we didn't . . .

That we didn't understand.

We couldn't just blame it on initial conditions?

No. That really requires a terribly delicate balance to get that set up right.

In stating the implausibility of that balance, did you conceive of lots of different possible ways that the universe could have been set up?

I guess I was thinking—again this might be an argument for a bouncing universe—that the universe bounces, and it is still closed, so that it collapses again. But it has to bounce to a quite reasonable size, and it can only do this in a very nearly flat universe.

Do you have any idea why people weren't asking these cosmological questions, like the flatness problem, earlier? You mentioned that people weren't taking cosmology as a whole as a serious subject. Do you think that might have been a reason?

It's a puzzle to me how cosmology got so separated off from the rest of physics. Here is all that matter in the universe, and it doesn't seem to bother anybody. It's here, but where did it come from? Questions of this kind just weren't asked. If you go back even earlier, these purely mathematical models—isotropic and homogeneous spaces—were purely mathematical exercises. Not completely so. There were certainly people like Robertson who were interested in the observational questions and observational support. One of the reasons that Bob Robertson left Princeton to go to Caltech was to be closer to the observations there.

Let me ask you a little about your reactions to some of the discoveries in the last 10 or 15 years. Do you remember your initial reaction to the inflationary universe model?

Not very well, I'm afraid. Also, I must confess that in the last 10 years or so I haven't really been keeping up too well with these things, so my present-day impressions aren't all that good. But it certainly is a very clever way of getting around some of the paradoxes that have been bothering me. Exponential growth is a great idea. The thing that I found a little hard to swallow was the delicate balance that's required to have the phase change held off long enough so that you can expand enough first.

Have you been interested in the work on the large-scale structure of the universe? I'm thinking of the de Lapparent–Geller–Huchra bubble-like structures.

Yes, I'm following that.

How do you react to that kind of picture? Does that alter your personal views of the large-scale structure?

Well, it's surprising to see a structure organized on such a large scale, as well organized as it is, with bubbles and sheets and so on. It does suggest that structure in a highly condensed, early phase played an important role.

Did it surprise you?

Yes, I would say it did. Although ever since Jim Peebles put together that galaxy map based on Shane's catalogue, I was impressed by the appearance of those filaments in there. They seemed real. I kept arguing with Jim that they were real, and he kept saying that they were a figment of the imagination.

So you always suspected . . .

So I always suspected that there was some structure there.

At this time, did you have any particular preference for a homogeneous universe versus an inhomogeneous one?

No. That the universe is organized as much as it is, and is as uniform and isotropic as it is, might also be an effect of the anthropic principle. If you get the thing too badly organized, I doubt the universe would be very hospitable—if, for example, some parts were collapsing before other parts could get started expanding.

Wouldn't it be all right if our local part of the universe was evolving at about the right rate? Would it make a difference to us what was happening in the other parts?

Well, if other big pieces that were close to us had collapsed completely, I think there could be some rather disastrous radiation conditions for us. I think you have to have a reasonably well organized universe.

Is this question of the organization something that still bothers you, or do you think that the inflationary universe model has answered it?

It still bothers me. I still am amazed at there being so much matter on such a grand scale and have this all come out of one explosion. I think there may still be a bouncing universe [laughs].

How does a bouncing universe solve this problem of the large-scale organization of the universe?

I'm not sure it does [laughs]. Jim Peebles has argued, I think quite correctly, that in the collapse phase, inhomogeneities grow so rapidly and so strongly that you can't come out with an organized state.

Do you think that we might be missing a big piece of physics in understanding these things?

Some new physics? I doubt it.

Do you think we just have to understand it in terms of what we know?

More observations would probably help.

I guess the inflationary universe scenario, if correct . . .

It certainly is impressive the way it takes care of the flatness and causality problems with one mechanism. The string business I don't understand enough to comment at all.

You mentioned the anthropic idea a couple of times. What do you think about some of the recent work that's been done on the anthropic principle? Your statement of the principle in your 1961 paper in Nature *was rather mild.*[8]

Yes, I would call it a very conservative statement. The really exciting one is

Brandon Carter's, which would have all of the physical constants adjusting themselves.[9]

Do you think that this could possibly give us some new insights? Do you think these are useful arguments?

First of all, I think that in the form in which I stated the anthropic principle, there isn't a lot of controversy, because it's a rather straight-forward question. The other form, I don't know. It would require quite a revolution in the way of doing physics. You have to understand these physical constants as functions of something if you're going to do it in a field theory way. Functions of what, then? Well, a universal scalar field is one possibility. If you have that, you could conceive of constants that would vary with the structure of the universe, and pick out of all these possible structures the one that you feel most comfortable to live in.

One of the things that I'm interested in is how scientists use metaphors and visual images in their work. Do you use visual images much in your own work?

I think in my research I use analogues a lot—not necessarily visual images but analogues. Something that you learned in high school electrical engineering might be useful in cosmology.

Have you found either analogues or images useful in cosmology in particular? I'm thinking of the expanding balloon analogy for the expanding universe.

I must say that I think always in terms of comoving coordinates. It's a great big, big girded structure with flexible girders that stretch.

Do you ever try to imagine the very early universe, other than with equations? Do you ever try to picture it?

No. I don't know enough.

Do you think that theory and observations in cosmology have gone in separate directions or have helped each other in the last 20 years?

I'd say that without the observations, the theory would have gone completely haywire [laughs]. Completely. That interplay has been very important. It's also been important to the observations. When you're doing experiments, you don't do any old experiment. You always have some kind of a theoretical model in the back of your mind. It might be crazy, but you have some kind of a theoretical structure that you want to test.

How comfortable do you feel with taking the big bang model and extrapolating it backwards to the first few minutes or to the first second?

A little uncomfortable, I must confess. It's a tremendous extrapolation. But not as uncomfortable as I feel about extrapolating the model beyond that. There's still one point in cosmology that I find very disagreeable, and that's the idea of time and space having no meaning up to a certain point and then suddenly appearing. A universe that is suddenly switched on I find highly disagreeable. I guess what bothers me is a sudden barrier, a discontinuity, whether it's in time or space—because I'm used to continuity. To have space exist on one side of a sheet and not exist on the other I would find most disagreeable.

A worrisome thing about the inflationary universe model is that we do have to take seriously some kind of physics back at an extremely early time.

Well, that's an extrapolation that is hardly warranted by the observations. I think it's a sheer fantasy.

And yet you're still attracted to the theory because of its explanatory power?

Yes. That may not be a sufficient reason.

Let me ask a couple of final questions even more speculative than the ones I've asked already. If you could design the universe any way that you wanted to, how would you do it?

Well, I think what I'd like to do—bearing in mind Carter's ideas about variable constants—is build in a scalar field that things could be functions of, so that you could have evolving constants, depending on the structure of the universe. I'd like to have the universe bounce and gradually grow in size, with each bounce contributing more particles and more entropy. I'd like to have the whole thing go in accordance with what the elementary particle people would believe, if they ever had a way of testing it.

Why does the oscillating universe model appeal to you?

You don't have to start.

Except from some initial quantum fluctuation?

Well, you have a space with some gravity waves running around in it. So you have a purely empty space to start with, nothing but gravity. And then, in a quantum fluctuation, a few particles appear and start a little thing [universe] oscillating. And not just one, but many, probably—which leads to many universes, each one doing its own thing.

So there would be some larger space, which would have different universes inside of it?

Yes. If they were closed, they'd be closed universes. That is to say, they'd be cut off from each other, but very nearly flat.

Why do you like the idea of having lots of independent universes?

You don't have to explain why only one. It also takes care of the anthropic principle.

There's some place in Steve Weinberg's book where he has this curious statement that the more we learn about the universe, the less it seems that the universe has a purpose.

I don't understand that, but it's certainly a philosophical statement.

When you say that you don't understand it, do you mean that it doesn't make sense to you, or that you don't know why Weinberg said it?

It doesn't make a lot of sense to me, because I'm not sure that it's the purpose of the universe to have a purpose [laughs].

So you have never personally worried much about this question of whether the universe has a purpose?

No. The closest thing to a purpose I can see is through the anthropic principle.

But the way that you have stated the anthropic principle, it doesn't require a purpose, does it?

No. It doesn't require any great revelation or any great revolution.

JAMES PEEBLES

James Peebles was born April 25, 1935, in Winnipeg, Manitoba, and is a Canadian citizen. He received a B.S. degree from the University of Manitoba in 1958 and a Ph.D. in physics in 1962 from Princeton University, where he was the student of Robert Dicke. Remaining at Princeton for his professional career, Peebles has been the Albert Einstein Professor of Science there since 1984. He is a Fellow of the American Physical Society, the American Academy of Arts and Science, and the Royal Society of England and is a foreign member of the National Academy of Sciences. He won the Eddington Medal of the Royal Astronomical Society in 1981 and the Heineman Prize of the American Astronomical Society in 1982.

Peebles, in collaboration with Robert Dicke and others, predicted the existence of the cosmic background radiation in 1965. In 1966 Peebles did one of the first detailed calculations of the cosmic helium abundance expected from nuclear reactions in the early universe. The agreement between those theoretical calculations and the observed helium abundance constitutes one of the major supports of the big bang model. In 1965 Peebles pioneered calculations of gravitational clustering of matter in an expanding universe and has become a leading theoretician of the "gravitational hierarchy" model of structure in the universe. Peebles has played a major role in bringing cosmology to the attention of physicists. His book *Physical Cosmology* (1971) injected physics into the detailed calculations of physical processes in the expanding universe as no other book before it. Tall and lanky, Peebles has not slackened his pace since the 1960s and seems to be completely absorbed by cosmology.

I wanted to start with your childhood and find out a little about how you got interested in science.

I've always been interested in mechanical things. I must have been heavily influenced by my father, who is also very good with his hands. He liked to build things, and I always loved to watch him do it. And I loved to build things on my own. I was never exposed as a kid to any real science. I read the occasional popular science book, and I loved *Mechanics Illustrated.* It wasn't until I got to college that I began to appreciate what physics was all about, and that was really an accident also. I started in engineering, where I think I could have happily remained and, who knows, made a bundle as a civil engineer or mechanical engineer. But more of my friends happened to be majoring in physics than engineering, so I switched over. No more compelling reason than that. But I then decided, once I was in physics, that it was great stuff and stuck there ever since.

When you were younger, before college, do you remember any particular books that you read that had an impression on you?

No, not a one. As I say, I did read the occasional book on popular science, but they were never of any great depth. So I can't say that I had any idea what science was all about before I got to college. In fact, I didn't know what engineering was all about before I went to college. I just wandered in. I came from a very small high school in which there was no guidance and not any appreciable amount of physics taught, nor much mathematics.

You probably weren't thinking about cosmology at all at that age then?

I had no idea what cosmology was.

Can you tell me about your undergraduate education?

At that time, the University of Manitoba was quite strong in classical physics but pretty weak in modern theory. So I came away from the University of Manitoba with a pretty good grounding in what I would need in astrophysics, in the sense that it was broad and strong on the classical parts but weak on modern physics. I didn't know any relativistic quantum mechanics, although I did know a fair amount of nonrelativistic quantum mechanics.

It was, as you can imagine, quite a shock to come from the University of Manitoba, as it was in the 1950s, to Princeton University—to come from being top dog in my class and getting all the honors to being totally bewildered and surrounded by all these people who knew so much more than I did. That shock lasted for maybe a year, until I managed to catch up and to find out what was going on. As is inevitable, my first interest was particle theory. It was the glamorous subject then, as it is now, and I started working in that direction. I was very inspired by some of the people here at the time. Murph Goldberger gave a brilliant series of lectures that I was totally overwhelmed by and totally grabbed by. So for some time I was dreaming of being a particle physicist. But then I ran into Bob Dicke, who had a weekly evening meeting on his research on gravity physics. I dropped in on it mainly because I knew some fellow graduate students in the group, but I soon came to attend because of the subject, which is fascinating. In gravity physics, one had to look at many different subjects, from the structure of the planets to the structure of the galaxy to cosmology. That's where I first began to see a little of what cosmology is and to become fascinated by it and gravity physics in general. So I deflected my interest and wrote a thesis with Bob Dicke on the possibility of variability in the fine structure constant. I placed empirical limits on how much alpha [the fine structure constant] could have varied through radioactive decay and the ages of meteorites. Then it was when I was a postdoc that he and I moved in the direction of cosmology as a way to do gravity physics. Cosmology seemed to me, at the time, to be rather a limited subject—a subject, as it used to be advertised, with two or three numbers—Hubble's constant, the deceleration parameter, and the density parameter. A science with two or three numbers always seemed to me to be pretty dismal. If that's all there were to cosmology, I wouldn't have found it very interesting. Of course, there was the other great topic, the debate between the steady state theory and conventional big bang cosmology.

Can you tell me a little about your thoughts on that debate.

Well, mainly I was scandalized that there should be such intense debate over what seemed to me to be such a vacuous question. At the time, I found that cosmology was not appealing because it based so much on an assumption that the universe is homogeneous, which seemed to me to be highly questionable. Of course, I hadn't looked into the observational basis for why one believed in homogeneity, which even then was not by any means insignificant. It seemed to me just unrealistic, on the face of it, to think that you could make a theory of the universe. I guess I remember first thinking that the assumption of homogeneity was unrealistic when I was preparing for general exams as a graduate student here. Among other things, you could anticipate that there might be a question in the relativity part on cosmology, and graduate students knew this and prepared themselves by learning about the standard solutions to Einstein's field equations for a homogeneous, expanding universe. So I dutifully learned them, but I remember thinking to myself, "Boy, this is silly. Who could imagine the universe would be so simple?"

That's what you mean when you say that you can't imagine anybody making a theory of the universe?

Right. The universe surely is a hierarchy of complexity. One could imagine making a theory of galaxies, a theory of clusters of galaxies, and so on, but to have a theory of the universe as a whole sounds to me more like a meta-theory rather than physics as I know it.

And yet that's what Einstein himself did in 1917.[1]

Yes, yes. In fact, I've often wondered how he could ever bring himself to make such a simple assumption, and I've wondered also why people like de Sitter let him get away with it. De Sitter was pretty cautious, in fact . . .

De Sitter argued a lot with him.[2]

Right, on that point. Other sensible astronomers and many physicists quickly accepted his idea as the way it has to be. The universe has to be so simple that we can analyze it in a one-dimensional differential equation—everything a function of time alone. Of course, Einstein had brilliant intuition, and he surely was awfully close to the truth—that's the way the universe looks.

So have you changed your view about whether it's silly to make theories of the universe?

I think it's *amazing* still, but I have to agree that it does make sense to make theories of the universe. They work so very well. First, I think the observations pretty clearly show us that in the large-scale average—on scales comparable to the Hubble distance now—the universe is remarkably isotropic. And more indirectly, but I think still pretty convincingly, the universe is remarkably close to homogeneous. So if one should believe general relativity theory, you have it. The universe is simple. It's expanding in a computable way—which is an amazing thing. I never would have anticipated it could have worked out so easily, but there it is.

Do you think the observations between then and now have made you stop thinking that it's a silly idea to make these theories?

Yes. Remember at the time I wasn't aware of the observations. I could have known then that the faint radio sources are quite isotropically distributed—that was known. I guess that was *the* main observational datum that showed homogeneity in the large scale. There were the galaxy counts, but they were always very questionable—still are—because it's so hard to make the k-correction and to correct for local mass concentrations, the local galaxy concentrations. Given that, it's hard to argue that we have much evidence of homogeneity there. Isotropy, of course, is wildly improved in the last 20 to 30 years. But, as I say, even in the late 1950s I could have looked at the radio source counts and I could have been impressed at how isotropic they are, but I didn't know about them.

You mentioned that Dicke's evening meetings got you very interested in relativity theory. Do you remember when it was that you got interested in cosmology in particular?

Yes, I guess I can pretty definitely date that. It was the famous idea of Bob Dicke's to look for the microwave radiation left over from a hot big bang. I can't give you a year. It would be around 1964. It was in the summer, I do remember, a very hot day. We met in his usual evening group but with a small number of people. For some reason, we met in the attic in Palmer lab. It was really ridiculously hot, I remember. He explained to us first why one might want to think that the universe was hot in its early phases. His thought at that time—and one that he does keep returning to—is that the universe might oscillate. And an oscillating universe does require something to destroy heavy elements so one could start again with hydrogen. The way you destroy heavy elements is to thermally decompose them in blackbody radiation. So he explained to us then why you would like a universe that's filled with blackbody radiation. He explained to us how this blackbody radiation would remember its thermal spectrum as the universe expanded and cooled off. He explained to us that a good window for looking at this was around 3 centimeters wavelength, and in pretty short order he had Dave Wilkinson and Peter Roll looking into the design of a radiometer to look for this radiation. I remember his off-handed but very inspiring remark to me, "Why don't you go think about the theoretical consequences?"—which is a sign, I guess, of a great physicist. You can throw off these remarks that can move mountains. And it certainly gave me an awful lot to think about. He and I wrote a paper—I think before the discovery of the microwave background—in which we expressed many of the theoretical ideas that had come up as a consequence of thinking about the microwave background.[3] I guess that was my first paper in cosmology, and it was really a lot of fun to write it. We could see so many things to be done.

At that time, do you remember having a preference for any particular cosmological model, say, open versus closed or homogeneous versus inhomogeneous?

My reaction there was consistent with my initial reaction to cosmology as a subject. Again, I thought, suppose someone told me whether the universe was open or closed. I think it would be a great let-down, because what would I do with that knowledge? I've always assumed that we won't know whether the universe is closed or open until the information becomes, so to speak, irrelevant because it will have been enmeshed in a much larger picture in which it's obvious that the universe has to be the way it is on other, deeper grounds. So I could never bring myself to get very excited about the question of open versus closed. It always seemed to me to be really a side issue. You'll find that in the book I wrote a few years afterwards, *Physical Cosmology*, I was very cautious about this question and didn't address it in any detail because it didn't seem to me to be very interesting. In fact, I think that's still the case. Now, with inflation, we have a reason, of course, to look for open versus closed, and within inflation suddenly this question becomes very dramatic and very compelling. So now, we can see that if we knew that the universe was *not* cosmologically flat, it would be of overwhelming interest because it would knock down a beautiful idea of inflation.

Did you give much credibility to the steady state universe at this time, in the mid-1960s?

No. I must confess I hadn't looked at it closely. I had heard a talk on the subject. I remember being amazed that grown people could get excited about such speculative ideas. By the time the blackbody radiation came along, it became clear that if this radiation was found, then, of course, it had to be an expanding universe. Shortly thereafter, the radiation did seem to have been found, so the steady state theory very quickly drifted away before I had a chance really to think about it long enough to decide whether or not it would have been a theory I would have fought for.

Do you remember what your motivation was in your work on the microwave background?[4]

Well, as I say, it was very direct motivation from Bob Dicke—"You go think about this." In response to whether I found that an exciting challenge, yes, I certainly did. I could think of lots of physics to be done here. If you have radiation, you have radiation drag; you have thermal history, so you can think about reaction rates; you can think of turning a plasma into atomic hydrogen, into molecular hydrogen; you can think of turning a gas of protons and neutrons into helium and so on. So that was fun because it was cross-sections; it was thermodynamics; it was something you could compute and relate to observations. So I got very excited about it very quickly.

Did you buy Dicke's theoretical motivation, that is, the oscillating universe idea?

No. I was willing to go along with it, but I couldn't get excited about the idea. I think it seemed to me then, as it does now, to be a possibility that we'd better pay attention to because it is an alternative to the now standard lore of inflation. And since inflation isn't all that firmly established, we'd better consider the alternatives, and that certainly is one of them that I would want to think about. But, in fact, I have never gone beyond that view—that it is an alternative and we better bear it in mind. I don't think I've ever written a paper exploring the possibilities of an oscillating universe. I have no idea how you invent physics that will make a universe oscillate and maintain its homogeneity—that's a real trick. It's an attractive idea, I must say. It's not as elegant as the inflationary picture, but I think it could be borne in mind.

I think I recognized pretty quickly that one could not have a universe that oscillates indefinitely into the past because one would have an entropy catastrophe. You would make entropy on each phase of the oscillation, but since the universe now has a finite amount of entropy, then one knows that it could have had only a finite number of previous oscillations. I think I remember computing fairly early—but perhaps after the paper you've mentioned—that the universe could have oscillated 100 times up to the present, starting from dead cold and with the present content of baryons. That's the number of oscillations that would be required to turn the radiation from stars into the energy density of the radiation from the microwave background. Of course, one recognizes that if you have only 100 oscillations, that makes the universe exceedingly old indeed but nonetheless a finite age.

I think it is true to say that Dicke was motivated by this thought that a universe that's oscillating at least pushes back the crisis of initial conditions from the standard big bang case—which is perhaps a step in the right direction.

Can you tell me when the flatness problem, which you posed with Dicke in 1979 in the Einstein Centenary volume, first occurred to you?

Again, to be honest, it didn't occur to me. It occurred to Bob Dicke. You know it's hard to remember when you learned to tie your shoes, and it's hard for me to remember when I didn't know and believe that as a good argument. I think Bob Dicke must have been presenting it to us back in the days when we were in Palmer Lab and met in the evenings in these gravity research group meetings. It always seemed to me to be a very obvious argument, and I was amazed when it suddenly became canonical conventional wisdom because the argument hadn't changed any from one that Bob had been giving for years. It was an argument of reasonableness and of coincidence, but it was never a geometrical demonstration and, of course, still isn't.

It seems to me that for the argument to be compelling, you have to imagine that the universe could have been made in many different kinds of ways, with lots of different initial conditions.

Yes, and of course that recalls to us another of Bob Dicke's ideas—the anthropomorphic universe idea [the weak anthropic principle].[5]

Yes, but he explained that in a much more natural way. Whereas the flatness problem is something that poses a puzzle if you can believe that our initial conditions had to have been rare, out of a very large number of possible initial conditions.

Let me see if I understand what you're getting at. There are two parts to this. One part is the coincidence argument that if the universe has an appreciable cosmological content or if it has space curvature that's appreciably contributing to the expansion rate, then there was a special epoch at which this contribution to the expansion rate became comparable to the contribution by the mass density. That special epoch then exists, and there are two problems with that. First, initial conditions in the very early universe, whenever classical physics became applicable, had to have set an enormous time scale, which seems a little difficult to arrange. Second, we have to have come on the scene just as the epoch reached that preferred time. Then you have to ask, could the second be only a coincidence or could it be such an unlikely event that this characteristic time, if it exists in fact, is very far off in the future. In that case, the space curvature cannot have been important, and lambda cannot have been important, and we arrived on the scene at some randomly chosen time so far as the evolution of the universe is considered. So, if space curvature is negligibly small and lambda is negligibly small, then we have a problem indeed. How did the universe arrange these initial conditions?

That's the version of the problem that I've been discussing.

Of course, we have a problem no matter what the space curvature is, since the universe is so dramatically homogeneous, and we know that it began as separate pieces.

Yes, that's another problem.

Right. So, all along there was a deep problem—how did the universe get to be so homogeneous? And that certainly is something that worried me a lot through the years. I've written papers on that, and I must say that the inflationary idea certainly is a brilliant way to solve the problem. A separate problem is what is the space curvature value now and could it be appreciable in its contribution to the expansion rate compared to the mass density's contribution? Another aspect of the Dicke argument is that you would be surprised to find that space curvature is appreciable now, since that would require that we came on the scene just as space curvature was becoming important, which is a coincidence argument. It seems to me that the second argument that space curvature has to be negligibly small is a weaker one because we had to come on the scene at some time or another and that meant that there had to be some sort of configuration of events and any specific configuration of events always is wildly improbable. How do you judge which is more improbable—that we should have come on the scene now or at some other time?

That anthropic element was not a part of the way that you stated the flatness problem in the Einstein Centenary volume.

Now, by flatness puzzle, you are not referring to the homogeneity puzzle. That's a separate puzzle.

*No, not the homogeneity puzzle. I mean that for omega to be so close to 1
now, it had to be incredibly close to unity at either a Planck time or 1
second. That argument by itself, without the anthropic principle,
apparently has a lot of power to some people.*

Yes. We know it has a lot of power because very quickly the argument that
space curvature must be negligibly small now [or omega very close to
unity] was taken up by the community, and I think in many circles it is now
considered to be gospel—a geometrical demonstration of the way things
must be—which I think is overstating the case. It really does boil down to
a coincidence argument, that we came on the scene as omega started to
drop away from unity. Actually, one can approach this from another
direction and ask what the observations say. As you know, there is still the
terrible problem that it's awfully hard to find this extra mass that's required to make omega equal to unity—which is something I tend to pay
more attention to than these philosophical arguments.

*For some reason, although these puzzles had been around for a long time,
your article in 1979 seemed to ignite the community. Do you have any
opinion as to why those arguments suddenly took hold at that particular
time?*

I've often wondered why the arguments took hold so rapidly from that one
article, since these arguments all were known—at least to many professional cosmologists—well before the article. I have the feeling that the
ground had become fertile for these arguments, that particle physics was
coming with the ingredients needed for the inflationary concept. I suppose you could argue that the particle theorists needed some philosophical motivation for their ideas on inflation and here they were, ready-made,
at just the right time. I could imagine it's as simple as that.

*Do you remember any of the initial reactions of people to the flatness
problem after your article in 1979 came out? I know that some people
dismissed the argument on the grounds that the universe is as we observe
it, and it makes no sense to talk about a collection of possible universes.
So it definitely polarized the community.*

I don't know that it polarized the community at that time. Perhaps it did,
because the arguments were made visible. Certainly responses of the sort
you mentioned were familiar to me. I'd heard them earlier many times
from sensible people. So I can't say that I noticed any difference in

response to these arguments before and after the publication of the paper. The only think I can recall noticing is a gradual drift toward acceptance of the arguments. That was surprising to me, as it became such a strong current of opinion—these arguments had to be the way it is. It would be fun to ask Bob Dicke how seriously he took these arguments when he presented them in the 1979 paper. I thought they were good arguments, but I never felt that we were laying down the law about the way the world has to be.

Well, you weren't giving a physical explanation. You were presenting an argument that there needed to be some physical explanation.

Right. And of course at the time I was pretty convinced by the argument, say, that omega really ought to be unity if the universe is at all rationally constructed. It still seems to me to be a good argument for omega to be equal to 1, but I don't think I thought then, and I don't now, that it was a demonstration that omega has to be unity. What I found a little surprising was the fact that many people took these arguments to be demonstrations for the way the universe had to be—that omega had to be unity.

Do you remember you initial reaction to the inflationary universe model when you heard it?

Yes, I remember great skepticism because of a simple point. In the original inflationary picture, one goes through a phase in which the universe is almost Einstein–de Sitter. One knows that an Einstein–de Sitter universe is invariant not only under translation in space and time but also under change of velocity. It seemed to me to be an amazing thing that the universe could have made the phase transition back to a Friedmann–Lemaître phase in such a way that all the parts were moving in the same direction. How did part A and part B *know* to be moving relative to each other such as to have the general expansion we observe if locally they have no signal from the geometry as to how they should move? At the time, I was stuck on that point. I gradually came to see that it is not a problem, and at the same time, I could see how beautifully the inflation model solved what seemed to me to be the essential puzzle—how did the universe get to be homogeneous. On the basis of those two realizations, I was glad to adopt the inflationary picture as a good possibility. I was not—and still am not—convinced that it *has* to be the way the universe started, but I certainly had to agree that it was a wonderfully elegant idea and so certainly should be pushed harder.

I'm personally amazed at how rapidly and how widely it caught on.

Yes, but that's the way physics operates, isn't it? In the absence of any other idea, a good idea will capture the field. And before inflation, we really had a blank. One could extrapolate the conventional models back in time to a singularity, which is a very ugly thing. So we all knew that we had this horrible problem, and I think it's no surprise that the first idea that came along that seemed to make sense would capture the field and become the canonical, standard model. Of course, it doesn't mean that the idea is right. It means that we didn't have any options. So I'm not surprised. I think we should be careful. I think there's a reasonable chance we've been led down the wrong path. It certainly has happened before. I've heard comparisons of the steady state idea with the inflationary idea, and indeed they have some similarities. In both cases, we had problems—in fact, the same problem. What do you do about initial conditions? In the absence of any very hard evidence—in fact, in the absence of any great movement in the canonical model—a new idea came along and because it was new and provocative *and* accounted for some of the known problems, it captured a lot of attention. Certainly the steady state theory did in the 1950s.[6] It soon became at least a co-equal in credibility with the expanding universe model, and that wasn't because of any observational evidence as far as I know. There were a few bits and pieces of evidence that the steady state theory might do better. One had a time scale problem with the big bang model. But I think sensible people didn't think the time scale problem was really serious. I think it was more a question that it was a new idea, not manifestly wrong, and worth exploring. And I think we saw much the same thing happen with inflation. A new idea came along, solved some known problems, so people jumped at it and started exploring it.

You mentioned that both models had in common some kind of a treatment of the initial conditions. Do you think physicists like to avoid specifying initial conditions?

I remember a remark of Eugene P. Wigner many years ago in which he said, "We used to think that the interest was in the physics, the laws of physics, and that initial conditions could be left to engineers." I think that is an attitude that pervades much of physics. I don't find the same attitude nearly as strongly in astronomy, interestingly. I don't find there the nervousness with initial conditions that I do notice among many physicists.

Why do you think that is?

Well, as Wigner remarked, one thinks of the interesting part of physics as being the laws of physics, and the initial conditions as being the province of engineers who set things up.

Then why have we gone to so much effort in both the steady state and the inflationary models to get around having to specify the initial conditions?

I guess we didn't trust the engineers [laughs]. I would draw another analogy to steady state, and that is that the invention was not motivated, in the first instance, by observation. For example, when Lemaître was re-introducing the expanding universe idea, I think he was strongly driven by the observations—by the redshift, the receding nebulae.[7] So I would say that here is an example where someone was led to a revolutionary picture by fairly direct observations that prodded him in that direction. I don't think there are any observations that prodded people into inventing the steady state theory. That was just a beautiful idea. I think it's true to say that there were no very close observations that led to the inflationary concept. That was theory, a heavy dose of it. Well-motivated theory, to be sure, and I think it could very well be right. In fact, we have another example of that in Einstein's introduction of the homogeneity assumption in cosmology. He certainly wasn't motivated by any observations—quite the opposite. He deliberately ignored them. So that path can work. Of course, it's not guaranteed to.

Getting to some recent discoveries, do you remember your initial reactions to the work of de Lapparent and Huchra and Geller on large-scale structure?

Yes, that was a case of the curtain opening. We had before then hints that galaxies tended to be in sheet-like distributions, but I was very nervous about believing those hints because I was so aware of the tendency of the eye to pick patterns out of noise. In fact, I wrote some pretty vitriolic papers with examples in the past of how astronomers had been misled by just this tendency. When I saw the Geller et al. map, I was just flabbergasted. The window had been cleared and one could see the 3-dimensional distribution, and there it was—linear. So I accepted it, I think without regrets. It certainly helped that at just about the same time we saw the redshift maps of Martha Haynes and Riccardo Giovanelli, which looked awfully similar. So there it is—reproducibility. What more could you want?

Did this change your thinking about the large-scale structure?

It certainly did. Previously, I had in mind the notion of a rather chaotic process that would lead to large-scale structure without much pattern formation. With these linear structures, it was clear you need a pattern-forming mechanism, which is not an element I'd ever given much attention to before. So, yes, I was quite strongly deflected in the way I thought of theories of the origin of large-scale structure.

You were already accepting the possibility of large-scale structure to begin with.

Oh sure. Large-scale structure, but not with a *pattern* in it. With the Geller et al. map, it became clear you need a way to form not only large-scale structure but also linear patterns in that large-scale structure, which seems quite a trick of nature to me. I'm still not confident we know how these patterns were produced or the origin of large-scale structure.

Can you think of any other example that has really altered your thinking in the last 10 years? Have I missed something that was a big watershed for you personally?

No, I think not. Of course, the dramatic watershed was the discovery of the microwave background and then very slowly the evidence accumulating that it does have the spectrum you would want for blackbody radiation. I couldn't give you a date at which I had decided that the spectrum had to be blackbody and therefore that this radiation had to have come from the early universe. Certainly that didn't happen very quickly because, although the long wavelength part of the spectrum was shown to be pretty close to Rayleigh–Jeans within five years, it was considerably longer before we knew that the spectrum turned over, broke away from the Rayleigh–Jeans power law at short wavelengths as it ought to. I guess that it became clear around about 1975 that the stuff was really pretty close to thermal blackbody. That would fall just outside your 10-year interval. In the last 10 years, the two big steps forward have surely been the inflationary concept and linearity. And I don't know how much the former has altered my thinking. I've been very excited with this concept. I'm willing to pay attention to its predictions, but I don't feel bound to those predictions. And I certainly am a little skeptical that those predictions are even *right*. As for the linearity, the redshift maps in general have been very important as revealing in more detail *what* the large-scale structure is, and the dramatic thing has been the revelation that linear structures are so common.

I'm interested in how scientists use metaphors and visual images in their work. One successful image in cosmology has been the expanding balloon model. Do you remember when you were first introduced to that image or metaphor of the expansion of the universe?

No, I don't. I suppose it's like many things—I must have heard it as a graduate student and slowly become aware that it is a very useful picture, not only for me but also as a means of communicating to others what one has in mind. I suppose it might have first become apparent to me that this is a very powerful metaphor with the discovery of the microwave background because a lot of people were puzzled. If there was a big bang, and it produced radiation, why don't we see this radiation streaming away from the site of the big bang? I remember Arno Penzias was very puzzled about this, and I remember telling him about the balloon picture. He said, "Oh, all right. Sure, I understand now." So I was very familiar with it then and certainly I found it awfully helpful.

Do you use visual images much in your own work?

Yes I do. I tend to think visually, I believe, rather than in equations. I don't know—how else do you think, besides in images? Perhaps some people think in words. I can't imagine thinking in words. For example, when I got into the game of studying large-scale structure, I pictured the distribution of galaxies on large scales as much like the distribution of surface height in water in a choppy pond. I was very interested to know what would be the characteristics of these waves on large scales. In the case of the choppy pond, one has a physical interpretation. This is a result of a competition between wind and dissipation. I wondered if there might not be some similar case for the large-scale fluctuations on the distribution of galaxies. It was strictly a metaphor that drove me to think about galaxy space distributions.

Have you ever tried to picture the very, very early universe?

Yes. All I get is a very hot, hot region [laughs]. Of course, we think of these density fluctuations that must have been present at fantastically low levels, so. I try to think about that, but it doesn't really grab me too much. I suppose we don't *need* to have much of a picture of the early universe if we can believe the conventional ideas. All of the length scales we're interested in were enormous compared to the horizon in the early universe. So apparently we have a very simple situation. Take what you see now, extrapolate back with linear perturbation theory, and you've got it—until you get into quantum fluctuations. I don't know how I think about quantum fluctuations.

How do you think that theory and observations have worked together in the last 15 years in cosmology?

And to what extent have they gone their separate ways? We're at a very interesting time. I think theory has not been very strongly constrained or guided by observation. That's because, at least in large part, the observational situation has been pretty shaky, pretty confusing. But that situation is dramatically changing for two reasons. People are getting very detailed models of galaxy formation now. For example, the canonical cold dark matter, scale-invariant model makes a large number of predictions that I think are starting to become firmed up and testable. At the same time, people are starting to observe galaxies at high redshifts. It's impressive that galaxies at a redshift of 1 are now observed and their spectra are measured. That's a galaxy at an appreciably younger age than you see now, and so you can infer quite a bit about how galaxies have been evolving and how galaxies must have formed. I think we're going to see an interesting clearance of theories in the next 5 years.

Do you accept this whole idea of extrapolating our theories back to the first few seconds or so of the universe?

It's awfully brave, isn't it? On the other hand, it worked so wonderfully well with nucleosynthesis—unless, of course, it *didn't.* Perhaps it was an unfortunate accident that the naive calculation worked so well. You can imagine that physics back then was wildly different from what we suppose, and by some other accidental route we ended up with the observed abundances. I guess I'm loath to think that the universe would have been so unkind as to give that a possible chance. I feel fairly comfortable with the notion that the physics of the universe when it was 1 second old can be traced back from the physics of the present epoch. If we go back another factor of 10^{10} in expansion to approach the epoch of inflation, then I feel very skeptical that we can know enough physics to be very confident in predicting what can have gone on. I have no idea how that situation can be improved, because it's going to be very difficult to have an experimental check of physics at those energies. Actually, one nightmare I can imagine is that as particle theory advances, particle theorists will hit upon a convincing story—convincing to them—of high-energy physics, from which they derive a cosmology of the early universe, of which there are no observational tests aside from the standard ones that, say, the universe is homogeneous. We knew that already, so it's not really a prediction. Or that space curvature is negligibly small. Well, we didn't know that already, but a lot of us were hoping it would be true. So, if out of this complete theory of particle physics and complete theory of the early universe we get *no* predictions other than things we already knew or were hoping for, will we be entitled to think that we have a physical theory here? How will we know it's right? It would be very frustrating. Of course, nature could have presented us with this frustration many times previously in the history of physics. It hasn't. Always it seems that as physics has advanced, we've been able to test. And the experimental contact has always seemed to be rather convincing. Consider, for example, such a spectacular thing as the discovery of the W and Z particles. It's just a remarkable thing. So I have hopes that it will continue that way and that, as our ideas mature and develop, we will be led to predictions that are sufficiently startling and different from what people had expected when they invented the theory that the predictions will be testable and will be considered convincing if they work. And so the subject will develop.

> *Let me end with a couple of questions even more speculative than the ones I've asked you so far. If you could design the universe any way that you wanted to, how would you do it?*

I would make omega close to unity, zero space curvature, and zero cosmological constant. I would make the universe out of particles we can see, that is, baryons, but then I would be a little embarrassed because I'm not so sure that my universe is going to agree with the real one. How do you mean this question? Are you asking how am I betting the real universe is constructed?

> *No, you can design the universe any way that you want to, and it doesn't have to look at all like our universe does.*

Okay. I have to admit that whoever designed the universe did a good job in making the universe homogeneous, because it gave us a kindergarten problem to work—everyone can analyze the evolution of the homogeneous expanding universe. Also, an expanding universe has a lot of attractions in the sense that we can start from something compact, where it's easy to imagine that the material is forced to be almost pure hydrogen, which is very useful for making stars. As the universe expands, it's easy to imagine how this material got collected up into galaxies, which are very useful for making stars and collecting the debris from stellar explosions to make planets to make people to look at the universe. So I approve of an expanding universe.

You like the idea of having people.

I like the idea of having people around, yes. I hope there are even a few tribes in the universe—in case we happen to knock ourselves off. I dislike the idea of making the dominant component of the universe some material that is undetectable. So I would make the universe out of baryons—of course, with radiation. It's awfully useful to have that radiation not only to control galaxy formation but also to give us—the inhabitants of this universe—some evidence that the universe really is there and expanding. As I say, I wouldn't put in a cosmological constant because it's ugly, and I wouldn't put in any space curvature because it's ugly. I'd have an Einstein–de Sitter universe. I don't really care whether it recollapses at some distant time in the future or keeps expanding forever. One shot is good enough for me.

Somewhere in Steven Weinberg's book, he has the statement that the more we learn about the universe, the less it seems that the universe has a purpose.

Oh. That seems an awfully egotistical remark.

Have you ever thought about this question?

Yes, in fact, I remember that comment, and I remember puzzling as to why he made it. I think he must have been feeling a little down that day, perhaps a little tired, a little discouraged, a fight with his wife. I remember being surprised to see it. I have never demanded that the universe explain to me why it's doing what it's doing.

Have you ever worried about or wondered whether it has any kind of purpose?

That's such an anthropomorphic remark, and not in the sense of the anthropomorphism of Bob Dicke, but I think in a much more naive sense. I'm willing to believe that we are flotsam or jetsam—I'm not sure which is the appropriate term—in a much larger scene. And I'm willing to believe that in the lifetime of the human race, we won't discover the full meaning of this larger scene.

CHARLES MISNER

Charles Misner was born in Jackson, Michigan, on June 13, 1932. He received his B.S. degree from the University of Notre Dame in 1952 and his Ph.D. in physics from Princeton University in 1957, where he was the thesis student of John Wheeler. Misner taught at Princeton until 1963, when he joined the department of physics and astronomy at the University of Maryland, College Park. He is now professor of physics. Misner has had visiting positions at the Institute for Theoretical Physics in Copenhagen, Brandeis University, the California Institute of Technology, Cambridge University, and Oxford University. He is a Fellow of the Royal Astronomical Society and of the American Physical Society and is a member of the American Mathematical Society and of the Philosophy of Science Association.

Misner's research interests include theoretical studies of general relativity, quantization of general relativity, cosmology, and science education. In 1967 Misner began a series of pioneering studies aimed at explaining certain properties of the universe as the result of known physical processes, rather than ad hoc initial conditions and assumptions. In 1969 Misner posed the influential "horizon problem," which raises the question of why the universe seems to be much more uniform than can be accounted for by the exchange of heat and homogenization following the birth of the universe. Misner is the coauthor of a widely used textbook in relativity and cosmology, *Gravitation* (1973). For the last 10 years, Misner has devoted himself mainly to teaching and to science education, particularly using computer-aided instruction. He has written on the relationships between science, philosophy, and theology and has been a consultant, since 1987, to the Committee on Human Values of the National Conference of Catholic Bishops.

I wanted to start with your childhood and ask you if you can remember any influential experiences, particularly any that got you interested in science?

The first thing I particularly think of was getting a prize in a science fair when I was probably in seventh grade. I had been playing around with a chemistry set, and I had already been using my spending money from delivering magazines or newspapers to buy extra gadgets for it. When my mother complained about the holes in my clothes, I did an experiment on how various kinds of cloth reacted to various kinds of acids.

I'm sure that didn't give her too much comfort.

No. My father was an electrical engineer, and he certainly taught me a little bit about practical electricity and fixing things around the house. I can't remember a time when I didn't know how to repair a plug. I went to parochial school in seventh and eighth grade, and we would get sent off across town to a public school for shop and mechanical drawing and things like that.

Did your father encourage your interest in science?

Yes. My father and my mother were both intellectuals, readers, and so forth. So there was general encouragement, but I don't remember specific things.

From your mother as well as your father?

Oh, yes. She was a school teacher. She taught me first grade before I entered regular school, so I was a year ahead of myself in school.

In addition to your experiments with plugs and electrical apparatus, do you remember any books that you read that had an impact on you?

At the grade school age, I don't remember any books. The other thing I played with was cameras. During third or fourth grade, I bought a camera.

How about in high school. Do you remember any books you read then?

I guess there were two. I dug up a college chemistry text and started reading that because I was interested in chemistry, and that contained an integral sign. I ran around trying to see if anyone could tell me what that was. Roughly, none of my high school teachers could tell me what it was.

Could your father tell you?

I didn't ask him. When I did discover what the integral sign was, probably from the chemistry teacher, then I started talking to my father, and he dug out the old calculus book that he had used when he was in college in 1910 or so. I started studying that. There was no calculus taught in high schools at that time.

Did you have any concept of the universe as a whole in high school?

No. I don't remember that. I was interested in chemistry. I was interested in electricity, which I didn't realize was physics. I probably owned a radio amateur's handbook and learned about theories of complex impedances and things like that, but I didn't know it was physics.

Had you decided that you wanted to be a scientist before you started at Notre Dame?

Yes. Probably by the time I was in grade school winning these chemistry prizes. By the time of the science fairs, I'm sure I knew I wanted to be a scientist. I don't recall that there was a serious question.

How did your parents feel about that?

They were quite happy. They would have been happy if I had been a priest. It was high on their priorities to have some children go into religion, but they were quite happy with my being a scientist. My dad helped me with my chemistry work. There was a big workbench he had built for himself that I took over. I then built a whole set of shelves to put my chemistry things on. So there were general signs of approval that they were happy to see me get involved with what interested me.

Did you have a mentor at Notre Dame?

Arnold Ross. He was then the chairman of the math department at Notre Dame. Another friend of mine, Tom Day, who I met at Notre Dame, actually was trying to read tensor analysis and Einstein's general relativity in high school. He ran into Arnold Ross and found out that Ross would encourage people to learn what they could and not take the standard courses. So I went and talked to Ross, and he got me out of analytic geometry and into a real calculus course. I was then a chemistry major, but I got switched to second year physics. In chemistry, in the first semester, they were doing qualitative analysis, which I didn't find at all interesting— mainly, I think, because you rushed through the lab hours so quickly that you never had a chance to figure out what was going on. You just had to follow instructions, get results, and be on your way. So that was the end of chemistry. I knew I had a lot of talent in mathematics, but I decided that I preferred something a little closer to the real world, and I started in physics.

You mentioned that you had a friend who knew something about relativity theory. Did he get you interested in relativity theory?

No, I didn't take any relativity until I was a senior, and then I took a graduate course in relativity at Notre Dame and got a good, solid start in it.

Did the relativity course have any discussion of cosmology?

I would think that it did, but I can't remember being struck by it. It was a one-semester course, and it was just one of the graduate courses that I was taking at that point. I don't remember any cosmology. I don't remember cosmology playing a big role until rather well into my thesis work at Princeton. Even there, it was never the prime target of what I was doing.

Let me ask you about your work at Princeton. Did you have any mentors there who had a big influence on you, as you did at Notre Dame?

Yes. I should mention that there were other people at Notre Dame. I got sort of adopted by the physics department and given a part-time job. There were a handful of people that I knew very well, including this one particular professor, Walter Miller, who I had for several courses and who was involved with the van de Graff machine there where I did a lot of my work. I had really good courses there, so that when I arrived at Princeton, I felt that I hadn't missed anything.

Who did you start working with at first?

I spent my first year with Art Wightman. I did some projects with him on double beta decay. The only course I remember taking was Valentine Bargmann's quantum field theory. I took very little in the way of course work. They didn't demand a lot. You just had to pass the qualifying exam, and I had had all the standard graduate courses at Notre Dame. So I chatted around with both math and physics students, read books and studied in the library, worked with Wightman, and tried to prepare myself for the qualifying exams in the spring, which I passed comfortably. Then I began seriously worrying about a topic for a thesis.

Tell me a little about your thesis work.

That really quickly came down to a question of working with Wightman or John Wheeler. I may have taken Wheeler's relativity course by then. Basically, the deciding point was that Wheeler was full of enthusiasm, and he thought every idea you had was wonderful. He really set far off goals to solve all the problems of the universe and thought it shouldn't be all that difficult. Whereas Wightman felt morally obligated to point out that his students had not gotten their theses done very quickly. There was someone there who was finishing his seventh year on his thesis, while the average graduate student took five years. So, although I thought he was doing very fundamental things, and I had a lot of math background for it, I chose Wheeler because it would be more fun and I would get finished quicker.

What was it like working with Wheeler?

He had regular meetings in his office of the students he was working with, and we would discuss everything in relativity. I spent a lot of time in the library initially and went through the literature and made myself a card file of essentially the work going on in relativity. I continued to study mathematics. Wheeler was interested in wormhole ideas. I had had point set topology at Notre Dame, and my Princeton roommates were mathematicians. I knew enough topology that I could do the mathematical work.

Did he talk to you at all about cosmology at this time?

I don't recall that there was a cosmology interest in the first years. It seems to me we got involved with neutron stars, wormholes, and eventually black holes, before cosmology came into the game.

Did cosmology come into the game while you were still at Princeton?

Yes, it did. Sam Treiman was interested. Treiman also knew about the black hole paper of Robert Oppenheimer and Hartland Synder.[1] He mentioned that to me once, while I was working with Wheeler. I think he also mentioned to me this peculiarity of some molecular spectra that were showing signs later recognized as the 3-degree microwave radiation. He was aware of that as an anomoly. This was in the 1950s, long before the work of Robert Dicke and Arno Penzias and Robert Wilson. So those ideas popped up, but I was not interested enough in cosmology to jump on them and say, "Hey, I should think about this."

My guess is that my serious interest in cosmology can be dated (much later), from an American Mathematical Society Summer Seminar at Cornell in July and August 1965. I was on the organizing committee and during the conference heard rumors of the microwave background. I called Peebles, got the story in outline, and then arranged to get the lecture schedule changed enough so that he could come and describe his work with Dicke and the Penzias and Wilson observations. I felt immediately that this was a turning point in the study of cosmology. His talk already asks, "What limits can we place on the uniformity of the universe if the observations establish that the radiation is isotropic?"[2]

Going back to Princeton and the 1950s, had you read enough cosmology to know about the big bang model and open versus closed universes and that kind of thing?

Yes. That was floating around. In the 1954–1956 period, I most likely heard Fred Hoyle defending the steady state theory by producing elements in stars and in supernovae.

At this time, did you have a preference for any particular cosmological model—steady state versus big bang, or open versus closed?

I was aware of steady state, and that didn't sound like physics to me. There were too many ad hoc explanations. They had to invent something [continuous matter creation] for which there is no other application, and it didn't seem to fit comfortably into the rest of physics.

You didn't like that?

I didn't like that because, on the microscopic level, there didn't seem to be any way for it to happen. I think I was also aware that there was a cosmological constant floating around, and I would have been happy to see that be zero just on the grounds of simplicity.

What about the question of open versus closed? Did you have any inclinations there?

Not strong inclinations. I might have somewhat preferred the closed model. Wheeler quickly jumped on the closed model because he felt it solved the problem of boundary conditions. Otherwise, there might be too many solutions to the equations and too much ambiguity. Wheeler was very much committed to the closed model on aesthetic or philosophical grounds. I can't recall that I had any preference at that time.

Moving to your own research, do you remember what motivated you, beginning around 1968, to look for physical mechanisms for producing the observed isotropy of the universe?[3]

It was certainly the microwave background radiation. Before 1965, in the standard cosmological models, the assumption of homogeneity was just because the data was [only] good to about 30 percent, so there was no use playing with more refined models. At that point, that assumption was simply a mathematical simplification—saying that we're going to treat a homogeneous universe because anything else would be wasted effort in the absence of data. But after the microwave radiation came out and began to show this uniformity to a tenth of a percent around the few arcs of the sky that were surveyed, then it became a serious problem. Things that you don't understand can be constant to 10 or 20 percent, but one percent requires an explanation.

In your mind, could the explanation simply have been that the initial conditions were isotropic?

No, that never seemed like an explanation to me.

Why not?

I don't know. It's not clear that a postulate of isotropy at the beginning is better than a postulate of isotropy now. There was nothing to get your hands on. I guess it was like postulating the creation of matter. You can, of course, throw in anything that solves your problem, but it's got to make sense from more than one viewpoint before it's a real explanation.

In your work at this time, looking for a physical explanation for isotropy, did you also group that problem with the homogeneity problem?

Yes, because homogeneity is necessary for isotropy. That was understood at the time. The cosmic background radiation could look isotropic if it was all carefully focused around us, but since Copernicus, we haven't been very happy saying we're at the center of the universe.

I really got deep into this when I went to Cambridge in 1966–67 on a NSF postdoc. Peter Strittmatter and John Faulkner were playing with some recent observational data that bore on this isotropy question. It was not microwave radiation. I think it was a question about quasars. The question was raised whether the isotropy of the universe bore on this data, and I quickly produced a calculation using all the relativity I knew, and then I realized I had the tools to tackle this other isotropy problem. So I began to get interested in the general isotropy problem. Also, a fall semester seminar series on cosmology in 1966, in which Dennis Sciama's group and others participated, had kept problems of cosmology continually in mind.

Were other people at this time raising this issue as a problem—that you needed some physical explanation to explain the observed isotropy and homogeneity?

I don't know. I'm not aware of it.

So on the theoretical side, you weren't particularly influenced by anyone else? It was your own idea?

That is my recollection, yes.[4] [The horizon paradox is clearly stated in my 1969 *Physical Review Letter*.[5] The 1968 *Astrophysical Journal* paper is concerned with explaining or deriving isotropy (without assuming it a priori), but does not identify the horizon paradox.[6] My main qualitative point was to ask that physics not just find the cosmos consistent with the laws of physics, but also try to show that no very different cosmos was allowed or was plausible. I was trying to change the *goals* of scientific cosmology from describing the universe to explaining it. The horizon paradox must have congealed in my mind in the spring of 1969. It does not appear in my notebooks around November 1968, when Wheeler passed on to me some reports of related "mixmaster" work by Belinsky and Khalatnikov.[7] Their work, like mine, was unpublished at this time. I presume that the horizon paradox arose in my mind that spring and provided enough excitement that I could publish a short description of it and the mixmaster solution—which I hoped would contribute to solving the paradox—in the short format of *Physical Review Letters*. My hope was that this kind of pancaking of the universe in the mixmaster model would wash out the horizons, which I considered a serious defect in the standard model. The mixmaster model was later shown not to help the horizon paradox.[8]]

Do you remember the reaction of the community to the mixmaster model?[9] Were other people also hopeful that this would solve a serious problem?

Well, I think Wheeler was. Wheeler was very excited about it and very happy about it. I don't remember other particular interactions, except with the Russians. Essentially, they had done the mixmaster at the same time, and I just couldn't understand their work.[10]

I gather that they weren't specifically trying to solve the horizon problem with their work. Is that right?

Right. I don't think they had that in mind at all. They were trying to solve the singularity problem. There was a time when they were claiming that the initial singularity was not generic.[11] In their mathematical solutions, the universe was doing this whole series of bounces that never actually became singular, although the sequence extrapolated to a singularity. Of course, at about that time the singularity theorems came out and proved that was not correct [that a singularity was unavoidable].[12]

When did you first hear about the flatness problem?

Dicke came up with the "Dicke paradox." This probably was in the fall of 1969, when I was a visiting professor at Princeton and attended the Dicke/Peebles seminars in Palmer Lab. The Dicke paradox was that the universe is too close to being spatially flat. I have a distinct impression that I was wandering around the old Palmer Lab building trying to understand whether it was fair to think about the universe in terms of initial conditions, the way you would think of throwing up a ball—because there are constraints in the Einstein equations that play in that game somehow. So it wasn't clear to me that the energy could be variable the way it is in the analogous Newtonian differential equation.

So you thought that the energy may not be a result of initial conditions but of some other constraint?

Yes, I just couldn't see how to play with those equations, and so I didn't come on board thinking that paradox was serious until the inflationary models came out. Later, I developed a strong preference for the flat universe, feeling that the Dicke paradox suggested it. The key point for me was that inflation offers an explanation. Even if it's not the right explanation, it shows that finding an explanation is a proper challenge to physics.

But if you go back to when you first heard about the flatness problem, did you take it seriously?

I took it seriously enough to worry about it for a while. I can remember it's causing me to spend some hours chewing it back and forth. Then I guess I just dismissed it. I decided that I couldn't take it as seriously as Dicke did because I wasn't sure that the energy in the Einstein equations was a free parameter. Somehow you seemed to have the choice of zero or plus or minus 1. There didn't seem to be the continuous parameter, in my view of it at that time. I think later one sees that there is a continuous parameter, like the equation of state of matter. You can run the universe in such a way to set these things up, but I didn't appreciate that. So then I put it aside.

You said that your view of the flatness problem changed as a result of the inflationary universe model?

Yes, the inflationary model was crucial. It was not crucial that the inflationary universe be right. What was crucial was that the inflationary universe provided an example that turned the Dicke paradox into a standard physics problem. By proposing a certain dynamics, you could solve that problem, explain it. Dynamics is an explanation to me, whereas a fiat that the universe starts out homogeneous and flat is not an explanation.

So having an explanation of the problem made it a real problem for you?

That's right. It could be solved correctly or incorrectly, but once you have seen one example of an *attempt* to solve it, you could feel that this is real physics, and if that attempt doesn't work, some other one could.

And once you decided it was real physics, you took it seriously as a problem that needed a solution?

That's right. Then I took it seriously as a problem that needed a solution. I would say even more than that. First, it allowed me to understand this question of whether energy was a free variable in the initial conditions. It became free by changing the dynamical laws through the equation of state in early times. Inflation seemed to show that there were "dials" you could turn in the equation of state that would lead to consequences that made this thing intellectually controllable, as distinct from being somehow built into the Einstein equations that the universe has to be closed, and therefore the total energy has to be zero.

You mentioned a little earlier that once the inflationary universe model came out, you became strongly attracted to a flat universe.

Yes, because now I take the Dicke paradox more seriously than the inflationary models. The inflationary models show that good physics probably is capable of producing an answer to this problem. The present models may not explain all observations and may need some changes. The elementary particle physics may be sufficiently different that the explanation will be sought elsewhere. But now I take the Dicke paradox seriously. There's no particular reason to think that just now, at the present age of the universe, when man has evolved, this number [omega] is about to deteriorate [deviate by a factor of two or more from one], and we're going to fade away from being on this borderline. So, we're on the borderline for dozens or hundreds of decades of expansion into the future. That doesn't say that omega is 1 exactly, but it says it is 1 to a precision that we are not normally prepared to deal with. Either that, or there is something very seriously wrong with our cosmology, and something is dramatically overlooked if that number is not 1 right now.

If you think it is very close to 1, then how do you reconcile that belief with the observational evidence that it is 0.1?

Well, I haven't followed the observations closely enough to criticize them, but roughly I have a feeling that the luminous matter is, of course, very small [as a fraction of the total mass], and the quantity of nonluminous matter that is detected gravitationally keeps changing, depending on the scale on which you try to find it.

If I were to tell you that most observers who work on this believe that the observed value of omega, including the nonluminous matter, is about 0.1 or 0.2 at the most, what would you say to that?

I would say, "Keep looking for ingenious ways to hide matter, maybe in exotic particles." Of course, I would also say to the theoreticians—the experimentalists sometimes have qualitative ideas that are more imaginative than those of the theoreticians—"Look for ways in which our picture of cosmology could be so dramatically wrong." If omega is not 1 [for example, if omega is truly 0.1], the mechanism that set it close to 1 in the old times must have known some number that correlates with the evolution of man. You either have to find that mechanism that, either by chance or nonchance, had some relationship to the present age of the universe, or you've got to find the missing matter.

What is your personal view as to which of these is more likely?

More likely, there is some invisible matter spread out wider than our present gravitational sensitivity to it.

Is that because you have a difficult time imagining the physical process in the early universe that could have known the age of the universe now, the epoch of man?

That's right. Yes.

We've talked a lot about the inflationary universe model. How speculative do you think it is? Do you remember how you first reacted to the inflationary universe model?

I can't remember, but my guess is that I liked it. It sounds great. Then and now, it has this slight drawback that it relies on physics at unreasonably high temperatures. It used to be that particle physicists in the 1950s laughed at relativists, saying, "What are you doing playing with numbers that don't make any difference? In quantum gravity, the distance scale is 10^{-33} centimeters. Nothing is known below 10^{-13} or 10^{-14} centimeters, so it's nonsense to talk about those extrapolations." Of course, now the particle physicists have their own motivations to extrapolate back that far and are doing it almost as wildly as the relativists would, or more so. I am a little hesitant to believe that you could skip 10 orders of magnitude in energy with nothing interesting happening in particle physics and still get the right answer. Not that it's never happened in physics. Maxwell's equations extrapolate the 30 orders of magnitude from the proton to the galaxy with no need for modifications. So, it's not impossible, but one's a little cautious.

So you were interested but cautious.

Yes, interested but cautious. However, I was roughly convinced that something like inflation would probably work. If that's not the answer, then the real answer is probably not too different. It wouldn't surprise me if the real answer was inflationary, but that the actual mechanisms leading to the inflation might be different.

Why do you think the community has been so enthusiastic about the model? Do you think it's for the same reasons you were?

Yes. And the fact that the particle physicists began to realize that there is not that much flexibility in building models of the universe. When people were not in the game, they would look at cosmology and think that you can do anything. But then when they tried to start making the models themselves, they saw there is a lot more limitation. You don't have that much freedom. Cosmology was not just pure invention, where you could throw out models everywhere and take them seriously. The other thing was that there is a nice long list of observational problems that inflation solved. It's not just that inflation solved one thing, but that it solves 6 or 8 things. Some of them are statable in terms of the standard model, like the Dicke flatness problem and the horizon problem. Some of them are only statable in terms of higher levels of elementary particle speculation, like the monopole problem.

You mentioned to me a few months ago that you consider the flatness problem to be in the same category of importance as Olbers' paradox?

That's right. I think *this* time, people are taking it seriously, whereas Olbers' paradox was never taken seriously until Hermann Bondi and Thomas Gold used it as a pedagogical device to discuss cosmology. It was never taken seriously before that, because there was always an out.

Let me ask you about an observational discovery. Do you remember when you first heard about the work of Valerie de Lapparent, Margaret Geller, and John Huchra on the large-scale structure?

By that time, I was not active in cosmology anymore. So I was intrigued, but it was not something I was going to jump on professionally. I had a general tendency that once things got away from the simple mathematics of general relativity to real astrophysics—like with black holes, once the research got away from idealized black holes to the accretion disks, x-ray sources, the plasma, and a lot of the detail—it just wasn't the kind of physics I was more efficient at than other people, and so I left it to the experts. It's getting that way in cosmology.

There is a whole group of observations of a similar character—the large-scale streaming motions, and so forth—that have shown that the universe seems to be more inhomogeneous and anisotropic than we thought. Has that kind of idea altered your thinking about the big bang model at all?

Well, just in the sense that it leaves a big mystery as to how the microwave radiation can be very homogeneous on every scale looked at so far. I haven't put the numbers down to determine whether there is a flat contradiction, but that is one of the things that one would worry about. The other is simply that the inflationary universes do lead over to all these wild ideas of parallel universes and things like that, which now become mathematically not so wild.

What does that have to do with this inhomogeneity?

It is a question of whether there is *simple* homogeneity within which there are perturbations, or whether there is a sort of fractal organization of irregularities, which keep showing up on all possible scales, including those much larger than the observable universe. That sort of picture is, to me, an a priori plausible picture. But then there is a serious technical problem of fitting that kind of over-broad picture with the observational uniformity of the microwave and the [local] irregularities [we see]. That's a technical problem that will be very challenging and very important. It's not at all clear that it will be solvable in any one of these broad frameworks that we just sketched.

Have there been any other developments in the last 10 or 15 years, either in observations or in theory, that have had a major impact on your thinking?

One is the Dyson article in the *Reviews of Modern Physics* on the distant future.[13] I think that is a tremendously interesting and insightful idea, which is made more so by the inflationary universes. Being convinced that the Dicke paradox is solid and significant, I expect a very flat universe, which means that I expect a very long future.

But you could also have a very long future in any open universe, without having omega so close to 1.

That's right, but there was no compulsion to be in an open universe. Now there is a compulsion.

So calculations of the distant future have more relevance.

Yes, they have more relevance because I am more convinced now that omega equals one than I was about any definite value of omega prior to the inflationary universe models. So, now it looks like we should deal with that long future. Dyson put all these ideas together with calculations and imaginative ways of making estimates. I summarize his conclusion by saying that the universe will last long enough for civilization to rewrite its literature infinitely many times, even though it gets colder and slower. You see, if the universe gets slow too fast, it won't be able to do much. Dyson says that even though physical processes are getting slower, time is expanding in front of us so rapidly that you can write your literature infinitely many times. I think that is a vision of our place in the universe that is very inspiring and very dramatic. I'm surprised nobody is paying much attention to it. It's not available to the general public at all.

I don't know whether the general public cares about what is going to happen in the very, very distant future.

They care about what was in the past. The big bang is on everyone's lips. If you have a talk or a public lecture on the big bang universe, you will get crowds. People are very intrigued by the beginnings of the universe. I'm surprised that this glorious picture of the future doesn't even attract physicists.

Let me take a detour here. Do you use mental images and pictures at all in your own work?

I think so. I am much happier with things that are geometrical. I suppose when I think of a closed universe, there is a picture of a sphere in my mind. But there is also so much backlog of mathematics that one has worked through that may be an important part of the picture also. In mathematics, I never get terribly comfortable and easy with pure algebra, because somehow it doesn't have a visual lead-in. I've found that there is a kind of pleasure I get from working in geometrical terms—which are not always pictorial but somehow have a sense of pictures behind them—that is not the same as I would get in other areas.

Do you think that theory and observation have worked harmoniously in cosmology in the last 10 or 15 years?

Yes, it seems to me that observations provoke lots of theorizing. Of course, it's a different game from tabletop physics. The ratio of observation to theory is very different. As we mentioned before, people are beginning to appreciate some of these very qualitative things, like the absence of monopoles, as important observations. You have to be imaginative about what constitutes an observation to get anywhere in cosmology. I think some of the evidence will be consistency relationships. It will be that all of physics ought to eventually hang together in one solid picture. In fact, that sustained a lot of people who were working in relativity for years when there were no experiments.

What do you think are the outstanding problems in cosmology?

The outstanding problems are certainly the large-scale problems, where the answers will ultimately come—just by checking the fine detail. As I said, I haven't been close to the day-to-day problems for 10 years. But the broad-scale problems are to situate something like inflation within a more solid framework of elementary particle physics. There are questions of really understanding elementary particle physics and not being quite so speculative when we make models of the very early universe. That includes, ultimately, all the questions of the Planck length and the quantum aspects of the beginning of the universe. The Hartle–Hawking idea that maybe quantum mechanics provides the initial conditions of the universe, allowing you to sidestep the singularity problem, is a very intriguing kind of thing.[14] That should continue to receive attention. And there are all these questions of 4 dimensions. Are we going to make a theory of a higher dimensional universe, which condenses into 4 dimensions of space and time plus a bunch of other dimensions that are essentially the graph paper on which gauge fields are plotted?

> *For the next question, you may have to put your natural scientific caution aside a bit. If you could design the universe in any way you wanted to, how would you do it?*

I never have thought about designing the universe. I am interested in the *question* of the design of the universe. I have published papers on philosophy and cosmology and theology.[15] I do see the design of the universe as essentially a religious question. That is, one should have some kind of respect and awe for the whole business, it seems to me. It's very magnificent and shouldn't be taken for granted. In fact, I believe that is why Einstein had so little use for organized religion, although he strikes me as a basically very religious man. He must have looked at what the preachers said about God and felt that they were blaspheming. He had seen much more majesty than they had ever imagined, and they were just not talking about the real thing. My guess is that he simply felt that the religions he'd run across did not have proper respect, or proper dignity, for the Author of the universe.

But then there is this other question of more straight philosophy. Is the design of the universe distinct from its enactment? Are the laws of physics necessary? I think in the time of the Enlightenment, people felt that if you think hard enough, it would be clear what the laws have to be—Descartes certainly felt that way. And the shock of quantum mechanics and relativity eventually, I believe, convinced people that even though Einstein did a lot with pure thought, that pure thought was not likely to be adequate.

> *Because we hadn't anticipated quantum theory and relativity?*

Yes, right. Physics probably did need a nudge from experiment, however philosophical or theoretical you wanted to be, to put you on the right track. It might be basically because quantum mechanics was inconceivable in the nineteenth century. That is to say, all of human intelligence and culture had not had the necessary experience so that these thoughts could actually appear in anyone's head. The whole evolution and the experience of a civilized culture was necessary before you could develop these ideas, and now we can think them. So we need these nudges. But there still is the question: could there be a different set of laws that was as good a theory of physics? It's not clear to me that we can say anything is wrong with the classical Einstein–Maxwell equations except agreement with observation. They don't agree with observations, but is there some philosophical viewpoint in which they are improper physical theory?[16] They have an initial value problem, they have lots of solutions. You could make models of the solar system by replacing each sun and planet by a black hole; they would run around and do Keplerian motions. So there is a lot of physics in this model. I don't know, on philosophical grounds, a check-off list for what constitutes a good physical theory. Other than comparison with experiment, is there any internal property that this theory doesn't have that a real theory ought to have? Maybe there is a distinction between a conceivable theory and an actual theory. If so, what can you say about the actualization? Do we just blame that on God or can we get more insight? My guess is we blame it on God and say there are things that we can think about but can't do.

If you were allowed to conceive of a theory yourself, what would you do?

The universe I see is always more beautiful and preferable to any I could have previously imagined—the more details I see of it. So in that sense I like the present universe. If I wanted to put that into a phrase, I would say "a universe which is inexhaustibly intelligible," where you could keep understanding things and the game never gets boring.

There is a place in Steven Weinberg's book The First Three Minutes *where he says that the more the universe seems comprehensible, the more it also seems pointless.*

Yes, I come down on just the opposite side of that. I'm impressed with the beauty and intelligibility of the universe. We would have to get into a whole other thing about the meaning of truth, which I have written a little bit about.[17] I don't see the universe as pointless. You might call Newtonian theory a myth in that we know what it's good for and we know its limitations. It's not so much of a myth now as it was in Newton's time, when people were unaware of the limitations. In that same sense I think there are truths of religion that are real truths, but that are also myths— myths in the sense that we will not want to change them when we understand things more deeply, but we *will* understand things more deeply. For example, Newton's theory was once understood and believed totally, and now it's understood and used and provides us with a grasp of nature, but we have some feeling that there are other things beyond it. My feeling is that in religion there are very serious things, like the existence of

God and the brotherhood of man, that are serious truths that we will one day learn to appreciate in perhaps a different language on a different scale. We will probably always continue to teach them in the traditional ways—and think of them like Newtonian mechanics: you don't want to play baseball with quantum mechanics. So I think there are real truths there, and in that sense the majesty of the universe is meaningful, and we do owe honor and awe to its Creator. With this Dyson future, I don't see anything wrong with imagining that civilization will succeed and evolve so that intelligent, responsible beings discuss physics or what comes after, long after the temperature has gone down and the heartbeat is once per 10 billion years. The activity will continue apace and be more glorious, and we're part of it, helping to produce it. I think there is a lot of meaning in the whole operation.

JAMES GUNN

James Gunn was born on October 21, 1938, in Livingston, Texas. He received a
B.S. from Rice University in 1961 and his Ph.D. in astrophysics from the
California Institute of Technology in 1966. His professional career has included
work as a senior scientist at the Jet Propulsion Lab at Caltech and as a professor at
both Princeton University and Caltech. Currently the Eugene Higgins Professor
of Astrophysics at Princeton, he is a member of the National Academy of Sciences
and won the 1988 Heineman Prize of the American Astronomical Society.
Gunn's research interests include astronomical instrumentation, the structure
and evolution of galaxies and clusters, and star formation.

Jim Gunn is one of a small fraction of physicists and astronomers who are
outstanding in both theoretical and experimental work. Among Gunn's contributions to observational cosmology are his design and construction of highly sensitive instruments to electronically record light from telescopes. Gunn was the
author, along with David Schramm and Gary Steigman, of an influential theoretical calculation in 1977 limiting the possible types of subatomic particles by
consideration of nuclear reactions in the infant universe and the observed cosmic
helium abundance. At Rice University, students still use the phrase "gunning an
exam," meaning to get a perfect score, although the origin of the expression has
long since been forgotten. Considered exceptionally talented by his colleagues,

Gunn seems to be always in a hurry. He is married to the astronomer Gillian Knapp.

I wanted to start with your childhood and how you got interested in science. I've heard many stories about how you built things when you were younger. Could you tell me a little bit about that?

My dad was an exploration geophysicist for Gulf Oil when I was a little kid. It was during the war. He was the chief of an exploration party, and you couldn't get parts for anything. So anything that broke, he had to fix or make another. We stayed in one place only six months at a time, and we carried around with us this machine shop that was in a trailer. From the earliest time I can remember, I was around power tools, and my dad was making things and fixing things. So it was very natural for me to grow up doing that kind of thing. It was when I was about seven or so that I became interested in astronomy. I don't quite remember what touched it off, but there was a little book that my dad bought me called *The Stars for Sam*. There were various science books for kids called *The Ocean for Sam*—a whole series. That really turned me on, and it just sort of took off from there. My dad had taken some astronomy courses as an undergraduate and still had his textbooks around—just descriptive kinds of stuff that a kid could easily read and more or less understand. It wasn't but a year or two until I graduated to reading that kind of stuff about astronomy.

That's before the age of 10, you think?

Yes, well before the age of 10. It was also a natural that I build telescopes. So about that time I started building some telescopes, and I built several by the time I was 11 or 12. It seemed a very natural thing to keep on doing because it was the thing I loved the best.

What size were these telescopes?

They started out being pretty small. You'd just buy lenses from what then used to be Edmund Salvage Company—little ones and mostly refractors. It wasn't until I was in high school that I built my first reflector—four and a quarter inch. I guess I was in junior high, actually. And it wasn't until the middle of high school that I started doing optics seriously. I made an eight inch mirror from scratch. That was the biggest telescope I built before I went pro. It was pretty ambitious. It worked, though. It's amazing.

You mentioned this book, The Stars for Sam. *Do you remember any other books or any particular scientific ideas that impressed you?*

The book later that really caught my fancy was Fred Hoyle's book, *Frontiers of Astronomy,* and the notion that we're all made out of the stuff that spewed out of stars. That kind of thing was not discussed in these old textbooks of my dad's because they had no idea how the heavy elements were made. Hoyle is a very persuasive writer, and that was a very cohesive view. The idea, to a high school kid, that one could make a cohesive *picture* of the universe and why things were the way they are is an exciting thought. It's something that you don't think about when you're that young—that you can more or less sit down in your armchair, literally in Fred's case, I think [both laugh], and write down a theory of the world. It's pretty powerful stuff, and it really caught my fancy.

At this age, say 10 to 15 or so, do you remember reading or hearing anything about cosmology? Did Hoyle talk about cosmology?

Yes, Hoyle talked about cosmology. The whole thing was, of course, couched in the steady state idea and how things could come about in the steady state. I also read George Gamow's books, and it was clear even then that there was some fundamental dichotomy in the world, although Gamow's books were written before the steady state became current. Hoyle kind of soft-pedaled standard Friedmannian cosmology because he thought it was beneath contempt. But it was clear that there was some great dichotomy of thought at that point. I was also somewhat aware of the kinds of tests that one might make to distinguish them [the steady state and big bang models]. At that time, the microwave background wasn't known, so the understanding of things was really pretty primitive compared with today. I'm sure it's primitive now also, but even compared to now it was primitive.

Do you remember having a preference for any particular cosmological model at that time?

I liked the steady state, and I don't know why. I guess just the idea of permanence. It's something that's hard to say. I have always had a philosophical predilection for universes that at least last forever. I don't particularly care if they have been in existence for any amount of time or not. It's probably one of these subtle biases that affects what one does, because I've come down fairly heavily on that side of the coin when it seemed to matter. I'm not sure it matters anymore.

Can you tell me a little bit about your undergraduate education at Rice?

Yes. I went to Rice for several reasons. The main reason was that my parents were not terrifically well-off financially, and at that time Rice had no tuition. I also got a quite nice scholarship for support. I haven't really regretted going there at all. It's a very good school. Rice had no astronomy department, for which I have been eternally grateful since.

So you were forced to take physics?

I was forced to take physics, right. I took a double major in math and physics, and I think that was one of the wisest decisions I made—not to worry about astronomy until later. It wasn't clear until the very last minute that I was actually going to go and *do* astronomy, because I was taking physics fairly seriously. I applied to many physics graduate schools, and also just on an odd chance I applied to Princeton, Chicago, and Caltech in astronomy and was admitted. And just in April of my senior year, I decided, "By God, this is what I really like; this is what I'm going to go do"—even though I had *absolutely* no idea what astronomical research was like. Zero. I didn't know what professional astronomy was.

Well, you certainly had been an amateur astronomer.

I had been an amateur forever, that's right. I read lots and lots of books, but I still didn't have a very clear idea about what a professional astronomer *did*, whereas I had a pretty good idea of what a professional physicist or mathematician did. I had actually been more active in mathematical research than in physics.

Once you decided to go into astronomy, and you entered graduate school at Caltech, do you remember how or when it was that cosmology in particular attracted you?

From the very beginning. In fact, one of the main reasons I went to Caltech was that H. P. Robertson was there, and he was certainly one of the leading lights at the time. Now in the summer between the time I left Rice and came to Caltech, H. P. Robertson was killed in a car accident. With his passing, there was not even a relativist at Caltech, much less a cosmologist. That was something of a blow, but Caltech was a sufficiently exciting place with other things going on that I didn't mind particularly. I proceeded to attempt to educate myself a little bit about relativity, and then in my second or third year, Frank Estabrook from JPL [Jet Propulsion Lab] came and started teaching a relativity course, and so I learned that.

Just back-tracking a little bit, you said you went to Caltech already with the idea of working with Robertson, who worked in relativity theory and cosmology. Do you remember how it was that you were initially attracted to cosmology?

I think through reading Hoyle and Gamow. It was just the thing I wanted to do. It's what every kid wants to do who thinks about it at all, you know, to understand the World with a capital W.

Well, you could think that you just wanted to go study how the sun works . . .

Yes, yes, yes, and there are people who do that. I don't understand them [laughs]. But actually during my senior year at Rice, Bob Kraft came around and gave a colloquium, and I spent quite a number of hours talking with him. I then got involved with him almost immediately after going to Caltech on some things as far away from cosmology as you can imagine, namely studying the atmospheric abundances in F stars because that's what he was doing. So my first year at Caltech, I got involved in observing. I went to the 60-inch telescope and got the spectra of F stars. I got busy on other things almost immediately and didn't really notice the lack until I had time to sit down and think about what it was that I was doing. So the loss was a disappointment, but not particularly a painful one. Caltech keeps you busy, as you know.

You said you took a relativity course with Estabrook. Did you then start finding more opportunities to do relativity and cosmology?

Yes, I did. I started thinking about cosmology in some detail and doing little calculations. I got interested during my second year by talking to Guido Munch about the statistical distribution of galaxies. He had done a similar thing for clouds with Chandrasekhar. It was some years before James Peebles's work on the clustering of galaxies. That project in my third year turned into a thesis. I counted galaxies on plates and did correlation functions and stuff like that. I tried to make models of the evolution of structure, without much success.

Did you have redshift data then?

No, this was just 2-dimensional. This was long before you could even get redshifts for most galaxies. 3C295 still had the record [the biggest redshift], and these were mostly objects fainter than that.

You were not doing the observations yourself, I take it.

Yes, I was. I got the plates at Palomar. I counted them, which is something I never want to do again. Then I put the stuff in the computer, computed the correlation functions, and did comparisons with the theories.

So you were already doing both theory and observations at this point.

Yes, right. Coming up from being an amateur and building instruments and such, I was already well into that at Caltech. I made several things while I was there. It never occurred to me to proceed in any other way. If you need data, you go out and get it. I still think that's the best thing to do.

At this stage of your career in graduate school, do you remember having any particular preference for one cosmological model versus another?

Oh, just the old preference that had been there for a long time, for things that went on forever. I was never convinced of the sort of neatness of compact models, which are closed both in space and time. It seemed to me rather a waste.

Let me ask you a little bit about your instrument building. A lot of people say that some of the most important things that you've contributed to the field are not to be found in the journals but are the instruments you've built. How do you go about deciding what instruments to build and how to design them?

I am nearly always driven by results I need. That's been true in almost every case. I want to do something. A lot of the things have been done by the pressure of wanting to get the spectra of ever fainter galaxies at bigger redshifts and, by the by, finding those galaxies at big redshifts and being able to set them on the spectrograph. So most of the stuff that I've done in the last 10—no, more than that—15 years has been instrumentation aimed at that.

When I first came back to Caltech in 1970, image tubes were just becoming off-the-shelf things that one could buy and think about seriously using. Bev Oke and I got involved almost immediately in this survey for distant clusters, which still goes on. That was begun with Schmidt plates at the 48-inch telescope, but it was clear even then that one needed to get better pictures with the 200-inch, and the way to do that was with image tubes. So the first things I built at Caltech were a couple of image-tube cameras.

Then along about 1973 or 1974, 2-D electronic devices were becoming available, and the first really major project that I did in my professional career was building this SIT [silicon-intensified target] vidicon spectrograph. I spent a couple of years, essentially full-time, building that instrument. I didn't do very much science during that time. I learned an enormous amount of electronics—analog electronics and digital electronics—my first real experience with digital electronics. This was 1974–1975, I guess. That was also the time that minicomputers were becoming widely available, and so from the outset we designed this thing as a computer-controlled device. It was my first experience with computer control. So those two years were very intensive, in an engineering sense. I did essentially all the design for the instrument. Looking back on it, you could say it was two years gone, but I learned an awful lot, and it has come in very handy since.

Why would you say it was two years gone?

I didn't do any science. So there were almost no papers. I then used the spectrograph quite extensively. It was not as successful as it might have been because the detectors were not as good as they might have been. It did a number of things very well, but not the problem I particularly designed it to do—to detect very faint things, because it was limited by its own inherent kind of noise whose properties were not easily understood.

Then along about that time it became clear that CCDs were the wave of the future, and so it was not long after then that I got involved with [James] Westphal and the JPL CCD effort, and all the instruments I designed after that used CCDs. That helped solve the noise problem, because CCDs really are essentially ideal devices. So the next thing I built was this thing called PFUEI, which is a combination camera-spectrograph that would do almost everything I wanted it to do, except the field of view was too small. Then I designed Four-shooter to get around that.

There are very few young astronomers these days who are getting hands-on experience both with building instruments and with actually looking out of the telescope. Do you think that kind of experience has given you any special feeling for the objects that you study?

I think probably the contrary. The problem is that my efforts are so fragmented between doing various things that I probably don't have time to understand the theoretical things as much as I would if I just did theory. One thing it *does* do—and there are two sides to this coin as well—is that it does give me a pretty firm handle on what to believe and what not to believe of my own results and other people's results, because I feel that I know the limitations of instrumentation very well. I've noticed in the last few years—I guess one always gets negative as one grows older—that I believe less and less [laughs]. Whether that's healthy or not, I'm not entirely sure.

> *My impression is that there is a new generation of observational astronomers who have much less hands-on experience with a telescope and instruments than you have had.*

Yes, yes. I think that's not very helpful. It's fostered by the National Observatory System and such, in which you go to a telescope twice a year to use an instrument whose development you have not been involved in at all and that you learned about from a book. It's almost impossible to get a real feel for what the limitations of that instrument are. I don't know where it came from, but there's a famous saying among instrumentalists that the problem with this instrument is that it always gives an answer [both laugh]. So you take home a piece of data, but you don't really have a very good idea of what it means, or to what extent it's to be trusted. I don't know what to do about that, because as universities become poorer and poorer and observational astronomy gets centered at fewer and fewer higher- and higher-powered places, there's just not much way to educate people to do instrumental stuff. Not very many people are interested in the first place.

> *I'd like to ask your reactions to some of the discoveries of the last 10 years. Have your broad views of cosmology changed as the result of your work or other people's work?*

Not so much as a result of my own work, but certainly as a result of other people's work. I think that if you're anywhere close to the game, you *cannot* have the same view of the universe now as you did even 5 years ago. If anything like inflation is right, the universe is just a very, very different place than any of us visualized a few years ago. And the questions that were burning issues then are by and large not very interesting anymore in the global sense.

> *They're made irrelevant?*

They're made irrelevant, and the kinds of questions that you ask are entirely different. It's not completely clear to me that inflation is right, and there's certainly no external evidence that it is. But it's such a pretty idea and fits things together so well that it . . . it just enforces an entirely different kind of mental outlook on cosmology than before. I think it's *very* important still to go on with the classical tests because that's one test that you make. I mean if omega is not 1.0000, then something is screwed up, and we have to know that. We have to be able to test whether general relativity is the right theory of gravity, which is also something you can only test, I think, with global tests. But the reason for doing those tests is entirely different from what we thought the reason was a few years ago.

You said that no one doing cosmology could keep the same views in the last 5 years. In terms of the change of viewpoint in yourself, has any kind of intellectual struggle gone on? Have you found any discomfort or hesitation, or did you just naturally glide from the old questions to the new questions?

A lot of people who work in cosmology have, I think, basically religious beliefs. I have had mild philosophical preferences but never very strongly held views. It seems to me that the universe is the way it is, and we'd better accept that. So it's not been any problem. In fact, it's been an enormous pleasure, because all of these locked doors that were present a few years ago now seem to be unlocking. And even though I'm not working in the forefront of early universe theory, just the idea that these questions can be answered is enormously exciting, and all very positive. I wouldn't have changed anything for the world in the last 5 years. The big problem, it seems to me, is that there is such a plethora of possible theoretical frameworks at the moment and no way of testing them. The subject is sort of running open loop. That's not very healthy from a purely scientific point of view. It's very exciting, but from the point of view of trying to learn the "truth," I think we've taken a large step backward, and that's just going to be the way it is, I think.

Do you remember, when you first heard about the inflationary universe model, what your reaction was to it?

I thought it was absolutely insane; it couldn't possibly work because these bubbles couldn't possibly ever get together. That objection turned out, of course, to be true. But that soon settled out. The idea was so pretty that I was immediately taken by it, but I realized there were big problems. Other people realized there were big problems, and now there are several ways around those problems.

Let me ask you about the flatness problem, which I guess got widely broadcast in the article by Peebles and Dicke in the Einstein Centenary book in 1979. Do you remember when you first heard about the flatness problem?

I don't know. It had been kicking around for many, many years before that, because I used to give popular talks, and I talked about it in the early 1970s, and it wasn't my idea. So I think it's true that that article really brought it to the forefront, but I don't even remember where it came from. I think for anybody who really seriously looks at Friedmann, Robertson–Walker metrics, it's a sort of obvious thing. I'm sure I read it somewhere and then went and looked at it. But I don't remember where I got it from.

When you first started thinking about the flatness problem in the early 1970s and lecturing on it, did you consider it to be a serious problem or did you think that it was just due to a random initial condition of the universe and not worth a physical explanation?

No, it seemed to be worth a physical explanation, but it's intimately tied with the causality problem. There were clearly things about the initial conditions that we didn't understand and that was something else. I mean, the fact that it required such exquisite tuning was maybe no more bizarre than the acausal nature, in general, of the initial conditions. So I guess I didn't worry about it as a special problem and I didn't attach a name to it. But it was certainly just one of the set of worrisome things about Friedmann cosmologies.

Looking back, one thing that I don't really understand was why I was not persuaded immediately that omega had to be 1. Because it would be easy to say that the initial conditions have to be such that the energy is *zero*, and that is somehow easier to swallow than to say that the energy is epsilon. But I was not persuaded of that argument, and now I cannot go back and reconstruct why that was so. I guess I was already thoroughly persuaded that the observations fairly convincingly told me that it wasn't so, or at least that the density was smaller than the critical density—something which I'd become less and less convinced about as I grew older.

When you gave talks about the flatness problem, did you describe it as a serious problem with the standard picture?

No, I think I just described it as a curiosity.

As a curiosity?

Yes, but not necessarily not a serious problem, because again the standard theory didn't address the issue. The issue was somehow in the initial conditions, and the initial conditions were already a mystery. So it was one of the couple or three curiosities about the initial conditions that one wondered about but was not capable of *addressing*. We didn't have the tools to do anything with it. *I* certainly didn't have the tools to do anything with it, and it didn't seem to me that anyone did.

So you sort of quietly suffered with the problem.

Yes, suffered with the problem.

When you first heard about the results of de Lapparent, Huchra, and Geller on the large-scale structure—or I guess they were extensions of similar results earlier by other people—do you remember how you reacted?

My reaction actually was that they had terribly overinterpreted their data. The data were very pretty, but at about the same time, and actually somewhat before, the Cornell group [Haynes and Giovanelli] had published their 21-centimeter radio stuff on the Perseus–Pisces region, which didn't look like the Huchra–Geller stuff at all. So here was another piece of the universe not very far away that did not have this property at all, so it seemed to me that . . .

They were imagining things?

. . . that they were imagining things. And I think it's still quite possible. In the *first* place, the Cornell stuff was for a 3-dimensional volume. Margaret's and John's stuff was for a 2-dimensional *slice*, and you can easily be fooled by 2-dimensional slices. And even if they were right, it was clear already that it wasn't universal, because here was another piece of space that didn't look like that. So I thought immediately that the idea that things were on these sort of regular bubbles had to be wrong, and it was just a statistical fluke that they happened on such a volume. I certainly thought Margaret's notion that it said immediately that gravitational effects were ruled out was completely off the wall. Margaret usually doesn't say off-the-wall things, but I thought this was off the wall.

Before seeing these results, did you have any particular notions of your own about how homogeneous the large-scale structure was?

Yes, because I had sort of believed the Peebles party line that the 2-point correlation function . . .

Died off at distances larger than about 20 million light years.

Right. So one expected the universe to be pretty homogeneous on those scales, and I was *very* surprised. But I was already surprised by the Cornell results in which that didn't seem to be true. It didn't take much thought to realize that the 2-point correlation function was not a very good tool for exploring that kind of structure. And there we are. The question still remains: what scales are really relevant on large scales, and we don't know.

Has your opinion about the de Lapparent–Huchra–Geller stuff changed since your initial reaction?

I still think they overinterpret the data a bit. It's clear in their larger surveys that things are not quite so regular as they would have had you believe from the first ones. And *still* their region of space appears qualitatively rather different from the Perseus–Pisces region, which says that there are differences from one place to another, which is another way of saying that there is really large-scale structure although it's not a question of the density here being different. It's a question of the *kind* of structure here being different.

Is it true that your view about the large-scale structure has undergone a change?

Oh, yes. It certainly has. Before that stuff, one had *inklings* that there were these structures on much larger scales than you liked to admit, but it was really *that* and the Haynes–Giovanelli stuff that put the nail in that. I think Haynes and Giovanelli should share as much credit, actually, as John and Margaret.

You said that earlier you had sort of gone along with Peebles.

Oh, yes. I mean, it was clear. You're confronted with the results. That's what the universe looks like and you can't ignore it.

One of the things I'm interested in is how scientists use visual images or metaphors. Do you find that you use visual images much in your work?

Oh, quite a lot. That also is something that happens more and more as you get older and are less able to figure out Christoffel symbols, differential forms, and such—you rely on mental pictures. But I always have, to a very large extent. That's just the way I think. Unless I can make an image for something, I don't really feel that I understand it. There are *dangers* in that, because you can often make images that are not right. I think a lot of people manage to do that. It's very powerful. I have never been a very high-powered technical person. I've never been able to push indices around and such very well, and it's been a very powerful tool in the past, being able to visualize things geometrically. It's one reason that I liked the

Misner, Thorne, and Wheeler book so much because it's Johnny Wheeler's approach to the world and I find it enormously attractive. Also it's Dick Feynman's approach to the world.

Do you ever try to visualize the big bang?

I used to—still do to some extent. I have all the naive wonders about where it happened and when it happened. No matter how much you know, and know those are silly questions, they're still sort of there.

What do you picture when you think about it, or maybe you don't picture it?

Well, I doubtless picture it, but I don't know whether I can describe what I picture. I don't know. I don't think I can answer that in any reasonable way. I sort of make a movie—always going *backward*. I never go forward in time, always backward.

Let me ask you a little bit about theory versus observations. There's certainly been a lot of controversy or conflicts between the two in cosmology in the last 10 years.

It's a very complicated topic. I think that the interaction has not been either very strong or very happy. There are two problems, one in the theorists' camp and one in the observers' camp. The theorists, especially in the last few years, have had this problem that the multiplicity of ideas has become so large that in some ways it has affected the seriousness of the field. I don't know whether I can make that clearer, but there are so many ways that things could be that the fact that the observations appear to be at variance with one or the other of them is not . . . The practitioners have so many things at their fingertips that knocking down one of them doesn't make very much of an impression. In fact, it doesn't make enough impression for them to abandon *that* one, much less any of the others, right? [Both laugh.] The problem with the observations is, as always, that cosmological observations are always right at the hairy edge of the possible. Observers tend to overinterpret their observations. Theorists tend to overinterpret the observations even more.

Why do you think people tend to overinterpret?

I think it's just ordinary human zealousness. You get a result, and you *want* to believe the result, even though you know in your heart of hearts that it's a little dicey, maybe even a little more than a little. A lot of the stuff about the large-scale structure, from the "observational" point of view, I think, is not very well substantiated by the data. And yet it drives the theorists into feeding frenzy, because they have various things to say about it. So there's a great deal of Brownian motion in the field as a result: theorists trying to explain the results of less than completely exemplary observations; observers going out getting observations to test various theoretical ideas and coming back with results that, one way or the other, are not necessarily believable. It's something that will settle out, I'm sure. The *correlation* of what is really true about the universe and the set of notions that we *think* are true about the universe I think is not very high at the moment. We don't understand the dynamics of galaxies very well. We don't understand the velocities of galaxies very well. We certainly don't understand lots and lots of things that the old-timers would worry a lot about, like the effects of obscuration on a galaxy, for instance. People now apply these cookbook things to their catalogs. It makes a lot of difference when you're doing things like large-scale screening.

You think that's because the big ideas have gotten almost within reach, and so we're just reaching a little further . . .

That's right. People are reaching further without really carefully going over the ground. It's forgivable, I think. It's human. But I think that ground is going to have to be tilled a little more before we can really believe a lot of the results that come. In a way, it's unfortunate, because it must in some sense slow down real progress. It's much better to build in a systematic way, but it's *very* much more exciting to do what people are doing.

Once in a while some really unusual idea, like the inflationary universe model, will come out—which probably couldn't have been gotten to by a very slow, linear process.

I think that's absolutely right. We all know that progress comes in those two forms, and the important thing to do is to balance them in some way. I think that "galloping ahead" is the mode that we're in now, perhaps unfortunately. But it's a very exciting thing to do, so people are going to do it. You can't change the way people work.

I'm almost at the end. Are there any additional points you'd like to make?

Nothing really comes to the top of my head. I'm very worried about research in general. These are not scientific remarks, they're sort of political remarks. There has always been this question of the conflict between little science and big science. I think that our lifestyle—the way we've been doing things—is very, very much in danger, and I see almost no way around it. I think the collapse of NASA is kind of a paradigm for what's happening to science in this country in general. I *worry* about it, and I think that people should be more worried about it than they are. I haven't any idea what to do. The whole idea of success-orientation of projects, I think, has now caught up even the ground-based projects. People have these grandiose ideas to build telescopes whose technology is completely untested. They go out and they scrape up enough dollars to do it, but not enough dollars to have any contingency. I can see several disasters on the horizon in the field in general. So I think we're very healthy at the moment, but I see that we might not enjoy that health for very long.

Let me ask you to take a step back and be as speculative as you wish and put your scientific caution aside. If you could design the universe any way you wanted to, how would you do it?

I actually can't imagine a much prettier way to do it than the current notions of inflation and the idea that the universe is Infinite with a capital I in a way that we couldn't think of before. [I like the idea] that there are regions of the universe far away that are very, very different from ours, and that reality will come crashing in on us in some exponential future, and essentially every experiment has been tried somewhere in the universe, perhaps even to the extent of changing the physical laws, because the symmetry may have been broken differently in different places. That is just such an immensely appealing notion. It's *chaos,* but it's such an immensely appealing notion, that it should have been done in a way that lets everything happen the way it is without any forethought or any need for special this, that, or the other. It explains in a natural way the *seemingly* special things about our universe without imposing any overall plan on the thing. I find it philosophically just enormously satisfying. It worries a lot of people, I think, because since the horizon is the horizon, there are questions about this big universe that we can't ask in principle. That would have bothered me perhaps even a few years ago, but it doesn't bother me now.

What's changed in your thinking?

Oh, just this *notion* [of inflation]. That was unthinkable. Well, not unthinkable, it just hadn't been thought of a few years ago. We knew we couldn't see all the universe, and *my* own notion was that the universe was infinite and I had the temerity to think that the rest of the universe was just like the part we *see,* because the part we *see* is homogeneous and isotropic. And that's just absolutely crazy. It's counter to every kind of Copernican development that has come in the world, and this is kind of the ultimate Copernican idea, that not only are *we* of no conceivable consequence, but even our *universe* is of no conceivable consequence. That's very pretty.

> *Somewhere in Steve Weinberg's book is the statement that the more we learn about the universe, the less it seems that the universe has a purpose. Have you ever thought about this question of whether the universe has a purpose?*

That's a hard question. I'm not a religious person. I have never been since I was a kid. I toy with the idea that the purpose the universe might have had is the development of things that can understand it. I find that a very attractive idea, but I can stand back and realize that I find it an attractive idea because I'm a human being, and I like to think that we're important. So I've never been persuaded of its correctness. It's not, I think, at *variance* actually with these new ideas, because intelligence probably can take as many forms as the universe can take. So I think that one could entertain this notion even in a chaotic universe of the sort that is envisaged today, but certainly one is not compelled to. I'm not quite sure what my own feelings on the subject are.

JEREMIAH OSTRIKER

Jeremiah Ostriker was born April 13, 1937, in New York City. He received his B.A. degree from Harvard University in 1959, followed by a Ph.D. in astrophysics from the University of Chicago in 1964, under the direction of S. Chandrasekhar. After a postdoctoral year at Cambridge University, Ostriker joined the faculty of Princeton University, where he is now professor of astrophysical science and director of the Princeton University Observatory. In 1972 Ostriker was awarded the Warner Prize of the American Astronomical Society and, in 1980, the Russell Prize of that society. He is a member of the National Academy of Sciences. Ostriker is married to the poet and critic Alicia Ostriker.

Ostriker's work has spanned a wide range of topics in astronomy and astrophysics, including the evolution of globular clusters, high energy astrophysics, black holes and neutron stars, the origin and evolution of galaxies, and the large-scale structure of the universe. In 1973, working with James Peebles, Ostriker argued that observed rotating galaxies would fly apart unless held together by a large halo of unseen mass. This was one of the first modern predictions of the existence of dark matter. A year later, in 1974, Ostriker, Peebles and Amos Yahil compiled data from a number of different previous observations showing that the amount of cosmic mass detected by its gravitational influence was about 10 times as much as the visible mass. In 1981 Ostriker and Lennox Cowie proposed that galaxies might be formed from huge explosions that compress gas, rather than from the slow gravitational condensations of small inhomogeneities, as generally believed. Witty and quick on his feet, Ostriker is constantly erupting with new theories and often involved with half a dozen collaborations at the same time. Some people regard Ostriker's greatest strength as his ability to leap from one theory to the next, without clinging to any so much that he cannot let go.

I wanted to start with some questions about your childhood and how you got interested in science. Do you remember any particular books that interested you?

I actually can remember. My portrait was painted by Raphael Soyer when I was around 4. I remember it vividly because I remember the book my mother was reading to me. It was an elementary science book that had dinosaurs in it and things about the earth and what makes plants green. That was what she chose to read to me while I was sitting for a portrait being painted.

Do you remember the name of the book that your mother was reading you?

No, but I can see it. I know what the cover looked like. It's a strange thing. So that just reminds me that I was always interested in science. Then, I think in high school, the first astronomy books I read were those of Jeans and Eddington. What made a big impression on me was that from fairly simple physical principles, which could be understood by a high school student, you could really calculate things that were interesting. Most of the things that are written for that age now focus on personalities, because they think that's what students are interested in. But what was interesting to me was how, with basic physics that I could understand, you could see that a star had to have the mass it had. You could calculate what the galaxy weighed. You could do all kinds of things from first principles. In some cases you needed observation, but in some cases you could almost show— as Eddington attempted—that a star *had* to have a certain mass. That was very, very impressive to me.

Were there any cosmological ideas that you came against at that age?

I don't think so. Not that I can recall now. Cosmology at that time really meant something else. If you look at Jean's early book on cosmogony, it really talks about our galaxy and formation of the solar system, things like that. I don't think Hubble and the expanding universe became the subject of semipopular writing until after my formative years. So I can't think of when I first learned about all that. It was pretty late. I wasn't sure I wanted to go into astronomy until very late.

Was that in graduate school or college?

I majored first in chemistry in college, because that was what was interesting to me in high school. Then I switched to physics. I had one astronomy course in college, but it was so terrible that I had to petition to get out of it, because whoever was teaching it didn't really understand such things as why the moon didn't fly away from the earth. It was just awful. That put me off for a while. Then I went to work after college—I think partly to avoid the draft because there was a draft at that time—in solid state physics for the U.S. Government in Washington.

Before going to Chicago?

Yes. There was an interesting thing that was very influential to me at that point. There was a *Fortune* magazine article—strange to think that that was influential. I think there was a series at that time—"The Sciences." It had interviews and good pictures. There was a specific article on astrophysics. I remember Jesse Greenstein was in it and Chandra [Chandrasekhar] was in it. I don't remember right now the other people. I actually have it at home. I saved it. Chandra, in particular, made a huge impression on me—just his presence. I read his *Stellar Structure* the year between college and graduate school. It was a Dover book and it was cheap. So I thought, "Well, I'll just read it and derive everything in it." I think that was partly influential in my going to the University of Chicago. I picked that place partly on that basis.

Let me go back just a little bit to some earlier days. Do you remember thinking at all about cosmology or about the universe as a whole?

I probably did, but I don't think that there was anything very definite that I can recall. I had no serious preconceptions or even intense curiosities that I now recall. I remember when the steady state theory came out, I read about that. I read Bondi's book, probably in college. I was not persuaded by the logic [of the arguments for the steady state theory]. It might have been true, I thought. But I remember his logic was that everyone postulates a big bang and that puts the breakdown in physics all at one point in time and space, and there's nothing different about steady state; it just has a breakdown in physics at every point in space and time. I thought that was just a rhetorical trick. Of course, the steady state theory could have been true. It had a certain attractiveness.

What about it appealed to you?

Just as we're probably all brought up to think how foolish it was to have a man-centered universe, any scientist is brought up to think that anything unique about his place in space and time is probably an artifact of either his imagination or the measurements. We're bred to have a bias against anything that shows things to be man-centered. In a big bang universe, we're at a privileged point because we know we couldn't have existed in the very distant past or the very distant future. And that seems strange. So there's some attractiveness in the uniformity [of the steady state].

My own favorite models—which I've never done any work on, but, if I understand correctly, some of the chaotic inflationary models are beginning to look like—are models in which every individual universe finds itself in the big bang, but globally it's a steady state. You have universes sort of budding off.

This is Linde's stuff. [1]

Linde's, yes. I met him when he visited Princeton, and it was very interesting talking to him. He started out in philosophy. Aesthetically, that's a model I actually had long before. Of course, I didn't know the physics for it. Who did? But that is, in some sense, aesthetically the most attractive model to me. I can't tell you why.

I have not thought that cosmology should be an area where there's belief, even a very small component. So when people ask me, "Do you believe in this or that," I find it hard to understand the question. "Do you at Princeton"—they say you, the plural you—"*believe* that omega equals 1?" I say, "I don't know. Omega's a number you measure." "Don't you have any *philosophical* belief about it?" they ask. I say, "Absolutely not." I have a kind of aesthetic preference sometimes for that, but I never seriously looked into it. My own thought on cosmologies is just do the measurements and find out.

You mentioned that when you went to Harvard you didn't have much of a positive experience with astronomy. When did you decide that you wanted to go into either astronomy or physics?

I had changed effectively to a major in physics by the time I had finished Harvard. Ed Purcell was probably the best teacher I had there. He made a big impression on me, as he did on many people. The physics department there was very cold, I thought. I was pretty discouraged by the time I left. One of the things that cheered me up was in my senior year I designed and organized a poll of all the physics majors at Harvard. I asked them various questions about how they liked majoring in physics at Harvard. The first fact was the tremendous fall-off in number from the sophomore year onwards. The second thing that came across was how they had gotten discouraged with themselves and the subject and had become demoralized. People were going in very optimistic about how exciting it all was. The third thing was, to a person—I think it may have been to a man, I'm not sure if there were any women—they thought it was *their* fault. No one questioned the system and thought there might be something wrong with the way they were being taught or the social organization of things. When you see all the data in front of you, you realize it can't be all the separate individual things about why this one didn't study hard enough, and that one had trouble with his girlfriend and so on. You have to look and see whether it's something in general. It actually bucked me up a little bit.

I think it was the year in between college and graduate school that I decided to do astrophysics. I remember thinking—it was one thought that was extremely definite that I had at that time—"Well, I like mathematics and physics, and I'm going to end up making mathematical models for things because I'm pretty good at it, it's fun, and there's some demand for it. Twenty years later, I might be telling Colgate Palmolive how to put the toothpaste in the toothpaste tube most efficiently by mathematically modeling it, or I might be looking at the interior of a star or a galaxy. The equations may be the same, and the work might be intellectually stimulating in exactly the same ways—to solve those equations and figure out how to do these things best. But one's more interesting than the other. If I'm going to do the same kind of work I might as well do it in an area where the application is to something which has a little glamour for me."

Do you remember why astronomy had a particular glamour for you at that time?

It's something that's so obvious I can't answer.

Was it more glamorous, say, than particle physics?

I don't think I considered particle physics. Astronomy was sort of outside of us and bigger and away from our messy lives. I did a lot of reading in other areas. I probably spent more time on literature than I spent on science. I married a poet. The fact that, as an astrophysicist, you get a perspective on humankind that is . . .

Broader than the grocery store?

Yes. Sweating on this little grain of spinning sand. That appealed to me. I'd always liked writers who took a cosmic perspective. Montaigne, who liked to look at the Europeans from the point of view of the Chinese. So the idea of having a broad perspective was very attractive to me. It still is. Now that argued maybe against biology and human sciences. It didn't particularly argue against particle physics.

You've worked in many areas of astronomy, but do you remember when cosmology in particular attracted you?

I've kind of backed into working in that area. I worked in a lot of other areas first, and still work in some other areas. There was no decisive moment. I think it was a time of opportunity, when it looked like there were a lot of interesting problems. I can give you a very specific example that makes the point clearly. All the classical cosmology was done with the point of view of getting two numbers, omega and H_0 [the Hubble constant]. You need standard candles and so forth. If you found the standard candles changed, you had to make corrections in the cosmological calculations. There were huge programs everywhere studying all of these things. Beatrice Tinsley made the very simple contribution of pointing out that as galaxies age, they become fainter. If you don't take that into account, you get the wrong value for H_0 and omega. It could be seriously wrong. She gave a lecture here at the Institute [Institute for Advanced Study in Princeton], and she demonstrated, I thought quite conclusively, that galaxies would change, would get fainter.[2] In an elliptical galaxy, if there are no young stars being formed—and we don't see young stars in them—then just at the rate the old stars die, the stellar population will change a few percent per billion years and get fainter. If you don't take this into account and you use elliptical galaxies as standard candles, which is what everybody had done, you get the wrong result. She got an omega of something like 2, I think, by just taking the result of Allan Sandage and making this correction.

I remember thinking to myself, and I said this at this talk, "I'm not a cosmologist, but I've worked on dynamics of galaxies. I bet I can think of several dynamical effects that can make the galaxies become brighter by that much—which isn't to say that your correction isn't correct, but there are probably others as well, which will have the opposite signs. If this one changes it in one way, they will probably change it in the other." And then I worked on it, with Scott Tremaine, who was my graduate student at that time. We worked out dynamical friction of satellites spiraling into galaxies—the fact that the galaxy M32 was falling into M31 and the Magellenic clouds into our own Galaxy. You can calculate the rate at which galaxies will get brighter by their smaller companions falling into them all the time. Typically, it's a few percent per billion years and it goes the other way [from the effect that Tinsley calculated]. That's an example of how I got into it. I thought, gee whiz, if this is going to change the universe from being open to closed, I could think of something else. I think we titled the paper "Another evolutionary correction" because the presumption is that there are still others.[3] But she had made the point. This wasn't criticism of her. She had made the point that the standard candles did change and you better pay attention. So this was more along the same line.

> *Going back to the early 1970s, do you remember having a preference for any particular cosmological model, such as open or closed or homogeneous or inhomogeneous?*

No. Scientists have followed their own biases, and my principle bias at the time was being contemptuous and intolerant of all of these people who *had* specific models. How could they be so certain when the evidence was as confusing and inconclusive?

> *Theoretical as well as observational?*

Yes. Only a small class of models was being examined. If you look historically, almost all of the models at any given time that people have are *wrong*. So there's no particular reason why they shouldn't be at this time, and why should scientists be so *stupid* as to not realize this? I had made myself obnoxious in a small way by making these remarks. I still do that to some extent. There was a meeting on large-scale structure, and in my summary talk I said: "Well, on the basis of history, I think it's extremely probable that most of the models we're currently discussing are wrong. What one wants is discriminants that talk about broad *categories* of models, rather than attacking or defending this particular little one."

> *You think that's a mistake we're making now, focusing on narrow models rather than talking about broad categories?*

Well, if I can be slightly arrogant on this, I think science progresses because you focus on *definite* issues. You take the most predictable models and you work out all the details. You either falsify them, or they're satisfactory and then you go on. So it does pay to do this. But, to some extent, it's something the troops do. It troubles me when I see people who I think should be slightly above the fray and should be thinking about the broader issues . . .

Going back in the trenches.

Yes. There's a belief—sometimes it seems to be the simplest way—the Manichaean logical fallacy of dividing things into *twos*. It seems to be easier for us and so you get to the point where you think if you can think of an argument against someone else's, then it's an argument *for* yours. Whereas, in fact, the truth is probably off in some other dimension [laughs] and neither one of you are in . . .

In good shape.

Yes. So I don't think I had any particular preference for any particular models early on. And I don't now. I've thought then and I think now that there are missing ingredients, essentially. There's no way of making the puzzle if you don't have all the pieces. So insofar as there's a real chance that some previously unexamined possibility from particle physics, or something that we haven't even thought of yet, is *the* critical element, there's no chance that any of what we're doing now is right. Of course, it could be that we now *have* assembled all the pieces. The opposite fallacy is the way the Greeks did it, where everything is always up in the air, and you're always at liberty to invent new hypotheses, throw out old ones, and not work out anything in detail really. So that's the argument for focusing on fairly narrow models. It's the way science has progressed in the past. By and large, I think that's the proper way of proceeding. Rather than just invent new physics all the time, see where the old physics gets you. But I just think it's gone almost too far in some cases of cosmology, where people are agonizing over the minutiae of a very small fraction of the possible models where they have been casting their net.

Have your own ideas about cosmology changed any in the last 10 years?

I wrote a paper with James Peebles on the size and mass of the universe, where we said omega was 0.1.[4] That still seems to be all right. At that time, it was considered very high because we were using massive halos to get it up to 0.1, which was about a factor of 5 higher than people had gotten previously. I still haven't seen any numbers for omega larger than that which are at all convincing. So in the question of how much mass there is in the universe, my ideas haven't changed very much.

Would you say that, based on this, you think that we have an open universe, or do you not go that far?

I don't really go that far, because I can think of ways of separating out the baryons from the dark matter.

When you say "omega," do you mean just the baryonic omega [the contribution to omega from baryons only], or are you talking about the total omega?

Let me put it this way. Dark matter is used in two opposite senses, which is extremely confusing. It's used for matter that we know is there by measuring it dynamically. So in this case, we know there's more than we can see. Dark matter is also used in the opposite sense, in which they say omega is 1, and the difference between omega being 0.1 and 1.0 is dark matter. That's matter which we've *no* evidence for.

When you first talked about your hydrodynamic model of galaxy formation, do you remember how people in the community reacted to it?

Well, the first and most significant thing that I think I've done in this field is the whole dark matter business and massive halos. That to me was more shocking.

OK, let's go back to that.

On the massive halos, there were two papers, which, first of all, were confused in people's minds, although I thought they were fairly clear. One with Peebles said that the picture we have of the galaxy—*inside* the galaxy—must be wrong, and the spherical component must be a larger fraction of the total.[5] It didn't change the mass of the galaxy at all. It just said that, if the galaxy's going to be stable, it can't be the flat disk that we "see" in other galaxies like NGC 4565. You have to have some significant fraction of the matter hot [in a halo]. As far as I know, I don't think anything major has happened to change that overall conclusion. All the models now have that. The other thing was to just look with what seemed to be a cool eye, in a nontheoretical way, at the accumulated observations of masses of galaxies.[6] By various determinations, they really seemed to indicate that there's a lot of mass at large distances from the galaxies. Fritz Zwicky had said the same thing for clusters of galaxies.[7] But, at that point, the idea seemed to be true on the larger scale. Then, on the smaller scale, galaxies had a well-defined mass and a Keplerian rotation curve. What we did in our paper was to show that as you go to larger and larger scales in between these two scales, the mass just keeps on increasing more or less linearly proportional to the radius—until you get up to mass–light ratios for galaxies which are comparable to what we find in the clusters—which indicated that the total extent of radius and mass of galaxies was probably 10 times the visible one.

And that surprised you?

It surprised me. But it just seemed to me what the data said. It wasn't a theoretical argument. I had no axe to grind. The first thing was that people reacted to that with *very* strong hostility. I couldn't see particularly why. It was just a fact. The other thing is that there was what seemed to me willful confusion. For example, to mix up the two papers and say, "But it doesn't show anything. It just shows that more of the mass is in the spherical part rather than in the flat part." Or for people on the other part to say that Ostriker and Peebles have shown that the stability of galaxies requires that they be much more massive, which we didn't show at all. So both the proponents and the opponents seemed to just not *understand* it. And the level of hostility was, I thought, fairly extreme. Not in my department. In fact, Martin Schwarzschild showed me that he had a rotation curve for M31 [the Andromeda galaxy] 20 years before and argued the same way that the mass is much more extended than the light, which was really a pioneering paper on that.[8] But in some sectors of the astronomical community, I think it's fair to say that it was greeted with hostility.

You didn't have any ideas or any opinion about why people were hostile to this?

I think it was partly the reaction to new things. In some cases it was people who were sort of sorry they hadn't done it themselves, I think.

When was this in time relative to Vera Rubin's work?

Oh, it was way before. It was at least 5 years before, I think. Vera, in her early paper, said she was confirming what we had done.[9] Vera's work provides the best evidence for it. But what we had used were binary galaxies, rotation curves of galaxies, satellites of galaxies—a whole slew of different things. I think it was just the idea that if you're told that everything you've been doing is only . . .

Ten percent of what's there?

Ten percent of what's there. Then I found the reaction changed. So it was a kind of a sea change from reaction with hostility to having forgotten and denying that we did it. I kept on running into people who would explain to me that I was probably too old to realize that things had changed and that galaxies had massive halos and things like that. Students would tell me this [laughs]. I think that was my first subject in cosmology where I ran into people's *very* strong feelings.

I remember there was one meeting where Agris Kalnajs showed some rotation curves for galaxies and showed that you could fit them with just the matter that you could see.[10] I think he had found 4 galaxies. But everyone came back from that meeting and said "See. We don't need massive halos. There's no more dark matter. At last we can get rid of it. It's all wrong." And there was a big sigh of relief. I think he never published it, but I remember there was a mini-flap, so that if you follow the sociology of these things . . .

On the basis of one report, unpublished, people were willing to . . .

Yes. For a period. It astonished me. I said, "But this is just 4 galaxies. He hasn't published the data on which it's based. What did he really say?" But apparently he gave a very convincing talk. And clearly people wanted to hear it, at that time. So that was one example. I think now it's fairly much accepted that there's a good deal of dark matter in the outer parts of galaxies.

On the explosions [Ostriker's hydrodynamic model of galaxy formation], the reaction varied.[11] Yakov Zel'dovich was extremely positive. Or rather, it was mixed. I heard from someone who visited—I can't remember if it was Rashid Sunyaev—who said that Ostriker has killed our pancake model. I was pleased that he had taken it so seriously. I didn't think that I'd killed anything because I had and still do see it [explosive galaxy formation] as an amplifier, essentially, even if things started some other way. And the origin of those perturbations could have been dark matter pancakes. But he at least took it very, very seriously. I think the prevailing view in the United States, in the West, has been benign neglect. "Oh, he's just carrying on."

So to some extent I've worked in the wilderness on this subject for some time. Fortunately, I do enough other things as well. I thought, "Well, it's either right or wrong, and time will tell." I'm not in an insecure position, so I can afford to do it. I tried to be wary of a phenomenon I notice in other people's careers. There are many people who do a number of right things. They do something, it's right, people recognize it, and they're applauded and they applaud themselves, and they go on and do something else right. They go on until they do something wrong. Then, when they do something that's wrong—and, of course, in any career people do things that are right and wrong—people object to it. Being so clever, they can defend themselves against these objections. But they're now defending something that is wrong. You forget everything else they've done because they've spent the last third of their career defending an increasingly untenable position. It's a scientific equivalent of the Peter Principle, but it's not their incompetence but their competence that does them in. I think with Robert Dicke and the Brans–Dicke theory and Fred Hoyle and the steady state theory, that really happened. In the eyes of many people, their previous accomplishments are overshadowed. So with these cautionary tales in front of me, I've always made a policy of not defending against attacks. Just go on, do something else. I've wondered on this explosive thing whether I'm doing the same thing. I think as long as I continue working out the aspects of it, rather than defending against attacks . . . And if I see something wrong with it, I'll just drop it because I don't feel that I have that much invested in it. For the present, it keeps on showing up surprising things that are attractive. We certainly did predict that things would be on bubbles.

Let me ask you about the bubbles. When you first heard about the work of de Lapparent, Huchra, and Geller, do you remember how you reacted to that?

I thought, "Oh, that's nice." That was it. It was not conclusive because in fact you can get things like what they observed in other ways. But it's certainly good news rather than bad news. At least it's consistent with the type of things I had been working on. I have noticed that the approach to the explosive amplification picture has changed from completely ignoring it to putting it in a footnote in papers to considering it among the possible viable alternatives. And that's as much as I would say.

Do you remember when you first heard about the flatness problem and how you thought about it?

Now, let's see, that's the problem that if omega is 0.1 now, then it was very, very close to 1 for a very long time. I guess it's one of those things that I thought I always knew, but I couldn't have always known. Since I'm here and Peebles and Dicke are here, they probably mentioned this to me long before they wrote it up in 1979. It's conceivable that Jim Gunn told it to me as well, or that I thought it up for myself. As far back as I can remember, I've known this argument. The most likely thing is Peebles mentioned it to me sometime, and I said, "Oh yes, I guess so."

When you first began thinking about the flatness problem, did you take it as a serious argument, in the sense that a physical explanation is required, as opposed to random initial conditions that produce an omega close to one.

It didn't demand an explanation. I thought it was a good argument, but the idea of extrapolating our present universe back to one second is not all that convincing to me. I thought then, and I think now, that we can extrapolate it back to a few minutes because the light element nucleosynthesis in the big bang works out all right.

Even if you go back to a few minutes you're going to find a very delicate balance.

Sure, but it [the deviation of omega from 1] isn't a smaller parameter than lots of other small parameters that you see around. What I remember thinking—I once worked it out—is back to nucleosynthesis, which is as far back as I think our extrapolations are *tested*, it's a small number but it's not unbelievably small.

Let me give you another example from cosmology just to make the point. The fundamental thing people believe—and people have tried since Einstein—is that they can calculate *everything* from G [the gravitational constant], h [Planck's constant], and c [the speed of light]. The way I think about that is this. On one side of the equation you have the electron mass, which you're going to calculate, but it's not a primitive number. On the other side you have the mass unit that comes out of the Planck mass. In between you have a factor of 10^{-22}. So you have to calculate then, by pure thought, a dimensionless number, which is 10^{-22}. It's not zero. It's 10^{-22}. I think if you're going to have a fundamental theory that involves only h, G, and c and are going to calculate the mass of the electron, then we absolutely assert that nature can contrive a dimensionless number of 10^{-22}. I've talked to Ed Witten about this. He says that kind of thing happens, you have e [the exponential constant] to minus a big number. So if that's the case, on something as fundamental as that . . .

You see no reason why nature couldn't have contrived to form the universe with a very fine balance between kinetic and gravitational energy?

Yes. I think you can invent inflationary models that do exactly this. I didn't find the fact that a number is small, therefore it has to be zero, an absolutely compelling argument, because I didn't see why we couldn't have a small number. It may be difficult for me to imagine calculating it, but since I can see that we're trying to do that in other domains, this didn't seem to be any worse.

Some people react to this argument by saying that it doesn't require a calculation at all, that the universe is as it is. We go out and measure omega and don't ask questions about why it is as it is.

I didn't take that view. For example, I argued with Peebles for years on the horizon problem on that.

Did the horizon problem bother you?

It always bothered me.

OK. So this bothered you in the same way then?

Yes, but I assumed that a physical theory would provide an answer. If you have some very special conditions, I would presume that they are determined by a well-defined physical theory, rather than just being a coincidence. That coincidence strikes me as being very unaesthetic. I've always assumed that omega is determined by a physical theory, not by "initial" conditions.

How did you react to the inflationary universe model when you first heard about it?

I thought it was a clever idea.

Why do you think it caught on so widely and so quickly?

I'm not competent to say. I think people *like* things that come from particle physics. And there were real problems that it addressed. The other thing is, philosophically I think, we don't like singularities. In physics we're taught not to like singularities. If you get a solution that has a singularity, you throw it out. You're taught that in college physics. You keep the solution without the singularity. So therefore, insofar as inflation addressed the singularity that was built into big bang theories and made a little more complicated and plausible picture of it, it was attractive. Also with the flatness and the horizon problems, it was addressing questions that in the standard cosmologies were just left hanging there in a bit of an embarrassing way.

One of the things I'm interested in is how astrophysicists or physicists use visual images in their work. Do you use visual images a lot?

Yes. There was a volume in honor of Stan Ulam, who's really a mathematician. He commented that physicists always think first in pictures. I think there's nothing I've ever done where I don't think that way.

You think that helps you work on problems?

I wouldn't even make it as weak as that. It's essential. That's what I do. In fact, in every single thing I think I've done, I can first just close my eyes and think about it until I can sort of *see* what happens, and then I can work out the equations later. It won't always prove that I was right. But then I work out the equations and see whether it comes out. I think that's just the way my mind works. I visualize things first.

Let me ask you to take a big step back and maybe think a little more speculatively than you have. If you could design the universe any way that you wanted to, how would you do it?

Well I like Linde's ideas, insofar as I understand them—where you have lots of firecrackers going off.[12] It's globally steady state, but locally everyone finds themselves in a big bang universe. I don't know why that is aesthetically appealing, but it always is.

So you would make something like that?

Yes. I like things that are hierarchical also. It's less boring. The steady state always seemed to me an extremely boring universe.

So Linde's model has some of the attractive features of the steady state but . . .

But doesn't have the boring features. It's like, would you rather live forever or go through cycles and be different people at different times and have many different lives? I think most people would say that they'd rather have many lives than live as the same person forever.

Someone mentioned to me that you made a comment along the lines that theorists who have a broad background in the humanities and literature have a deeper reservoir from which to produce new ideas. Could you elaborate on that?

I didn't say exactly that. I said something that's close to it and that has struck me as something very interesting. Where do scientific ideas come from? It struck me as uncanny how many ideas that we've had on a mythological literary basis have turned out to be right. In some crude sense, right. An example is the Lucretian idea of atoms. It is the idea that we're still with to this day. You can ask what evidence did he have for it. None. I can't think of any evidence. It's turned out to be an extremely useful idea—just way beyond what anyone would have thought, even though we've had to go on to further and further levels, and we've never found the ultimate particle. Another example could be that Judeo-Christians—and it goes back to the Babylonians—have a big bang cosmology. We can think of many examples, including literary ideas, that existed for black holes in the past. Now there are many different possible explanations for [this unexpected scientific correctness of mythological thought]. One is the Nostradamus effect. When things come out right, you notice it. So you pick out the ones that are right. But my argument against that is that the a priori probability of any completely ad hoc idea being at all correct is so tiny that *even* if we notice them, it's amazing to me that *anything* the Greeks said on the basis of no evidence would be such a powerful idea. So the Nostradamus fallacy is one explanation, but I don't think that it is sufficient.

Another possible interpretation is Fred Hoyle's panspermia idea that there was a race of very intelligent beings, and they understood all of these things, and we come from them. So it's built into our genes—all the answers to these things—because they were previously known. I think in the past you could have said that was probably absurd, because Jungian memory is impossible, but we now know that animals do have genes, and DNA can really control a lot. I just don't think this is too plausible. So let me ignore that one. Another one, which the deconstructionist approach would take, is it's all just metaphor. There's no "reality" there anyway. We just pick and choose among these things on a basis that's more or less fashion. Lucretius isn't any more correct than Aristotle. The atomic theory of matter isn't any more correct than the phlegm theory. But the current state of science adopts this model or that model. This picture is, I think, a false reading of Kuhn. I reject that. I'm a realist. I think you do the experiment and the needle points. But that's a possible interpretation.

Now, having rejected all of those explanations, I may not come up with the right one, but here's the one that I've thought of. It's an appealing one, and it's very open-ended to me. We have a store of images, which I don't think are unique to a civilized, literate person. Probably the literate person has more of them than the less literate person. But we absorb them somehow or other, and they provide the basis very often for the models that we're willing to explore and entertain. Some of them we can falsify, and some of them survive. Now what that means implicitly is that, had we come from another cultural tradition that had other models in it—not that we would disprove things that we now know to be true—but, rather, there are other phenomena, which we're not even investigating now, but that are quite important, which we would then be able to address. Because you just don't even address things if you don't have the mental pictures for them. And as our cultural experiences becomes broader, as more different cultures are brought essentially into the world of science . . .

You get a larger store of these pictures.

Right. And we may be able to address problems that we've just ignored. I'm continually struck in the history of science by things that are *perfectly* obvious, but that couldn't be seen. You know the famous example of the Crab Nebula. It wasn't seen anywhere in Europe. The monks have no records of it. Many other cultures, including American Indians, Chinese, etc., have them. That was because for the Europeans, there were "fixed stars." My guess is that if you have an idea of fixed stars and you see something that isn't, you just assume, "Well, that's some atmospheric phenomenon." You ignore it. The monk doesn't put it in his day book. And so you can't study things because in some way you reject the reality.

So when people ask about such things as extrasensory perception, I always think that there are things that we think are part of our senses now that would have been extrasensory perception 50 years ago. One example is that they now know that people can tell when other people are looking at them, in part. I put a screen between you and me, and I turn my head this way and this way and this way, and someone tests whether you can tell. First of all, it's statistically significant that you often can, more than random. Second of all, it depends on how far away we are, which is good because most of the false effects don't scale with distance properly. Now I think there's a simple explanation. We all give off a lot more infrared from the front of our face than from the sides, and we have infrared detectors, because we can tell if we turn a light bulb on with our eyes closed. When the person turns their head, then they think that others are looking at them, and if we use aluminum foil rather than silk . . .

So there's a physical explanation for this.

Which is infrared. But people didn't know about infrared. So they could have done the same experiment, proved it the same way, and it would have been an "extrasensory perception" detection 50 years ago. My guess is that there are a lot of phenomena that, if we don't have the mental apparatus to notice—like new stars or the like—we don't see.

I'm astonished looking back over the history of astronomy, because this happened in *my* time. Probably the most significant thing in cosmology that happened in my time you didn't mention. That is, if you look at cosmology from the 1920s to the 1950s, what everyone studied was the large-scale structure of the universe—omega and H_0, the Friedmann models. And the structure within it—galaxies, clusters of galaxies—was interesting only insofar as the objects could be used as standard candles or as standard meter sticks. And no one *asked,* where did these things come from? You can't find papers on it. It only came up really—because this really did happen in my scientific lifetime—when people were using these as standard candles. They said, "Look, galaxies weren't there at a z of 1000, and they are there now. So therefore they were made in between now and then. If they were made in between now and then, they were changing during this interval. If they were changing within this interval, then they're not good standard candles." But then you say, "My God, forget about the standard candles. Where did these things come from?" And, thus, the whole thrust of cosmology has changed to understand the origin of stars, of galaxies, clusters of galaxies and large-scale structure. There's much more work done on that now than in learning H_0 and q_0.

Could that be because it's a harder problem? With H_0 and q_0, at least you could set up the problem, as Einstein and other people did.

Well, it's not as if we've solved the one before . . . I think it's amazing that for 40 years, no one asked those questions, which have got to be obvious questions. That things were uniform early on, how did structure develop?

I'm sure you could find smart people who worried about galaxy formation in the past, but there was very little work. So that's an example of a blind spot. My guess is that there are many glaring blind spots now, which will be glaring in the future—things in astrophysics and other areas where people will ask why didn't they examine so and so, when the data was all around them and they weren't looking for it. I think in some cases, at least, it's because we don't have the equivalent of a Lucretius. We haven't been set up for it. Where we have, the ideas have conditioned us for the kinds of questions we're going to ask and the kinds of models we're going to apply.

There's some place in Weinberg's book where he makes the statement that the more we learn about the universe, the less it seems that the universe has a purpose. Have you ever thought about this question of whether the universe has a purpose?

Yes, and I find myself mystified by the question. If we go back to what Lucretius said the purpose of the universe was . . .

He didn't talk about a purpose. In fact, the whole point of the atomistic picture was to rid us of the vagaries of the gods.

That's the point. It's not something new. I think those who want to make much of *us*—the forked radish, the bipedal ape, etc.—and want just to find *us* writ large, and somehow or other find a significance in human terms, are always struggling to find . . .

To find a purpose?

Yes. But I've always thought the enterprise was essentially quixotic.

VERA RUBIN

Vera Rubin was born on July 23, 1928, and was raised in the Washington, D.C., area. Her degrees are all in astronomy. She received a B.A. from Vassar College in 1948, an M.A. from Cornell University in 1951, and a Ph.D. in 1954 from Georgetown University. Her master's thesis examined evidence for bulk rotation in the universe, a thesis that helped stimulate Gérard de Vaucouleurs to claim evidence for the "local supercluster" of galaxies. In her doctoral work, which was advised by George Gamow, she conducted one of the early investigations of clustering of galaxies. Since 1965 Rubin has been a staff member in the Department of Terrestrial Magnetism of the Carnegie Institution of Washington. She is a member of the National Academy of Sciences.

Beginning in 1951 and picking up again in the 1970s, Rubin was one of the first to document the peculiar motions of galaxies not following the universal expansion of the big bang theory. In 1978, during detailed investigations of the rotation of galaxies, Rubin and her colleagues found persuasive evidence for the existence of dark matter in galaxies, which confirmed earlier theoretical work of Ostriker and Peebles. Throughout her career, Rubin's scientific work has centered on understanding galaxies, both as individual entities and in groups.

Rubin has served on several committees on the status of women in astronomy and physics and is presently a member of the NAS committee on human rights.

She has recounted that she integrated the Palomar Observatory in 1965 by becoming the first woman permitted to observe there. She is married to the physicist Robert Rubin, and they have four children, all of whom have doctorates in scientific fields.

Could you tell me about any influential experiences you had as a child, particularly anything that got you interested in science?

My father was an electrical engineer. He's presently 92 and still could be holding down a job. He had a very analytical way of looking at things, and I enjoyed that very much. I think that was a large influence. My childhood bedroom—if childhood could be about 10 years old—had a bed which was under windows that faced north. At about age 10, while lying in bed, I started watching the stars just move through the night. By about age 12, I would prefer to stay up and watch the stars than go to sleep. I started learning, going to the library and reading. But it was initially just watching the stars from my bedroom that I really did. There was just nothing as interesting in my life as watching the stars every night. I found it a remarkable thing. You could tell time by the stars. I could see meteors. My parents were very, very supportive, except that they didn't like me to stay up all night.

They knew you had an interest in this?

They knew I had an interest in this. A few years ago, I met a friend of my mother's whom I do not know at all and hadn't seen for 50 years. She said to me, "The last time I saw you, I was picking your mother up at your house. As your mother went out the door, she yelled up the stairs, 'Vera, don't spend the whole night with your head out the window!'" So yes, they knew I was doing this. But the net result was that when there were meteor showers and things like that, I would not put the light on. Throughout the night, I would memorize where each one went so that in the morning I could make a map of all their trails. I really don't think my parents would have protested as much as I then thought they would.

Did you get encouragement from your mother as well as your father?

Yes, from both of them. My father helped me build a telescope, which was really a total flop, but was sort of fun. I ordered a lens from Edmund's and got a cardboard tube that linoleum came rolled on. We were living in

Washington, in the city, and you could see the stars during the night in Washington in the early forties. So I went downtown on a bus and got a linoleum tube, which I brought home and turned into a little telescope. I tried to take some pictures, but none of it worked because the telescope didn't track.

> *Do you remember any books that you read that had an impact on you at this age?*

I read a lot of books, such as James Jeans's book, *The Universe around Us* and Eddington's early books. But I was already hooked. It really came from the sky. In the late 1930s, I remember, there was an alignment of 5 planets. That impressed me. I didn't realize at that time how likely such a thing was. Then there were several auroral displays. It was those things that really captured my interest. It was the visual experience more than what I read in books.

> *Did you come across the concept of the big bang or the birth of the universe in your reading?*

That's hard for me to remember. I probably did, although it may have been a little later. George Gamow was in Washington. Ralph Alpher was writing his thesis under him. I remember going to a talk by Ralph Alpher on his thesis work.[1] That was probably early in college. So I knew those ideas at that time.

> *Did you know when you were in high school that you wanted to go into astronomy, or was that later on?*

Yes, by high school, I knew I wanted to be an astronomer. I didn't know a single astronomer, but I just knew that was what I wanted to do.

> *Did you have a sense of that as a career possibility?*

Yes, I knew about Maria Mitchell, probably from some children's book. I knew that she had taught at Vassar. So I knew there was *a* school where women could study astronomy. So, yes, it never occurred to me that I *couldn't* be an astronomer.

> *Is that why you went to Vassar, because of Maria Mitchell?*

Yes. That and a lot of other reasons. I needed a scholarship, and they gave me one. I didn't apply to many colleges. There were not an enormous number of colleges where a woman could study astronomy. But I knew about Vassar because of her. This is off the subject, but one of the things that I have attempted to do in my life, and I clearly haven't succeeded, is to make the story of Maria Mitchell as well known as the story of Benjamin Franklin. She really is a great heritage of the American scientific scene. When Vassar was founded in 1865, she was the first professor of astronomy. She was already well known from doing astronomy.

Who was influential for you when you were at Vassar?

Well, Maud Makemson was the director of the observatory. I learned a lot of important, fundamental astronomy, but even there the idea of being an astronomer was not especially encouraged. The feeling was that there were very few observatories and very few astronomers needed. I wasn't discouraged at all, but I can't say that I was overwhelmingly encouraged. There was a lot of encouragement there for science in general. There was a lot of physics and mathematics. But astronomy, even there, was a tiny department.

Tell me a little bit about your stay at Cornell.

I got my master's at Cornell. I got married when I graduated from Vassar and entered Cornell. Actually, I had been accepted by Harvard. I have a letter somewhere from Donald Menzel saying, "Damn you women," handwritten across the bottom. This was in response to a letter I wrote saying that I wished to withdraw because I was getting married and going to Cornell. He scribbled across this very formal letter, thanking me for letting him know, something like "Damn you women. Every time I get a good one ready, she goes off and gets married." My husband was already in graduate school and getting his degree under Peter Debye. I think it's fair to say that, although he has probably been my strongest supporter, we never even considered doing anything else but my joining him at Cornell. So I entered the Cornell graduate school in 1948 and got a master's while he was completing his Ph.D. That was even *less* of a department than Vassar. It almost didn't exist. There was one man. He had been a navy navigator during the war, and he was not very sympathetic toward anyone being in astronomy. He actually called me in when I arrived and told me to go find something else to study. He said that they didn't need astronomers, and I wouldn't get a job and so forth. There was a second person in the department, Martha Stahr [now M. S. Carpenter]. She may have been the only woman faculty member at Cornell at that time. The place was almost exclusively male. I have since read in Margaret Rossiter's book that at that time there was one assistant professor, so I presume it was she.[2]

Thomas Gold and Edwin Salpeter came somewhat later?

That whole development in astronomy came later. Cornell was a very exciting place because of the physics. Philip Morrison, Richard Feynman, and Hans Bethe were there and I studied under all of them. Feynman was on my committee. He took off for Caltech about a week before my master's orals, and I added Morrison to my committee. The astronomy department really was an undergraduate teaching department. But Martha Stahr had come from Berkeley and knew galaxy dynamics. I learned a lot of galaxy dynamics from her. It was in picking a master's thesis that I chose to work on the question of whether there were really large-scale motions of galaxies as distinct from the conventional Hubble expansion.

Let me ask you about your Ph.D. work at Georgetown.

It's very much related to the master's work, because when my husband finished at Cornell, he took a job with the Applied Physics Lab in Washington. He shared an office with Ralph Alpher. It was through that contact that George Gamow, who was then a professor at George Washington University, heard about my master's work and called me and talked to me about it. It was through those discussions that I ultimately wrote my thesis under him.

Maybe I should go back and ask you some questions about your master's work. You worked on the rotations of nearby galaxies and on a search for a universal rotation?

I did not do the internal rotations of galaxies. At the time, there were something like 108 or 109 galaxies with known radial velocities. That's a number that seems incredibly small to us today. I just asked the question whether, once the expansion had been taken out for those 108 galaxies, there was a large-scale systematic residual motion. The only large-scale motions I knew about were rotations, because I knew the galactic dynamics. So I just attempted to take that whole formalism and apply it to the motions of the galaxies. But at that time, I had done no work on internal motions.

Did you have any cosmological motivations for working on the bulk rotation of systems of galaxies? What really got you into that?

The only motivation that I can point to is just plain old curiosity. That really has motivated an enormous amount of my work. In retrospect, I think part of it was that since I came from such a nontraditional background, I didn't know what everyone would have said, such as, there weren't enough galaxies and the velocities probably weren't good enough. I used magnitudes to get distances, and the magnitudes probably weren't good enough. But I think that because my background had been so nontraditional, it was just a question that seemed worth answering.

Why do you say that your background was nontraditional?

Well, I had three years of Vassar, a couple years at Cornell, and I probably had not met a single person that you would call an astronomer. By nontraditional, I mean it wasn't Harvard and it wasn't Princeton and it wasn't California, where working astronomers and really great people did their work and taught their students, and therefore, the students probably tended to come out in a more traditional mode. Even if it hadn't been taught in that way, they had absorbed certain attitudes. They knew what people were working on. I just hadn't been through that. So I call it, in that sense, a nontraditional background. And truly, even to the present day, I really don't consider myself a cosmologist. I'm just curious about these things and try to find answers.

You presented this master's thesis work at a meeting of the American Astronomical Society in 1950. Can you tell me how that was received?

I could write a book. There's more than you want to know. I had a child, our first son, who was born the 28th of November. We didn't have a car. December in Ithaca is very snowy. My parents came up from Washington and drove us to the meeting in Philadelphia. This was their first grand-child. My father has since said that he aged 20 years on that drive through the snow with his first grandchild. I think it's a correct statement that I did not know one person at the meeting. I was nursing a child. I was not a member of the AAS. I walked in that morning, gave my talk, and left. That was the extent of it. So that's the background of giving this paper. I put this in as nontraditional. I didn't really know what I was getting into. It was a 10-minute talk, which was not an awful lot. I submitted it under the title "Rotation of the Universe." Well, I know what I would do now if a 21- or 22-year-old, someone who you had never heard from, submitted a paper called "Rotation of the Universe." I've reread that paper in the last couple of years. The abstract appeared in *Astronomical Journal*.[3] In it, I say what I did and it's perfectly respectable.

This was based on your own observations?

No. I did no observing. It was *impossible* for a normal person to get observations. Let me back up a minute and say that while I was working on this, it was known that Milton Humason had a whole set of radial velocities, and Martha Stahr wrote to him asking if these could be made available. I think the letter came back saying that they would be published very soon, and I would have them. They were published in the Humason, Mayall, and Sandage paper in 1955 or 1956, so that was 6 years later.[4] There was also talk from Princeton that Kurt Gödel was working on rotating universes, and therefore I should wait. So from both the observers and the theorists that were in contact, there were reasons why I shouldn't do this. You had to have access to one of the largest telescopes to get radial velocities. So this was just data analysis. But to get back to this abstract. At the end of it, I actually evaluated terms like the radius of the universe, the shear, the energy—things that are only a mild embarrassment now. But the thesis itself, I think, is a respectable piece of work. Once a year or so someone like Jim Peebles asks for a copy. I have xeroxed it half a dozen times. It's totally out of date, because the data were not really good enough to do it. So I came in and gave this talk. Then a variety of people got up and made their comments, none of whom I could identify because I didn't know who they were.

What were the comments like, do you recall?

All but one of the comments were of the form that you just couldn't do this for a variety of reasons, comments that I think I have probably hidden in the back of my mind. Martin Schwarzschild got up, as he does; he's so nice to very young students. He said all of the things that one would like said— that this is an interesting thing to do, that the data probably are not good enough, but that it was an interesting idea for a first step. So he made me feel somewhat less than mashed to the ground. Then when it was over, Dirk Brouwer, who was the editor of the *Astronomical Journal*, came up to me and said, "We can't publish a paper that's called the 'Rotation of the Universe.'" He changed the title to the "Rotation of the Metagalaxy." I think that's what it is. It was published with all the other abstracts. The paper itself was rejected for publication by both the *Astrophysical Journal* and the *Astronomical Journal*.

When you said that all but one of the comments were of the sort that you can't do this, do you think that people were saying that you can't do this because the data is not good, or do you think they were saying you can't do this because the whole concept of a bulk rotation is ridiculous?

I can't remember. I'm sure that even if they were saying the first, what they really meant was the second—that the whole concept of these very large-scale motions was just ridiculous. I think that's fair. I think that was the general belief. But it's too long ago for me to really remember what was in people's minds.

> *At this time, when you were doing your master's work, were you familiar with the different types of cosmological models, say steady state versus big bang, or within the big bang, the different types of models such as closed versus open?*

I certainly knew of the big bang model. But I really didn't do the work in the framework of any cosmological model.

> *Did you have any preference for any particular model?*

No, probably not. Well, certainly probably for the big bang over the steady state, but that was really just prejudice more than anything else.

> *Let me ask you about your work at Georgetown.*

That was an interesting place, and it probably deserves more credit than it's gotten. I studied radio astronomy under Hagen, who was at NRL and had just done some early radio astronomy work. I studied spectroscopy under Kiess from the Bureau of Standards, and I did some work with Charlotte Moore Sitterly on spectra.

As I said, Gamow contacted me because of my master's work. He was to give a talk at the Applied Physics Lab, and he wanted some details of my master's work, which I gave to him. Then I asked him if I could come hear this talk, and he said no, because wives were not allowed in the Applied Physics Lab. It was a classified place, and there was security. It was through that contact that we started talking. I spent about one day at Georgetown on a thesis problem that Carl Kiess had given me. It had to do with faint lines on the solar spectrum. By the end of the day, I decided that wasn't the thesis problem that I wanted to work on. Gamow had an interesting question. He wanted to know whether there was a scale length in the spatial distribution of the galaxies. He just posed this question to me, and I decided to work on it, and that's what I took for my Ph.D.[5]

> *So it was fairly vague . . .*

Yes, it was very vague. He wanted one number. He wanted the scale length.

> *That doesn't direct you much on how to obtain that number.*

No, it doesn't direct you at all. But that's the kind of person he was. I knew what I was getting into. He just threw out lots of interesting questions. I ultimately performed a 2-point correlation function on the Harvard galaxy counts to do this.

This must have been one of the earliest ones.

Yes. Like the large-scale motions, this also pre-dated the current interest in the problem by about 20 years. In this case, at least I could say I entered it in total ignorance of how to proceed. I had to learn from scratch how one could go about doing this.

I am curious why you chose not to continue your work on the large-scale motions, on the universal rotation.

That is an interesting question. I think there are two answers, and they both contribute. First of all, the only real contribution would have been observational. What would have been required would have been more galaxies and better velocities and magnitudes. I had two children by then. I think the honest answer is I knew that I just couldn't do things like that. There was no way I could get myself to an observatory and gather data. It was totally out of the question. So there was never any thought of doing that. The second answer is that although several times in my career I have found myself in relatively controversial positions, I really don't enjoy it. For me, doing astronomy is incredibly great fun. It's just a joy to get up every morning and come to work. In a sense, the heated controversy really spoiled the fun. I mean people were really *very* harsh. Maybe one learns to take this. I'm not sure you do. My way of handling that, in every case, has just been to go off and do something very different. It never occurred to me to continue on that work. Probably 10 or 15 years later, I started thinking about ways in which I could return to this earlier problem.

After dropping this problem in your lap, did George Gamow interact with you very much?

Very much is probably the wrong word. My husband really has been an enormous support, especially in the early days when I knew no astronomers. He listened to everything I had to say. He's a very high-powered, skillful mathematical physicist. Although I have no idea of how I came to the formalism I used in my thesis, I'm sure talking to him was one of the main ways. Then, ultimately, I did get much of my direction from François Frenkiel, who was a hydrodynamicist. He was a refugee, and he also worked at the Applied Physics Lab. All my contacts really were through my husband. I think in rather short order it became clear that one way of doing this problem was to take the Harvard counts of galaxies along the sky and apply the 2-point correlation procedure. My whole thesis consisted, virtually, of this one calculation, which people would now make on a computer in minutes. I had a desk calculator, and I did virtually all of my work at home at night because I had two kids.

Do you think that you absorbed any of the style of doing science that George Gamow had?

No, I wasn't smart enough. Mark Kac, in his autobiography, says that there are two kinds of geniuses. There are the kind of geniuses we would all be if we were very, very smart and knew what we were doing. Then there is the kind of genius in which there is no way your mind would think that way. I would put Gamow in that class. He had an incredible mind. He was very curious about how the universe worked. And he had a lot of good ideas about how the universe worked. He had no interest in how you got the answer. In fact, in most cases, I think he would be totally incapable of getting the answer.

Oh, really? So he was not technically proficient?

He was not technically competent at all. He couldn't spell; he couldn't do arithmetic—I may be exaggerating a little bit. But he could pose the questions that no one else thought of asking. He was incredible when it came to giving ideas. He did this throughout his whole lifetime. He wrote a postcard to Walter Baade, after a meeting that I had been to, saying, "Tell me where the stars leave the main sequence, and I will tell you the age of the cluster." He seemed to be the first person who understood what that meant. But I had been at that meeting with him, as a graduate student, and he embarrassed me no end because he would fall asleep and wake up and ask questions that I considered stupid questions. His behavior was unconventional. And then he would just understand things that no one else had understood. So it was fun, but I'm not smart enough to do science that way.

What motivated you to start doing your work on the rotation of individual galaxies?

Oh, that I remember. That's current enough, because that's really the early 1970s. Again, there were two reasons. One was that I had come to work here, and Kent Ford had built a very exceptional spectrograph. He had a spectrograph that could do things that no other spectrographs could do, and what you need for a program like that is a good spectrograph. Then, of course, the other side of the question is what you are going to look at. In the late 1960s, when I came here, we did what everybody else did who had their hands on a good spectrograph, and that is we looked at quasars. Kent and I would go out to Lowell Observatory or Kitt Peak a couple of times a year, and we would get spectra of a few quasars. Like everyone else's, they were sort of crummy spectra. It was hard to see what you had. By this time, I knew other astronomers. I would get calls from friends, Margaret Burbidge and Maarten Schmidt, saying, "Have you observed thus and so? Do you have a redshift? If you don't, I'll go get one." It was all done in a very friendly manner. But, again, it just wasn't the way I wanted to do astronomy, because by and large, these people had much more telescope time than I did. If I had a spectrum and wasn't sure what the redshift was, I was either put in the position of having to tell them what I *thought* it might be or giving up on that spectrum. After about a year or two, it was very, very clear to me that that was not the way I wanted to work. I decided to pick a problem that I could go observing and make headway on—hopefully, a problem that people would be interested in, but not so interested in that anyone would bother me before I was done. I chose to study the rotation of M31, and that was what really started that work.[6]

> *You told me a negative reason why you did that. Is there a positive reason why you chose to do that particular kind of work?*

It was my old interest in galaxy dynamics. Before the 1970s, since large telescopes were so few—although, by the early 1970s, good image tube spectrographs could make small telescopes behave like large telescopes—much of astronomy had to operate in a realm in which much was inferred. You observed a few facts. Astronomers have been incredibly clever, throughout the history of astronomy, at *inferring* what they would see if they had a slightly bigger telescope. I really think that's almost the history of big telescopes. Then you get the big telescope, and you go back and you see that what you thought you'd get is, in fact, correct. But then you take the next step, and you make another inference. So people have inferred what galaxy rotations must be like, but no one had really made a detailed study to show that that was so. Inner parts of a few galaxies were pretty well known from the work of Margaret and Geoff Burbidge, but outer parts were not.[7] So, partly, it was my old interest in galaxy dynamics, and then also the realization that with the instrumentation I had available to me, I could really do this. And truly, it was this other aspect of picking a program that I thought was a valuable one to do, but wasn't so in the forefront of astronomy that everyone was doing it. It's only fairly recently that I realized I must just like doing things that other people are not doing. Partly because of the way I get to a telescope, which is relatively seldom.

When you first began finding evidence for dark matter, did it come as a surprise to you?

It's very hard to tell. I think I learned slowly. Well, I guess the answer has to be yes. Of course it was a surprise, because if it hadn't been a surprise, we would not be sitting here talking about it. Kent Ford and I would take about 4 spectra a night. That's about all we could fit into 2 or 3 hours. We would take turns guiding the telescope. I would develop the plates. He built the instruments, and I sort of did the science, but we always observed together because we both liked to. I do remember my puzzling at the end of the first couple of nights that the spectra were all so straight. My first ideas were, by today's ideas, just totally wrong. The first thing that came to my mind when I looked at these very straight spectra was that there must be some kind of feedback mechanism. If the stars got too fast, they were slowed down, and if they got too slow, then they were speeded up. It just didn't look like a random occurrence. The idea of a distribution of matter that would just give you that velocity distribution really didn't enter my mind at first. So it just shows that your intuitive ideas, the first thing you think of, is apparently just irrelevant. I was really thinking more in terms of observables than the distribution of matter.

When you did realize that it meant something about the distribution of matter, do you remember, as you began talking to people about this, what the reaction was?

The reaction was two-fold. In fact, historically, we've left something out. After I finished the early M31 work, which was in the early 1970s, then I went back to the large-scale motion problem in the mid-1970s.[8] My going to the rotation curves was, again, to get away from the controversy of the large-scale motion research. Therefore, I really loved it, because the rotation curves were so flat.[9] Observationally, it was such a nice program. All you had to do was show someone a couple of spectra, and they knew the whole story. In a sense, it was a wonderful observing program, because when you ask about people's reactions, there was never any doubt on anyone's part that these rotation curves were flat. You didn't have to show them measurements. You didn't have to argue. All you had to do was show them a picture of the spectrum.

Well, they could have doubted the data.

No, no one did. It just piled up too fast. Soon there were 20, then 40, then 60 rotation curves, and they were all flat. My recollections are that no one doubted the data. And it was just a joy to have that kind of a program, after a program where you had to go through deep analysis and everybody doubted the answer. I think the fact that the rotation curves were flat was doubted by no one—at least in my presence. The interpretation was more complicated. I think many people initially wished that you didn't need dark matter. It was not a concept that people embraced enthusiastically. But I think that the observations were undeniable enough so that most people just unenthusiastically adopted it.

So you felt in this case that you didn't have to go out and promote the meaning of the work?

Not at all, because there already was a theoretical basis. Ostriker, Peebles, Yahil, and others had come out with very good ideas that dark matter ought to exist, independent of observations, for other reasons, for stabilizing the disk.[10] The ideas had been around for a while. The observations fit in so well, since there was already a framework, so some people embraced the observations very enthusiastically.

Let me turn now to your reaction to some outstanding cosmological problems. Do you remember when you first heard about the horizon problem?

I remember hearing about it in the mid-1970s, because I spoke at a summer course at Erice, and Roger Penrose was also teaching. I sat in his courses because I had so much to learn. In fact, my husband and my youngest son came over toward the end because we were going to climb Mt. Etna. I even insisted that they sit in, saying, "It doesn't matter whether you understand what he's saying. It's just such remarkable stuff that you ought to be exposed to it." So that would be the mid-1970s. Earlier than that, I don't have any clear recollection.

When he talked about the horizon problem, what was your reaction? Did you regard it as a serious problem?

I'm an observer. I really feel very much that I'm an observer, and I tend to relate what we know about the universe to what has been observed. So my feelings about cosmology, I think, are probably much more loose and relaxed than those of many people. Maybe this comes from starting with

Gamow, from my early work. I think many of the models are brilliant, and some of them probably have some parts that are right. But, personally, my attitude towards many such theories is that we're still groping for the truth. So I don't really worry too much about details that don't fit in, because I put them in the domain of things we still have to learn about. I really see no reason why we should just have been lucky enough to live at the point where the universe was understood in its totality. I think the best thing I can say is that I didn't worry about it any more than any other facet of details that don't seem to fit. As telescopes get bigger, and astronomers get cleverer, I think all kinds of things are going to be discovered that are going to require alterations in our theories. The horizon problem doesn't exactly come into that kind of situation, but I think science consists of just continually making better and better what has been usable in the past. So all I can say is I would put it in the domain of something that needed attention.

Did your view of the horizon problem change any after the inflationary universe model?

Oh, I thought that was pretty clever. I thought that was fine. I liked that very much.

Let me ask you about another problem, the so-called flatness problem.

That I consider a real problem.

Do you remember when you first heard about that?

Much later. A few years ago really.

Did you regard that as a serious problem when you first heard about it?

Yes, I did and I still do. I don't know whether you're going to ask me what I think omega is?

No, I won't ask that.

Well, then let me tell you. I see no reason why omega should be 1. I really don't know why the universe should be that finely balanced, and that comes from the observations.

When you say you see no reason, you mean that the observations say that it is not 1?

That's correct. The observations tell us that omega is 0.2 or so, and all kinds of theorists can tell me that 0.2 is awfully close to 1 and go through the flatness arguments. I do see that as a problem. I don't understand that at all.

When theorists say that if it's 0.2, then it should be 1, do you accept that argument?

And therefore think it's 1? No, not at all. I believe the observations, and I believe the observations are telling us that it's 0.2, and I'm not convinced by arguments that tell me that 0.2 is so close to 1 that it has to be 1. No. I think there is some kind of a problem there, and it has to be solved, but I'm not convinced. I don't understand what the solution to the flatness problem is going to be, but it is not yet so troublesome to me that I'm willing to say that omega is 1. I think it's fun. I sort of enjoy it. I often ask myself what's cosmology going to be like in 500 years. I presume they'll have solved the flatness problem, or they'll know that omega really is 1 in a way that will satisfy an observer and so forth. I think if there were no problems, it wouldn't be much fun.

When you talk about solving the flatness problem or finding that omega is equal to 1, do you mean reconcile the fact that omega is 0.2 with the reasons that theorists think it should be 1?

That's correct.

So that's your version of the flatness problem?

That's exactly right. That's how I see it at the present time. But I'm very willing to admit that if we could really understand, if we could learn about the distribution of dark matter or how much dark matter there was or what it was, then there might be good observational evidence that omega is more than 0.2. So part of the problem might go away. I don't know.

So part of your view of the flatness problem is that it may be a problem that theorists have? I want to be sure I understand clearly what you are saying.

That's correct. That's really what I'm saying, that somehow or other the theory has put them in this position of saying what they do, and that something may be missing. I remember, in the early and middle periods of the large-scale motion problems, some of the people I admire most telling me that you can't have large-scale motions because any irregularity since the early universe would be damped out. They gave me all these reasons, which impressed me—indeed really terrified me—as to why you couldn't have large-scale motions. But if you ultimately get to the point where everyone believes that there *are* large-scale motions, that that is what the observations show, then the theorists just have to tell themselves that they have missed something, and *they* have to go back to their drawing boards and fix up their theories. That's fine; that's progress.

> *What is your view of the inflationary universe model? Do you take it as a working hypothesis? Do you think it's speculative?*

I would take it as a working hypothesis. No, I don't think it's highly speculative. I don't think it's much more speculative than the big bang, and that in its time was awfully speculative. I don't know enough physics, really, to know how far back you can trace the physics. I feel comfortable, as most physicists would, extrapolating way, way back—whatever numbers people give you—10^{-46} seconds or something. I'm willing to look at any model that people want to put in, because I don't know any way to refute those models. The simple, conventional big bang was a very general framework that described only a few things, and now Alan Guth and others have put details into that era. I look upon it as a working model. However, let me just go on and say that, for an observer, it doesn't make a lot of difference at the present time because, except perhaps for the cosmic background radiation, we have very few observables from that time.

> *When you say it doesn't make a difference, do you mean that your observing program and your projects are not going to be determined by the physics of the early universe?*

I don't mean that they're not going to be determined by it, because in fact if someone was clever enough to think of some observations that might be very relevant, they might be directly determined. But by and large the observations I make, and the observations that most observers make, are much closer, much younger. We're not observing in that domain. I guess we all hope that we will learn something that will be relevant, that will tell us about that era.

> *In your view, why has the inflationary universe model gained such wide support?*

I don't think I'm capable of answering that question. I don't know.

> *When you first heard about the work of de Lapparent, Huchra, and Geller, how did you react to that work?*

I thought it was spectacular.

Did it present any conflicts in your own view of things?

None whatsoever. Of course not.

Yes, because you've done related things.

Yes, that's right. In fact, the only thing that bothered me about it was, with these wonderful wedge diagrams, some people didn't understand that they weren't really looking at spatial diagrams.

But velocities.

Yes, velocities, because I did worry about large-scale motions. So I even tried figuring out how you could turn them into a spatial diagram. But I thought they were lovely.

> *Taking into account a large body of work besides the Geller, de Lappa-*
> *rent, Huchra work—your own work on the large-scale motions, the work*
> *of the Seven Samurai,*[11] *and all of that work that has shown that the*
> *universe is more inhomogeneous than might have been present in simple*
> *models—has that altered your view of the big bang model at all, or of the*
> *validity of the assumptions of the model?*

It certainly has convinced me that we're not living in a homogeneous, isotropic universe. I mean these things that I really *suspected* in the back of my mind, I can now say publicly. I'm not sure the Robertson–Walker universe exists. I can think of more questions to ask because of what they've done, which go more in the direction of making things more inhomogencous, and I've at least asked some of my theorist friends some of them. No, it hasn't concerned me about the big bang—maybe because I just don't put my mind to it. If someone came out with a different model that could incorporate such large-scale inhomogeneities, I would be delighted to see it, but until then I will just live with the big bang model. I'm not quite sure what you mean. Are you asking is it all gravity?

> *I'm asking whether you feel like some of these things threaten the big*
> *bang model, in your own opinion.*

I wouldn't use the word "threaten," because I don't believe—or I hope 500 years from now astronomers still aren't talking about the same big bang model. I think they won't have done their work if they are. And so these are the kinds of observations that will get some theorists thinking about how to incorporate them. I still believe there may be many *really* fundamental things about the universe that we don't know. I think our ignorance is probably greater than our knowledge. I wouldn't put us at the 50-50 point of knowing about the universe. So how correct the big bang model is is not a worry of mine at all.

> *Besides the few things that I have mentioned, has any other development in the last 10 or 15 years changed your thinking very much?*

I've been very impressed with the Davis, Yahil, Strauss work in attempting to start from the distribution of galaxies and predict what motions would be.[12] That's on a smaller scale, within some tens of megaparsecs or something, but I like that approach very much. That seems to me a fundamental way to study the relatively nearby universe. I've been very impressed with their diagrams. I'm not sure they're *right*, because it's not clear that the IRAS sample that they use is a complete sample. In the past, we've used galaxies as test particles in the gravitational field, and we've asked how fast are they moving, and what does that mean. But to start with a distribution of matter and to predict what the galaxy should do is somehow, to me, a very satisfying way of doing this problem. I've been impressed with that. Other than that, I guess I would say I've been impressed with how little we still know. The Seven Samurai have 300 galaxies.[13] The number of galaxies whose distances we know well, independent of the Hubble expansion, is beginning to get near 1,000. It's incredibly small. There is a lot to do.

> *I'm interested in the extent that scientists use visualization in their work, whether it is important to visualize a problem that you are working on. Does visualization play any role in your own work?*

I'm not sure I understand. I have to see my spectra. I mean I look at spectra. When I come back from an observing run now, when everything's on tape, the first thing I do is make a picture of each spectrum. Also I go observing with a picture of each galaxy.

> *You mean an actual photograph of the morphology?*

Yes. So at this very low, fundamental, observing level, sure. I sometimes ask myself whether I would be studying galaxies if they were ugly. I really do, and I'm not sure. I see ugly bugs. My garden is full of slugs. I sometimes think, well, maybe if I started studying them, they wouldn't appear to be so ugly. I battle the slugs because they ruin the flowers. I put that at the other extreme. I think it may not be irrelevant that galaxies are really very attractive.

How do you think theorists and observers have worked together in cosmology over the last 10 or 20 years? Do you think it's been a successful working relationship or do you think one has gotten way ahead of the other?

I don't know. Maybe I'm not a good person to answer that, because I really tend to work pretty much alone. I personally don't often interact with theorists at all. I don't see anything wrong with the way things have gone. I think it's hard to point to many observations that have been made because theorists said we should. Now maybe you could argue that the dark matter has been an example of that. But at the time it wasn't.

It sounds like in your work that was not the motivation.

No it wasn't at all, and I don't know why. In retrospect, I can't remember theorists standing up and saying, "Galaxies have heavy halos, so go find rotation curves because they'll be flat." I just thought of something. When you asked me what the response to the flat rotation curves was, I remember that the first set of galaxies we observed were very high luminosity galaxies. We published an *Astrophysical Journal* letter.[14] I had forgotten that the response I got from many people was, "Okay, but that's because you did the high luminosity ones."

The selection effect. Of course, it's hard to explain even for the high luminosity ones.

That's right. But their point was: when you go to the lower, or the lowest luminosity, they will all have falling rotation curves. I really had forgotten that. So it wasn't so well expected that people said, "Okay, that's what they predicted and now you've found it." It was amusing, because by then I had been convinced they were all going to be flat. In contrast, let me say in all honesty that when we did the M31 rotation curve and that was virtually flat, we weren't smart enough to make a great thing about it. We live in a diverse world, and some people like to think and some people like to observe. We interact in some places and, by and large, I think it helps us both—the observers and the theorists.

What do you think the outstanding problems in cosmology are?

Well, certainly, what the dark matter is, how much there is, how it's distributed. Hopefully, some of that will teach us about the conditions of the very early universe—whether it was hot dark matter or cold dark matter, if omega is just 0.2, whether it's all baryonic. I think those are fundamental problems. I think the question of whether it's just gravity moving things around, if these are large-scale motions—these are probably all related questions. Probably the one that we're most likely to be able, I hope, to address soon would be what dark matter is and what the distribution of it is, whether it's distributed like the galaxies or more uniformly. It's hard to think of good observing programs that will do any of that.

> *You've made some comments bearing on women and science. Do you think that your experience in science has been different because you are a woman rather than a man?*

Of course. Yes, of course. But I'm the wrong person to ask that question. The tragedy in that question is all the women who would have liked to have become astronomers and didn't. For those of us who have been successful in doing science, clearly the problems haven't been so great that we couldn't overcome them. By and large, if you ask a set of successful women, their answers would have to be that whatever the problems or differences were, they managed.

> *What do you think are some of the problems that prevented other women from going into science?*

I think probably it's the way we raise little girls. It happens very early. I think also it's what little girls see in the world around them. It's an incredible cultural thing. I have two granddaughters. One of them—her mother and father are both professionals, her aunt and uncle are professionals—said her toy rabbit was sick. Her uncle said, "Well, you be the doctor and I'll be the nurse, and we'll fix it," and she said, "Boys can't be girls." And her mother realized that she never had seen a doctor who was a woman. By the age of 2, she knew that men were doctors and women were nurses. So you may talk about role models and think about colleges, but this happens at the age of 2. It's a very complicated situation. I'm not sure how you handle this. I think it sets in very young.

Somehow or other, you have to raise little girls who have enough confidence in themselves to be different. I went to a D.C. public high school. I was very, very interested in astronomy, and I just could keep myself going by telling myself that I was just different than other people, that they just had different interests than I did. I had a physics teacher who was a real macho guy. Everybody loved him—all the males. He did experiments; he set up labs. Everybody was very enthusiastic. I really don't think he knew how to relate to a young girl in his class, and it became a terrible battle of wills. He never knew that I was interested in astronomy;

he never knew that I was interested in science. The day I learned I got my scholarship to Vassar, I was really excited because I couldn't go to college without a scholarship. I met him in the hall, and probably said the first thing I had ever said to him outside of the class, and I told him I got the scholarship to Vassar, and he said to me, "As long as you stay away from science, you should do okay." It takes an enormous self-esteem to listen to things like that and not be demolished. So rather than teaching little girls physics, you have to teach them that they can learn anything they want to. When I was at Vassar, I sent off a postcard to Princeton asking them for a catalog of the graduate school. Sir Hugh Taylor, the eminent chemist who was then the dean of the graduate school, wrote me a letter saying that since they wouldn't accept women, they wouldn't send me the catalog. Some things are better, but a lot of them are not. My daughter is an astronomer. She got her Ph.D. in cosmic ray physics and went off to a meeting in Japan, and she came back and told me she was the only woman there. I really couldn't tell that story for a long time without weeping, because certainly in one generation, between her generation and mine, not an awful lot has changed. Some things are better, but not enough things.

> *This is a speculative question, so you may have to put aside some of your natural scientific caution. If you could design the universe any way that you wanted to, how would you do it?*

I don't think I'm smart enough to design the universe. I used to think I could design a woman's plumbing. I used to think that if I had to design a woman's plumbing, I could have done a better job. But the universe, I couldn't do it. I couldn't do it. I don't have that kind of a view.

> *There is a place in Steven Weinberg's book* The First Three Minutes *where he says that the more the universe seems comprehensible, the more it also seems pointless. Have you ever thought about this issue about whether the universe has a point or not?*

Oh, off and on. And do I think it has a point? I think I would agree with him. I think the laws of physics being what they are, galaxies, stars, planets come into being, and supernovae come into being, and people come into being, and evolution is remarkable. I think it is remarkable. But I think it is a game, an amusing game. And some of us happen to be here, and we happen to have children, and they happen to be who they are. I don't think there is an enormous point to it all. But, for some of us, attempting to understand this universe is important and a major part of our lives.

EDWIN TURNER

Edwin Turner was born May 3, 1949, in Knoxville, Tennessee, and attended the public schools in Raleigh, North Carolina. He received a B.S. from the Massachusetts Institute of Technology in 1971 and a Ph.D. in astronomy in 1975 from the California Institute of Technology, where he was the student of Wallace Sargent. Turner was a research fellow at the Institute for Advanced Study and an assistant professor of astronomy at Harvard before going to Princeton University, where he is now professor of astrophysics.

Turner's current research interests include galaxies and clusters of galaxies, quasars and active galactic nuclei, the cosmic X-ray background, image processing, gravitational lenses, and statistical techniques. Among Turner's contributions to cosmology are measurements of the masses of small clusters of galaxies and one of the earliest extensive computer simulations of galaxy clustering. Turner has also been a leader in the observation and interpretation of gravitational lenses, which are large masses lying between us and distant astronomical objects that distort the light of those objects. Turner's friends are often surprised by some of his nonscientific interests, which include hard rock music, baseball, Tai Chi, and Japanese poetry. A student of Eastern thought, Turner has remarked that Western religions envision two worlds, the physical and the spiritual, while Eastern religions see just one. Turner and his wife have two sons.

Let me begin by asking you some questions about your childhood and anything that you remember that first got you interested in astronomy.

Not unlike lots of astronomers, astronomy's always been something I've been interested in. When I was grade-school age, I read a fair number of books about it. I don't remember any specific ones. I don't have a good story, like I read this book and was really turned to the subject, but I was very interested in it and read lots of books. There's a planetarium near where we lived at the University of North Carolina—the Morehead Planetarium—and my parents used to regularly take me to the shows there. I remember that as always being a very exciting expedition. I'm sure a lot of the lectures and exhibits there had a serious impact. I guess the *earliest* connection that I can make to astronomy was when I was about 3 or so. Since I remember it from a young age, it must have made some sort of impression on me. I had polio, and for a number of years after that I was fairly sick. I often had to wear a lot of orthopedic contraptions while I slept, and during the summer, particularly, it would be very hard for me to sleep, with all this junk. So I can remember on a number of occasions, on hot summer nights, my mother would take me outside and lay out a blanket on the lawn. We'd lay down and look up at the sky. I guess the idea was just to get me distracted enough to go to sleep. I don't remember if it worked, but in any case we would lay down and look at the sky and talk about it at some preschool level. I have no idea what we said, but I have a clear memory of that, and I suppose that's when I first became aware of the sky as a kind of particular object or place. It may have had nothing to do with me later being interested in astronomy, or maybe it did, I don't really know.

Could you tell me a little bit about your parents, before we go on?

Sure. Neither of them is in science or any technical field. Neither of them, in fact, has a college education. My mother graduated from high school. My father didn't because of World War II. He went into the service instead of completing high school. They both, on the other hand, have always been very supportive of and interested in intellectual kinds of things. My mother, particularly, has always been quite interested in science. She had an ambition to be a chemist when she was in high school. As far as their effect on me—there are many deep ones, of course, as with any parent and child—but as far as astronomy goes, they were always very supportive of it. They never regarded it as an unwise, eccentric, or crazy ambition. For instance, when I was admitted to MIT and was trying to decide whether I should go there, there were a number of people, including the minister of our church, who advised them that they shouldn't let me go up to the North to go to school. This was a small-town, Southern mentality. It would be just as good to go to the universities in North Carolina. And there are very good ones there, of course. But my parents were very supportive and fended off all that pressure. And they made significant financial sacrifices, as middle-class parents generally do to send people to college.

I remember that once before we talked about your childhood, and you said that you were very impressed by questions of scale and the distances to the stars. Could you describe that?

Certainly. One of the things that I remember being most grabbed by when I was a kid was the size of the universe. And it still is what I like in astronomy—to work on the far-away, on the quasars mostly these days. But when I was a kid, in particular, I was quite gripped by just the scale of the universe and how big things are. For instance, if the sun were a grapefruit, how big is something else. I was generally interested in the idea of big numbers and large scales. I used to spend a lot of time thinking about that kind of thing. I remember once I found some popular book that had a table in it of fairly nearby stars, with their names and the distance in light years and probably some other information. I made a little table to go along with that, in which I found out what historical event was going on when the light that we were seeing from that star first left it. I was also interested in history then, so that was the connection between the two. Because of the broad luminosity function of stars, the bright stars we can see are a wide range of distances from us. So with just naked-eye stars, these tables ranged over modern history.

Were you a teenager at this time?

I was probably in junior high. Twelve or thirteen would be my guess.

So you were well aware of the concept that when you looked further into space you're looking back in time?

Yes, which is a neat concept and one that grabbed me. I think one of the things that particularly attracted me about astronomy then was the feeling that *we,* stuck on this tiny and obscure and very limited planet, living only for this short little time, embedded in both a universe big in space and a history long in time, can *learn* about all of these vast reaches of space. We can learn about the stars, and we can learn about the past by looking far into it. I'm sure I went around explaining to people how you could look back in time. I probably abused many of my peers in those days with little mini-lectures on the subject, but I thought that was particularly interesting.

At this age, before going to college, do you remember having a preference for any particular cosmological models?

No, I don't think so. I was aware, just from reading, that there were different cosmological models. I was aware of the contest between the big bang and the steady state theory, which was pretty well dying off in professional circles at the time I was in high school.

Within the big bang models, did you have any particular preference at this time for an open universe versus a closed universe?

Before I went to college, I think I would say probably not, or at least I don't remember one. Fairly early, I developed an attachment to the open model, but I would guess that was after I went to MIT. I'm not particularly sure. As best I can remember, the big cosmological issue in my mind in high school was the steady state versus big bang. But at that time I think I was more focused on noncosmological parts of astronomy, partly because in high school I was an amateur astronomer. I did the usual rite of passage of grinding your own mirror and making a little reflecting telescope and being in an amateur astronomy club. Of course, then you tend to get focused on things that you can see with instruments like that. Cosmology did not seem like a big part of astronomy to me at that time, I think.

Tell me a little bit about your undergraduate education at MIT.

I took a physics degree there. MIT doesn't have an undergraduate astronomy program. While I was at MIT, it was pretty clear in my mind that I wanted to do astronomy. By that time—or soon after arriving—I understood that that meant getting a Ph.D. in physics or astrophysics, and so that was more or less my goal. I got as much astronomy as I could. I remember particularly a seminar given by Phil Morrison and other faculty members that was centered around various aspects of the Crab Nebula. I remember that being a particularly exciting course. I took a course in solar system astrophysics from Irwin Shapiro that I remember enjoying. But most of the coursework was physics oriented. When I was at MIT, I also sought out opportunities to get involved with astronomers, and I got jobs working with the X-ray group at the Center for Space Research. I actually had some summer job at American Science and Engineering, which used to be one of the big centers of X-ray astronomy in Cambridge. I did my senior thesis with Phil Morrison on a theoretical topic that had to do with cosmic rays. So I was kind of looking for opportunities.

Did you know at this time that you were going to go into cosmology or were you that specialized yet?

No, I don't think so. I became more interested in cosmology later. I remember reading Dennis Sciama's book.[1] I read that and got a lot more interested in the open versus closed issue and so on.

Let me ask you about that—the open versus closed models. You mentioned earlier that sometime during this period as an undergraduate you had a preference for the open universe. Could you tell me why you had that preference?

I think there are two reasons—indeed, by which I still feel moved in that direction. On the one hand, I've always thought that a one-performance-only universe was aesthetically pleasing: a kind of a universe that is born and then goes through all of this evolution and then just sort of fades away. I don't know why, but for some reason I always thought that was an appealing image, rather than the time-symmetric ones, which most people find aesthetically more pleasing. I think it was less a mathematical aesthetic and more an emotional one. So I always liked that. The other thing I liked is that I was aware that this was an experimental question. If you looked out and added up the matter that we could "see," so to speak, in the stars and galaxies and interstellar medium, you didn't get nearly enough [matter to close the universe]. So the evidence looked like it was open. I have to be careful not to project back my current vocabulary and ways of thinking onto that time, but I think it was clear to me even then that if the universe were closed, that implied that we were somehow missing most of the game. It was hidden or there was something out there that was very much different than what we could directly observe, and I thought that was unappealing. Also I think it's a little bit scientifically improper to make the argument that is sometimes made that a closed universe would be nicer; therefore, the stuff must *be* there. It seems to me that if the evidence was that it wasn't there, it wasn't there. It should just be an empirical question. But also I didn't like the idea that "everything you know is wrong," which is the title of some Firesign Theater album. Firesign Theater was a 1960s San Francisco comedy group. They once had an album called *Everything You Know Is Wrong*. I think that's a worry that astronomers have to have all the time, or cosmologists particularly.

Tell me something about your graduate school career at Caltech.

I went to the astronomy department at Caltech thinking I wanted to be a theorist, but was soon disabused of that idea. I became interested in optical observational astronomy. Wal Sargent was my thesis advisor, and I also worked with him before I started my thesis. He was probably the faculty member who had the most direct influence. Also, to a considerable extent, I think of Jim Gunn as a kind of informal secondary advisor and I saw a lot of him. During my first year or so at Caltech, I think I really wasn't very happy. I didn't like it very much and considered leaving.

Why was that?

A wide variety of reasons. Partly, I think the coursework load was very heavy. By the summer after my senior year at MIT, I'd been spending a large fraction of my time on research, or programming at least, working for somebody doing research. Going back to the one-problem-set-a-week-in-three-different-courses grind reminded me of the early MIT years, which I didn't like very well. I was more or less a hippie at MIT, I think it would be fair to say. Caltech was not very conducive to that sort of thing. So there were a lot of changes. Also, I'd been married just before I came out to Caltech, and my wife couldn't find a suitable job for the first year, so she wasn't very happy. All of those things made it a grim year. But by the second year, she'd found a job, and I got to start doing some research, and after that I enjoyed it a great deal.

What about your thesis?

My thesis was on dynamics of binary galaxies and groups of galaxies, which I got into after having been at Caltech for a couple years.[2] I'd been much more focussed on extragalactic astronomy, and I looked around and tried to see if there was some topic that seemed like a fundamental topic in extragalactic astronomy that had been neglected for a while. I decided on binary galaxies. There had been some classic work by Page, mostly in the 1950s, just using photographic spectroscopy.[3] It seemed to me that it was then possible to do a much better job on that. It attracted me because it was what I thought of as a fairly fundamental problem in extragalactic astronomy, a statistical problem, and I rather liked statistics. One of the issues that interested me was that a lot of the discussion of dynamical masses was tied to rich clusters, but rich clusters are actually a rather rare thing. Only a few percent of galaxies are actually in rich clusters. So it seemed to me that something on the masses of galaxies in more common systems would be valuable. Of course, the thing that soon came up was the issue of massive halos of galaxies, although I didn't know about it when I started the thesis. At that time, a pair of papers, one by Ostriker and Peebles and the other by Ostriker, Peebles, and Yahil, suggested that galaxies might be just some bright tip-of-the-iceberg, some luminous bit of stuff embedded in some much larger and more extended mass distribution—a dark halo.[4] One of their key bits of evidence for this was that they plotted various mass determinations of galaxies as a function of the distance over which this mass had been determined. They had a nice linear relationship, which showed the mass going up and up as you measured it on a larger and larger scale. They had a variety of points using different techniques. There was one point that was way off, which was Page's binary galaxy point. It was an order of magnitude below the line. So suddenly, when they published these papers and when Jerry Ostriker began going around and promoting this idea of the massive halos, insisting that people take it seriously, this discrepant point suddenly became the focus of a lot of attention and interest, which made my thesis seem much more fashionable.

Did you have any speculation as to how this was going to turn out before you got your results?

Yes. Allan Sandage asked me about that, I think, on my first observing run on the mountain. He said, "Well, what answer are you going to get? You should always have some answer in mind before you start." I was relieved to hear him say that, since I did think that I would find low masses [for the galaxies in binary systems]. I thought this dark massive halo stuff was probably wrong.

Did you have reason to think it was wrong? Does it go back to that earlier philosophical feeling that what you see is what you should get, or were there other reasons?

I think partly that, and partly some more technical reasons that just had to do with the fact that I could see various weaknesses with some of the other points that had been plotted. I was quite reasonably heavily influenced at that time by the Gott, Gunn, Schramm, and Tinsley work, which argued that the universe was open by a wide margin.[5] At that time, people tended—improperly, I would say now—to equate the idea of dark, massive halos with a closed universe. The idea seemed to be, "Well, if there's all this dark matter, then that was probably just what was needed to close the universe." Of course it doesn't follow. There's plenty of room for dark, massive halos without closing the universe. But the Gott, Gunn,

Schramm, and Tinsley stuff influenced me. And also the fact that I thought Page had already done it. I thought I could do it better, but I didn't think that it would change the result by an order of magnitude. So I was quite surprised when I did the analysis and the point moved up, bang on to the Ostriker and Peebles line. It was not the result I expected, but I was happy enough with it. My expectation wasn't so strong that I just published it and forgot about it. By the time I finished my thesis in 1975, there was still lots of debate. I remember defending it [the dark matter hypothesis] and supporting it.

Did your own result here alter your belief in an open universe at all?

No, I don't think so—which maybe just reflects how illogical one's beliefs on these things can be. I had just become convinced that there was 10 times more matter than we could see, but it didn't move me much on the question of whether there might be a hundred times more than we see—enough to close the universe. I think the defense I made, the scientific reasons, were mostly of the Gott, Gunn, Schramm, and Tinsley type—the deuterium and so on. So I thought there was at most 10 times more matter. I thought it was probably baryons in dark stars of some sort. You could put a pretty consistent picture together at that time. An omega equal to a tenth basically fit pretty well everything from clusters of galaxies to binary galaxies to deuterium. The only real fly in that ointment was Jim Peebles's covariance function work, which he claimed gave evidence for a large omega, an omega equal to 1 or greater. I remember conversations with various people at that time in which the opinion was expressed—and it was certainly mine—that *other* than the covariance function result, everything seemed to be consistent with an omega of a tenth. I think it was by this time that I'd become relatively obsessed with the open–closed issue. Obsessed isn't exactly the right word, but that had become a sort of a major focus or organizing theme, and that really led into the work that I did right after I got my Ph.D. on N-body simulations of galaxy clustering.[6] That work was an attempt to see if the covariance function argument was really a strong one.

Do you remember when you first decided that you might want to do that work?

I know where it started; it was with a colloquium that Jim Peebles gave at Caltech around my last year there as a graduate student. He showed a movie of an N-body simulation that he and Ed Groth and perhaps other people had made.[7] It was just an early, perhaps the first, example of these now-standard kind of simulations. As part of a talk on the covariance function and galaxy clustering, he showed the movie and put up the pictures and said, "Now this really has just gravity in it. Doesn't it more or less look like the galaxy distributions we see in the sky?" It was just a kind of audio-visual enhancement of his show. But people had a lot of questions about it, and it seemed to Rich Gott and me, talking about it afterwards, that this was a potentially very powerful tool for studying large-scale structure, because you could put in whatever physics you thought might be going on. You could simulate observations of the universe and compare them to the actual observations and test models in a fairly direct way. That was the germ of it. Rich and I had some discussions that maybe we could write a big N-body program, but then Rich went off to Cambridge, England, as a postdoc for a year and met Sverre Aarseth. Sverre, of course, is one of the world's greatest experts on N-body codes and already had some big ones. So that started a collaboration in which we did many N-body simulations. We tried in particular to see whether you could tell the difference between an open and closed universe from a covariance function. Our conclusion was that you couldn't, which was reassuring. The "cosmic astrology" notion that you can read off the future of the universe by looking at the positions of the galaxies in the sky just turned out to be wrong, as far as these simulations went.

I seem to remember that the three of you had nicknames for your collaboration.

Right. We had a standing joke. Since we were doing these simulations, we used to say that the collaboration was a simulation of God, Mother Nature, and the astronomer. Rich Gott was God because he generated or prescribed the initial conditions. Sverre Aarseth, who had the big numerical code, executed the laws of nature. [He was Mother Nature.] And I then took the tapes that Sverre had generated and calculated simulated observations. I did things to the simulations that were analogous to what astronomers do to the sky.

I wanted to talk to you a little bit about your reactions to some of the new discoveries in the last 10 years. Have your views of cosmology undergone any major conceptual changes in the last 10 or 15 years?

[Pause] I guess not. In some ways I am influenced by the inflation stuff to take the larger omegas a *little* more seriously, but I don't think I'm as moved by that as most people are. When I do papers, I tend to use omega of 1 as a convention, just because I think that seems to have become the accepted convention. With these parameters that we really don't know, I think it's very useful if everyone settles into some convention, so I'm willing to go along. My own work has not been focused on the omega question in the last 10 years, at least not nearly as much. I think I'm a little more cynical about the probability that we will really have any clear-cut answer soon, or inside my career.

You mentioned the inflationary universe model. Do you remember what your reaction to it was when you first heard about it?

I thought it was extremely clever and a big surprise. I was not expecting that someone would come up with such a clever new idea about the early history of the universe that would help so much with the causality problem and other things that one is aware of—these little uglinesses of the early big bang. I also thought it was very speculative. I guess when I first heard about it I thought what would have happened with the inflationary universe was that it would have been put aside, so to speak, as a clever little speculation, or clever idea, that would always be mentioned when one discussed the early history of the universe, as a possibility that might well have happened. I would *not* have expected what did happen, which is that it became the foundation of a whole renaissance in the study of the early universe. The whole thrust of cosmology in the last 10 years has in many ways been exploring the ramifications of that idea. I would have thought that it was too speculative and too hard to check for that to have happened.

Do you have any personal opinion as to why the inflationary universe model caught on so dramatically and universally?

First of all, I think it does have the property that it offers explanations for things that are otherwise hard to explain, and it is very clever. But the other reason, I think, is that it is a paradigm, if I may use that word, or a model that allows one to do lots of cute calculations. It's a theorists' gymnasium, so to speak, where one can go, and there are lots of nice problems to do and to be worked out. It wasn't obvious to me when the theory was originally described that that would be the case. But I think that does tend to happen when some new idea comes up that has lots of problems or aspects of it that can be worked out. Naturally people do. There's nothing wrong with that. But I think it often gives the *impression* that the theory is likely to be right, or that the field is more important than maybe it is. It's just the fact that there are nice calculations that you can do. In other words, the theory of star formation is obviously incredibly important for astrophysics, but you don't really have theorists flocking on to that, just because there's not all that many well-defined calculations and problems you can set yourself and solve.

I can think of analogies. The field I've mostly been working in—gravitational lenses—is a bit like that. There's been a huge number of theoretical papers on that. There are probably 20 theoretical papers on it for every gravitational lens candidate that's been found in the sky, and the reason is that there are a lot of nice problems that no one had thought of before that you can work out. Black hole theory in the 1960s was also like that. It involved techniques that allowed one to calculate all sorts of detailed properties of black holes, even though at best maybe we had a little indirect evidence for one.

But I think it often misleads people in their assessment of how likely something is to be true. If you made a great pile of all the papers written on inflationary cosmology, it would be very impressive, and there's some kind of subconscious impression that forms in the community that this must be right. A more external rationalistic approach would say, "Well, what are the hard pieces of evidence that inflation actually occurred?" I wouldn't say there's any evidence against it, but there's not a whole lot for it. I still regard it as something that could easily be right or could easily be wrong. You shouldn't be surprised either way.

Let me go on to another theoretical idea. Do you remember how you first reacted to hearing about the flatness problem?

I probably first heard of it when I was still at Caltech, probably from Jim Gunn. I remember somewhere or the other hearing [John] Wheeler mention it in a lecture as well. I think I thought of it as one of those things that's a little hard to know what to do with. It seems curious but maybe it's just the way things are—like the sun and the moon being the same angular size in the sky, which they are incredibly precisely. That's something that just seems like it must be an accident, given our understanding of things. You're a little uncomfortable to see such a strange fluke, but since it's only one thing and there are many things you can notice in the world, you're not uncomfortable enough to really worry about it. The flatness problem did not bother me a lot. It was something that, if someone could come up with an explanation for it, great, but otherwise I don't think I would have lost any sleep over it. I guess if someone had insisted at that time, before I heard about inflation, that I take it seriously, I would have said, "Well, it's something that we may understand some time in the far future when we understand quantum gravity or God knows what. It may be that it's some result of some early physics that we don't know."

Has your point of view about the significance of that problem changed any? Do you view it as something that's accidental or something that requires a deep physical explanation?

Having had an explanation for a while, it now seems more important to have it explained. As merely a practical matter, I don't think that inflation is likely to be replaced as our best consensus, or best guess, at the early history of the universe *unless* it is replaced by another theory that can also explain the flatness problem in some way. I think that we won't give up inflation for some other theory unless the other theory can also cover that base. One could imagine, although only barely, I think, that some experimental test of inflation would prove that the universe had never inflated. If you had some empirical evidence of that, I guess you might be forced to go back to throwing up your hands. But I have the feeling that in a search for some replacement theory, the flatness problem would be taken seriously. I really don't quite think that it has to be, in the sense that if one wanted to, one could go back to the kind of pre-inflationary idea where you blamed a lot of things on the initial conditions. Initial conditions in this context are a very special kind of idea or thing, and it's not very aesthetic to imagine that the universe had some special initial conditions. It's certainly much more pleasing to have some theory to explain it, but it doesn't strike me as intolerably ugly to just have to say, "Well, the fact that anything exists at all is so strange in a way, that maybe the initial conditions are just very special in some way." As far as the practical evolution of the field, I think it will continue to play a very big role.

How did you react to the results of de Lapparent, Geller, and Huchra on the large-scale structure of the universe?

The main reaction I had was that it seemed like a fairly convincing indication that the old gravitational hierarchy—the amplification of small, initial fluctuations by gravity—was not the whole story. I don't really think that even now that result has been fully appreciated. It seemed to me such an unexpected result.

Were you expecting to see things much more homogeneous?

Yes. I just wasn't expecting to see anything with such a large amplitude, and so regular, on large scales. On the basis of the covariance function work, one would have thought the fluctuations should be small. I guess I was heavily influenced by lots of N-body simulations, including our own, never having seen anything like that. Of course, in the years since, people have wiggled around in various ways to try to produce something that looks a bit like that [the observed bubble-like structure]. But I think my reaction, which was not much different from that of most people's, was one of considerable surprise. And I think that probably means that at least on large scales, there is some process going on that we don't understand yet, or at least there is no single good candidate. Jerry Ostriker's explosions hint in such a direction,[8] but I think that the observed structure makes the whole vocabulary with which we used to discuss galaxy clustering seem to have missed the boat. Clearly, there must be some mapping of those old concepts onto the new picture, but I don't think we really know what it is. The new observations illustrate the impotence of the covariance function, if you will, when applied just to galaxies distributed on the sky, because this structure is clearly not a small perturbation or some slight effect but some dominant kind of phenomenon going on that we just missed entirely. I don't think it's clear yet what it will mean, but I do think it means that all that work that Peebles and Groth started, that Aarseth, Gott, and I did, and other stuff [N-body simulations] that a lot of other people did better probably is not relevant. We may have learned some physics from it, but I don't think we learned anything about the universe from it.

> *Does that bother you at all, that maybe a lot of work that has been done has been irrelevant?*

In a sense, yes. I would say that it has given me a considerably more modest or pessimistic view of what we can hope to accomplish in cosmology and astronomy. Especially on these very big questions. It seemed to be the antidote to the cosmic astrology dream of reading off the fate of the universe from the positions of galaxies, because the new observations of large-scale structure showed how far one could go wrong with a few simple, pretty ideas and some data. Everything was fine and very nice in a theoretical way, and then the data reared its ugly head. Another discovery, which may seem totally unrelated but about which I feel much the same way, was the Voyager pictures of Saturn's rings. Here is an object that has been known since antiquity, that has been studied by astronomers forever, that is nearby—basically a point mass and gravel. If there is any astronomical object of any significant complexity that you could argue we ought to be able to understand, you might have thought we could understand the dynamics of Saturn's rings. People have reflected radar off the rings. We've got all sorts of data. It's a reasonably simple situation, and we've done centuries of work on it. And the pictures were full of surprises, just totally unanticipated, a major phenomenon, again not these tiny little effects. Then you ask yourself how well do we understand something with the complexity of the universe as a whole, based on a handful of observational facts. How well do we understand a quasar, when we can't resolve it? It's just a point of light.

This gets back to your earlier suggestion that there's a chance that we're going off in totally the wrong direction.

Something like that makes me think there's a reasonable chance we're all wrong. As far as the large-scale structure is concerned, it quite discourages me. I haven't worked on this since. If I had an idea, I would, but it seems clear to me that all of the lines of thought I was pursuing are not on the right track, and I have not been particularly persuaded by any ideas that have been put forward since then. So I've more or less stopped working in the field on the basis of the surprising results. [I think there is a particularly good chance that our theoretical ideas qualitatively underestimate the complexity of reality. That is what usually happens when there is little data. We make a model that is no more complicated than it needs to be to explain the data, failing to realize that nature most often realizes complex situations even in apparently simple situations, as in the case of Saturn's rings. People with training in physics, like nearly all astronomers and cosmologists, are particularly likely to fall into this trap. The fundamental physical laws have a subtle and beautiful simplicity about them, symmetries and so on. Physicists learn, or are taught really, to respect or even venerate this quality of elegant simplicity. Looking for it is even a tactic which has paid off in the search for new fundamental laws. However, I feel that this approach may backfire dreadfully on us in our attempts to guess at the *actual realizations* of the simple laws in the physical universe, to guess at the early history of the universe and so on. In fact, I think that the ability of the universe to manifest such astonishing complexity as we see around us on the basis of profoundly simple fundamental laws has a kind of beauty of its own, which may just as well apply to cosmology as to planetary ring systems.]

I'm interested in how scientsts use metaphors and visual images in their work. Do you use visual imagery in your own work?

Yes, I think so. Certainly imagery. I don't know about metaphors.

Have you ever tried to picture the universe during the time around the big bang, around the beginning? Do you have any image of that, or do you just never try to visualize that?

No, I think about that some. The trick there is to not think about it as something you're outside of, looking at, a little thing going "bang," but to think of something you're *inside* of.

Are you able to do that?

A bit, yes. I imagine the atoms or the nuclei or the quarks or whatever it is, depending on the epoch, as like galaxies. You can imagine yourself sort of sitting there on some quark or on some point in this early fluid and having the density dropping incredibly fast all around you and things shooting off. Or, alternately, during inflation you think of everything shooting off but the density *not* dropping somehow. It seems peculiar. I'm not sure if this is the kind of thing you're looking for, but when I think about the early universe, I generally think of it with the same picture in my head that I use for the universe now, except to replace the galaxies with—oh, I don't know—hydrogen atoms if you're talking about the recombination epoch, or nucleons if you're talking about nucleosynthesis, or, I'm not exactly sure what, some other particle if we're talking about an earlier epoch. And you think about forces other than gravity playing a role, but it's a kind of scaling metaphor or something like that. So I'm much more inclined to think of it that way—and to try to get students to think of it that way—than to think of it in the more obvious way of, here sits the universe a centimeter in diameter and I'm over here somewhere.

As both an observer and a theorist, how do you think that theory and observation have worked together in cosmology in the last decade?

I don't know if I'm a theorist. I often say that the secret to my career has been that the observers think I'm a good theorist and the theorists think I'm a good observer. In my own mind, I think I'm an observer and not very much a theorist, but I like the theory. I think there's a real problem. The theoretical work and observational work in cosmology are diverging to a considerable extent. We're getting theories that are tied by more and more extended and tenuous chains of argument or calculations to things that we can directly observe. I think there's this worry, or nightmare, that we could end up conceivably at some future time with several different scenarios for the evolution of the early universe—which perhaps are *very* different physically—but with no testable consequences.

How do you feel about using our general theoretical models to extrapolate backwards 10 or 15 billion years to the first nanosecond? Do you think that we're more or less justified in doing that? Do you think that's a good thing to do?

Oh, yes. I think that it is. What I was saying almost sounds like a criticism of the theorists, but obviously any opening that we have we should pursue because there's no telling what we'll find at the end. Conceivably, there'll be some very elegant way of tying the theory back to the observations, or even if there isn't, it's still valuable to do the theory. I would represent it as our best guess. But I'm very distrustful of our ideas getting too far removed from being checked empirically. I just think the chances of going wrong are very large. There's nothing to do about it, I guess. The universe may have "forgotten" how it formed, so to speak, at least within the limits of what we can find out with our current abilities to tell what's out there. Then it becomes a little like history. This is another subject I'm interested in. History is very frustrating. There's just a finite amount of information about what Alexander the Great did and thought, and there are no doubt very important facts about great events that we're interested in that just were not recorded; they're not passed down to us, and we're not going to find out. We're just not.

Let me ask you a question that will require you to take a big step backwards. If you could design the universe any way that you wanted to, how would you do it?

I may take longer to think about that than the time we have. The changes I would be most inclined to make are not of interest to astronomers and scientists I suspect. One could make lots of improvements in the world on how people behave.

You mentioned earlier that you would prefer that there not be a lot of dark matter.

Certainly, I would have the universe be rich and complex and accessible. I guess those are the ways I would describe it. Accessible to observations, and, even better, accessible to interaction. It would be nice if the speed of light were a lot higher relative to the size of the universe. Another way of saying that is that it would be nice if the universe were younger or the energy requirements were such that chemical rockets were practical vehicles for traveling to the stars. Obviously, if you start fiddling around with these things, you'll soon have no life left. The astronomical universe has this annoying property of being like a bakery filled with delicious goods but you're outside with your nose pressed against the glass. You can see it, but you can't touch it. [Also, I would like a universe with richness and complexity, which is clearly what we've got to a considerable degree. Nevertheless, the more the better. For instance, I much prefer theories in which the dark matter is some sort of population of dead stars, a Population III in current jargon, which formed early in some unknown way and environment, influenced other components of the universe, and had some evolutionary history, some current set of complex properties and so forth. Theories in which the dark matter is taken to be merely a sea of some sort of exotic identical elementary particles, weakly interacting, hardly participating, with properties so simple that they can be derived from first principles in a few lines, strike me as very dull, a real waste of potential richness for the vast majority of the stuff in the universe. Perhaps this example is just a sort of turf war. I'd rather see the dark matter in the astronomer's domain than the physicist's, but I'd still stand by the general point of preferring richness and complexity.]

So you'd like to be able to get around to the stars.

Sure, I'd like to be able to get that barrier down. I'd like to be able to look at a quasar as closely as we looked at Saturn or better. Maybe this would make science too easy. But I am inclined to think that astronomy is a little too hard. One can take various philosophical attitudes towards the nature of the truth and so on, but if we take the most simplistic view of an external reality that has some truth value, and consider that our job as scientists is to try to get hold of that, I think that a hard-nosed rationalist might conclude that the job is too hard for us. There is much interesting and true stuff about the universe that we will never learn, or not in a foreseeable time. As I said, much of what we currently believe may well be wrong, or at least we won't be able to find out whether it's true or not. And I think that's very disappointing.

So you'd like to see a universe that was accessible.

I don't know if that's really how you change the universe, but in the context of being a professional astronomer, that would be an attractive change.

Let me end with one final question. Somewhere in Steve Weinberg's book The First Three Minutes *Weinberg says that the more that the universe seems comprehensible, the more it also seems pointless. Have you ever thought about this question of whether the universe has a point or a purpose?*

Yes, I guess a little bit. I remember when I read that, I was very attracted by it. It was intended, I think, to annoy the reader, but in at least my case it failed. I think that my understanding of the point of the universe—if I had to give it a name—is probably close to the kind of Buddhist concept of something that just is. Does the water *mean* to catch the reflection of the moon? It just *is*. I think trying to impose reasons on it, or points, is too anthropomorphic and just a dream.

SANDRA FABER

Sandra Faber was born on December 28, 1944, in Boston. She received her B.A. degree in physics from Swarthmore College in 1966 and her Ph.D. in astronomy from Harvard University in 1972. While conducting her doctoral research, Faber lived in Washington, D.C., and worked with Vera Rubin at the Carnegie Institution of Washington. Since 1972 she has worked at the Lick Observatory of the University of California at Santa Cruz and is a professor of astronomy there. In 1986 Faber won the Heineman Prize of the American Astronomical Society. She is a member of the National Academy of Sciences.

Faber's scientific interests include stellar spectroscopy and stellar populations in galaxies; the formation, structure, and evolution of galaxies; the study of clusters of galaxies; and cosmological applications. Her contributions to cosmology include the discovery of a new method for determining the distances to galaxies called the Faber–Jackson relation and, more recently, the investigation of large-scale motions of galaxies not following the simple expansion expected for a homogeneous universe. A modest person by nature, Faber is especially gracious in giving credit to others.

I wanted to start with your childhood. Do you remember any particularly influential experiences you had as a child?

I remember looking at the night sky a lot—just visually, with star charts and a pair of binoculars. I had a big chart from the American Museum of Natural History. It was bulky; it was hard to use; it was something that unrolled and rolled up again. I remember spending evenings looking at the sky with my dad. He was a civil engineer and was interested in science as a kid. And he always encouraged me. He found the binoculars for me. But I was the kind of kid who liked omnivorously almost all kinds of science. Rock collections, fossils. I like leaves, I like plants, I like biology. Almost everything. I didn't really focus until I got older.

Was your mother interested in science as well as your father?

Not at all. Well, let me qualify that. She was very interested in medicine, and she followed that. I think if she had been growing up today, she might have become a doctor. She renounced career interests and things like that in favor of family responsibilities. But all her life, she did follow medicine, and she knew a good deal about it.

Did you build any things, any science projects as a child?

No. I'm very bad with my hands. I'm not an instrumentalist now. I'm a remarkable bull in a china shop in any kind of laboratory. In fact, I really think twice before I pick up a screwdriver, because I'm liable to do something really stupid. One of the most embarrassing failures of my childhood educational career revolved around science projects, which inevitably turned out badly. My talent isn't along those lines.

Do you remember any popular books in science that you read?

Yes. I read popular books by James Jeans written in the 1920s. My godfather, who was also an engineer, had collected some popular writings from the earlier part of the century. I should say, by the way, that there are almost two generations between me and my parents. I was born when my mother was 42 and my father was 45, and I was the only child. And so their friends were also older. Their education came from the beginning of this century rather than the middle of the century, as it would have had the age gap been more normal. So, that's by way of explaining why my godfather had in his collection books by James Jeans in the 1920s. So I read some of those. *Stars and Their Courses* is one I remember the best. I think, though, the most influential book for me from that period was Hoyle's famous book *Frontiers of Astronomy,* which had a big influence on so many people. I read it when I was in high school. It dealt with steady state theory, and the aesthetic beauty of the steady state theory really appealed to me. I was quite convinced after reading his book that that had to be the proper explanation for the beginning of the universe.

Did he mention the alternative, the non-steady model?

Yes.

So, among the two alternatives, you personally preferred the steady state?

Yes, I strongly preferred it because I thought, aesthetically, it was compelling. It definitely got rid of the question of the beginning by saying there wasn't one.

And that appealed to you, to remove the concern of the beginning?

Yes. It was my first exposure to something which has happened to me a lot of times since. It's a phenomenon that happens repeatedly in physics, astronomy, and cosmology. One poses what appears to be a perfectly reasonable-looking question, and then one finds that the answer is really not to answer the question as originally understood but to show that the question somehow was flawed, that there was an implicit assumption that wasn't correct.

In this case, what would the question have been?

What is the beginning of the universe? Steady state answered it by saying that there wasn't a beginning, so we don't have to answer the question as posed. We replace it with yet another view. I no longer believe in steady state, and the demonstration that steady state was wrong was a lesson to me. It showed me that there's more to doing physics and astronomy than just aesthetics. But I still do find myself strongly influenced by aesthetic principles, as do most physicists, I think. My enthusiasm about inflation is an example of that.

Did you know by the time that you went to Swarthmore that you would want to be a scientist?

I wasn't sure that I could be anything, because I was quite confused at that time about how I could be a woman and, at the same time, pursue these questions that I thought were really interesting. I just was taking one day at a time. A very clear path was open to me by education. One went to high school, then one went to college, then one could see going forward to graduate school. I could see that path fairly clearly. Beyond that, the future was a total mystery to me, as to whether or not I would have a job and work, whether or not I would get married and have children. I just had no idea what I would be able to do.

Do you remember why it was that you felt confused about what the future would hold? Was it a message you got from your parents or your teachers, or was it a broader message?

I think it was a broader, diffuse message. Just simply because at that time, say in 1960 or 1962, there weren't very many women scientists, and there were even fewer who had what might, at that time, have been termed a normal family life. A woman scientist in the 1940s and 1950s, in general, was a single woman.

Did you know of some of those women?

No, I didn't know any of them. The only working women I saw whose lives interested me at all were school teachers in high school. But I didn't think that I wanted to be a school teacher in high school.

So you had no role models whatsoever?

I really had none. The first role models that I began to see were women professors in college, and specifically there was a woman astronomer at Swarthmore, Sarah Lee Lippincott. She was, on the one hand, an inspiration. She was a woman who was professionally employed as a scientist and as a researcher too. She wasn't a faculty member. Her job in life was to do astronomy in the observatory. But, on the other hand, I felt that I wanted to go beyond her position because she had never gotten a Ph.D., and that prevented her from becoming a faculty member or an astronomer at a larger, more forefront institution. I thought she was a wonderful person. But I also felt that I wanted not to stop at a masters degree as she had done.

Why was it important to you to get a Ph.D.? Why was there something more than just the research that was important for you to strive for?

I felt that if I had a masters degree, I would be limited to a small institution, and therefore I would have limited access to tools. I also felt very strongly that going on and doing a Ph.D. would add greatly to my education. A seasoning, actually. But to return to your earlier question, when I said that I was confused, it was really this diffuse message that I couldn't clearly see people out there who had gone in the direction that I wanted to go. But I must say that as far as specific messages were concerned—with either my mother, my father, relatives, or teachers—I had fantastic encouragement from all of these people. That kind of encouragement was certainly responsible for my having gone along as far as I did. So there was a conflict. My way of dealing with that was simply to take one step at a time.

You mentioned Sarah Lee Lippincott at Swarthmore. Tell me a little about the influence she had on you.

One thing that really impressed me was the fact that she was very independent. She's the first woman I ever met who was in that situation. All of the women in my family were either married or divorced and living within a family situation. She lived by herself. And, on one occasion, she invited me over to her apartment and fixed dinner for me. It was very pleasant. I saw that she had made a life for herself. She had a place to live that she had created. She was a photographer and had published a book of photographs of Philadelphia. She had a world for herself outside of astronomy, and she was doing this completely by herself.

This is a very hypothetical question. Do you think if she had been a male astronomer, who also had the same personal relationship with you, that you would have been as inspired to go into astronomy?

I might have been, because, in fact, she was not the only one at Swarthmore I interacted with. There were two astronomers at the observatory there who influenced me a great deal. The other one was Peter van de Kamp, and he obviously was a male. He was a fantastic and very positive influence on me as well. Clearly, I didn't look at van de Kamp and say he was a remarkable person because he has a career and a family and a house—the sorts of reactions I had to Lippincott. But he certainly inspired in me a great love and interest in astronomy. The observatory was a very happy place. There was just a handful of students who were interested in astronomy, and it was a very small, congenial atmosphere, a very good place to get one's basic foundation.

When you studied astronomy at Swarthmore, did you also study cosmology?

Unfortunately, I didn't. I think,, partly, this is one of the bad things about learning things in a small liberal arts college. The astronomy that was done was focused on nearby stars, and the astronomers there didn't really know much about the wider issues of cosmology. I didn't care too much about that at the time, because I felt it was time for me to be learning fundamentals of a classical sort. And, even more than astronomy, I studied physics in the physics department. However, it would have been good, in retrospect, had I had some exposure at that time not only to classical cosmology, which I found easy to learn, but more importantly to particle physics, the physics of the early universe, which I find extremely difficult. Most of that I have learned completely on my own, since graduate school. I didn't get that in graduate school either. It's been hard for me to gather whatever understanding I have. I didn't have much of a foundation laid in my formal education. It's been a real handicap.

When you went to Harvard to do your graduate work, did you get an introduction to cosmology then?

Not really. I did take a course in general relativity, but it was a very classical course that just dealt with the equations and a few applications to the expanding universe and black holes.

Did you study cosmology any on your own at that time?

No, I didn't. At that time, if you had asked me what I was doing, I would have said I'm interested in galaxies. There was a lot to be done just in that field, having to do mostly with stellar and gas dynamics and so on. It was clear to me at that point that I was better suited to be an observer than a theoretician. And so I was pretty fully occupied mastering the techniques of observational work, data reduction, learning how to program a computer, which I had not done in college. So that was a new experience.

Do you remember any people who were particularly influential?

I had a strange career at Harvard, in the sense that I had by then gotten married. Andy and I had become acquainted at Swarthmore. My graduate career was affected by his problem of serving in the Vietnam War. He was a member of that cohort of men who were permitted to have one year of deferment in graduate school but no more. This was a transitional policy. So he took one year of study in a masters program of applied physics at Harvard. I'm a year older than he is. At that point, I had completed two years of course work at Harvard, which, as in most places, is all you need. Then you go on to do your thesis. He had to leave graduate school and either serve in Vietnam or find a job that would give him a deferment. He had previously worked in underwater acoustics and now was able to find a job at the Naval Research Laboratory in Washington. I was not prepared to consider staying on at Boston and conducting a long-range marriage—

which would have been possible to contemplate had we had more money to allow airfares and frequent visits. But we had very little money, so visits would be quite infrequent. That seemed very unpleasant. So since I completed two years of course work, I didn't really have to be in residence at Harvard. And they were quite flexible with me. I wangled an office at the Naval Research Laboratory, amongst the astronomers there, who didn't do anything related to galaxies at all. They were doing things like measuring the temperature of Venus and studying water masers. So at least there was some astronomy, but nobody to talk to. After a year and a half there, I was invited to take residency at the Department of Terrestrial Magnetism, with Vera Rubin and Kent Ford, with whom I had worked one summer before. And I finished out my graduate thesis work there. That was a very *good* atmosphere because they were very interested in galaxies. Overnight, I had found a very pleasant and productive niche.

I guess Vera Rubin was extremely influential?

She was more influential than anybody else. I like to call Vera my de facto thesis supervisor. My formal supervisor was John Danziger. He did a very fine job with me. He was a wonderful facilitator. Whenever I needed something from Harvard—my permission to go to Washington or travel money—he was right there with incredible support. On the other hand, he was a stellar astronomer and not too able to guide somebody who was doing a thesis on galaxies. Why was he my advisor? Well, he was the only optical observer at Harvard at the time.

Was Vera Rubin interested in cosmological applications at that time?

No, she wasn't. But she was on the verge of thinking about things that were to have a profound cosmological application. I like to recall fondly one day when I was doing my thesis work there, she got a phone call from Morton Roberts, who was at the National Radio Astronomy Observatory. He said, "Vera, I have some very strange results on rotation velocities in M31, and I'd like to come up and discuss them with you." He came up the very next week. I remember sitting around a table, and Mort Roberts was showing Vera this graph of rotation velocity versus distance [from the center of M31]. His new radio results went 50% farther out and showed the HI there going at exactly the same velocity as the inner stuff. Mort was clearly very excited about this. I said, "Well, so what? Fifty per cent farther out is only 50% more mass. It's not a big change in mass." He said, "You don't understand. There's no light there." And I remember being profoundly unimpressed, I think partly because I was already convinced that there was something totally crazy about mass measurements in astronomy. My view then was that somehow these velocities didn't mean anything, that they were wrong, for some reason.

I remember why this was. My thesis project was a photometric study of elliptical galaxies. I was interested in whether or not the properties of double galaxies were more closely correlated than pairs picked at random. It was a crazy idea, in light of hindsight, but it had led me to go through all the galaxy catalogues—by hand of course, because they weren't on tape at that time—and look for doubles. That brought me right up against the problem of when do you pick a double on the basis of proximity as opposed to velocity differences. I had found many doubles that had big velocity differences, and when I just calculated characteristic masses for them, the masses were impossible. Of course, I was aware that people were measuring masses of clusters and finding excess mass. So I attributed this velocity discrepancy in doubles to the same phenomenon—whatever it was. I knew that people argued about it. I knew it was a bottomless pit. Nobody had ever solved this problem. And I just didn't want to bother with it. I didn't want to think about it. It didn't seem to be a question that was ripe for solution. When Mort Roberts came and showed yet one more crazy velocity, my attitude was, "These velocities are never right. Don't bother me with that—we don't understand this. Why are you making such a big deal? It's just one more example of this problem" [laughter].

Do you remember what your principal motivation was in deciding to work, with Robert Jackson, on the velocity dispersion of elliptical galaxies?[1]

Yes. I remember very well. My thesis was concerned with the spectral properties of these galaxies, and because I was limited to a small telescope, my measurements in the thesis work were rather coarse.[2] When I came to Lick, the image dissector scanner had just gone into operation, and it was clear that people could obtain much higher resolution spectra of galaxies. So the very first thing I wanted to do was to take these spectra and see what they looked like in full detail. I was very interested in making radial measurements as a function of distance from the centers of these galaxies to find out how the spectral properties change as a function of distance. However, when I looked at the spectra, any idiot could see that some of these things had broad lines and some had sharp lines. So I took a quick detour, because there had been almost no work done on measuring velocity dispersions, and it seemed like this was something that had just fallen my way that I should follow up.

Did you have any idea that measuring these velocity dispersions would lead to some interesting relationship? Or were you motivated first just to measure them?

Well, as I was taking the data, I could not help but notice that the more luminous galaxies had broader lines. So, when we worked up the data quantitatively, it was with the intention of finding out how good that relationship was. Of course, it turned out to be a pretty good relationship. The next question you might ask is, "At that time, did you have the faintest idea of how you might explain something like that?" I viewed the work we were doing as very similar to what might have been done at the beginning of the century when you plotted, purely empirically, luminosities of stars versus their surface temperatures—to make the first H-R diagram. I'm sure the people who did that knew that they had stumbled on something very fundamental. But for a long while people didn't really know why there was that relationship. They knew it was a key to explaining things. Similarly, I felt that the relationship we found was a key to explaining the origin of galaxies. But I had not the faintest clue at that time of any theory that might explain it.

Let me go on to some much later work. Do you remember your own personal motivation in beginning the Seven Samurai project—the project to measure distances and velocities of a sample of elliptical galaxies?[3]

Yes. I seem to have a knack for doing things for the wrong reasons, and this was no exception. As a result of collaborating with Roger Davies and Roberto Terlevich and Dave Burstein previously, we thought we had discovered that the scatter in the Faber–Jackson relationship was real, and that there was a second parameter involved. And furthermore—and this is where we went wrong—we thought there was a correlation with ellipticity in our sample [a correlation between luminosity and ellipticity].

Now, if *that* had been true, it would have been possible, in all probability, to say something about the true shapes of elliptical galaxies. That seemed to us to be a rather exciting prospect, since that was just about then coming to the fore as a major question. We thought we would do a much larger sample, and in the end, perhaps be able to tell you whether a round galaxy was round because it really was round or whether it was elongated and seen face on. In the light of a subsequent sample, we know that there's no correlation there. But if you read our NSF proposal for that work, there's a great deal of energy devoted to explaining why there might be more than one parameter and why the second parameter might be related to galaxy shape, and therefore, people should go out and collect bigger samples to study elliptical galaxies. And somewhere in there, in like two sentences, it says that after studying all these properties maybe we can develop a better distance indicator. Maybe we can somehow tune-up the original Faber–Jackson relationship and collapse the scatter once again. And if we can do that, then we propose to study deviations from the Hubble flow. But the whole thrust was on the galaxies; it wasn't on the Hubble flow.

When that work got started, did you expect to find significant deviations from the Hubble flow, or were you sort of surprised in that result?

We were completely surprised. We were shocked, elated; we didn't believe it. One day we believed it, the next day we didn't believe it. In the meantime, Mark Aaronson and his group had published their 1982 analysis of the Virgo infall into the local supercluster.[4] So we were very interested in trying to find out whether we could confirm their measurement of Virgo infall.

Was that infall relative to the microwave background radiation?

Well, looking back on their first paper, they really didn't emphasize that the local group was moving at 600 kilometers per second with respect to the cosmic microwave background.

So, they were just measuring relative velocities?

They were just measuring relative velocities. Now any thinking person— and they probably realized this themselves—would have realized that their whole system was translating rather rapidly with respect to the cosmic microwave background. But I must say that I wasn't really a thinking person. In fact, it's amazing that as I look back on it, I was wading through all the icky parts of data reduction, just keeping my eyes focused on day-to-day activities, not thinking about larger implications. I knew those two facts. I knew that the local group was moving at 600, and I knew that the Aaronson analysis had shown nothing very interesting in the local group coordinate frame. That meant that all the galaxies must have been moving rather rapidly [relative to the mean rest frame of the universe]. But it took me a long time to put those two things together.

Do you think that you didn't see it because you weren't looking for it?

One of my faults as a researcher is that I do not think all the time. I spend large amounts of time just dealing with trivia—going observing, doing all the myriad trivial things that you have to do to produce data. And then, suddenly when I get the data, I think. I have a very productive period of thinking, and then months can go by after that in which I hardly think at all. So, I was in my nonthinking mode here [laughter]. I was just wading through this big project to get all this stuff done, and that's all I was doing. There were a couple of things, though, that began to point in this direction. One of them was a statement by Donald Lynden-Bell, who contributed a great deal to this project. He said: "I think we ought to conduct all of our analysis in the microwave background frame." Up to this point we had never really said in what frame we were working. He said, "The natural frame is the microwave frame, so why don't we do that?" We were standing on the street corner in Hanover, New Hampshire. And right then I could see: "Hey, that's going to put 600 kilometers a second on all of our velocities. That's interesting." I got very excited and said, "Donald, that's a brilliant idea. Why hasn't anybody done that before? Obviously that's what we need to do."

After we adopted that as the fundamental frame, the person who really first saw in our data that there was an area that was doing something totally bizarre was Dave Burstein. Dave was always out in front of the rest of us. He kept all of these big data files. He would punch in the numbers and run a quick plot, while the rest of us were still cleaning up the loose ends. And at our next meeting, in Pasadena, lots of us were doing deadly, dull, trivial little things and getting programs to work while Dave was making these plots. "Look at this," he said. "There's a whole region in Centaurus, and it's moving at 1,000 kilometers a second." Another tendency of our group was that nobody wanted to talk about things until it was the right time. If I was doing some little operation that was dull and boring but had to be finished, usually I'd say, "I'll think about that later. Don't bother me with that now." Because sometimes it didn't turn out to be right. It might go away in two days, and why should I waste time thinking about it until we were really ready to think about it? But he was right about that one. He kept bugging us about it. As the weeks and the months went by, that region got more and more interesting. It wouldn't go away.

Has this work raised any concerns for you about the validity of some of the assumptions of the big bang or the interpretation of the cosmic microwave radiation? This is one possibility that you mentioned in one of your papers.[5] *How do you feel about that now?*

Well, I think we mentioned that purely for the sake of intellectual honesty. I personally don't think that's the right approach. I tend to think the cosmic microwave background really is the right rest frame [of the universe].

Do your results have implications for the degree of homogeneity in the universe? If there were huge bulk flows, wouldn't that call into question the assumptions of large-scale homogeneity?

Not necessarily. The challenge is to find a theory in which one can simultaneously match the scale of velocity flows we observe here with the cosmic microwave background fluctuations. I think that we don't know yet whether or not that's possible. I tend to think that it is possible. For example, there is a student at Santa Cruz, John Holtzman, who has calculated motions and fluctuations for a hybrid theory, which combines cold dark matter with a fairly light neutrino. That looks like a theory that on paper can match any constraint. So I would make two statements at the moment. First of all, even if you take our results here at face value and interpret them naively, it's not at all clear that there aren't theories that are consistent with both streaming motions and any other observational constraints like the microwave background. Secondly, I think that what we are learning is that comparing these streaming motions with theory is a very subtle business. We haven't seen the end of how best to do that yet.

So your faith in the big bang model has not been shaken?

It's not shaken. No, I don't really think it's going to be that bad. It's going to be an interesting constraint, but I don't think it's going to be earth shattering.

Let me ask you, do you remember when you first heard about the horizon problem?

No. It would probably have been a few years after graduate school.

When you did hear about it, do you remember whether you considered it a serious problem?

Yes, I did. I believed that it was telling us about a very fundamental problem in our way of looking at the initial conditions.

You mentioned earlier that you like the inflationary universe. Why do you think the inflationary universe model has caught on so widely?

Because it solves these three or four classic problems of conventional cosmology. The horizon problem is one, the flatness problem is another. Those really speak to me. One conventionally mentions the monopole problem, but, not being a particle physicist, I have a dimmer appreciation of how serious a problem that is. But they do say it's quite serious.

Do you think that the inflationary universe model has a good chance of being correct, or do you think it's too speculative?

I don't know anymore what people mean when they say "the inflationary model." I'm kind of thinking that inflation is a process—which is likely to be important, but exactly at what phase in the evolution of the universe, I can't say. And I don't have an opinion on whether there are multiple inflationary periods. I think inflation is a general concept that we need to use.

How do you reconcile the amount of matter needed for inflation—omega equal to 1—with the observed value of omega? Does it bother you that those two values are in conflict?

It bothers me intensely. The only way out of it, I think, is that there are baryons in the voids.

You make the dark matter with baryons?

No, I mean that there is *both* dark matter and baryons in the voids, in the normal proportion of about 10 to 1. I like to think that the universe was well mixed in its early stages. We know that only a few percent of the baryons have shown up in galaxies. My guess is that the other baryons—perhaps 5 times as many baryons—are out there not in galaxies, probably in voids.

I'm trying to find out how you reconcile these things. Are you sufficiently impressed by the inflationary universe model that you are willing to take on faith that there are these baryons or missing mass in the voids, even though this matter has not been currently observed?

Well, I never believe things perhaps as strongly as other people believe them. So I would just like to say parenthetically that if you made me bet and put money on something today, that's where I'd put it. But my world would not fall apart if it turned out otherwise. I was one of those who voted for omega equals 1 at the dark matter conference. Scott Tremaine took a poll in the audience, and he asked people to vote in three categories. This was at Princeton two or three years ago. He asked the audience to vote on the value of omega—those between 0 and .9999; .9999 to 1.0001; and over that. Almost nobody voted for omega much larger than 1. I think the rest of the audience split about 50/50.

Do you remember when you first heard about the flatness problem? You said a few years after graduate school for the horizon problem.

Yes, about the same time.

Did you also consider the flatness problem to be a serious problem?

Yes. I did. In fact, I mentioned that at a talk that I gave at Swarthmore, a Sigma Xi lecture. But I didn't consider myself a cosmologist at that time. I considered myself to be someone who was interested in the structure of the galaxies. And I knew that they were related. But I didn't have much familiarity or any real contact with the bigger questions of cosmology.

Did you have any opinions about how the flatness problem might be resolved?

Yes, I've actually had an opinion all along on this. And I still hold this opinion. Again, this is something I wish I knew more about. But at some dim level, I perceive a relationship between doing cosmology and trying to answer the questions of cosmology, and Gödel's theorem. It just seems to me that trying to explain the universe viewed as a closed system, like a closed mathematical system, is incomplete. I don't think it will ever be possible on logical grounds to explain everything about the universe in a completely self-consistent way. So, to arrive at the properties of the present universe, I think it will always require dragging in some kind of information outside this closed system. Now that extra information could take many forms. I think of myself now in terms of the ancient Greeks. What they saw as their universe was the sun, and the planets, and the earth. And they might have said, "Well, why is the earth the way it is, or why is the sun the way it is?" We now know that the right answer to that question is to say that the question doesn't make any sense. The earth is the way it is because we happen to be on it. This is the old anthropic way of answering cosmological questions. I really think that we will probably find that our universe is the way it is to some extent just because we're in it. Implicit in that is the notion that there are many different kinds of universes. One of the questions I am interested in, although I have no idea of how to address it, is: What is the total possible range of parameters you could think of for something that you might call a universe? What can be different? Probably things like omega can be different. Can the mass of a proton be different? Can particles be different? Can space and time be different? I'm interested in those questions, but I frankly have no idea of how to address them.

Do you remember when you first heard about the results of de Lapparent, Geller, and Huchra on the large-scale structures, or the results of Haynes and Giovanelli? Did those results surprise you?

No, they didn't surprise me. They didn't surprise me because I had gradually begun to become more familiar with our local territory, just from looking at the spiral samples. And gradually it has been dawning on me over the years that the galaxy distribution is very, very inhomogeneous. When I looked at their picture I could see voids that were comparable to the voids that I already knew existed. Now the other interesting facet of their picture, of course, are the thin walls. I *am* worried about that because they're making their map in velocity space, and we were simultaneously showing that regions can have big peculiar velocities.

So you don't really know whether the velocities correspond to distance.

I don't really know. On the other hand, you could imagine velocity fields that are such large scale that they merely distort the picture but don't fuzz it out. I don't think that their results are necessarily inconsistent with what we're finding, but the interpretation of the location and sizes of their structures might need some refinement. At the same time, I think most people believe that their first slice was more striking, as far as the sharpness of its edges, than the subsequent slices that have been made. So we probably don't have a fair sample yet of how sharp these boundaries are.

Yes. I would assume from your earlier comment—that the large-scale streaming motions are not really inconsistent with the big bang picture— that you would say the same about the Geller–de Lapparent–Huchra work, that those inhomogeneities do not necessarily conflict with the homogeneity in the microwave background.

I believe that to be the case. But I think in the case of their work, it has not yet been demonstrated. I know that in the case of our work there are calculations that show that is true [that you could have a model that will satisfy both constraints].

Besides these examples, have any other developments in cosmology in the last 10 years or so changed your thinking about cosmology in major ways?

Actually, I think you haven't mentioned the one that's been most crucial to me. That's this whole picture of the growth of gravitational instabilities. I used to read papers on galaxy formation in the 1960s and 1970s. None of them made any sense to me whatsoever, because the authors—maybe they had it in their own minds—didn't succeed in putting in the paper a *picture* that I could make in my mind of what was happening. I think very geometrically, and I can't evaluate a model unless I can generate a mental movie of what's happening and what's moving where. None of those early pictures of galaxy formation models were able to make me generate my movie. And this has all changed for me now with James Peebles's work, which however is not very visual. But people have built on his concepts. The White and Rees paper was a profound influence on me—galaxy formation due to the collapse from dark matter halos.[6] Then the N-body simulations, which actually showed us how things were going. That I think has been very influential for me in developing intuition about these things.

Would you say that has really changed your thinking?

I can't say that it has *changed* my thinking because my thinking back then didn't exist. It's given me the crucial conceptual peg to hang my hat on— the whole scenario of fluctuations coming out of the big bang, being modified according to some transfer function, and then after recombination growing into the structures we see today. I don't bet much money on too many things, but I'll bet you that basic picture is correct.

Let me come back to the issue of gender in science. Would you say that your experience in science has been any different because you're a woman?

Yes. I would say it's probably been somewhat different. On the most mundane level, I've had less time to do science than most of my male colleagues. That was really a factor when my children were young. I came to Santa Cruz in 1972. I really didn't publish a paper aside from my thesis work until Bob Jackson and I published the work on velocities. That's a gamble. Three years or so. People wouldn't survive very well with that kind of hiatus now. I felt acutely aware of it at the time. I felt very badly I wasn't getting any work done. It was hard. I had my first child nine months after I came here. So that's been a definite handicap.

I think, though, I've had a lot of opportunities. I've been appointed to lots of committees, national committees, and other things because somebody wanted representation from women. Committees have played a very big role in my education. You get to meet people that way, and you review all kinds of scientific issues. I think astronomy should ideally be one of the most synthetic subjects. It's hard to do effective astronomy if you're too specialized. Ideally, astronomers should have a very global view. You don't know if the solar neutrino problem is going to have something to do with the big bang. Who would have guessed 15 years ago that that might be a very intimate connection? So you like to stay as broad and as flexible as possible. Moving on the national scene and serving on committees gives you a way of keeping in touch. I think therefore I've had an advantage.

Let me end with a couple of philosophical questions. If you could have designed the universe any way that you wanted to, how would you have done it?

Gosh, has anybody given an interesting answer to that? I don't know. I have to beg the question because, again, I echo the theme that troublesome answers to questions usually mean the question is flawed in some way. I'm sorry, but this strikes me as a flawed question.

Well, you don't have to answer if you don't want to.

The reason the question is flawed, I think, is that it is trying to put human values on a question that is intrinsically totally unrelated to human values. And that's why I'm having difficulty with it.

Okay. That leads naturally to the last question that I have. There's a place in Steven Weinberg's book The First Three Minutes *where he says that the more the universe is comprehensible, the more it also seems pointless. Have you ever thought about whether the universe has a point or not?*

Yes, I've thought a great deal about it. And my answer here relates to what we were saying before about why is the universe the way it is. Let's use the same analogy. Why is the earth the way it is? To me, the questions are very, very similar. I don't believe the earth was created for people. It was a planet created by natural processes, and, as part of the further continuation of those natural processes, life and intelligent life appeared. In exactly the same way, I think the universe was created out of some natural process, and our appearance in it was a totally natural result of physical laws in our particular portion of it—or what we call our universe. Implicit in the question, I think, is that there's some motive power that has a purpose beyond human existence. I don't believe in that. So, I guess ultimately I agree with Weinberg that it's completely pointless from a human perspective.

MARC DAVIS

Marc Davis was born on September 8, 1947, in Canton, Ohio. He received his B.S. in physics from the Massachusetts Institute of Technology in 1969 and his Ph.D., also in physics, from Princeton University in 1973. In 1975 Davis joined the faculty of the astronomy department at Harvard University. Since 1981 he has been a professor of astronomy and physics at the University of California at Berkeley. Davis won the Pierce Prize of the American Astronomical Society in 1982.

Davis has contributed to both observational and theoretical cosmology. His research interests include the distribution and clustering of galaxies and the evolution of large-scale structure in the universe. In the late 1970s, collaborating with John Huchra and others, Davis started the original Harvard-Smithsonian Center for Astrophysics redshift survey, which compiled redshifts, and hence distances, of 2,400 galaxies. Hoping to develop theoretical models for the observed galaxy distribution, Davis then moved to Berkeley and began working on computer simulations of galaxy clustering. In 1983 he and his co-workers demonstrated that "hot dark matter" models of galaxy clustering have serious difficulties. With Simon White, Carlos Frenk, and George Efstathiou, Davis has done influential work on modeling the evolution of structure in a "cold dark matter" universe. Davis is a tennis player and plays to win.

I wanted to start with your childhood. Do you remember any particularly impressionable experiences?

In terms of my interest in science, I recall vividly attending a NASA exhibit in Cleveland. I was raised in Canton, Ohio, about 60 miles to the south. I'd been interested in science as a young child. My parents took me to a big NASA exhibit for the public. I must have been 10 or 12. They had lots of material to give away, and they had a big chalked sign on a blackboard saying, "More material to anyone who knows what an ion is." I answered the question of what an ion was, and they gave me a pile of material. I have a very vivid memory of a chemistry set, as a young child. I kept trying to make explosives, but it was a very safe set and I couldn't make anything explode.

Tell me a little bit about your parents and what they did when you were growing up.

My parents encouraged me to read. I used to go to the library a lot. Neither of my parents had gone to college. My father and mother were raised in a poor family. They're second-generation immigrants. My mother had been a very good student in school. We used to read a lot of books together and talk about books all the time. When my father was a student, he had to work part-time and help support his family and had to go to school only part-time. He never had the chance to go to college, but he was very enterprising and was a businessman, a wholesaler. He started his own business, and it grew over the years, eventually employing 40 or 50 people.

What about your mother?

My mother worked for him as a bookkeeper. She was always working, but usually at home. For a long time, the only office for the business was in the living room of our house. When I was a very young child, my parents owned a truck and no car. My father bought things and sold them, using the truck. I was told that he would be gone most of the week and only come home on weekends.

Do you remember any books that you read at that age that were memorable?

Well, I must have been older. I read a lot of science books at the eighth- and ninth-grade level—books on relativity and on physics and chemistry. In the fifth grade, I had an influential teacher, Mr. Friedman. We had just moved to a different school district. He saw my interest in science, and he gave me a chemistry book to read, a high school chemistry text. I read that over the summer and thought it was great. That was my first serious science book, I suppose. I must have had a strong interest before that. I do recall very vividly that later in high school, George Gamow's books, *One Two Three . . . Infinity* in particular, were really influential. And I was quite intrigued by *The Creation of the Universe*. But particularly *One Two Three . . . Infinity* was just spectacularly great for me. It really made me decide that I wanted to study physics.

Did it make you interested in cosmology in particular, or just physics?

That one turned me on to physics. I think all the relativity books that I looked through made me particularly interested in gravitation and cosmology. Even in high school I thought I was interested in relativistic questions. I wasn't an amateur astronomer. I never really knew anybody who knew anything about telescopes. Maybe I would have been interested in telescopes if I knew they existed. That part I've never experienced—building my own telescope, for example, which a lot of my peers have done.

Had you been exposed to theories of cosmology in any of your popular readings at this age?

Well, certainly *The Creation of the Universe* was about a theory of cosmology, but I don't recall understanding, or even reading in any detail, more than a few buzz words—big bang versus steady state.

Do you recall ever thinking of the universe as a whole at this age?

Yes, I definitely did, particularly after reading relativity books. I get confused as to when I read what. I read Bertrand Russell's *The ABC of Relativity*. That was pretty amazing. I still have a few of them around. Let me see if I can see a date on it. [Davis looks on his office bookshelves and finds *The ABC of Relativity* by Bertrand Russell.] Here it is. It's got my home address in Canton, Ohio, so I must have read this in high school. I think I have *One Two Three . . . Infinity* here also. Yes, it also has my home address in it.

And you have these books here with you in your office right now.

Yes. On occasion, I've gone through my various collections of books and sorted them. These went into my office. I've had them here as a reference to undergraduates who come by and want to learn more science at an introductory level. I'll recommend books like this to them. These are still really good.

Let me move on to your undergraduate education at MIT. Do you remember any particularly influential experiences there?

The whole experience at MIT was staggeringly influential to me because I had come from a modestly sized town in the midwest and was somewhat unusual in my interest and talents for science. My school didn't have that many peers to talk to—no one at all with my interest in science. Then I go to MIT and discover that the entire class seems to be made out of class valedictorians like I was. I remember the introductory week at MIT, when the freshman dean said, "Look around the room. Half of you scored 800 on the advanced math aptitude test." I came from a school where I didn't even have preparation to take that test. I had to take the intermediate math test because I didn't have an advanced course in high school to prepare me for it. So I felt very bad—thinking here I was, perhaps in over my head. And MIT was definitely a lot of work. It was not like high school, where you could just breeze through without worrying about anything. MIT required a serious effort. But it was a fantastic environment to be with that class of peers, people who really were interested in science, people who were really cultured, who knew about music. I didn't know a damn thing about classical music when I got to college. I didn't know anything about anything, it seems, compared to what all these kids knew. I felt I had a lot of learning to do. It was an incredible experience. It required a lot of effort, but I managed to make my way through.

Did you take any courses in physics or astronomy that got you particularly fired-up?

Well, practically the whole curriculum. I started off as a physics major and, course after course, I was really fired-up—even the introductory freshman physics with 600 students in it. I remember Professor [Anthony] French, who was very, very good when we were just learning calculus and mechanics. I had a very inspirational course with Ray Weiss, who taught the first introduction to electricity and magnetism. He was a great lecturer. There were any number of really good courses in physics.

Did you know at this time that you wanted to become a scientist?

Oh, that was clear. I knew before I went to college that I wanted to study science, and that's why I chose MIT.

Were you still interested in gravitational physics?

I wanted to remain open. I thought I was interested in physics. I didn't know enough to specify my interest beyond that, but it always had been clear that I was more interested in gravitational physics than solid state physics, for example. So I felt that was what I would probably end up doing, and it is what I ended up directing myself to. At the same time, though, my parents were clearly struggling to pay for my being in college, and they wanted me to work while I was there, which is only fair. The university paid rather poorly, and I was able to work outside for a software firm, a computer firm called Adage, Inc. I became a systems programmer and made a lot more money than I would have made working in the dining hall. That took a fair amount of time. I was working as much as 20 hours a week in my junior and senior years.

> *Do you have any opinion why gravitational physics would have appealed to you more than solid state physics at this age?*

I felt that it addressed large questions that I had read about as a child, questions related to great philosophical issues. In my humanities, I minored in philosophy. I had read philosophical discussions—Kant, for example—on the early notions of the universe. That topic used to be the domain of philosophy. These were large questions that always appealed to me. How can a high school kid who doesn't know any science at all say he wants to study solid state physics? He doesn't even know what solid state physics is. It's a much more technical subject that requires knowing the minutiae.

> *Let me ask you a little bit now about your graduate work at Princeton. Can you tell me about the influences that you got there?*

By the time I had gotten through my senior year at MIT, I was doing pretty well. My grades were good, and I applied to a bunch of graduate schools. When I went to Princeton, I encountered, once again, a culture shock. I thought I knew some physics, but here the class was full of people that were just unbelievable. There was Claudio Teitelbaum, there was Jacob Bekenstein, there was Bob Wald. There was Claude Swanson, who has since left physics. He was a friend of mine from MIT, and he was reading about Banach algebras and all this stuff. I didn't know what the hell he was talking about, frankly, and these guys just left me in the dust. Partly, I felt that here I was paying the price for the fact that I had worked while at MIT and hadn't taken quite so many graduate physics courses as some of my peers had taken. That put me at some disadvantage, particularly in theory, where people seemed to really have a head start.

Is that what you wanted to do—theory?

I wasn't sure. I never could quite decide what I wanted to do. I thought I might want to do theory, but I like tinkering with my hands as well. Because I was interested in gravitational physics, I started to associate myself with the gravity group. That was Bob Dicke's group at the time, and, of course, Dave Wilkinson and Jim Peebles were major players. I started to work with Dave and Paul Boynton. Bruce Partridge was also there.

Sam Treiman was a particularly effective teacher. He was amazingly brilliant as a lecturer. Even though I thought I knew statistical physics, taking a course from him was just wonderful. Then he taught a course in field theory that was just great. He just made it all seem so simple. Of course, then you get out of the class and you discover that you can't calculate anything, but you sure could understand what he was talking about.

Did you take any astronomy courses?

I took relatively few. I sat in on an essentially shotgun seminar with Lyman Spitzer, which was a bit intimidating because I really wasn't up to speed with the subject. Martin Schwarzschild was teaching a stellar interiors course, and I sat in on a seminar with that. During my first year, Jim Peebles was offering a course in physical cosmology, which led to his first book.[1] At the time, I sat in on it a bit, but I was so busy with other courses that I didn't take it regularly. It was really geared for more advanced students.

Didn't you ultimately do your thesis with him?

No, I did it with Wilkinson. People think that I worked with Peebles, but the project with Peebles was later. So there were some really great people around. You start to associate with these really high-powered physicists, and it's hard to maintain an air of superiority like you could in high school. So it was good. It was humbling and I worked hard.

At this time, was cosmology particularly attractive to you, or was it just another possible discipline?

It was another possible discipline. It hadn't branched out to the point that it has now. Jim Peebles certainly made it interesting. His course was great;

his book was terrific. He was truly inspirational. He basically invented the subject, for the Westerners at least—he and Zel'dovich. My thesis was inspired by some crazy paper he had written with Bruce Partridge about how primeval galaxies might appear in the sky.[2] Looking back on my thesis, it's almost embarrassing. I did a project with Dave Wilkinson.[3] I built some hardware to search for faint, fuzzy extended objects on the sky. We took it to Hawaii. It was nice to go to Hawaii and use the telescopes. We searched in blank areas of the sky, looking for faint, fuzzy objects. We didn't find any. Of course, in those days we didn't have any CCD detectors or anything. We had a photo-multiplier. It's ludicrous, the limits that we could set then compared to what people can do now. The Partridge and Peebles models are no longer considered relevant classes of models for forming galaxies, but I got a thesis out of it and went on to do other things.

> *As a graduate student, do you remember having a preference for any particular kind of model, that is, open versus closed or homogeneous versus inhomogeneous?*

Well, the question that had just been settled five years earlier was the hot big bang versus the steady state. Of course, being at Princeton there was no question that the hot big bang was right. So that was over. It was interesting to see that cosmology, because of the microwave background radiation, had now advanced to the point of having data. It was a real science. It wasn't just philosophical speculation. It rapidly became apparent, as Wilkinson, Partridge, Boynton, and their students did one experiment after another, that the radiation was quite isotropic.[4] I remember the first several attempts with Dave's students. They couldn't detect any anisotropy at all. Eventually, they succeeded in detecting a dipole anisotropy after a lot of hard work. So the evidence was clearly pointing toward an isotropic universe, and by the Copernican principle, unless we're at a special place in that universe, you would infer that it's homogeneous as well. Of course, that's the simplest universe, the simplest mathematical model. On the question of open versus closed, much of the data we use now was somewhat in existence at that point. The evidence was that the universe was open, based on masses of clusters and groups of galaxies. Even though the data is greatly improved in the last decade or two, the conclusions have basically remained the same.

> *Did you accept that or did you think about it?*

I thought about it. We thought about it a lot. I had some preference for a flat universe or closed universe. I've always had trouble visualizing an infinite, open universe and still do. I understand what it's saying, but somehow an infinite number of galaxies seems a little bit strange. But that was just a problem that I had.

So you thought that maybe a higher density might be discovered in the future, or something like that?

We always knew the standard arguments that you can recite today. The omega estimates of masses of clusters were sensitive only to the cluster component and were insensitive to a uniform component that could be there for reasons unknown. It still could. The other test, which at that time was considered more promising—the curvature tests at large scales—were variously debated one way or the other. Allan Sandage wrote these elegant papers in the early 1960s that I read closely.[5] We all followed the data coming from Palomar with intense interest—the question of magnitude–redshift tests and all that stuff. Then Jerry Ostriker and Scott Tremaine, and also Beatrice Tinsley at Yale, started to calculate evolutionary corrections of one sort or another.[6] By the time I left graduate school, it was clear that it was hopeless. The evolutionary corrections were going in both directions. They didn't know which one was greater amplitude. There were too many astrophysical parameters. That cast a pall over global tests. That must have been 1973 or so when those tests were really in question.

Peebles and I then started working on these questions. I had come to Peebles when I was searching around for a thesis topic, and he had tried to turn me off by saying that what the world needs is good data. It doesn't need another crummy theory. That has always been his attitude—good observations are worth more than another mediocre theory. I've been told that by quite a few people. I remember John Bahcall had a similar conversation with me once.

It's remarkable that Peebles would say that, given that he has been one of the leading theorists in cosmology.

Well, there are two reasons for that. One, the subject is data-starved, and good data is worth its weight in gold. Secondly, in all honesty, I think that Jim would prefer not to have too much competition from other theorists. He liked it when he had the field all to himself, and he did such a good job of selling it that he had all these young turks come and be with him. He never considered me as competition because he knew I was primarily an experimentalist. But I think he felt a lot of other people were competition.

Maybe I'm misreading him. In any event, regarding other tests, Jim and I started to follow up some kinetic theory calculations that he had done while he was on sabbatical at Berkeley. He had showed me a notebook of calculations he had done related to what is known as the BBGKY hierarchy. He had thought about applying it to cosmology, and he and I started working on that. We wrote a couple of long papers, one infamous one in *Astrophysical Journal Supplement*.[7] We had a lot of flak for what we said in that paper, and I think fairly so, but that got me a little inspired that maybe you can infer omega in rather different schemes than the way we had always thought of—virial analyses in isolated groups. At that point, we realized that we had some interesting statistical tests related to velocity fields and relative velocities of galaxies in space, and it sure would be nice to have some data to test this with.

Which leads to the redshift survey, which I was going to ask you about.

Yes, I had this theoretical interest that led me into redshift surveys. It was becoming clearer that these statistical procedures could be very powerful, but they just needed better data.

Did it occur to you to get N-body data, or did you just immediately decide that the real data was by far preferable?

The N-body data was cooked. You knew what omega was. In fact, I don't know when it first occurred to me that the N-body data would be a good way to calibrate the sky. You really wanted to have both. In any event, you clearly needed to look at the sky in some detail. Margaret Geller, John Huchra, and I started talking about this at length. We did some analysis with the Shapley–Ames catalog, which is a whole-sky survey of about 1,100 galaxies to the 13th magnitude. We started looking at the 3-dimensional distribution of galaxies there. We didn't do very much with it. It was clearly not a very fair sample because there were three times as many galaxies in the north as in the south.

So you were using other people's data?

That data had just been made available by Sandage and Tammann. So we were using that data. We studied correlation functions and wrote a paper but didn't do too much.[8] I remember a workshop we once had at the Radcliffe Quad, where Bill Press had this very interesting way of describing the available surveys of redshift space. He said it was like a mace—a mace being an iron ball with spikes—and the height of the location at any angle is the depth of the survey information at that region. A few regions had been surveyed deeply, but most had been surveyed very shallowly. I think at the time he gave that talk—I remember Jim Gunn being around then—we were already getting started to get on with the CfA [Center for Astrophysics] survey.

So you had decided then that you were going to collect your own data?

Yes. I had been at Harvard for about a year, and I knew that I could count on a six-year term hopefully without too much problem. I had the time to do a big project and decided that it wasn't worth fiddling around with just a few observing runs. We would do something serious. John Huchra and I were chatting with everybody to try to get some money. I had gone out to Mt. Hopkins to use their spectrograph and see what it was like, and it was a disaster, totally useless. It was clear that we were going to have to do a lot of work. Here is where the Center came into its own. It was a project that we were able to do at Harvard-Smithsonian because of its size and resources. It would have been impossible at any normal university setting.

We wrote a proposal to the NSF for $300,000. I remember it was only when George Field and Alexander Dalgarno got on the phone and badgered Jim Wright, or whoever it was, that I got $20,000 out of them. Herb Gursky had just become associate director of the optical and infrared group. He was used to spending money because he came from the x-ray division. So he didn't think anything was amiss when we said we needed lots of bucks to build a real computer-driven system and we needed the resources and manpower to make it work. Fortunately, Tom Stevenson, who had written the drive and control system for the Multiple Mirror Telescope, was available to help with our system software. I latched onto John Tonry, whom I'd known at Princeton. He was a math student at Princeton and then went to Harvard physics. He was interested in the project from the beginning.

I had met Dave Latham when I first went to Harvard. He was a photographic expert, but no one was utilizing his talents at the time, and he got interested in the project. The Smithsonian put an enormous amount of money into rebuilding the spectrograph there. We had all kinds of problems; we had to fix up a lot of it. Dave Latham took responsibility for building the image tube package which we needed for this. I thought the most suitable device was the photon counting Reticon systems that Steve Shectman had built. One summer I went out to his lab and spent the summer copying his electronics. There were wires running all over underneath the circuit boards and various components removed and replaced.

I'm no expert at any of this so I didn't do the neatest job in the world, but it did work.

I wanted to ask you one more question about the CfA redshift survey. You thought it up, designed it, and got it running. Why did you then move on to other things? Why didn't you keep pushing it beyond the initial survey?

We had a great time doing the survey, and we learned an enormous amount about the large-scale clustering of galaxies. When I wrote our series of papers, I started to compare the data with the best available N-body simulations, since it was clear that they were a useful tool for comparing to these data.[9] The models were so bad in comparison to the observations that it was clear that the data was way ahead of the theory. As I had been interested in theory anyhow, I decided it was time to focus a little more on theoretical questions. If we repeated the survey, we could work three more years and double the database, but I didn't think it was going to change qualitatively. I think I've been proven wrong by some of the results that Margaret Geller, Valerie de Lapparent, and John Huchra have shown. They've shown that the clustering is even a little more dramatic than we thought it was. They have better signal-to-noise on the structures, and you're really quite struck with how empty and round some of these things are. But, in any event, I felt qualitatively—and I still more or less believe it—that we had a pretty good feel for what the clustering was, but we had a very poor intuition about what it meant. I was therefore inspired to try to focus on that.

I want to move to your reactions to some recent discoveries. Do you remember when you first heard about the horizon problem?

Oh, I probably didn't hear about it in any serious way until I took a relativity course—probably in graduate school, in 1971 or so.

Do you remember how you reacted to it when you heard about it? Did you regard it as a serious problem?

I never paid great attention to it because it was clear that in some sense it was an initial value problem, at least from all the theory that we knew of. I didn't have any basis for understanding how I could argue one way or another about initial values. I didn't know how to address the question scientifically. It seems a little bit bizarre, but so be it. That's about as deep as I went on it.

*Did your view of the horizon problem change any as a result of the infla-
tionary universe model?*

I have to say that I was so impressed with the inflationary model because it
had promoted the horizon problem to a tractable problem. I was so
amazed. I thought, "My God, this actually can explain the horizon problem." That sold me immediately. Here we had a real explanation for these
things that we didn't think could be addressed at all. In a sense, when
people ask about the role of God in science, scientists frequently answer
that God is just beyond. He was pushed out of the solar system when
Newton made celestial mechanics tractable with his inverse square law.
He was pushed back to initial conditions when the big bang model came
around, because we didn't need God once the initial conditions were set,
but we sort of needed Him to set them up. Now inflation has pushed Him
back to the Planck time. Maybe if Linde's ideas of eternal inflation are
correct, and ideas about higher dimensional spacetime are correct, then
God's in 10-dimensional space–time or something.[10] But the progress of
science pushes back these questions that you think aren't even address-
able. That's what is so impressive—when you can actually push back your
ignorance to a point where you can address a question that you didn't
think was in the bounds of science at all. So that was a real step forward.
Although it's not proven that the inflationary universe model is right, it
certainly has a strong appeal to all of us.

*Why do you think the inflationary universe model has caught on so
widely, given that there is so little direct observational support for it?*

I guess it's because it solves the horizon problem and flatness problem.
Perhaps it solves the flatness problem too well, but it seems to solve it in
what appears to be a dynamical and physically intuitive method. I think
what people like is the *results* of inflation. They like what it does, that we
have a dynamical explanation of these amazing observations. I think a lot
of us would be happy to use whatever mechanism is shown to work. It
doesn't have to be inflation, if you can think of something else that
produces these good results. So far no one has, but maybe they will
someday.

*Do you think that the inflationary universe model or some version of it
has a good chance of being correct?*

Yes, although to say that authoritatively probably requires field theory
credentials that I don't have. I know that while the particle theorists are

impressed with the results of inflation, as is everyone, they are less impressed with the means by which the results are obtained. There is no fundamental scalar field that anyone has ever seen, and that's what you need. It has to be weakly coupled in an extremely strange way. You need rather refined parameters in order to not overproduce fluctuations when you end inflation, and simultaneously get to a sufficient temperature to do baryogenesis, which must occur after inflation. So the fine-tuning problem in the current inflation models is pretty extreme and not considered very compelling by any particle theorists. That's just the state of affairs. It's not satisfactory. I don't know how we're going to make progress. I'm not going to make progress on those questions.

Do you remember when you first heard of the flatness problem?

Oh, it must have been about the same time as the horizon problem. It might have been earlier, because in popular books on general relativity, they talk about the universe expanding or contracting eventually, so the flatness problem comes in there. I must have read about that.

Was it stated as a problem?

No, it was stated as options. As a dynamical instability, it was not appreciated. That wasn't impressed upon me until later. My first recollection of it was the analogy that Jim Peebles used to make with Bob Dicke, namely, that the flatness problem is like the problem of a pencil standing on its point.

Did you consider it to be a serious problem?

I did. I considered it to be a very serious problem and my heart was not sold on the famous paper by Gunn, Gott, Schramm, and Tinsley, in which they presented four lines of argument for an open universe [with omega equal to about 0.1].[11] I knew the arguments, but, philosophically, I wasn't convinced. Peebles and I at the time were arguing that the shape of the correlation function dictated against an open universe.

When you say that you took the flatness problem so seriously that you were not persuaded by the Gunn, Gott, Schramm, and Tinsley article, does that mean that the flatness problem made you think that omega was equal to 1?

Yes. That's right. I had a reputation, I think, of being one of the people trying to defend an omega of 1.

Did you ever wonder how it was that the universe began as an omega-equal-1 universe? Did that concern you, or was that just another question of initial conditions?

In the same class. The reason that the flatness problem wasn't wholly compelling was that we couldn't justify why omega started off at one in the first place, so we couldn't really justify why it had decided to deviate from one a few Hubble times ago. It's just one of those things. Unless you have a dynamical argument, you're arguing about nonphysical questions. You can't address the question.

If you took omega equals 1 as an implication of the flatness problem, did you then worry about where all of that extra mass was?

Yes, I definitely worried about it. It was clear that we had a lot of loose ends, that we simply didn't understand how we could make these mistakes unless there were things like Peebles was suggesting—dead galaxies out in space. In his first book, I think he mentions dead galaxies. I don't think he'll talk about that anymore, but you have to hide the stuff somewhere. He talks about all the ways to hide it. It had to be something we didn't know about. Relativistic background sounded really favorable for a long time.

So you were willing to entertain the possibility that there was all of this stuff out there that was unmeasured?

Yes, I was willing to entertain that.

Based upon this theoretical argument.

Yes, the theory was a strong influence. I had no data to support a high omega so it had to be a theoretical argument or a philosophical argument that drove it.

What would the philosophical argument have been?

I guess the uniqueness of an omega of 1. If it's not 1, then why isn't 10^{-6} or something completely different? Why is it so close? That sort of thing. Why are we so special?

When you first got whiff of the results of de Lapparent, Geller, and Huchra, which came out of a program that you had created yourself, were you surprised?

Yes, I have to say that I was a little surprised that they seemed to find a case of such clean delineation of on or off voids, on or off filaments. That didn't actually square with my perception of what I had seen in all the different slices of the CfA survey. I think, in fact, the reality is that [their first survey] was a fairly exceptional slice of the universe, that a more typical slice of the universe is muddier.

We will know more in the next few years.

Well, there is a lot of data that is available from them and from people such as Giovanelli and Haynes. We finished our southern sky survey, which just was published this year.[12] Generally, we don't see quite such compelling on and off slice distributions. Rather, we see what appears to be a heterogeneity of structure. In some cases, you just see a loose agglomeration of stuff with no well-defined structure to it at all. I think the reality is that once we have put all these data sets together, we will perhaps be able to characterize the frequency with which you see such clean structures as that [the de Lapparent-Geller-Huchra survey], and it isn't going to be all that high. So I was surprised that they found such a good one. I have to say that I felt that the data got to be overinterpreted, because they started to argue for bubbles. We had actually looked at lots of regions in space. We had looked at bigger volumes than they have, and we haven't seen things that I could convince myself were bubbles in the sense of surrounding 4π enclosures. I immediately liked the term "sponges," when Rich Gott made it up. I think that's a fair description. We have interconnected voids and interconnected filaments. That means that nothing is completely surrounded, and that I think is a better representation.

Do you think that these structures suggest that the universe may not be homogeneous enough to satisfy the assumptions that go into the Robertson–Walker–Friedmann models. Are you worried at all?

No, I'm not.

You still think that rough homogeneity is a good approximation?

Yes, and the reasons for that are more than just philosophical desire. There are constraints. If inhomogeneity persists on too large a scale, you start getting fluctuations in the microwave sky, which we don't see. If the inhomogeneity is gravitationally attracting, you would expect large-scale coherence in the velocity field—more than is observed. We are working now on a paper, in which we computed velocity correlation functions, and the coherence length is not all that excessive. In fact, it matches the cold dark matter models, believe it or not. The Seven Samurai amplitude is very large.[13] That's hard to match with cold dark matter. But other catalogs of velocity fields, such as the Aaronson, Huchra, Mould study of the local supercluster flow field, have the same coherence length as the Seven Samurai, but not the same amplitude.[14] There are other data sets, such as the IRAS velocity field, or IRAS gravity field, that I've been working on with Michael Strauss.[15] It matches the Aaronson, Huchra, Mould velocity field very well. None of these things have a coherence length that's excessive. So I think there are real tests now. Jerry Ostriker and I apparently disagree completely from the same data. I think the data supports the notion that we are approaching the Friedmann limit [of homogeneity].

So your faith in the big bang has not been shaken by any of this?

No, I'm not willing to claim, as he apparently is, that the dipole anisotropy may not, in fact, be a result of motion. It may be intrinsic. I don't think we need to take such a Draconian step just yet.

Do you think that the theory and observations in cosmology have worked well together in the last 10 or 15 years?

Oh, absolutely. I have enjoyed it immensely, because I was fortunate to be in the field before it became so popular. We've seen the situation shift from one side to the other. In a sense, cosmology was originally thought to be a science in search of two numbers [H_0 and q_0]. Now all this industry of people searching for fluctuations has kept an army of theorists busy trying to calculate the expected small-scale and large-scale fluctuations in various models. It has already ruled out several classes of models. The large-scale structure dovetails in nicely with this because it provides normalization to the amplitude of fluctuations expected. That's observational input that the theorists absolutely need. So the interaction is quite good. Now this interaction with particle theory is just amazing. In fact, that there is any interconnection at all took us all by surprise. Before the invention of inflation and grand unification, none of us thought too much about the intimate connection that now exists. However, I think the particle physics connection is dropping by the wayside. The particle theorists have their problems with grand unification and supersymmetry and superstrings. The astrophysicists are worried about the details of galaxy formation and whether the cold dark matter model works or not. They're not communicating as much as they were a few years ago.

What do you think are some of the major problems in cosmology right now?

On the observational side, we still need to understand the Seven Samurai result. That's very ill-explained. The data is not secure because the different groups do not agree on the flow field in detail. That has to be developed. In the next few years, I think we are going to see an enormous growth of knowledge in the flow field given by peculiar velocity studies using Fisher–Tully or Faber–Jackson relations, or improved relations. That will finally allow us to measure a gravity field. That will be great, because then we'll have real information about mass distribution and not just the galaxy distribution. Another serious question is whether there is any evidence that this bias mechanism works. There are observational consequences of our explanation for natural bias. For example, we would expect weak, small galaxies to be less biased than rich galaxies. Some of us have written papers saying that we can see it in various data sets. Other people have criticized those data sets for various incompleteness problems and have said their own data shows no such effect. It's just this year that this controversy has really been coming to light. That has to be resolved because the bias predictions are very clear. There has to be something out there that's clustered, but not as clustered as the galaxies. If that's wrong, then the model's basically irretrievable.

You can't explain omega equal to 1?

No, it isn't going to work. So that's fundamental. There is another experimental question—find the dark matter. No physicist worth his salt is going to like this until you can actually discover this dark matter in a laboratory. It's got to be there in some form or another, but unfortunately, there are too many candidates and none of them are well enough motivated. Theoretically, I guess the particle theory question is very serious. I hear lots of problems with all the supersymmetric particles. They're all in bad shape. None of the inflationary models make sense. None of the dark matter candidates are particularly compelling. It's not a good situation. We think—prejudicially—that the cold dark matter model has provided a good motivation for looking at galaxy formation with real initial conditions that you think make sense. The technical challenge is to take these initial conditions and try to form galaxies with real physics. This will probably have to be done numerically, but it's going to be extremely difficult. There is too much physics that happens all at once to likely lead to believable results. But work will pick up on that in the next several years as people really get geared up for it. We'll see a lot of output on that in the next few years. I don't know if it will lead to anything.

Let me ask you to take a big step backwards. If you could have designed the universe any way that you wanted to, how would you have done it?

I know what I would definitely have done. I would have provided a noncompact dimension—or some mechanism—that would permit you to do time travel, because here in astrophysics we see all these staggering things. I liked Carl Sagan's recent novel, *Contact,* where they managed to travel to the center of the galaxy by going in some sort of space warp. I would like something of that sort. The fact that we can't travel faster than the speed of light is obviously essential to make physics work. It wouldn't be a very sensible universe if it were otherwise. But our inability to get to astrophysically interesting sights and see an accretion disk around the black hole or any such things is really disappointing. We can speculate on all this, but we can't get there. I fear that we're never going to be able to travel the galaxies. I'd like to design a universe that would permit that.

There is a place in Steve Weinberg's book The First Three Minutes *where he says the more the universe is comprehensible, the more it also seems pointless. Have you ever thought about whether the universe has a point or not?*

I try not to think about the question too much, because all too often I agree with Steven Weinberg, and it's rather depressing. Philosophically, I see no arguments against his attitude, that we certainly don't see a point. To answer in the alternative sense really requires you to invoke the principle of God, I think. At least, that's the way I would view it, and there's no evidence that He's around, or It's around. On the other hand, that doesn't mean that you can't enjoy your life. I feel very privileged as a physicist. I think about the question a lot. What is the point of most people's lives? They make a living and do some work that's moderately interesting, but physicists get to do these amazing things. I'm very exhilarated when I wake up in the morning because we can go back and knock on these questions a little more and eventually make a little progress. That's got a real point. I'm very enthusiastic and excited by it. I'm sure that many other people in different lines of work have similar good feelings about what they do, but being a physicist, I think, is very special.

MARGARET GELLER

Margaret Geller was born December 8, 1947, in Ithaca, New York. She received her B.A. in physics from the University of California, Berkeley, in 1970 and her Ph.D. in physics from Princeton University in 1975. Geller was a thesis student of James Peebles. Following her degree, she took a postdoctoral position at the Harvard-Smithsonian Center for Astrophysics. From 1978 to 1980 she was a senior visiting fellow at the Institute of Astronomy at Cambridge University. She is currently a professor of astronomy at Harvard University and an astrophysicist at the Smithsonian Astrophysical Observatory.

Geller's research interests include the origin and evolution of galaxies and the nature of the galaxy distribution. She and John Huchra are frequent collaborators and have written over 30 papers together. Since the early 1980s Geller and Huchra have jointly led the Center for Astrophysics redshift survey. They are extending the original survey of 2,400 galaxies to 15,000 galaxies; over 10,000 have already been measured. Working with Huchra and Valerie de Lapparent in 1986, Geller discovered that some galaxies appear to be located on the surfaces of bubble-like structures, about 100 million light years in diameter.

Geller has an unusually strong capacity for 3-dimensional visualization, having been trained in geometrical perception as a child by her father, a solid state physicist at Bell Labs. She maintains a strong interest in design and says that had she not been a scientist she would have chosen a career in the visual arts.

I wanted to start off asking some questions about your childhood and how you got into science. I know that your father is a scientist. Do you remember any of the early influences that made you decide to go into science?

Yes, I think I was interested by my father's encouragement. He bought me a lot of toys that especially had to do with geometric things. I used to build solid shapes and play with all kinds of toys that really improved my 3-D perception. I think that is a skill we acquire, that most people actually don't have. I was trained by my father in that regard, whether I was going to be a scientist or not. Then I was interested in mathematics. I was very bored in school. In fact, I often made excuses not to go to school. I would pretend I was ill and stay home, because I just really didn't like the elementary school too much. My parents realized pretty quickly that I was not ill and decided that I could stay home provided I studied something. I was allowed to read or whatever. When I was in grade school, in fact, my father encouraged me to learn algebra on my own. He had old algebra books by Hall and Knight, and I would work through the problems instead of going to elementary school. I think I was absent about a third of the time from school. Fortunately, my parents were educated enough that they could really educate me. I was interested in mathematical things at an early age, but I was also very verbal and I liked to read. I read voluminously. My mother really encouraged me more in the arts, and my father was always encouraging me in the sciences. He worked at Bell Labs, and he used to take me to his lab. I would sit there and measure X-ray diffraction photographs. People would come around and say, "Ah, that's cute," but it made me feel very important. At that time there were two women scientists at Bell Labs, Betty Wood and Vera Compton, and I think somehow that registered on me—I don't know exactly how—but they would always take special care to come around. I was somewhat aware that there was something funny about being a little girl interested in mathematics. I got the message from my teachers. One in particular gave me a very hard time, which I think was one of the reasons I didn't like to go to school too much.

You got the message that girls should not go into science?

Oh, yes. From this one teacher in particular I certainly got it. She would ridicule me, sort of as a girl, not about what I did, but she would make me feel awkward. And I was an awkward kid. Then, of course, other kids would join in. It was quite subtle, but I realized it; I wasn't stupid. But I was very much encouraged by my parents, and then I went to a very good high school where I was able to take examinations and advance to higher level mathematics courses.

Would you say a little bit more about playing with toys and how that gave you a feeling of 3-dimensional visualization?

My father is a crystallographer. Not only is he interested in crystallography, but he is also interested in design—the design of furniture and buildings. I was fascinated by that, although I never got included. He had an attraction for any kind of toy that had anything to do with geometry. I don't know what company they were made by, but there were toys where you could connect flat shapes up with rubber bands to make solid figures. He bought me that, and he'd explain to me the relationship between things that I built and things in the world. For example, I'd make a cube, and he'd explain to me the relationship between that and the structure of table salt. And I'd make an icosahedron, and he'd explain how you see that in the real world. He would talk about filling space and which figures could fill space and which couldn't. I guess one of the things that attracted me to it was that my father was making his way in science. He was under a lot of pressure, and I didn't see him that much, but he was always ready to talk to me about these things, or to play, to show me how they worked. He would show me and I remember not understanding. He would say things—this structure was this or that—and he'd show me the beautiful models of the crystal structures, and I wouldn't be able to see it. But he would say it again and again and again. And finally, I *would* be able to see those spatial relations.

You'd be able to visualize it in 3 dimensions?

Yes, I would be able to visualize in 3-D. And I realize now—I've talked to lots of people in science—that very few people have that ability. It's very rare. But my father really trained me as a child, and I really learned it. I had lots of things to help me learn it. And it's related to what I do. In fact, I think one of the reasons I was able to recognize what we had [in the data revealing the bubble structure of the distribution of galaxies] was that early training. In a way, it's kind of funny, because in a funny sense my father and I do the same thing. It's really kind of strange. After the redshift survey results, I had several talks with him about how we're essentially both doing geometry, but on very different scales.

When was it that you decided you might want to be a scientist?

I guess when I was thinking about going to college. I thought I wanted to be a mathematician, but I was disillusioned from that pretty quickly. I had the idea for a long time, and it was clear to me it might be an interesting thing to do. As I said, I'd go to my father's lab, and he made it very clear to me what was involved in what he did. And I liked other people there. The lab was a very exciting place at that time and had a very different atmosphere from what it has now. People were very encouraging to me, by and large. I got the idea that science was an exciting thing to do and that there were people who really enjoyed doing it and that it could be rewarding. It's not that I didn't have doubts. I had doubts for a very long time about doing science, even after I got a Ph.D. And I never thought about doing astronomy. I thought I wasn't the least bit interested.

When did you decide to go into astronomy and cosmology?

It was really almost an accident. It was essentially when I finished undergraduate school. At that time pulsars had been discovered, and it was clear that there were interesting things happening. First I thought I would go into mathematics, but I realized very quickly that I wasn't creative in mathematics. It was clear to me that I could do it, but I didn't know the questions to ask. And so, encouraged by my father primarily, I took physics and found that I liked it. Not only that, but there were lots of questions hanging around, and it wasn't that hard to ask questions, even though at that time I certainly couldn't answer them. I really was so heavily influenced by my father that I was thinking of going into theoretical solid state physics. I think it's probably fortunate that I didn't, because I would have been very much in his shadow. It would have been very difficult for me psychologically, I can see in hindsight. I was sort of afraid of astrophysics, because I felt at the time that it was the kind of thing where you could never be sure. And it *is* the kind of thing. In solid state physics you go in the lab, you measure something, you can touch what you do. But we're never going to know the answers to the questions that we ask in cosmology, because we're never going to touch these things. We're only going to have consistency checks. That sort of frightened me, the idea that it wasn't really the same kind of thing. But I was encouraged a lot by Charles Kittel at Berkeley, who told me that you should choose a field that is going to be most alive when you are doing it, and that astrophysics was a field that was just opening up, and it would be a good idea to do it, and I should go to a place where it is best done. I applied to only two graduate schools, Princeton and Berkeley. I more or less decided that if I got into Princeton, I would go there and do astrophysics.

You hadn't chosen cosmology?

No, I'd do astrophysics. I did get in and I went there. I was in the physics department, and Peebles was there, so I more or less decided at that point. But I have to say that I really didn't find it all that exciting until much later. I just sort of *did* it, and I got through as a student. I really didn't turn on as a graduate student, to say the least. Even for a while afterward, I had very grave doubts about whether I had even chosen the right profession, and I came very close to getting out many times.

Why did you decide to go into the large-scale structure part of cosmology?

I don't think it was that conscious. That was the option that was available, and I think even at the time I went into it I didn't really understand what profound changes were in store in our ability to explore it. I didn't understand for quite a long time how fuzzy the foundations in this field were, and what they were based on, and how little was really known.

I remember one of your very first papers with Peebles was on the question of the expansion of the universe, a very fundamental issue.

Yes, but that didn't really make me realize it. It was only much later, when I went to England in 1979–80, and I was asking myself the question. I wasn't that excited by the problems I was doing. I was working very much in the style that Peebles was, and I have to admit it wasn't my style. I guess Al Cameron actually put the bee in my bonnet. My office was down the hall from his, and he once asked me, "Do you do problems because you can, or because they're interesting?" And I was doing them because I could. I realized that if I was going to stay in science, I was going to have to make some changes and do problems because I was interested in them. Because for me as a person that was really the important thing. That's what really drives me. What surrounds me in my life just has to be interesting. So when I went to England, I realized that I had to spend some time deciding whether there were things that really interested me in this field, or, if not, what else I would like to do with my life. And so, although I didn't publish much of anything in the year and a half I spent in England, I read a lot about what was known about the large-scale structure. What I realized was that much of what was said in the literature and much of what people seemed to think was known wasn't known at all. It was known on the basis of one poorly studied system or none, perhaps. Virtually nothing was known even about the universe nearby. There were no data. It was induction based on one object.

And also theoretical prejudice, like homogeneity.

Yes, very much. Theoretical prejudice is still very important in this field. But, fortunately, it's become more and more popular to observe what the universe is like. So we have some hope that eventually there will be more interaction. But, in fact, right now there is really a remarkable separation between the route that theory is following and the observations we're taking. They are not really meshing as well as one might like to see.

> *When you were in graduate school and you first started working in cosmology, do you remember whether you had any preference for different cosmological models, say an open universe versus a closed universe, or a homogeneous universe versus an inhomogeneous one?*

I was very aware that Peebles had a prejudice for a flat cosmology because it was elegant. He had this view that it was the elegant thing—why should the solution be anything else, and it was so close. And I was aware that [John] Wheeler was saying that the universe is the way it is because we're here.[1] I guess my reaction to that was that it was rather silly, because in my view the universe is a physical system and we're eventually going to measure its properties, and they're going to be what we measure. You could ask, "Why is the fine structure constant not exactly the ratio of integers?" Well, it isn't. Maybe it would be nicer if it were, but it isn't. It's something we measure. It seems to me that there are questions of science and there are questions of philosophy, and they're different questions. The question of what the mean mass density of the universe is is a *measurable* thing. It's a question of science and until we measure it, we don't know it. So that was my attitude about it. You might say it's not very imaginative, but I think there are things about which one can be philosophical, where you're not going to have a constraint, but this is something for which we're eventually going to have a constraint.

> *There have been a lot of changes in cosmological thinking over the last 10 years, both in theory and observations. Have any of your perceptions of cosmology changed over the last 10 years as a result of work that either you've done or that other people have done?*

Partly as a result of working with Peebles, I had this prejudice for the homogeneity and isotropy assumption, but I have certainly begun to wonder. One of the things that's very sobering about the surveys of the universe nearby is that the structures, at least as traced by the galaxies, are as big as the surveys. I think that really leads one to question. If we find much bigger structures, we're really going to have to question those fundamental assumptions, regardless of the smoothness of the microwave background. If we *find* the structures, if we *see* the structures, there is something wrong somewhere, and we're going to have to start looking at those basic assumptions. But I had certainly never questioned that and I've begun to. I think that there is a lot of resistance . . .

Is there any resistance in your own mind? I know there's resistance in other people, in the community.

Oh, I think it's exciting to be able to question it [the standard cosmological assumption of homogeneity and isotropy]. There are people who questioned it before. Gérard de Vaucouleurs has persistently questioned it, and he has an uncanny way of being right about these things.[2] It's exciting to question those things because it means we're making progress. I think in a field like this, you're very unaware of the enormous number of assumptions you make in order to do a problem. And it has turned out in my own work that much of it is based on questioning the tried and true assumptions, and they fortunately never turn out to be correct.

It sounds as if you're not particularly committed to any of these assumptions. I'm wondering whether you feel any intellectual struggle in yourself as you find new data that contradict those assumptions?

I don't know. I guess maybe I'm an iconoclast. I like to see the tried and true assumptions fall. I'm always looking to make the simple assumption, the tried and true assumption, fall. It's funny, but there's always this idea in physics that the simplest model should be the right one. But, on the other hand, these are really macroscopic systems. There are lots of phenomena and it's not so clear. You could say, "Should the simplest model of human behavior be the right one?" Well, there is no simple model of human behavior. And these systems really are complex. OK, they are physical systems, and they are far simpler than the interactions among human beings. But I'm not so sure that in this case the simplest models are going to turn out to be the right ones. The simplest models for large-scale structure formation, although people are holding onto them tenaciously, may not be right. A much messier model may be the appropriate one. It's going to be a long time before we know anything, and I'm sure we won't know until we *see* it. But when we see it, if it's a mess, no amount of arguing about simplicity is going to show us that that's correct, other than what we see.

So when your recent work in the last couple of years with John Huchra and Valerie de Lapparent seemed to be at odds with the accepted theoretical view . . .

Oh, I was delighted. I was really surprised by the result.

It surprised you?

Oh, absolutely.

Why, were you expecting otherwise?

Because I thought the universe was homogeneous and isotropic.

Why did you think that?

Because that's the prejudice of the field. That was the assumption. So I thought, "Well, OK, people have discovered the void in Boötes and maybe it was really unusual," just the way others were saying.[3] We were making a survey to constrain the existence of big structures and probably we wouldn't see anything. And presto, we did. And not only that. Not only did we see the big structures, but we saw that they were very sharply defined, which was another big surprise that people are *really* refusing to come to terms with. It's amazing. The refusal is really striking.

Because you can't explain the observed distribution of galaxies just in terms of gravity?

That's right. It appears that you can't, but people are really fighting it, trying to say that they have. It's interesting to see how this happens. And it's going to take a while to see if people want to argue that different data sets look different. I am not impressed by this argument, because I think they all look the same. No, I was really delighted. It's always fun to find that you don't really know very much. But it's not a surprise. If you think about the fraction of the universe that we've mapped out, it's really small. It's interesting because many people deny that they were surprised, even though it's obvious that everyone was. I mean there were people who essentially said their grandmother knew this. But nonetheless, it's very clear from the reactions of the community that they were surprised by these data. It's one of the surprising things about the whole thing, all the politics and how people react to it.

It's interesting to ask why it is that people were surprised. You said that you had bought into the standard assumption of homogeneity and isotropy of the universe to some extent.

Nobody had seen *sharp* structures before. There was the Perseus–Pisces chain. But this isn't that. There are many of these structures, and nobody has really seen anything so sharply defined before. Why should you expect to see such a sharply defined pattern on such a big scale? I think that was the thing that was so surprising about it. People had this picture of hierarchical clustering. You had this gloppy distribution of galaxies. But I think had we found *that* in the survey and written our paper, people would have said, "Yes, yes. This survey is a good thing. The people at the Center for Astrophysics should keep doing it with their 60-inch telescope." And nobody would bother to be racing to do the surveys, which everybody is doing now. The reason they're doing it is because there was a pattern, which everybody has recognized. That makes a big difference.

By pattern, you mean the bubble-like structures?

They were sharply defined. Okay, are they bubbles, are they sponges? But I think the fact that there is a real structure there that you can see is one of the things that has really motivated people. They want to not believe that it's really there; they want to believe it's atypical. Maybe it is. After all, our survey covered only a small piece of the universe; we don't know. But the thing that's attracted people is that the pattern is so striking. It really is.

And part of that is the sharpness of the boundaries?

Yes, part of that is the sharpness of the boundaries.

Which really suggests a qualitatively new phenomenon that's beyond just inhomogeneity, some new physical process.

That's right. You certainly see holes in simulations of galaxy formation models, but it just doesn't have the same appeal. It doesn't have the same coherence. If you look at the pattern in the redshift survey, you can put your pencil down and sort of connect all of the galaxies. You can also remember the patterns. Whereas when you see the patterns in the models, you can't fix them in your mind. You can tell that it just isn't the same kind of thing. It's not as coherent. Of course, that all remains to be tested quantitatively. I think it's going to be interesting to do that. Maybe the perspective we're taking on the basis of the observations isn't right, but I think there's just a resistance to changing, to really recognizing the impact of these big structures. They really *mean* that we don't know things like the mean density, we don't know the low-order moments [correlations] of the galaxy distribution. There was just a meeting in Rome where people like Peebles were insisting that we know them. I don't know how they can believe that.

> *Let me ask you about your reactions to a couple of important theoretical ideas in the last 10 years. Do you remember when you first heard about the flatness problem?*

Oh, yes. I heard about it from Peebles. In the early seventies.

> *Do you remember what you thought of when you first heard about it? Did you think it was something that was just a question of initial conditions, or did you think it might be a more profound issue that had to be explained by physical processes?*

I guess this is a view that is not widely held, but it's always seemed to me like an argument of religion rather than an argument of science. I have never understood this argument as an argument of science. I've heard it many, many times. I can repeat it. But I just can't see it as a scientific argument. Because the universe is *one* realization. It's *one* system. So how can you talk about a priori probabilities? We have one system. That's it. This is a system we're in and we're going to measure it. And so I don't understand it. I just have a really difficult time. The set of principles on which I operate to do science does not include that. I can understand it as a philosophical issue, but I can't understand it as an issue of science. It's like the issue of deus ex nihilo. It seems to me the same kind of issue. It just happens that our cosmological model is such that if you go back far enough, you get arbitrarily close to a flat universe. So big deal. I'm sure that if we ever understand what the structure of the universe is, there will be plenty of arguments about why it is that way. After the fact, it will certainly be easy to construct them. Of course, it's a challenge to construct them before the fact and to try to use them. There's a lot of feeling in science that one ought to be able to do things by pure thought, but that has rarely been successful in fact.

> *This might be a good place to ask you a question that I was going to ask you later. Do you feel that our current theoretical framework for doing cosmology—including the equations of general relativity and the willingness to extrapolate those equations backward in time—provides a useful context for interpreting observations?*

Oh, absolutely. Absolutely. We don't have that large a number of observations that support it, but so far things are consistent. So I think it is useful. It may be that at some time in the future, we'll find that it isn't. That's conceivable. But I think that its predictive power is very great, and it drives people to really test. It makes it very clear what the things are to test and what the things are that have to be measured. You *have* to have a theoretical model. You can't do science without prejudice. There are certain quantities that it's clear you have to measure. But when you go to do the project, you always have a prejudice. When we designed the redshift survey to look for large structures, I had a prejudice. It turned out that my prejudice wasn't right.

Is it possible that, if the universe were really very inhomogeneous on any scale that we looked at, the fundamental equations could be so misleading . . .

It's possible. I'm not really so expert at the mathematical aspects that I could say how large the inhomogeneities have to be or what their amplitude has to be on what scale in order to pose really serious problems. I think it's conceivable that could really cause questions. But so far the success of the hot big bang model is remarkable. It really is remarkable how it's withstood the test of time and how much of a guide it has provided. It's also sobering how difficult it really is to find out about cosmology itself. We always end up finding out more out about the objects in the universe than we do about cosmology. One of the serious problems is that the objects are complicated. We don't understand them very well and yet we have to use them as the measuring sticks for the cosmology. There's a conspiracy that the time scale on which they evolve is the same as the time scale on which the universe evolves. It makes it really hard to find out anything. But I think that we may. I don't think it's going to be as rapid as one might think. At every time, people think they are very close to the answer, and that's what keeps people going. It's not clear how long it will be. And as I said before, we'll never know for sure because we can't really go out and touch it.

Let me ask you about the inflationary universe model. How speculative do you feel that model is?

Well, any model is speculative until it's proven to be the correct one, right? But I think it's a really elegant model, and it does answer some profound questions. Of course, it replaces them with other profound questions, such as why is the cosmological constant so close to zero today when it was enormous in the past? So, as often happens in science, you answer some questions and you open the door to others, and sometimes it's not really clear whether you've answered more than you've raised. But, because of its elegance, there's certainly an enormous temptation to take the predictions of the inflationary model very seriously and to assume that omega must be 1. I think it has really affected people who work closer to the observations in a remarkably strong way, in that most of the models now assume that omega is 1. The frustrating thing from the point of view of somebody who looks at the data is that, at the moment, there is *no* direct observational evidence for omega equals 1. On the other hand, you can't prove that it isn't 1, because you have no good way of detecting the uniformly distributed stuff [matter that is not clumped, such as galaxies]. We can only place constraints on it, and the more and more uniform we force it to be, the harder and harder it gets, and one has to say what it [such uniformly distributed matter] is. In principle, there could be a revolution in both cosmology and particle physics if one of the particles that's suggested to be this stuff were discovered. That would be one way of really knowing the answer to this question. The astrophysical means are extremely hard. One of the reasons the astrophysical means are so hard is that we don't have good ways to measure distances. It's one of the reasons it is so difficult to get good constraints, especially on a large scale.

We can't really apply the virial theorem on that scale either.

No. As people look more and more, essentially they're finding that all the assumptions that were made about large-scale structure are not true. One after the other. Today we see sharp structures. Now it appears that there are motions [large-scale movement of galaxies not explained by the overall expansion of the universe].[4] It's always been assumed that there is no shear, that there are no large-scale flows. Well, if these flows are really there, we have to explain them. So one by one the simplifying assumptions are dropping by the wayside. What it shows is that this is a young field, and that we started out on the basis of no data, and now that we have data the picture is fairly rapidly changing. At least it's changing on a time scale of a few years. It's not changing as rapidly as I think some people would like to make it seem. But on the time scale of a few years, there are really changes.

Why do you think that the inflationary universe model has been so widely accepted? You said that it's even beginning to affect observers now.

I think that a lot of people have a prejudice for an omega equal to 1 to begin with. One of the things is that it appeals to the prejudice for simplicity and mathematical elegance that many scientists have. It's hard to resist that kind of appeal, and the arguments that go along with the inflationary cosmology sell it very well. And it also addresses some profound issues. People really try to make what we see now consistent with the solution imposed. If inflation said that omega was .003, nobody would pay any attention, because we know it can't be that low. We see more than that. I think it's because inflation predicts more mass than what we see, and we can't prove that the mass isn't there, and it has appeal on scientific issues. Also, it's been productive. It's made people think about a lot of other problems that they might not have otherwise thought about. Maybe it provides some answer as to how to get the right fluctuation spectrum. So it's really been a valuable thing. But I think that it's a mistake to be guided to the value of the mass density by theoretical prejudice rather than by observational constraints. You have to be very careful to distinguish between measurable quantities and those which aren't. To have such a strong prejudice about a measurable quantity—maybe it's valuable because it makes people push very hard on the data. It makes people ask a lot of questions about how good the constraints they have really are. They're not great, that's for sure. And it's made people make models that are actually more complex. The models where the galaxies don't trace the matter density on a large scale are actually more complex in some respects than those where it does. In some cases, it isn't really warranted by the data. On large scales, the data do not rule out models where the galaxies do trace the true mass distribution.

Yes, but you needed that property even without the inflationary universe model.

You need it, but not on such big scales. You need it on small scales associated more or less with galaxies and clusters of galaxies. But even in clusters of galaxies, it's consistent to say that the dark stuff is more or less the galaxies. In fact, it turns out that people have made lots of models for clusters that are either more concentrated or less concentrated than the galaxies. The best fitting models for the Coma cluster, where there is the most data, are the ones where the galaxies trace the matter distribution. Of course, it's not a best fit by a lot. But it still is the best. Nonetheless, because of this prejudice and also because we don't know what the dark stuff is, obviously we have to try all of the options. And that's valuable, because we really do have to consider that question. After all, there's no reason we should see all the matter in the universe, or that what we see should really trace where the matter is. So I think a lot of these things have led people to be more critical of the assumptions. It's ironic, because in some ways it's led people to be more critical of the assumptions they make when there is data and less critical of the assumptions they make where there is not data to constrain.

> *I'm interested in how scientists use metaphors and visual images. In cosmology, one powerful metaphor has been the expanding balloon metaphor for the expansion of the universe. Do you remember when you first were introduced to this metaphor? Was it in graduate school?*

No, I think it was as a kid, probably in one of George Gamow's books.

> *Did it help you visualize what an expanding universe was like?*

Oh, yes. I think that not all scientists have to have a visual image, but I do. I can't do a problem unless I have a visual image. That's how I solve problems. I have to have a visual model or a geometric model or else I can't do it. Problems that don't lend themselves to that, I don't do. That's where my strength is, it turns out. But I'm also generally visual. I'm very aware of visual cues in my environment. I have a good visual memory, and I am very observant. So it's not so separate from the way I am in general. I have an interest in the visual arts. I think that had I not been a scientist, I probably would have done something in design.

> *Do you remember, as a child or in high school, when you read Gamow's books about cosmology, did he talk about the big bang as being the beginning of this expansion?*

Oh, yes. I remember reading about the formation of the elements and so on in the early universe.

> *Did you take that seriously, that the universe had a kind of a beginning?*

Yes, I think I did. I took it pretty uncritically probably. As a kid, I was brought up to question things. I was brought up not to take anything at face value and to ask questions. But I think that something like that, I probably just took it uncritically, never dreaming that I would someday work on something related to it. I remember, quite vividly actually, reading that book. I remember being impressed by it, by the idea that the elements were so old and the universe was hot early on. I remember being quite amazed by those ideas. They seemed really remarkable to me, almost like a story.

Did they seem real?

I don't know whether they seemed real or not, but they had an appeal that was different from a lot of other scientific problems. They were sort of romantic. I remember it very well. I read lots of books about science, but it's one of the few that I remember. Gamow also wrote very well, in a very charming way, which probably appealed to me too. But I don't think that I ever have been that critical about the big bang model per se. In fact, I've just sort of gone along and accepted it like everyone else, or almost everyone else, except that I have certainly recognized that the number of observations that constrain it, that really support it, are very, very small. When you talk about them, you end up talking about things such as why the sky is dark at night and so on. It's really remarkable how simple the observations are. That's a reflection, in a way, of how little we really know about the universe.

I guess that's not too surprising considering that we're talking about something we think was 15 billion years ago.

That's right. And a system that's really enormous, where we don't explore it, and we only receive information, and it's very limited information. Cosmology is about as different as it can be from our laboratory sciences. It has a very different feeling, although I think that a lot of scientists who do cosmology feel the same way as scientists who work in other fields. That always surprises me. I think it's hard in this field to really have the courage of your convictions. You have to have a sense of humor about it because the likelihood of ever being right is so low. You can be sure that you have good data. But how sure can you be of your interpretation? Well, not very, ever.

Well, it is comforting at least to be sure of having good data.

That's true. That's one of the reasons I like to work close to the data. I think that it's unfortunate, but I recognize that interpretation is likely to change, and that you just do the best you can at the time, and you aren't really leaving something carved in stone. It's a very different kind of thing.

We've talked about a lot of different topics from the scientific point of view. I'd like to ask you to drop your natural scientific caution a little bit and just talk as an ordinary person. If you could have designed the universe any way you wanted to, how would you have designed it?

I don't know. I've never thought about it before, but had you asked me that before the redshift survey results, I probably would have said I would like the universe to have a pattern, although it didn't seem to.

When you say "pattern," you mean large-scale pattern?

Yes. I would like the universe to have large-scale patterns. I would like it to have structure, interesting structure.

Not necessarily orderly, but just structure?

Yes. Not necessarily too orderly, but structure. I like the idea of its having structure. I guess a lot of people find that objectionable. But I like that idea. I like patterns in nature. In general, they are things that I really find intriguing. So that's what I would have wanted. As far as the big bang is concerned, I don't know that I would have had the imagination to design something like that. I like the idea that the universe changes and evolves, and I don't think I would like it if it didn't. I would like that less. If you think of the things that go into the big bang model, one of the most important is that the universe evolves.

So you like that more than a static cosmology?

Oh, yes. It's more interesting. By the same token, I think patterns are more interesting than not. I like the universe to be as interesting as possible.

Do you think that philosophical considerations have ever had much of an influence on your work?

At some level, it's impossible to separate them because everybody has some set of principles on which they operate. People choose problems or find problems according to the way their mind is structured, according to their philosophy. I think that it's impossible to separate your philosophy or who you are from the science you do. I think scientists do have a style. It's something that's much neglected. I felt it very greatly because when I started out, I started out in the style of my advisor, and it wasn't a good style for me. It was not my style. I had to find my own style. So my philosophy clearly did influence what I ended up doing in science. It couldn't help but do that, and I think probably had it not, I would have gone off and done something else where who I am or how I am put together would have affected the work I did. And the politics of science and the structure of the scientific community influence how I present my work. So I think in every way those things really do affect me. And they affect what answer I expect. I never do a problem unless I have a prejudice about it, as I said. My prejudices can certainly be wrong, and even sometimes the answers I thought I have gotten have been wrong, in fact. But I've learned not to be so afraid of those, which I think is a very important thing. I mean it's another philosophical thing. In a field where you can't really hope to be right, how much do you worry about it? Well, you do the best job you can within the constraints at the time. You try, and if you are taking an approach which is different, which happens to be out of favor at the moment, what do you do? Do you fight it, do you stand for it or not? You have to have some courage in order to take a different approach. Maybe you're going to be wrong. But, you know, you don't die. And it's interesting.

You could just take the point of view that you're just finding out what is.

That's right. That's the point of view I take.

And do as little interpretation as possible.

Oh, I don't know. I like to interpret. It's a challenge to interpret the data and to ask what it means and what implications it has. It's not good to be wrong about it. On the other hand, the cost of not interpreting is infinite. The cost of not taking risks is that you don't really find out very much. It's hard to do anything exciting without taking any risks.

Let me return to something you got into slightly in the beginning—that's the question of gender and science. Often the world of human experience is divided into two classes: masculine and feminine. Some people consider the scientific endeavor to be a masculine-type activity, because there is a longstanding tradition of objectivity and not becoming emotionally involved with the subject and so forth.

Well, the first thing I think is that men are emotionally involved in their subject. So the basis of this idea is a complete myth. There are differences between men and women. Good heavens, it would be pretty boring without that. But as far as the *doing* of science, it's never been demonstrated that there's any difference in the ability to do science. There are subtle social pressures which often steer young girls away from science, and even steer women who made it farther along away from science. I think it takes a lot of stamina and strength to be a scientist, whether you are a man or a woman. It's a very rough profession. I think what happens is that there are a few more pressures on women that are not positive, that steer them out. But I don't think it's that women are less able. There are plenty of women who have shown that. There's not a genuine issue of whether women can do it. It's a question of whether society is ready to allow it. Unfortunately, things have not changed very much for women in science—indeed, in many professions—but they've changed very little in science. That has to do with the way universities are structured, and the way university hiring is done and the structure of the scientific community. There are also many social differences, because men have dominated science to such an extreme. Socially, the structure of science is not one in which it's easy for women to be immediately comfortable. I think women have to find a way to deal with it, and they are going to deal with it in different ways.

I guess you might say that a sensitive male scientist may also have to balance his family life and his working life.

Oh, absolutely. I think science is a very harsh profession. It's very difficult for people who are sensitive, period. Very difficult. It's not a particularly supportive profession. I don't think it's easy for anyone and, as in any profession, a lot of people have professional personae that they adopt. But for women, it's just more. It's just a little more and requires a little more. Also, as in other demanding professions, as far as family is concerned, it requires a kind of decision for women that men face but less strongly— that the time of making a career is the same time as having a family. The problem is compounded by the difficulty of getting a permanent position and so on. So that's another issue. It certainly affects men, but it affects women more. I also think one of the things that probably affects women in science—and in other things—is that women, even still, are brought up to seek security much more than men. So it means that most women may not take as many risks as men. To really be successful as a scientist or at any creative endeavor, you have to take risks. I think it's much harder for women in general. There are exceptions, of course. There are women who succeed in these professions. But I think it's consistent with this kind of encouragement, still in society, that women should seek some security. These things are very deeply culturally rooted and they affect professions that are tough and creative like science.

Somewhere in Steve Weinberg's book The First Three Minutes *he remarks that the more we understand about the universe, the more pointless it seems. Do you have any thoughts or reactions about that?*[5]

Yes, I guess I've always been sort of puzzled by it. I've read that statement, and I've never understood what he really means. Maybe it's because I have a different view of my life than he has of his. I don't know him very well, but maybe he's a person who thinks you live your life with a certain point, a certain goal and that everything has to have a point. I'm not that way. I guess my view of life is that you live your life and it's short. The thing is to have as rich an experience as you possibly can. That's what I'm trying to do. I'm trying to do something creative. I try to educate people. I enjoy encouraging people and meeting people. I enjoy seeing the world, and I have as many broad experiences as I can. I feel privileged to be able to be creative. But does it have a point? I don't know. It's not clear that it matters. I guess it's a kind of statement that I would never make. I figure, thinking in the small way that I think as a human being, well, okay, why should it have a point? What point? It's just a physical system, what point is there? I've always been puzzled by that statement.

JOHN HUCHRA

John Huchra was born on December 23, 1948, in Jersey City, New Jersey. He earned his B.S. degree at the Massachusetts Institute of Technology in 1970 and his Ph.D. in astronomy in 1976 at the California Institute of Technology, where he was the thesis student of Wallace Sargent. Since then he has worked at the Harvard-Smithsonian Center for Astrophysics. He is currently an astronomer at the Smithsonian Astrophysical Observatory and professor of astronomy at Harvard University.

Huchra's research interests center on extragalactic observational astronomy and cosmology. In 1979 Huchra and others improved a leading method for determining the distances to galaxies and produced a new value for the rate of expansion of the universe, significantly higher than the previous value and thus significantly decreasing the estimated age of the universe. With Marc Davis, Huchra was one of the pioneers in developing the Center for Astrophysics redshift survey of galaxies, producing 3-dimensional locations of galaxies in space. Huchra frequently collaborates with Margaret Geller. In 1986, with Geller and Valerie de Lapparent, Huchra discovered that some galaxies appear to be located on the surfaces of bubble-like structures, about 100 million lightyears in diameter. An observer's observer, Huchra has more hands-on experience with telescopes than almost anyone of his age. He is also considered to have an encyclopedic knowledge of astronomical facts.

I wanted to start with some questions about how you got interested in science. Were you interested in science as a child?

At a real early age. From the time I got involved in reading, I always had in my mind the idea that I wanted to be a physicist or a mathematician. I think the thing that sparked my interest in cosmology was a series of books that I came across when I must have been about 10 or 12 years old, in sixth grade. Books like George Gamow's *One Two Three . . . Infinity.* And a book by Fred Hoyle. I can't remember the title. It wasn't *The Frontiers of Astronomy.* Then there was also Gamow's *The Birth and Death of the Sun.* I read those books when I was in sixth grade and I said, "Gee, this stuff is interesting. It's what I'm interested in doing." I grew up just outside of New York City, in an area of the world where it's not possible to see the sky at night, except for those things that are flying at low levels. It wasn't a case of looking up that decided me.

It was reading those books?

It was reading those books. I have a feeling that there's a generation of young physicists, astronomers, and cosmologists who came across those books all about the same time, in the late 1950s. A lot of people saw them. They were very well written, compared to a lot of other things that were floating around at the time. There really wasn't that much in the popular science literature, unlike today where there's a vast mass. Those books stood out.

At this period of time when you were 10 to 14 years old, and you were thinking about going into science, do you remember whether cosmology had any particular appeal above other sciences?

Not particularly. I was interested in doing science, pretty much physics and mathematics, as I said. That's what I settled on. I went to MIT, and I started out in physics. I tried to double major in physics and math, which would have been possible except for a small thing called the Vietnam War. It shut down classes for about a year. For the first two years I was there, I was interested in particle physics, in the basic idea of trying to understand grand unified field theories and the like, which were also just about coming to the fore at that time. There was a lot of renewed interest in theory. At the time, the experimental part of it was pretty much dead. That was the thing that finally decided me not to do particle physics. I ended up getting involved with a laboratory research project when I was a junior at MIT. It was required that you do a lab project. I looked around at a variety of things and ended up in an x-ray astronomy lab, aligning detectors for rocket flights. That was sort of interesting, although at times a little dull. I had to make the choice then whether to get a math degree or physics degree. I decided to stick to physics because the last remaining course I needed to take in the math department was much too hard. I ended up doing a senior thesis on stars—in theory, basically on stellar pulsations.[1] I worked with Icko Iben, who is a very interesting character.

> *Going back to your childhood for a minute, do you remember having any ideas about cosmology or about the universe as a whole?*

Not really. The one thing that I picked up reading the literature was that there was a bit of turmoil going on in the ideas about the universe. I knew the words "steady state theory" when I was 12 years old and the words "big bang." I don't remember having any profound thoughts at that time.

> *You don't remember having any preference for any particular cosmology?*

No. Not at that age. In fact, I still try not to.

> *Let me go back to MIT. When did you decide to do experimental physics rather than theoretical?*

I didn't make the decision to become an experimentalist until after I had been at Caltech for a while. I should mention there's probably another point in my education at that time that had a fairly profound influence on me. As a freshman, I took a seminar with Philip Morrison. I loved that course; it was really fun. We played around with measuring distances inside the galaxy and talked about cosmology.

> *Then after MIT, you went to Caltech?*

The best way of describing it is that I wasn't sure what I wanted to do. I applied to graduate school and for some strange reason—I can't really remember why and I don't think I knew why then—I applied to graduate schools in astronomy.

> *You don't remember why you chose astronomy after majoring in physics?*

I think I had become disillusioned with the idea of becoming a particle physicist, maybe because I didn't think I was smart enough to be a particle theorist and because the experimental end of the field was in pretty bad shape in 1969. I'm not exactly sure what the reasons were, but I think I decided I wanted to go do astronomy—not necessarily cosmology, but astronomy.

But you were already very interested as an amateur in astronomy, weren't you?

No. I grew up in New York City. I never owned a telescope. But while I was at MIT, I did make an effort to do things like go out to Wellesley to use the telescope there. At MIT, I took Icko's course in stellar astrophysics. I took a radio astronomy course from Alan Barrett and Bernie Burke. And I took Irwin Shapiro's planetary astronomy course when I was a junior. By my senior year, I think that I was fairly certain that I was interested in astronomy. I still can't remember why I picked it. I also took a course with C. C. Lin on galactic structure. That too was a wonderful experience. I didn't like nuclear physics. But I loved galactic dynamics. I think that pretty much describes it.

When you went to Caltech, how did you move in the direction of experiment rather than theory?

Well, that's a funny thing. First, I didn't think I was good enough compared to the other people in my class at Caltech to do a lot of things. I figured I was a dummy. It wasn't entirely true in terms of grades. I had the best grades of most of the people in my class, but somehow I never seemed to quite catch the right spark. But there were one or two people who sort of took an interest in me as a student. I worked for Guido Munch. You go there on an NSF [National Science Foundation fellowship], but $200 a month when you're paying $125 a month rent means that you've got to find a research assistantship if you intend to survive. The first way I started getting paid was doing plumbing for Guido Munch. We were building a pressure-scanned Fabry–Perot interferometer, something I knew nothing about, but what the hell. I've got to learn sometime. And that was actually interesting. I learned a lot of things I didn't know and probably will never need again.

The astronomy department at Caltech was not rife with theorists at that time. Most of the theorists were in the physics department. The people on the faculty then whom you would most remember were experimentalists. Just being in contact with those people, and the fact that experimentation was something I had done before, working with the x-ray telescope during my junior year—I don't know—I moved into being an experimentalist. The theorists always seemed off in ivory towers in the physics department, far away from where astronomy graduate students were. If I remember correctly, there were very few astronomy graduate students that ever went in the direction of doing a thesis with people in the physics department. There were some physicists who came over to work in astronomy. The flow was usually one-sided and towards astronomy. My tendency was to stay in astronomy because I was shy and retiring. I didn't want to go too far outside the general vicinity.

I infer from what you're saying that you felt slightly intimidated by some of the physicists.

Yes, I think that's right. I was intimidated from the word go. I didn't understand how I had gotten into Caltech first of all. I was OK as an undergraduate in terms of my grades, but they didn't take a lot of students. A lot of the people I knew at MIT were applying in astronomy to places like Caltech. There's actually something that took place, a funny psychological thing. In the late 1960s, especially 1969 and 1970, it was my impression that because of the Vietnam War, many of the people who started out as physicists were swayed away, partly because of student politics. The Body Politic was very much antiwar. Getting involved in war or anything that might be construed as defense-related research was really anathema. I would say that out of my graduating class at MIT, of which there were about 80 physics majors, maybe half of them were applying to graduate school. Almost all who applied to graduate school did so in physics-related fields that could not be construed as being related to the defense industry. Almost half of those people applied to astronomy or astrophysics. A lot of those people were better than me, so I was a little surprised at getting into Caltech. I think that the only thing I might have had going for me was that I started doing astronomy before they did. Maybe that helped, I don't know.

Do you remember when it was that you got interested in cosmology?

Interesting question. I was working for Guido; that was my research assistantship. As the normal course of events at Caltech, you have to find a research project to do. I went up to Santa Barbara Street, which was then the headquarters of the Hale Observatories, and looked around to find somebody who might have something to work on. As I said, I thought I was sort of the low end of the scale of graduate students at Caltech. I was real shy. I felt that if I went uptown, I'd have a better chance of finding a research project. There were a lot of people from the Observatory who came down to Caltech and taught in the astronomy department. I ended up doing a research project with George Preston. My first year and a half spent doing research was on observations of peculiar A stars.[2] That was about as far away from cosmology as you can possibly imagine. That was sort of interesting. I wasn't really turned on. I had to hunt around for a thesis after doing that. I had taken a couple of cosmology courses. One was the extragalactic astronomy course, which was taught by three different people throughout the course of the year. They were on the quarter system, so there were three terms. Jim Gunn, Wal Sargent, and Leonard Searle taught sections. I decided that it would be interesting to try and do something with one of these guys. I ended up working for Sargent.

Did you start working on cosmology from the beginning?

Pretty much. Again, to pay the rent. I got a job doing the Palomar supernova search.[3] I started working with Wal and Leonard, who was a behind-the-scenes thesis advisor, on something they had started only a year or so before. That was the topic of these little blue galaxies, extragalactic HII regions, the question of finding more such galaxies, and starbursts and star formation in galaxies. This was not "cosmology" as you'd call it, but it's definitely a related topic. Are galaxies still forming now, and things like that. I must say it took me a long time to figure out exactly what it was I was trying to do. I started out with an idea, but the idea wasn't the idea of what the scientific problem was, but was rather the idea of how to approach an experimental problem. Another thing I remember, which was different from the normal course of events, was that I spent a lot of time taking courses that had nothing to do with astronomy or cosmology. I took some economics courses. I was taking them because it turns out there are interesting techniques in economics that can also be applied to stellar population synthesis, techniques that at that time no one in astronomy was dealing with, but that looked like they might be interesting— linear and quadratic programming techniques. Astronomy is one of those fields that sometimes picks up mathematical techniques 10 years after they've been discovered by the statisticians or the economists or whatever.

In your work as a graduate student, were you heading off in any particular direction in cosmology? The area that you're in now pays a lot of attention to individual galaxies. Can you trace that back to your work as a graduate student?

Not directly. Or I should say maybe indirectly. I was a dabbler. I was trying to do an observational thesis, to collect a body of data about a particular set of objects, to try to understand the properties of star formation in these objects, and to try to see if the sample, which somebody else has selected, could be used to answer a problem. It actually took me a long time to realize what the "right" question was. Now I know that figuring out the right question to ask is a very difficult problem. That may be the most important thing for a young scientist to try to realize. What questions should you ask? What questions can you answer? What questions do you have a hope of making any headway against? I didn't understand that quite at the time, but it came eventually. I did a lot of different things, because I was trying to do this assemblage of data on this particular topic and because the weather was terrible. I had a great reputation. They used to say, "What's the weather like at Mt. Wilson?" and the answer was "Huchra's up there," in which case everybody would know that it was snowing. Since things were going very slowly in that department, I had time to get involved in other projects. First of all, there was the supernova search. I was doing that as a job, but I got to learn a lot about the properties and the problems of finding supernovae and what that meant for stellar evolution and a variety of other things. By virtue of being at the telescope a lot, I had access, which is a very important thing. By being at the telescope a lot for *that* project, I discovered a comet.[4] I eventually became the master of the 18-inch telescope and more or less in charge of some of the instruments on the 60-inch at Palomar, which is also a good thing. I got a lot of hands-on experience. That's something that's very bad today. It's hard to get students hands-on experience with equipment.

There were a variety of other things that went on. I did some work on variability of Seyfert galaxies and looking at quasars.[5] That's again a connection through the supernova search. You develop a fairly good understanding of what's going on with the photographic plates when you have to take 100 of them a night, and develop them, and look at all the defects, by eye, bit by bit. That kind of painstaking effort is something that doesn't happen nowadays, but nonetheless is good for the soul and sometimes good for the mind.

> *In learning about various cosmological models—the open model, the closed model, and so forth—did you have any preference for a particular cosmological model?*

Made no difference to me.

> *You didn't have any preference?*

No. There was one other series of things I did that didn't take place during the actual course-taking part of being a graduate student—the theoretical seminars. I may have been scared of them in part, but I also found those tremendously useful. I learned a lot of things. You had a chance to interact with people. You were really forced to do the readings, to try to understand before you got there, so you didn't go in cold. It was a *bad* idea to go in cold, especially if your name got pulled out of the hat. In the course of about three years, we went through magnetohydrodynamics, supernova models, cosmological models, and quasars. Astrophysical theory at that time, aside from that branch of theory that was specifically associated with compact objects, really was pushing towards cosmology, cosmological implications. There was a lot of that, partly because the seminar leaders were Sargent, Gunn, and Peter Goldreich. Sargent and Gunn, of course, were cosmologists. So I think that also helped inform me.

Do you remember any major questions in cosmology that they were particularly concerned with?

No, not particularly. Both Jim and Wal are real generalists, and to some extent that's sort of what I am. Wal is more an observer-generalist, and Jim is a little bit of both. I ended up being somewhat like that. I do a lot of different things.

After your thesis at Caltech, were you then pretty sure that you wanted to continue working in cosmology?

After my thesis at Caltech, the best words I can use to describe my state was "scared shitless," because I didn't know if I could cut it, if I could think of new projects to work on, if I could come up with ideas that were doable and saleable. I had begun to get the idea that there are things you can do, and there are things you can't do, and there are things that are more interesting than others. I don't think it was really set in my mind that *I* would be able to do things.

I had actually started working on a few things when I left Caltech. There was a collaboration that was started by Trinh Thuan, who was a postdoc at the time, and Jill Knapp and Wal Sargent and I. Thuan, who had come from Princeton and who had been basically learning at the feet of people like Ostriker and Peebles and company, had come up with the idea that it wouldn't be a dumb thing to start trying to measure redshifts for a large sample of galaxies. Ed Turner had been at Caltech as a graduate student along with me, and J. Richard Gott had come through as a postdoc from Princeton for a couple of years. Turner and Gott had written a series of papers on field galaxies and galaxy groups based on a catalogue of galaxies, the Zwicky catalogue, but just on magnitudes and 2-dimensional positions, with no information on velocities.[6] So we decided—I think the motivator was really Thuan—that it wouldn't be a dumb thing to go out and measure the velocities. By that time, I was known as someone who could reasonably get data, and I spent a lot of time at telescopes. I was one of the remaining masters at the 60-inch spectrograph at that time. We started up this project. Wal and Jill were going to go to Arecibo and Greenbank to do 21-centimeter [radio wavelength] observations of spirals. Thuan and I—which meant me mostly—were going to go off and do the optical observations to get velocities for the elliptical galaxies in the sample. We got some telescope time at Palomar to start it up. There were also a bunch of proposals that had been accepted at Kitt Peak. Since this was a reasonably good proposal and could be done on a little telescope— that's the other part of the game—it got time. So even before I left Caltech, there was a project in the works that I hadn't really started, but nonetheless . . .

A project measuring redshifts?

A project measuring redshifts. That's right. In fact, the first three or four hundred redshifts that ended up in the CfA [Center for Astrophysics] redshift survey were measured as a part of that original program, at Greenbank and at Kitt Peak. Actually there may be even more than that, maybe more like 500, because I know I ended up getting about 350 galaxy spectra at Kitt Peak. I did all of the observing, which was both a bad habit and a good thing.

When you measure the redshift of a galaxy yourself, do you think you get any feeling for the galaxy that you wouldn't have gotten if you were just looking at data someone else had collected?

Yes. I have a long and somewhat simple answer to that. I've actually spent a little time looking back over why I think the way I think about things. What I've come to realize is that I have a world view. Everybody has a world view. I have one, and my world view about astronomy is one that has developed by virtue of *looking* at things. It's a visual world view, or however you want to put it. Looking at things like scanning photographic plates. Doing a supernova survey is looking at those plates. It taught me a lot—not a lot that I realized at the moment, but a lot that I now know because I've looked at the plates and *saw* what they looked like. I know very well what a cluster of galaxies looks like. I can draw you shapes of anything you're interested in. I have almost a catalogue of NGC galaxies sitting in my head, which sometimes surprises people. You don't need to remember it, but nonetheless it's up there simply because I've looked at all of these things. I've now spent a lot of time doing spectroscopy of galaxies and looking at galaxies. I think I've probably set aperture to [measured the spectra of] more galaxies than any other living human, maybe the sum total of all other living humans. That's probably not right, but it's awful close, it's awful close. As part of this redshift survey, we now have spectra of 20,000 plus galaxies. In addition to that, you have to remember I've also been involved on a couple other big survey projects of galaxies.

Can you say what your world view is, or is it just this big amorphous body of knowledge? Can you describe it a little bit?

Yes, I'll get to that. I've got this world view. As I've said, the world view comes from having looked at lots of different things. What I find that I do is whenever someone comes up with a new theory or idea, a new interpretation of observations, or new observations, I have a tendency to match those results against my world view, the accumulated body of knowledge, and ask, "Does it make sense? Does it fit?" So I will match ideas against my world view and see if they fit.

What are some elements of your world view?

Something that comes to mind as very important is the general shape of galaxies and what the distribution of different types of galaxies looks like. There's a game that's played a lot these days in cosmology that's called the N-body game. The N-body models have gotten a hell of a lot better, but the original game was played with point galaxies, all of which had the same mass. And the universe ain't like that. So, at least while that was the case, I could say, "Yes, these are real nice pretty pictures, but you know that's not quite right." There are a lot of things that go on because of the range of properties of objects that you certainly won't be able to see in the models. So the body of knowledge consists of things like knowing the range of properties in galaxies and some idea of what objects look like, what clusters look like, how galaxies are distributed in cosmology, having some idea of what the spectroscopic properties of galaxies and the range of those properties are. That has been a critical element in deciding whether some relatively recent work is right or not. Mind you, I have applied the world view on my own stuff as well. If I make an observation, I want to file it in this world view. If it doesn't make sense compared to what else I know, that usually means I have to go back and re-observe it or make checks to see if it is indeed correct.

Does your world view include any global qualities of the universe—for example, whether it's open or closed, or whether it's homogeneous or inhomogeneous?

Well, my viewpoint on that has changed as a function of time, as a function of getting more data. I would say that if you asked me what I thought of, say, the distribution of galaxies in 1975, the answer would have been: Well, there are clusters and then surrounding the clusters are field galaxies; maybe 10–20% of all galaxies are in clusters and the rest are uniformly distributed. That was the prevalent world view at the time, and I hadn't spent a lot of time looking at it myself then.

Did you accept that world view based on what other people were saying— for the lack of better data?

Yes, for the lack of better data.

You just accepted that.

Well, there are things you can accept, and there are things that you cannot accept. Then there are things that you *must* accept until you have more information one way or the other. That's sort of the way things go, at least in my end of the game. I have found—and this is not always easy to do— that for my role in doing astronomy or cosmology, it is exceedingly important that I not get too married to any particular theory. I can be married to the *data,* but I can't be married to the theory.

Why is that?

Because if I do, I almost always get something wrong sooner or later. It's just practical experience. If I'm too married to a theory, I get into trouble. I've done it once or twice, and it's not been a real good thing.

Can you give an example?

A classic example is this question of whether or not there are field galaxies. There is a real controversy. I had thought there most certainly were large numbers of field galaxies. When we actually did some analyses—Thuan and I and Margaret Geller and I—of how galaxies are distributed as a function of density, over density, in space, there is no easily identifiable field.[7] Most things tend to be associated—maybe not gravitationally bound, that's different, that's a *special* type of association, but "not uniformly distributed" is the magic phrase. And galaxies are not uniformly distributed. In 1975 I thought there were a lot of uniformly distributed galaxies. That's *not* the case.

Here's a better example, maybe for me really the turning point in my idea of why I should keep an open mind. At Caltech, I did some work with a lot of people at the Hale Observatory on Santa Barbara Street. The party line was that the Hubble constant was 50, 55 [kilometers per second per megaparsec], even back in 1975—some low number. When I was a graduate student, Brent Tully came through to give a talk on this thing called the Tully–Fisher relation. Allan Sandage got up, and he and Tully got into a great debate over whether this relation meant anything, because Tully was getting a number that was more like 80 for the Hubble constant, based on the Tully–Fisher relation results.[8] And Sandage wanted 50, 55. They really went at it in the colloquium, and it piqued my interest, you might say. It was interesting to watch the *political* process that was taking place, although I wouldn't have called it a political process at the time. One of the things I did when I came here was to try to think of a way of *testing* the Tully–Fisher relation, to try to answer some of Sandage's complaints against Tully and Fisher. Mark Aaronson was here, and we had been doing infrared photometry of galaxies at the time, since as an extension of my thesis I wanted to do infrared photometry of some of the starburst galaxies. But I also realized that if you could do photometry at two microns, you could get around one of the real big problems of the Tully–Fisher relationship, which is the internal reddening in edge-on galaxies. So I came up with the brilliant idea of correcting the optical colors of the edge-on galaxies by doing infrared photometry. Mark had the photometer. We tried to find a connection at Kitt Peak to allow us to get on the telescope to try this out, because we didn't have observing time for this project at the time. We talked to Jeremy Mould and Jeremy said, "You guys are stupid. Why bother to correct the optical data? Why not just do the infrared photometry for its own sake to try to do this project." And that's how the IRTF [Infrared Tully–Fisher] calibration was born.[9] So we went to Kitt Peak—Mark and Jeremy and I—to the 36-inch telescope. You can do cosmology on small telescopes. We played the game as straight as we could, measuring the magnitudes for these galaxies, and damn if when we were finished we didn't end up getting 70 for the answer. Not 50.

How did you feel when you came up with that result?

Scared. What did we do wrong? That's sort of the first thing. Because it went *against* my world view, which at that time had been based on convention.

And also, it made the universe younger.

Yes, all of those things that are not particularly pleasant when you've got that world view.

So you were scared at first.

I was scared that we did something wrong. We checked it and checked it and checked it six ways from Sunday. It was one of those crazy things. We couldn't make the results go away. I had honestly thought that when we started to do the experiment, we were going to take old Tully and Fisher and show that their stuff was garbage and that the Hubble constant was 50. *No.* That's not what we found. As I said, we couldn't make the results go away. We set out after that to do other clusters at larger distances. Same thing. We couldn't make the results go away. That was really the first time I realized that one had better not go into an experiment with really strongly preconceived ideas. Because if you do, you might miss something. We could have sat there re-doing the data until we *got* 50. And it would have been wrong. Fortunately, we didn't.

Let me ask you a little bit about the difference between theory and observation. Do you think that theory in cosmology has provided a useful structure for interpreting the data and extrapolating the data?

Oh, yes. People sometimes talk about the conflicts between theory and observation, and other people will pooh-pooh that idea. But I think it's there, and it's a very good thing. What I've found is that a good theorist should let his mind roam free, but periodically check back to the data—particularly through a good observer—to make sure it hasn't roamed too free. A good theorist must make theories that predict things that observers can check. There's a serendipity factor involved in the game, but by and large a good observer is someone who is making observations to try to check theories. An observer who just goes out and points a telescope at some random place on the sky because that might be interesting—that's not a real useful activity. That's like a theorist who decides that the real answer to the universe is that some aliens from planet Zork created it 16 billion years ago, or whatever. That's not a real useful theory. In fact, that's called a religion as opposed to a theory. So theorists and observers need to interact; they need to guide each other. Sometimes the theory paces the observations. That was certainly the case when we started doing the redshift survey stuff back in the mid-70s—the theory was way ahead. The theory of what there ought to be was way ahead of what the observations were. We had no data. Right now, temporarily, the situation is reversed. Right now we have more data than the theorists are happy trying to explain.

And the data you do have disagrees with the theories up to date. How do you feel about that?

Win a few, lose a few. In general, I think it's OK because theorists are generally not dumb people. They'll figure out a way of making it all work. But, in particular, right at this point of time, I'm worried that the theorists are wed to a model, that is, omega equals 1, closed universe, inflation. Or I should say flat universe, inflation, because it's not closed, which people forget. If they remembered that, they might not like it so much. Every theorist I know makes 99% of his or her models based on that assumption.

What do you think about the inflationary universe model?

I don't know. My impression is that taking the data as it exists at face value, without making assumptions that the data doesn't tell us anything about the universe, then the inflationary model must be wrong. If you just straight look at the data on things like mass density, which is a key parameter in this game, you do not get an omega equal to 1.

And you think that the theorists may be unduly wed to the inflationary model?

The people who have been making the N-body models are unduly wed to that model. Let's put it that way. And those are the theorists whom I've had most direct contact with recently. That may be a mistake. I don't think there's been quite enough effort placed on looking at alternative models, although a few people now—Peebles and a couple of other people—are beginning to really start thinking about open-baryonic or even open-cold dark matter universes.

Peebles certainly seriously considered open models before the inflationary universe model.

Well, he did, but there was a period of time—between about 1975 and 1980, before the inflationary model was first announced—during which the school of theory that Peebles and others were leaders in pretty much saw a closed universe, omega equals 1. And the reason is that they had to have galaxy formation.

You didn't need omega to be quite 1 for galaxy formation.

That's true, but you needed it to be a lot bigger than you found just measuring the masses of systems of galaxies. There's a fundamental assumption that goes in here. It may not be inflation itself, with which I see a problem now, but this particular assumption. The assumption is that the universe started out homogeneous. I have a sneaking suspicion—based on the data we've been collecting recently on things like the galaxy cluster distribution and the very large-scale galaxy distribution—that this assumption, namely homogeneity in the initial epoch, is one that is going to make *all* theories based on it wrong, or not workable. It may well be that something that comes out of these observations is the idea that we have to start out with inhomogeneous universes. It's an idea that's actually very old, but which was discarded—you know Charles Misner and the mixmaster universe.[10]

Yes, he wasn't able to make it work. Do you remember when you first heard about the flatness problem?

Well, yes, I heard of that problem when I first heard of inflation.

Do you remember how you reacted to it? Did you think it was a religious argument or a physics argument?

I thought it was a religious argument. I'm a card player. One of my vices. The probability of getting any particular bridge hand is damn small. But when you play, you get hands. So I've never been convinced by arguments that say, "The probability of omega being so close to 1 but yet not 1 is very small. Therefore, it must be 1." That doesn't wash. The first thing that came to my mind when I heard the argument was, OK, there is a probability that all of the air molecules in the room are up in the corner, and that's pretty small, too. But the probability that the air molecules are distributed in this room the way they are is *also* pretty small. Any individual realization of a universe or a physical system, if there are lots of possible states for that physical system, has a probability that's very small. So what else is new.

So this argument didn't sway you.

No. There are equivalent "religious" arguments on the other side. In balance, I disregard them both. To me, it wasn't a clean argument. The universe is what it is. Just because we can't *think* of a good way to form galaxies doesn't mean that galaxies shouldn't be there. In the course of developing a world view, in the course of looking at the universe, I've seen a lot of awfully improbable things. As an example: a couple of years ago we found a gravitational lens in the redshift survey.[11] It turned out that there was this quasar right smack behind the nucleus of a nearby bright galaxy— by right smack I mean less than a few tenths of an arcsecond, in fact right now the measurements are off about a tenth of an arcsecond. The probability of that happening, when you go out and calculate it, is getting to be on the order of the probability that omega is close to 1 —one part in 10^{15} or so. But there it is. It's there. That's an example. We *shouldn't* have found one [a quasar right behind the nucleus of a galaxy], but we did.

> *Do you have any personal view of cosmology? You were talking about a world view before, but from what you mentioned, the specifics dealt with particular features. Do you have any global world view that you might call personal?*

That's a tough one. I would say that if I took a look at my world view right now, I would favor—but hope I'm not wedded to—a universe that's of order 13 billion years old, a universe where the Hubble constant is of order 70 or 80, even though the measurements that we've been making recently give a higher number. I would like to see the number come down. I don't know if it will, but it might. There are things that need to be done to fix it up, things that will happen when the space telescope flies. I think there's a small but finite chance that we might have to reintroduce the cosmological constant, although right at the moment it's not required by the data. In other words, there is a difficulty, if you take things at face value, between the ages of the globular clusters and the expansion age of the universe. If the numbers stay the same and the errors get smaller, which is a possibility, and those ages do not overlap, then one is going to have to *do* something. The only way I know of solving that problem right at the moment is by sticking back in the cosmological constant, although I'm sure there are others. It's my impression that the limits on fluctuations in the microwave background are getting down at the level of embarrassing for forming galaxies. It's not the kind of thing where you've lost your pants and have to go running into the men's room quite yet. But the belt is loose or the fly is open and there may be some problem.

The large-scale structure business, the really large-scale structure business, is still very vague. The data is in bad shape—some data. It's got a lot of problems. I'm a data merchant, that's one of my stocks in trade, and there are things I trust. If you ask me the redshift of a galaxy, and I've measured it, I will tell it to you with a degree of precision that's pretty good. I *know* that. If you ask me if the catalogue out of which we pick distant clusters of galaxies is any good, I will say, "Probably not." In fact, I can offer you some proof that it's terrible, but I can't offer you a better catalogue yet. It's easy to make the negative statement, but improving the situation will take some time. So, in fact, one of the things Margaret and I and several other groups of people are working on now is to make some improvement on these galaxy and cluster catalogues so we can make stronger statements in the large-scale structure business. That will take another 5 or 10 years, but that's what we're up to. So the data that exists from not such good catalogues, which is to say, not such good basic data, indicates inhomogeneities on very large scales. We need to do it better. If we do it better, and the answer stays the same, you can guess what happens next. I think that we'll have really produced substantial modification in the current big bang or the current inflationary models.

> *Let me ask you to take a step backwards and leave your normal scientific caution aside for a little bit. If you could design the universe any way that you wanted to, how would you do it?*

Well, I'm an observer and not a theorist [laughter]. So I can design the universe any way I want to. Could I design it so that I could go off and be a novelist instead of an astronomer? If I were to come up with a design like that . . .

> *Yes, you could make that part of your design.*

The universe would be *exceedingly* simple, with galaxies that were all the same shape, size, mass, luminosity. It would consist of only baryons that emit light. These dark things are bloody hard to detect, a real pain in the tootsies; you have got to find some way of learning about them. It would be a steady state universe, I think, because that's the easiest one. It's a little weird, but nonetheless . . . Gravity would be a very weak force, or there would be something to offset it. It's not the easiest theoretically, but it's the easiest to visualize. No change. I'd be a very lazy universe maker. Possibly the first thing I'd do is make a universe that contained lots of planets with lots of life forms. That would make it more interesting for purposes that have nothing to do with astronomy. Also, get rid of the speed of light limit. Yes, I think I would certainly do that. I'm not sure what I'd do with the human race. That's a different story; that's a detail of the cosmological model. Right now, I'd design it with a ski chalet in New Hampshire.

STEPHEN HAWKING

Stephen Hawking was born in Oxford, England, on January 8, 1942. He received his B.A. degree in mathematics and physics from University College, Oxford University, in 1962, and his Ph.D. from the Department of Applied Mathematics and Theoretical Physics, Cambridge University, in 1966. He was the thesis student of Dennis Sciama. In 1974 Hawking was elected a Fellow of the Royal Society, one of its youngest members. In 1977 he became professor of gravitational physics at Cambridge and in 1979 Lucasian professor of mathematics. In 1982 Hawking was made Commander of the British Empire. Hawking shared the Wolf Prize in Physics with Roger Penrose in 1988. Hawking is married and has three children.

Hawking's research has centered on theoretical aspects of general relativity, particularly black holes, cosmology, and quantum gravity. Among his contributions to cosmology are his 1969 mathematical proof, with Roger Penrose, of the cosmological singularity theorems, showing that the universe must have originated in a big bang explosion even under much more general conditions than had previously been assumed. In more recent work, beginning in 1983, Hawking has attempted to formulate the initial conditions of the universe from first principles, the so-called no-boundary proposal. Hawking was the coauthor, with G. F. R. Ellis, of an influential scholarly book in cosmology, *The Large-Scale Structure of*

Space-Time (1973), and he wrote the astronomical bestseller *A Brief History of Time* (1988).

In the early 1960s Hawking became afflicted with an incurable and degenerative neuromuscular disease called amyotrophic lateral sclerosis. Although this disease has progressively incapacitated his body, his mind has remained unaffected, and he continues to work. Now unable to speak, Hawking communicates laboriously through a computer, which allows him to select one word at a time from a limited vocabulary of words, arranged in columns on a computer screen. On the walls of Hawking's office, beside the bookshelves and computer, are two large posters of Marilyn Monroe.

> *I first wanted to ask you what people and books influenced you most when you were a child.*

My father, who did research in tropical medicine, made me want to be a scientist, but I didn't want to go into biology because that seemed too descriptive, and not fundamental. So I decided at about 15 that I wanted to do physics. When I was an undergraduate at Oxford, I had to choose between research in cosmology or in elementary particles. At that time, in 1962, elementary particles seemed like botany. There were all these particles, but no theory.

> *When you decided to go into cosmology, do you remember any early ideas about particular cosmological models? Did you have a preference for one model versus another—for example, steady state versus big bang, or, within the big bang models, a closed universe versus an open universe?*

Before beginning research, I thought there must be some explanation, like tired light, of the redshift. And to judge from my crank mail, many people still think that way. But I realized that [explanation] wouldn't work, because the number of cycles in a train of light waves couldn't change. My supervisor was Dennis Sciama, and I was in the same department as Fred Hoyle, both of whom supported the steady state theory. Also, there were arguments that you got retarded potentials in electrodynamics only in a steady state. But I kept an open mind. I began working on the hot big bang model in 1964, and began to prove singularity theorems in early 1965.[1]

> *Are you concerned at all that your work on the initial state of the universe may not be testable by experiment or observation?*[2]

My work on the existence of a big bang singularity seems to be in agreement with all the observations so far.[3] My work on how the no-boundary proposal determines how the universe comes out of the big bang is still in a preliminary stage, but it leads to predictions, like inflation and a scale-free spectrum of perturbations, that are at least not inconsistent with observations.[4] I don't think you can completely test it by observation, any more than you can test the existence of a big bang singularity. But I think that the no-boundary proposal, like the big bang, will become one of the background assumptions of cosmology. There is not much alternative, unless you are going to suppose that God is sending messages into the universe.

> *Do you remember when you first heard about the flatness problem? Did you regard it as a fundamental problem with the big bang model?*

I was aware of it about 1967, although it was not called the flatness problem until 1981. But I wrote a paper in 1971, on the isotropy of the universe, which referred to it.[5] I think it was generally known long before inflation. In particular, Charles Misner was aware of it in 1967.[6] At that time, the only explanation seemed to be the anthropic principle. Brandon Carter discussed it in an unpublished paper in about 1970.[7]

> *Have the observational results of Geller, Huchra, and de Lapparent on the bubble-like structure of the distribution of galaxies, and previous work on the large-scale structure of the universe, altered your views on the basic assumptions of the big bang model? Has it shaken your confidence in the basic assumptions?*

No.

> *A* Newsweek *article last year quoted your saying that you avoid problems with a lot of equations.[8] Could you comment briefly on the role of visualization in your work. Is that a critical tool for you?*

I really have to be able to visualize a problem. There are plenty of people who can manipulate equations better than me, because it is difficult for me to write them down, and, even if I do, I don't have much intuition about equations. I do, however, have intuition about geometry.

> *If you could design the universe any way that you wanted to, how would you do it?*

It is like the anthropic argument: If I had designed it differently, it wouldn't have produced me. So that is a meaningless question. I'm prepared to make do with the universe we have, and try to find out what it is like.

> *There's a place in Steve Weinberg's book* The First Three Minutes *where he writes that the more the universe seems comprehensible, the more it also seems pointless. Have you ever thought about this question of whether the universe has a point?*

I don't feel like that. I think that human intellectual history is a record of how we have come nearer and nearer to an understanding of the order in the universe. I'm proud of our achievement.

[Addendum: The following question, which was posed to the other interviewees, was not asked Hawking because he had already addressed it in his book *A Brief History of Time*. His response is excerpted from that book.][9]

> *What is your opinion of the inflationary universe model?*

The new inflationary model was a good attempt to explain why the universe is the way it is. However, I and several other people showed that, at least in its original form, it predicted much greater variations in the temperature of the microwave background radiation than are observed.[10] Later work has also cast doubt on whether there could be a phase transition in the very early universe of the kind required. In my personal opinion, the new inflationary model is now dead as a scientific theory, although a lot of people do not seem to have heard of its demise and are still writing papers as if it were viable. A better model, called the chaotic inflationary model, was put forward by Linde in 1983.[11] This model has all the advantages of the early inflationary models, but it does not depend on a dubious phase transition, and it can moreover give a reasonable size for the fluctuations in the temperature of the microwave background that agrees with observation.

DON PAGE

Don Page was born December 31, 1948, in Bethel, Alaska, and grew up in Alaska, where his parents were elementary school teachers. He received his B.A. in physics and mathematics in 1971 from William Jewell College, a small Christian college in Missouri that his parents had attended. He completed his Ph.D. in physics at the California Institute of Technology in 1976. Page worked as a post-doctoral fellow with Stephen Hawking at Cambridge University during 1976–1979. In 1979 he went to Pennsylvania State University, where he was a professor of physics. Page is now Professor of Physics at the University of Alberta and is a Fellow of the Canadian Institute for Advanced Research. He and his wife have two sons.

After being trained in general relativity at Caltech, Page began applying quantum theory to problems in gravitational physics and cosmology and has collaborated with Hawking on a number of projects. His research interests include the thermodynamics of black holes, quantum gravity, and studies of the quantum state of the universe. Page is known for his ability to plough through long and difficult calculations, writing down one equation after the next until he arrives at the final result.

I'd like to start by asking you to tell me a little bit about your parents.

My parents were elementary school teachers. After they got married and after my father worked a year in Kansas City, he got tired of commuting, and they both wanted to go off somewhere. So they applied to the Bureau of Indian Affairs to go teach in some remote place. They got a position up in Alaska. In 1941 they went to Alaska with the goal of staying a year. But then the US got into World War II, and they were frozen to their jobs. I came along at the end of 1948. So, in a sense, I think when they went up there, they were almost pioneers. They had gone out to villages where there weren't any roads and weren't any telephones.

What kind of work did they do in Alaska?

They taught elementary school. The Bureau of Indian Affairs had schools out in these small villages, varying from 100 to 300 people. In the early stages, my parents had to do a lot of other things, too. In some villages, they were the postmaster or they would help if medical services were needed in the village. They might give shots and various small things, because we would be away from doctors. So they were quite independent. They were out there from 1941, and it was several years before they even left Alaska, just to see anything. They had fallen in love with it by the time the war was over, so they just stayed.

So you grew up there?

Right, I grew up there. I was in Alaska all my life until I went to college. Then my parents retired in 1972, after I'd been at Caltech one year.

Can you tell me a little bit about your childhood—what kinds of things you liked to do?

Of course, Alaska was good for various outdoor things. In the summertime, we would go boating. In most villages, my father would build a boat and we'd go with a motor boat. In wintertime, we'd go skiing. Occasionally— my brother and I had a couple of dogs which we didn't train too well—we'd go dog-sledding or have the dogs pull us on skis. In the summertime, we might go fishing or hiking up mountains and such things. Wintertimes tended to have fairly long nights. I did some photographic work and developed and enlarged photographs, and that took a lot of evenings in winter.

I remember in the earliest days, my mother was fairly mathematical. Both my parents had a college education and a little bit of graduate work beyond, but not enough for another degree. They taught me a lot when I was a child. I can still remember a big board, maybe about two feet by three feet, on which my father had written all the numbers out from one up to 100. I don't remember exactly when this was, but it was somewhere between the ages of 3 and 6. It's rather amusing. I still sort of visualize the numbers from one to 100 more or less how they're arranged on this chart. So my parents did a lot of work with me when I was a preschooler.

In mathematics?

Math was one of the things that I had an early aptitude for, and so I think they helped me some. Neither of them knew advanced math. Algebra, I think, was about as far as they'd gone. So, maybe, I got off on my own. I remember, in the fifth grade, I had a solid geometry book that I quite enjoyed playing around with and just reading. It was fun.

That certainly was not your fifth-grade textbook.

No, this wasn't.

Did geometry appeal to you particularly? That kind of structure, that style of doing mathematics, the lemma-proof approach?

That kind of geometry that I got in high school—proving things by angles and stuff—didn't appeal to me too much. When I got to analytic geometry, Cartesian geometry, that was much more appealing. I think in the eighth grade I knew the definition of an ellipse as points for which the sum of the distances from the two foci was a constant. I was trying to prove that that was an elongated circle, and I never succeeded. This was before I knew anything about analytic geometry. So I didn't have any real hopes of writing down equations, but I tried drawing lines and stuff, and I never succeeded in proving that. It fascinated me. In those days, they had these big paperback books called *Everything Made Simple.* I know when I was about 10, I was enthralled with chess, and I read *Chess Made Simple.* Then, in about the beginning of high school, I got some of the *Advanced Mathematics Made Simple* books. There was one about analytic geometry, I think, and then some other one about linear equations, using determinants to solve linear equations, and a few others of that sort. Most of those things in math I probably learned from books before they were really covered in school.

Do you remember any reading in science that you did at this time?

I do remember reading a certain amount, perhaps in chemistry—although that's a bit paradoxical because I never had a chemistry course in my life. One of my aunts gave me a chemistry set that I played around with a little bit when I was maybe 10. I do remember reading some books early on. At the early stages, I was probably more interested in math than science per se, although I did have a little bit of interest in science. Some of the things in nuclear physics sounded fascinating, although I was a little disturbed that the only application I knew was to make bombs.

Where would you have heard about nuclear physics at this age?

Probably from books, or maybe some encyclopedias. I do remember one article that was rather seminal in generating interest. The *Saturday Evening Post* had a series about the great ideas or something like that. I never really went back to find out exactly when that was, but I think it was around when I was 10. One of the articles I remember was on the origin of the elements, written by Willie Fowler.[1] That article made a deep impression on me—particularly the notion of hydrogen turning to helium in the sun and so on. When I was taking Fowler's course at Caltech, I remembered that article. I guess I was interested in a certain amount of science besides mathematics in the early stages, but I probably didn't actually decide to go into physics rather than mathematics until I went to college.

Before we leave this period, do you remember at this age having any cosmological thoughts?

It's hard for me to remember whether I had any cosmological ideas. I think I had my own theory of elementary particle physics. I thought everything was made of little tiny things. I called them "nuttons," and they were supposed to be little things that you couldn't possibly cut, but they could have any old shape. I think it was just an idealization derived from seeing metals that you couldn't do too much to, but that wear down. But as far as cosmology, I don't remember having too much of an idea, other than about the origin of the elements. I can vaguely remember reading some things about the big bang theory and the steady state theory.

Tell me a little about your college experience.

I went to William Jewell College, which is a small, Christian, liberal arts college in the Midwest [Missouri]. It was the one that my parents had gone to, and some of my aunts lived in the area, so in some sense it was more home than if I'd stayed within Alaska. If I'd gone to the University of Alaska, I would have had to live away from home anyway. For a college of that size—there were about 1,000 people when I was there—they had a good physics department with three physics professors. One of them, Dr. Wallace Hilton, was a very energetic teacher. He didn't know an enormous amount of theoretical physics, but he did a lot with acoustics and optics and lab work and really encouraged people to do things. Later after I'd left, he won the Oersted medal as the top physics teacher of that year. He was a great encouragement. He had programs of independent study and research, giving students one or two hours of credit just for spending Thursday afternoon working on something, on one's own. So he was a very enthusiastic teacher and would have signs around, like "Physics is fun." By the time I went to college, I thought I would eventually earn a master's degree. But I never particularly thought about going on for a Ph.D. Of course, at that stage, I didn't really know what aptitude I had.

Looking back now, I can see that it was much better for me to go into physics, although the physics I do may be somewhat mathematical. I'm not good at doing rigorous proofs or faking the style of pure mathematicians. Probably, at the beginning of college, I would not have realized whether my aptitudes were more in physics or math. At that stage, the direction I chose was probably more due to the better physics department there and the encouragement that I got. And Dr. Hilton used to embarrass me by always referring to me as Dr. Page. That was his way of telling me I'd better go on and get a Ph.D., since I had not particularly expressed whether I would or not.

Did you learn anything new about cosmology as a college student?

One thing I do remember is that between my sophomore and junior year, NASA had a summer institute in space physics, held at Goddard Space Flight Center in New York, not too far from Columbia University. Also I remember in my senior year I read an article or two by Kip Thorne. That developed at least a little bit of interest in black holes before I came to Caltech, though I'm not sure that it engendered so much interest that I knew for certain that I would go into relativity.

It sounds like you had a little awareness of cosmology, but not very much at this stage.

Yes. When I went to graduate school, at first I thought I was more interested in elementary particle physics. It was after I came to Caltech that I developed an interest in relativity. It's hard for me to remember precisely how early this was. I do remember that after the placement exams that we took early on, they had a party for us new students over at the Caltech Atheneum, and Kip Thorne was one of the relatively small number of professors who had enough interest to show up at this sort of thing. I remember having a nice conversation with him. I seem to remember I had a certain amount of interest in relativity after that, so maybe that was a turning point. It could also have been that I found particle physics somewhat confusing. There was such a variety of information, and I didn't see such a clear pattern in it. Another thing was that I did better in the relativity course than in the elementary particle physics course. So I had more of an interest of going into relativity. Then I started in Kip's relativity group after the first year candidacy exams.

You took Kip's relativity course, I guess.

Yes, I took that. I think Jim Gunn taught one semester of it. If I remember correctly, in the first term Jim Gunn was teaching the relativity course while Kip Thorne was teaching a high-energy astrophysics course. I certainly admired Kip's lecturing style and his clear expositions of things. And the relativity course became quite fascinating as well.

When you took the relativity course, do you remember at that time having a preference for any particular cosmological model, like open versus closed?

I don't remember an enormous preference. Of course, at Caltech there seemed to be more emphasis on the observations, which at that time favored an open universe.

And still do.

[Laughs] That's true. They still do. There wasn't evidence for an omega higher than 0.06, or whatever it was at that time. I took Willy Fowler's 8 a.m. course on nucleosynthesis and nuclear astrophysics. I remember at one stage he said that according to the Caltech religion, the universe is open, and according to the Princeton superstition, it's closed. I inherited a bit of that feeling of going with the observations more. I know, of course, that the Misner, Thorne, and Wheeler book, with some parts of the cosmology having been apparently written by Wheeler, strongly favored the closed universe. But I don't know that I had any strong prejudice. I suppose just by listening to what the observational results were and given the lack of evidence of other matter, I thought that it was more likely that the universe was open. But I think that was more a function of what I was hearing at the time rather than any particular prejudice on my own part.

Could you tell me how it was that you started working in cosmology?

Yes, it was a long route. After the first year, I started with Kip. I asked him whether he had some good problems with black holes, and he thought that so many people were working on black holes, maybe I should work on something else. He gave me some problems with the cylindrical universe, which I wasn't really getting anywhere with. Then I became interested in this superradiance that Saul Teukolsky and Bill Press were doing a lot of work on.[2] I tried to figure out how one would describe that quantum mechanically. It took me a while just to realize that it was simply stimulated emission. After realizing that, it occurred to me that there ought to be spontaneous emission. I remember we managed to get Richard Feynman to come over to Bridge Lab, and we wanted to explain this to him and see if it made sense to him. He came over, and he wanted to have a model for black holes with little vanes sticking out or something, so he could have a physical model for how these rotating black holes can amplify waves. I remember I was there, and Saul Teukolsky and I think Doug Eardley was there at that time as a post-doc, and you and Bill Press.

I remember those conversations.

Yes, I remember it was a lot of fun, because after we more or less convinced Feynman that there should be this application, then I think Saul Teukolsky asked him why the neutrino waves weren't amplified. Feynman quickly understood this and realized it was basically an exclusion principle—you couldn't get more than one neutrino in a state, so if you sent one in, you couldn't get more than one out. I do remember an amusing situation. When Feynman was at the blackboard, he was trying to draw some neutrinos going in, antineutrinos going out, and he was trying to figure this out at the blackboard. He wasn't too sure of exactly how to write that down. And he said, "I'm supposed to be good at these diagrams." Feynman had such willingness to talk to even green graduate students about this.

So this got you involved with putting quantum mechanics and relativity together, in a way?

Yes. I remember that a couple hours after Feynman left, I was trying to find some way to calculate this spontaneous emission. I discovered that Zel'dovich and Starobinsky had a paper in which they had a two-line statement predicting spontaneous emission.[3] Of course, I was a bit crestfallen that someone else had already predicted this effect. I got interested in calculating it, but I didn't really know how to do it formally. Then Stephen Hawking indirectly, I think, heard of some of this stuff through

Doug Eardley, who had told Hawking that I was working on this. Hawking had a chance to talk to Zel'dovich and Starobinsky in Moscow. He liked their idea but didn't like their derivation. So he sat down to do it himself and, lo and behold, he found that there was not only spontaneous emission of these amplified modes, but all modes produced particles. This was very surprising to him. This was his famous Hawking radiation.[4] Then, of course, we heard of this at Caltech and got his preprint. At first it was a little hard to know whether we believed it or not. So then I began doing some numerical calculations. When Stephen Hawking spent the year 1974–1975 at Caltech as a Fairchild Scholar, I got to know him. We ended up writing a paper proposing that people look for gamma ray bursts from primordial black holes that might be exploding.[5] Then I finished my thesis on calculating numerical rates of particles coming out from rotating and nonrotating black holes.[6]

So this was the start of your association with Hawking.

Right. In my last year, 1975–76, I was applying for a number of post-doctoral opportunities. I had Stephen Hawking as one of my references. Then he wrote me at some stage and said, "I've been writing letters of reference for you, but I may have a position myself." I went there in 1976 and was a postdoc. Then he got support for me for two more years.

You started working with Hawking on quantum gravity?

Right. It was quantum gravity more than cosmology. Then, after going to Penn State, I worked a little more on that, but I had a variety of interests. One thing that occurred after I got back to the U.S. related to Stephen Hawking's proposal that black holes might cause a loss of information from this universe. I was working more with black hole thermodynamics and various problems of quantum gravity until Hawking proposed the idea for the boundary conditions of the universe, at this meeting in the Vatican, I think in October 1981.[7] Then, in the next year, he and James Hartle worked out the mathematical details, which they published in 1983, in *Physical Review*.[8] I had a great amount of interest in that because I realized that one needs a lot more than just dynamical equations in physics. One also needs to have boundary conditions, particularly to explain such things as the second law of thermodynamics. It's not something that one can derive purely by laws as we know them. The dynamical laws as we know them are CPT invariant. So it would seem very difficult to get an arrow of time directly out of those dynamical laws. Maybe the new boundary conditions could help.

In this case, the boundary conditions should show why the universe started off in a state of very low entropy?

Right.

Were you thinking in terms of a beginning of the universe?

I've had a lot of interest in the second law of thermodynamics and the arrow of time—partly arising, I suppose, out of black hole thermodynamics and entropy. So I knew there was this problem of the arrow of time, even before Hartle and Hawking made their proposal. I can remember that Paul Davies had argued that perhaps inflationary universes might explain the arrow by winding up the gravitational field, although I was never completely convinced that he had explained it. It seemed that inflation *assumed* that you had a second law of thermodynamics. We had a bit of debate with a couple of articles in *Nature.*[9] These amazing facts about the universe—that it's so highly ordered, with very low entropy at the beginning, as well as its isotropy and homogeneity—seemed to impress upon me that one does need more than just dynamical laws. Therefore this Hartle–Hawking proposal had a great deal of appeal to me, and I perhaps felt like defending it, even before other people did.

How would you defend the philosophy of it?

Well, I just said that one needs something to specify the initial state of the universe. In physics, we wouldn't have a complete description until we had something like that.

Some people might just be willing to accept initial conditions as givens.

Right. I remember when I said this, Bryce DeWitt immediately popped up and he said, "You don't want to give God any freedom at all." So before I could think of an answer to that, Karol Kuchař said, "That's His choice." And I imagine that he meant that it's not really restricting God's freedom by finding out what laws He may have used.

Would you agree with Kuchař on that?

Yes, I think very much that I would agree. I am a Christian and believe that God has created the whole universe. Of course, as a physicist I'm trying to understand a bit more of how He did create it or in what state He's created it. I think these laws show the faithfulness of God and the patterns that He's used. But I don't believe at all that these choices were forced upon God by anything. Einstein said that the question he was interested in was whether or not God had any freedom or any *choice* in creating the universe. And it seems to me that you could never prove that He didn't have some choice, although I know a lot of physicists think, or seem to have the hope, that there might be only one consistent theory for the universe.

> *Do you think that there is only one consistent theory of the universe, that the universe can have only one set of physical laws?*

No, I think the universe could be vastly different. I can imagine all sorts of ways in which it could be different. In fact, when most of these people say that there might be only one consistent theory, I think what they mean is only one *fairly simple* theory that might be consistent. In principle, the whole thing could be quite different. I don't know why God's chosen to follow these laws. Of course, if the universe were completely chaotic, it would be hard for any intelligent beings to be created within it, and it would be very hard for the universe to have any order. On the other hand, it does seem to have a lot more order than would be necessary merely for our existence. So why has God chosen to make it even much more ordered than we might have needed? At the moment, I would even say I could allow all conceivable possibilities whether or not they allow life. Of course, we know if the universe were so chaotic that people couldn't be here, then we couldn't be here to ask the question.

> *In your program in quantum cosmology, when you're doing these path integrals over lots of different 3-geometries, isn't it true that you can interpret that operation as the consideration of many different universes, each with a probability amplitude?*

Right. Yes, I think in a sense when you do the path integral you get nonzero probability amplitudes for a whole lot of different possible situations.

> *If we imagine that there is this large range of universes that all simultaneously exist, do you think there's anything other than the anthropic principle that would make universes with life special?*

Well, let's see. Of course, the weak anthropic principle is in a sense saying that our observations have to be consistent with life.

> *Right, but what I'm asking you now to do is to take a step back—to look at the whole ensemble of universes.*

Right. So I imagine myself outside of our universe, which I do all the time. So I'm not restricted to looking at just those components [universes] in which life exists. I suppose God might be more interested in those components in which there is life, although I'm not quite sure how I would put that in operational terms. It's an interesting thing. I'm a conservative Christian in the sense of pretty much taking the Bible seriously for what it says. Of course, I know that certain parts are not intended to be read literally, so I'm not precisely a literalist. But I try to believe in the meaning I think it is intended to have. I know that a lot of Christians believe that man is the main purpose for God's creating the universe, but I'm perhaps a little less certain of that than many of my friends are. The Bible reports that God created man in His own image, but whether that means man is uniquely in His image, whether He could have created some other beings somewhere else that are in His image but quite different from man, I don't know. Maybe God even created some things that aren't life that might be said to be in His image. I don't know. The Bible doesn't really say much about that. I think the main point of the Bible is how He created man in His image, and man fell and rebelled against God, and then God offered the way back through Christ and Christ's death on the cross. I think it focuses mainly on man's relationship with God, without attempting to answer the questions of what other purposes God may have had in creating the universe. So I am a bit skeptical about applying the strong anthropic principle, that the universe *had* to have life in it, or that life was one of the necessary purposes for God to create the universe.

Do you think that the universe has a purpose?

Yes, I would say that there's definitely a purpose. I don't know what *all* of the purposes are, but I think one of them was for God to create man to have fellowship with God. A bigger purpose maybe was that God's creation would glorify God. I'm a little reluctant to say whether that's God's only purpose. Maybe I should say that I do believe the Bible is God's revealed word to us, so I think that's one purpose that has been revealed to us. We could just try to *guess* other purposes from some other means. We see such mathematical beauty and simplicity and elegance in the physical universe, in the dynamical equations that God's created here. I'm not quite sure how to tie those aspects of the creation into purpose. In some sense, the physical laws seem to be analogous to the grammar and the language that God chose to use. The purpose seems to relate more to other aspects of what is created. It's a bit like if you tried to analyze the grammatical structure of some of Shakespeare's writing, but you didn't look at all at what the plays meant, or what the story was there. There can be these different descriptions on different levels.

I want to ask you about your reactions to some discoveries that have happened in the last 10 or 15 years. Do you remember when you first heard about the horizon problem?

To be honest, I'm not sure that I do remember precisely when I heard that. I think it must have been sometime when I was doing research or something, and I read something about it. It might have been mentioned in courses, but I don't really remember its sinking in.

When you did think about it—whenever that was, late 1970s or whenever—do you remember whether you regarded it as a serious problem?

Yes, I thought that was a serious problem. I do remember one time—it was before inflation—I had dinner or breakfast at some meeting with Bill Press. I remember he made the remark that homogeneity and isotropy were not so hard to explain. You just have to say that the situation at early times was *smooth*. You use one word. So he was emphasizing that the real problem was to get the fluctuations for galaxies. Now, in a sense, he was taking more of the attitude that there was something analogous to the Hartle–Hawking proposal that would involve some special initial conditions.

Did you personally think that the resolution of this horizon problem, or large-scale homogeneity, might lie with the initial conditions?

Yes, once I thought about it. When I first took the courses and heard the problems, I don't know that I thought of it. But yes, I certainly came to realize that would be something for the initial conditions to explain.

Were you willing to just accept initial conditions that had this property?

I thought that it ought to come out of something fairly natural. I'm not quite sure what. A lot of these things were influenced by Roger Penrose's calculations of the huge entropy if the entire observable universe were an enormous black hole.[10] Although one might question some of the assumptions of such a calculation, such as fixing the energy and putting it all in a black hole and so on, Penrose's emphasis was on how special the initial conditions of the universe were. You get enormously big numbers for the possible entropy if the universe were in some different configuration that had a lot of gravitational entropy. Before these calculations, the question of entropy was always a little bit confusing because the early universe was rather hot and uniform and, *locally,* it was fairly near thermodynamic equilibrium. With nongravitational systems, thermal equilibrium tends to be a state of *high* entropy. So in that way of looking at it, you'd tend to think that the early universe had high entropy. Penrose is probably the main one who got me to see that the entropy of the early universe really wasn't that high [compared to what it might have been], because the gravitational field was quite smooth. So I felt that one needed to have some explanation for that. And admittedly, I wasn't quite as much convinced that inflation solved this problem as a lot of other people were.

I was going to ask you about that.

Maybe by the time inflation came around, I realized that this second law of thermodynamics question seemed to be a more special problem than just the homogeneity and isotropy. Those were special aspects, but these numbers that Penrose got from thermodynamic arguments were so huge. I don't know exactly how to quantify the homogeneity and isotropy. I'm sure they must be extremely special, too, but probably not *as* special as the second law. So it seemed to me that special, low-entropy initial conditions were required even to *have* inflation.

You mean you would have to have a direction of time already defined. Is that what you mean?

Yes, I think that in fact gets back to these arguments that I was having with Paul Davies. Paul Davies was arguing that you could wind up the clock of the universe and get it into a state of low gravitational entropy by having inflation, whereas it seemed to me that you would have to have low entropy even to have inflation.

So you felt like the inflation didn't really solve this problem, because there was a more fundamental problem that needed to be solved prior to inflation.

Right. I thought that the homogeneity and isotropy might come as a consequence of whatever it was that solved this bigger problem of the second law of thermodynamics.

Did you regard the inflationary model as a big success in some ways?

I perhaps reacted a little bit negatively towards it, because some people made what seemed to me slightly exaggerated claims about what the inflationary universe model solved. Alan Guth has really been quite fair in not making too many exaggerated claims. It didn't seem to me that the inflationary model solved the problem of the second law of thermodynamics. And if you could solve that problem, I was trying to think of some alternate ways that you could get the isotropy and homogeneity out. The monopole problem was explained well by inflation. I had some secret hope that maybe we could find some other solution to that problem. But it did seem that inflation solved the monopole problem. Now I believe that inflation is certainly a part of the evolution of the universe. I mean, it seems likely to have been part of the past history of the universe.

Because it's come out of quantum cosmology?

Yes. I still have some resistance to inflation. People talk about how probable inflation is. I've even been doing some classical calculations, with Hawking and on my own.[11] I found that you can try to calculate what the probability of inflation is with some very simple models. It turns out to be unambiguously ambiguous. That is to say, the number of inflationary solutions is infinite, but so is the number of noninflationary solutions.

Why do you think the inflationary model caught on so widely?

It does seem to solve some of these problems. Maybe I've been a little bit too much prejudiced against it. Maybe I've overreacted to some of the extreme claims for it, but it does provide a mechanism for making the universe large and fairly smooth. And it does provide a mechanism for amplifying some small quantum fluctuations into larger fluctuations that might be able to turn into galaxies and so on. Although even there you have to assume that these small perturbations start off near the ground state, which is something that can be justified by the Hartle–Hawking scheme modulo all the difficulties of really doing quantum gravity. Then, of course, it was another model that combined particle physics and high-energy physics with cosmology. Alan Guth started off in particle physics, and then he realized that some of the results there could have impact in cosmology. So the inflationary universe model is certainly a very important development. By the time inflation came out, Roger Penrose's arguments about entropy had caught my mind, and the second law of thermodynamics seemed to be a much bigger problem. Maybe the fact that I didn't think inflation solved that problem made me think it wasn't the solution of everything. But it's certainly an important contribution.

Another problem that the inflationary models are purported to solve is the flatness problem. One way of stating it is why was the ratio of kinetic to potential energy at the Planck time so close to 1?

Right. Or, as I sometimes say it, why has the universe grown so big and yet gravity's still important?

Do you feel the same way about the flatness problem—that it is one of those problems whose solution will come out of a pre-inflation period?

I think the flatness problem is something that inflation could well solve. I never did consider that to be as important as the homogeneity and isotropy.

Why is that?

I suppose just because it's one number. It's true that it's one number that's very finely tuned. Say, at the Planck time, this balance was to 1 part to roughly 10^{60} or maybe 10^{57}, whatever it is. So you might say that's extremely improbable, getting 1 chance in 10^{60}, but then [with] these other things that Penrose argued, you're 1 chance in $e^{10^{123}}$. It's extremely much less probable that you'd have a second law of thermodynamics.

Maybe I thought the anthropic principle might be sufficient to solve the flatness problem. It would depend on your a priori probabilities. I thought that was conceivable, whereas I did seem to recognize that it'd be very hard for the anthropic principle to explain the isotropy and homogeneity. Why would our existence depend on something way off somewhere else, something we are just now beginning to see, which could not have influenced us at all before now?

When did you first hear about the flatness problem? Was that later than the horizon problem?

Probably earlier, I'm not sure. Sir Bernard Lovell, I think, wrote something about how precisely was this balance between kinetic and gravitational energy. Of course, it also came out in the Einstein Centenary Volume.

But you're saying that it didn't strike you as a problem as serious as the horizon problem, because you thought that there was just one number that had to be fixed at the Planck time?

Yes.

And how did you imagine that number would have been fixed?

I might have thought there's a quantum ensemble of universes. There are all these different wave components and omega might have a whole lot of different values, and maybe life can only exist in the ones in which omega is fairly close to 1.

At this stage in your career, do you have any preference for any particular kind of universe, an open versus a closed?

Well, this Hartle–Hawking prescription so far has been given in the case of the 3-dimensional space, which is closed.

Other than the Hawking–Hartle prescription, you don't have any other particular preference?

Probably not. I do believe in inflation enough to have some prejudice toward thinking that omega is likely to be very close to 1, rather than say, around a half or whatever the observational result is. So, if I had to bet on it, I would put money on saying that more careful observations are going to find that omega is close to 1. It seems very odd that omega would be so close to 1 and yet just now, in the present age, be drifting off—unless the anthropic principle is the right argument for it. But somehow, if the anthropic thing were right, I'd expect omega to be more different from 1 than just a half. Maybe a tenth, or three.

Let me ask you one last question. We've already been pretty speculative, but if you could design the universe any way that you wanted to, how would you do it?

I never particularly thought about that, because I've been thinking about how the universe *could* be. If I was going to do things any differently, the only things I would have any idea about would be more on the human level. I mean, the Bible has some picture of heaven. One can imagine changing things like social injustices or eliminating diseases. But if I wanted to do this in some mathematical physics way, writing down a different Lagrangian, the trouble is that I have very little idea what the consequences of some of these equations are. If I tried changing the state or changing the boundary conditions, it's very difficult for me to visualize what the outcome might be. To be honest, I guess I never really thought about trying to design the universe. I suppose I'm a bit daunted by the task.

ROGER PENROSE

Roger Penrose was born August 8, 1931, in Colchester, England. He received his A.B. degree in mathematics from University College London in 1952, and his Ph.D. in mathematics from Cambridge University in 1955. After a number of temporary posts in the United States and in England, he went to Birkbeck College, London, where he was reader from 1964 to 1966 and professor of applied mathematics from 1967 to 1973. Since 1973 he has held the Rouse–Ball Chair of Mathematics at Oxford University. Among Penrose's awards are the American Physical Society's Heineman Prize for Physics in 1971 and the Wolf Prize in physics in 1988 (shared with Hawking). He is a member of the Royal Society. Penrose's interests include algebraic geometry, differential topology, plane tilings and quasicrystals, the theory of twistors, classical general relativity, and singularity theorems in general relativity.

While working on his Ph.D. in mathematics at Cambridge, Penrose came under the tutelage of Dennis Sciama and was encouraged by Sciama to pursue physics and cosmology. Penrose is one of the most mathematically sophisticated people ever to have worked in general relativity and cosmology. He is renowned for his discovery of Penrose tiles, which are two geometric shapes that can be fit together over and over to completely cover a 2-dimensional plane, without repeating any pattern. In 1965, with unusual mathematical techniques, Penrose found the general conditions under which a mass will collapse to form a black hole, the first of the so-called singularity theorems. In 1969 these ideas were adapted by Penrose, in collaboration with Stephen Hawking, to show that the universe as a whole probably originated in a big bang, even if the universe is not isotropic and homogeneous, as frequently assumed.

Penrose has strong views about physics, yet he is soft-spoken. On the wall of his office is a picture of a complex, 3-dimensional surface, with twists and arrows. Side by side is a beautiful and highly detailed drawing of a dinosaur, done by Penrose's son at the age of 8.

I wanted to start with your childhood. Could you tell me a little about your parents?

My father was a scientist. Both parents were. They both were medically trained. My father became a professor of human genetics, but with a considerable interest in mathematical subjects. My mother also had a definite interest in science. My father really didn't let her practice medicine, though. He found it too threatening, I think. Neither of them were terribly personal, but my mother was much more so. It was very hard to talk to my father about anything personal. We talked about science. We used to go for long walks. He would describe how things grow. It was fascinating. He was certainly a great inspiration to me on the scientific side.

Did you get any inspiration from your mother on the scientific side?

Not so much, but to some degree. However, I think I got a lot of understanding from her in other ways.

Was she supportive of your interest in mathematics and science?

Yes. But he was much more dominant. He worried when I elected to do mathematics that this was not an appropriate subject. Science was all right, but mathematics was something you did only if you were fanatical and weren't interested in other things. So when I wanted to go to the university and study mathematics, he got one of the lecturers there to give me a special test to see whether I could do it properly. I think perhaps he was hoping that I might fall down on this test. Unfortunately, I did do quite well!

Besides discussions with your father in science and mathematics, do you remember any influential books that you read?

I never read very much. Let me mention something else. I think I got a lot from my older brother, too. We were three brothers, and I was the middle. I also had a sister who came along much later. She was born when I was 13. My older brother was very precocious, and he was way ahead at school. He would sometimes explain mathematical things to me. My younger brother was more a threat in a different way. He was terribly good at games. Later on he became British chess champion 10 times. When we were young, it wasn't just chess, it was all games. We developed a sort of competitive attitude. But I didn't play chess much. My younger brother was too good already and so was my older brother, so I opted out.

Did you learn calculus before high school?

Well, let me explain. I was born in 1931. I lived in the UK until 1939. In 1939, just before the war, my family moved over to the United States. We were there for a few months in Philadelphia. Then we moved to Canada, where my father had a job as director of psychiatric research at the Ontario Hospital in London, Ontario. And so we lived the war years in Ontario. I had just moved over to high school for the last year I was in Canada. Then we came back to England, just after the war. My father had obtained a position at University College London as a professor of human genetics, or eugenics, as it was called until my father had the name changed.

In England, one would learn calculus at what I suppose you call high school. My father was so keen to get the thrill of telling me about calculus that he taught me just before I learned it at school. I'm sure my older brother, Oliver, knew about calculus a lot earlier, because he was racing ahead at great speed. I just learned it more or less at the time everybody else did.

Did you ever read anything about cosmology at this age?

Reading wasn't quite the thing I did. I never read much. Oliver read a lot. I did do things like make models—polyhedra and so on. They were usually geometrical models, which I made out of cardboard. There were lots of things I made. I was very interested in doing things on my own.

Did you talk about cosmology with your brother?

My father would have known more, but I'm not sure how much he knew on the subject. I'll tell you what did influence me, though. That was Fred Hoyle's radio talks. It was probably around 1950. I think it was when I was doing a maths degree at university. Those were absolutely fascinating. I was interested in astronomy. My father was interested in astronomy. He knew quite a lot about the stars and constellations. Also, he had a telescope, a nice brass telescope, a refractor. In fact, he and my mother went to see an eclipse once. They climbed up a mountain and saw the eclipse through this telescope, which had a special attachment for viewing the sun. That was before I was born.

When you heard Fred Hoyle's talks, did you have a particular preference for any type of cosmological model? I imagine that he was emphasizing the steady state model.

Yes. I didn't know anything about that at the time. It was a series of radio talks. They started off, I think, at the local level. First planets and then stars. Then he talked about cosmology and the steady state. And he definitely talked about this, because I remember his saying that galaxies, when they reach the speed of light, disappear off the edge of the visible universe. The idea was that when they exceeded the speed of light, the light wouldn't get here. I remember being very puzzled by this and drawing various pictures and so on. I went up to Cambridge because my father had this idea that if you wanted to be a mathematician, you had to go to Cambridge. I'm not sure if this was the occasion, but I actually tried to take a Cambridge scholarship exam, which I didn't get. I thought the whole thing was pretty ridiculous because I really wanted to stay in London. But it was during one of those visits that I first met Dennis Sciama. My brother Oliver was a good friend of Dennis's, and both were research students at Cambridge at the time. They used to eat lunch at a restaurant called the Kingswood, which is where I met Dennis. Oliver told me that Dennis was a cosmologist. So I said to Dennis, "Well, you're a cosmologist aren't you. Can you explain this to me about galaxies disappearing off the edge? It doesn't make any sense." I drew a picture and said, "Surely, you can always see the galaxy. You draw back your light cones, and any galaxy crosses the cone." Apparently he hadn't thought about this before and was very impressed. I think he asked Fred. Fred thought about it, and it turned out that what I was saying was right.

You thought about it in terms of light cones?

Yes. I don't think they [Hoyle and Sciama] were used to drawing pictures, whereas that was the sort of thing I did naturally, being geometrically minded. That experience was quite important to me, especially because I made friends with Dennis. When I did go to Cambridge as a research student, in quite a different area [in pure mathematics], I think Dennis felt it was his duty to look after me. He somehow took me under his wing and had lots of discussions about physics, cosmology, and all sorts of things. Although I was officially a pure mathematics research student at Cambridge, I learned an awful lot from Dennis—not just cosmology but physics generally, and a kind of excitement and enthusiasm for the subject that was very important to me. I think he was a tremendous influence, and I suspect this meeting in the Kingswood restaurant was important because it meant he paid attention to me.

I remember reading that you were sympathetic to the steady state model at this time.

Oh yes.

Can you tell me why?

It's very hard to disentangle these things, but it may partly have been Dennis himself. He was very enthusiastic about the merits of steady state, and that is probably what got me involved. But I think I also liked the idea of steady state for itself: a universe that always was there. This has an aesthetic appeal of its own. There was always a conflict in me because I also felt very strongly that general relativity was right and obviously a beautiful theory. Again, it was aesthetics to some extent. This conflict [between general relativity and steady state] was influential in the way that I thought about cosmology. In trying to resolve the conflict, I was finally able to see that it was irresolvable, in a geometrical way. This ultimately led to singularity theorems—by thinking about light cones and how they focused.

When you say that steady state had an aesthetic appeal, could you elaborate slightly on that?

I suppose it's partly the big bang, which was unpleasant in some ways. The idea that you have this singular origin seems to go against physics. It's where your view of physics goes wrong. If you have to have this singular state in the beginning, that's ugly. It's ugly because you don't understand it. Aesthetics has a lot to do with understanding. It's a difficult issue.

And the problem with a singularity is that you don't have any physics before then? So you have incomplete understanding?

Yes. It's somehow just magic. That magic might be beautiful, if you like, but if one is trying to be scientific, it is understanding that appeals. And here, at the singularity, you just have to give up, whereas steady state gets around that problem and gives you a kind of picture in which you believe that you can comprehend things.

In the early or middle 1950s, did you have a preference for a particular big bang model—say open versus closed or flat?

Not at that time, I don't think. I'm not quite sure what I really thought. But I suspect that I rather liked closed models, just for the reason that you feel you can comprehend the universe in its totality. It seems more pleasurable. If the thing is finite, then somehow it's more comprehensible.

If you are asking for my preferences now, I would say—for reasons that are indirect and not really scientific perhaps—I think I'd prefer the open models. So I think I would put my money on the $k = -1$ [open] models. Not much money. My reason is one that has to do with my feelings about physics.

Could you tell me briefly about that?

Yes. It starts with a long story, which I've written about, concerning the second law of thermodynamics.[1] The second law implies that the big bang has to have a very special structure. [The universe had to begin in a state of unusually low entropy.] It can't be just generic. This structure presumably has to come out of quantum gravity. But the special structure doesn't apply to the final state. The final state has to have high entropy and be very complicated. So the question is, what kind of simple initial state does one expect? I have this belief that complex analytic structure, or holomorphic structure, is a very basic part of what's going on in physics [and this structure should apply to the initial state of the universe]. [For mathematical reasons] this requires that $k = -1$. So that would be my reason for taking an open universe—in the absence of some other reason for believing in something else.

So that theoretical reason compels you more than the observations in your belief in an open universe?

Well, the observations would have done at one stage. When observers were saying that $k = -1$, I would sit back and say, "That's nice." But now, since we know there is dark matter, and we don't really know how much there is, it is quite possible that there's enough to close the universe. In fact, I imagine that, for various reasons, the observations may well not be able to distinguish $k = 0$ from $k = -1$ or $k = +1$ for a long while to come. One could attempt to distinguish by using the ideological argument about the holomorphic structure of the group. But these are not things I would attach a great deal of weight to. We must look to see what observations have to tell us. Except that the observations may not be able to distinguish the models, either. I suppose that what these arguments imply is that if people find that omega is very close to 1, we can't go and say that shows that k is 0, i.e., the universe is flat.

Let me go back briefly to your period of graduate education at Cambridge. You said that you were working at pure mathematics. Was it with Hodge?

I was a student of Hodge's, yes. For one year, before he threw me out.

Why did he throw you out?

Perhaps that gives slightly the wrong impression. He had a tendency not to want to hold on to his research students if he didn't think they were working in the sort of line that was interesting to him. Although I was working on a problem that he originally suggested, it took me off in directions that were not quite his interests. I decided that the problem he suggested had no solution, and I tried to explain to him that the problem had no solution. He was always terribly polite, although he didn't believe me. He never quite said so in so many words. He just said "Oh, that sounds interesting. Would you like to explain it to me?" And I tried to show him this funny notation I had developed. I was writing tensors in terms of blobs, with arms and legs. I developed this notation partly because Hodge's own lectures were so chaotic. I just tried to make sense of what he wrote on the board. So I wrote a great screed, a fat manuscript on all of this. As a young and green research student, one always had the feeling that the great professor would understand anything one could say. I wrote this all out, and it was totally incomprehensible to him. I think he thought I was a bit mad or something.

At what point did you decide that you might want to go into physics as opposed to pure mathematics?

While I was an undergraduate, when I was specializing in geometry in my final year, I went to lectures on quantum theory, just for fun. When I was at Cambridge as a research student, in the back of my mind I was still interested in physics. And Dennis Sciama very much stimulated my interests. He kept trying to persuade me, saying, "What are you doing with this pure mathematics nonsense. Come and work on physics and cosmology." He was always going on like that. And I think to some extent I was persuaded, even though I wanted to finish what I was doing. In addition to all this, I went to lectures in Cambridge in physics, not just pure mathematics. The two lecture courses I found most inspiring were not on pure mathematics. One was by Hermann Bondi on general relativity, and the other was by Paul Dirac on the quantum theory. In different ways, they were absolutely beautiful. They influenced me more than any pure course.

Let me move to the mid and late 1960s. Do you remember what motivated you in your work on the singularity theorems?[2]

One of the things that Dennis Sciama did when I was at Cambridge was to persuade me to come down to London and hear a lecture being given by David Finkelstein. It was on the Schwarzschild solution, extending through the horizon, and he mentioned the piecing together to form the whole Kruskal picture. I remember being struck by the fact that although you remove the apparent singularity at $r = 2m$ (the horizon of the black hole), you still had another singularity at $r = 0$. It seemed that you just pushed the problem somewhere else. When I got back to Cambridge, knowing very little about general relativity, I started to try and prove that singularities were inevitable. It seemed to me that maybe this was a general feature—that you couldn't get rid of the singularity. I really had no means of proving this at the time. But the only thing I did have was the mathematics I'd been playing around with before.

I should say this was not while I was a research student but later, when I was a research fellow. After being a research student at Cambridge, I went back to London. I had a job there working with Christopher Strachey on mathematical problems connected with computer systems. After that, I had my first academic job, lecturing at Bedford college in London. Then I went back to Cambridge again on a research fellowship. So the Finkelstein lecture was after I'd gone back to Cambridge. At Cambridge, I had decided I was going to start learning some physics, in a serious way.

I got myself out of historic order. When I was a research student at Cambridge, I had started thinking about spinors. I'd been mystified by spinors. I finally learned them properly from Dirac. I'd been intrigued by spinors, because my thesis work had been on tensor systems. The fact that spinors were important to physics was an additional intriguing aspect.

So you started working on the singularity theorems because you were stimulated originally from this lecture by Finkelstein?

That's right.

*Did you think originally that there would be any cosmological applica-
tion, or were you just thinking of black holes at first?*

I think both. I'm sure I had the cosmological singularity in my mind at the
same time, but I was aiming more at the black hole singularity, the
Schwarzschild singularity. This digression about spinors was relevant
because it was the only thing I knew, or thought that I knew, that perhaps
would provide a different angle. It was important for me always, if I wanted
to work on a problem, to think I had a different angle on it from other
people. I wasn't good at following where everybody else went. I wasn't the
kind of person who could pick up the prevalent arguments and knowledge
of the time. Other people were good at that. They could suck it all out and
put it together and make advances. I was the kind of person who'd have
some kind of quirky way of looking at something on my own, which I
would hide away and work at. So it meant that I had to have some way of
looking at a problem that was my own. And since I was thinking about
spinors very much, and I felt I got hold of the geometry of spinors, I
thought this has got to be my way of looking at general relativity. Perhaps I
could get some insights into this singularity problem by looking at it in a
spinor way. The singularity problem was the motivation for my thinking
about the spinor way of looking at general relativity. But I never got
anywhere with the singularity problem at the time. I just started developing the spinor method and writing everything I could think of in terms of
spinors. I was rather startled by how magnificently it all worked.

*After you and Hawking and Ellis applied the singularity theorems to
cosmology,[3] did you feel then, and do you feel now, that the observations
of our universe sufficiently satisfy the conditions of your theorems that you
believe there had to have been a singularity?*

I always felt that the required conditions were probably satisfied, without
knowing much about the observations. The words one uses here sometimes shift a little. When I first looked at the problem, I didn't like to use
the word singularity or think of the universe as singular, because I always
believed it just meant that classical general relativity has to be replaced by
some other theory. We shouldn't be throwing up our hands and saying
physics gives up. We should be understanding what physics is. And so I
would tend to say that you don't really have a singularity; what you have is
some change in our physics.

Well, if we can go back to the Planck time, let's call that a singularity.

Yes, okay. But then I think I did use the word singularity, because Stephen Hawking tended to use it that way. I'm not sure that his viewpoint was the same as mine, but I got round to thinking that way. I tended to say it's some new physics, maybe it's quantum theory. Quantum gravity is what I think I really said. But I did think that we knew enough to believe that is what the universe did.

Do you believe that now?

Yes. As you mentioned before, I was sympathetic to steady state for a long while. Then I began thinking about how it was really incompatible with general relativity. And then Dennis Sciama, who had been a strong supporter of the steady state in the early stages, swung around. He had been swung around by observations. I was impressed that he was swung around, having been such a strong supporter of the steady state before. The observations must have produced a strong counterargument. So I think it was indirect. My belief that the observations were powerful in support of the singular initial state [thus favoring the big bang model over the steady state model] was rather vicarious, through Dennis. Stephen Hawking was more cognizant of the direct observations and what the presence of the microwave background meant. I was prepared to believe that the universe was very close to a singular state at one time, from observations. One might then have believed in a previously collapsing phase, such as in an oscillating universe model. But the singularity theorems really showed that that couldn't happen unless one had a gross violation of energy conditions.

Let me ask you a question about the twistor theory.⁴ How have you felt about being the leader of such a nonconventional approach to space-time physics for over 20 years now?

I suppose I don't care about that. It's a complicated question you're asking. It's unconventional and it's not unconventional. It's unconventional because people don't know about it. A lot of it is mathematical techniques. Twistor theory is a way of formulating physics that is unconventional, but it doesn't change the physics. On the other hand, a lot of the motivation behind twistor theory is unconventional.

Could you describe that?

Somehow it's the idea that space-time isn't the primary concept, that it is a secondary concept, arising from some other, more primitive ideas—twistors and complex analyticity. So it leads one in unconventional directions.

I've read in one of your papers that you thought the primary ideas were just numbers, finite quantities.[5]

Yes. There are different motivations, which to some extent are at odds with each other. One of them is the finite discrete, which I was certainly very strong on in my younger years. And the other is the complex analytic. So I had these two things that have appealed to me from very early on. I think even when I was an undergraduate, the two mathematical fields which I liked the best were, first, those dealing with integers, mod two, just "on" or "off" states. And the second is the complex analytic, complex numbers, which have such a tremendous power and beauty of their own.

Would you feel that's primary to understanding physics?

Yes. It's mathematics first of all. But I think that you cannot separate physics and mathematics. One of the things that has always impressed me tremendously is how complex numbers are so fundamental to quantum theory. Complex numbers forced themselves into mathematics, even against the mathematicians' wills. Numerous mathematicians kept imagining that they didn't exist, but complex numbers kept coming back, and they became a powerful way of looking at mathematics.

So you are saying that ideas that are needed for physics also are particularly powerful in mathematics.

Such ideas seem to be underlying physics. It's not simply that we have a powerful body of mathematics that's used in physics, but somehow it's already sitting there. One sees this in Newtonian mechanics, in Maxwell theory, differential forms, fields, quantum theory, complex numbers, quantum field theory. This seems to me to be no accident.

Let me change the subject. One of the things we're interested in is to what extent scientists use metaphor, visualization, and pictures in their work. This seems to have been a strong characteristic of your work, your space-time diagrams[6] and your geometrical tilings.[7] Could you tell me a little bit about the role that pictures play for you?

They've always been extremely important in my way of thinking. Mathematicians certainly don't think the same as each other; some are much more geometrical, some much more analytical. This impressed me very much when I first started as a mathematical undergraduate student. I thought they would be more uniform in their thinking. I found myself to be much more geometrical than my mathematical colleagues.

Do you every try to picture the big bang?

Oh, yes [laughs].

What picture do you form?

One uses many pictures. Certainly if I wanted to visualize something 4-dimensional—and I often have to try and do that—I might use lots of different descriptions, each of which would be only a partial description. I might try to think of the big bang rather physically. That would be imagining what it was like to have everything all scrunched up together. You asked me about reading, and one thing I did read was Gamow's Mr. Tompkins. He did make some attempt to picture what the universe was like at the big crunch. So, I have tried to imagine myself in a situation in which the universe is recollapsing. How helpful that is to me, I'm not sure. Or I would tend to think of space-time pictures. All these things together are helpful. It's useful to try to get a physical feel for things. Or you try to imagine yourself near a black hole and watch it evaporate. But that has to be coupled with a much more rigorous mathematical way. But it's still often visual.

Let me turn to your reactions to some recent developments. Do you remember when you first heard about the horizon problem in cosmology?

I suppose I had vaguely thought about that for a long time. Whether I worried about it is another question.

Do you remember at all what you thought about it in the early days?

The first time I began thinking about this in connection with the sort of thing that people now worry about—in terms of inflation and so on—was when George Sparling was talking to me about symmetry breaking. The question was: if one believes the standard models, then how is it that if you look at two distant quasars in opposite directions, they both seem to have broken their symmetries in the same way [thus having the same physical properties]. My reaction, being unconventional in other ways, was completely the opposite from other people's reactions. It's a problem. If you take the standard view, you have to believe that those two quasars were causally connected in some way.

I see. You are talking about a different version of the horizon problem than I usually think of, which is in terms of uniformity of cosmic radiation or temperature, but this is basically the same problem.

It's the same problem.

Let's talk about the uniformity problem. When you first knew about that, or thought about it on your own, did you regard it as a serious problem with the big bang model?

I can't quite remember. I remember other people worrying about it. My reaction was not being quite sure about what they were worrying about. I suppose it's partly that we don't understand what the big bang was anyway. I know what I think now, but you are asking me a historical question and I'm not quite sure. I think I probably wasn't worried enough about it. I remember Dennis Sciama talking to me about this.

What do you think now?

I think it's all coupled to the second law of thermodynamics problem. The horizon problem is a minor part of that problem. It's not the big problem and, in a sense, that's what I always thought. Now I think I have a more coherent picture of why it is not the big problem.

And you think that when we understand why the universe began with a relatively low state of entropy, we will understand the horizon problem?

That's right. It's one part of the initial singularity and why the initial singularity is so special. People nowadays, over the last several years, have tried to make inflation solve the problem, which I just don't think works.

Before you tell me why you don't think inflation works, let me ask a sociological question. Why do you think that the inflation model has caught on so widely?

I think it's like so many things that are fashionable. I suspect there are many factors, and I wouldn't know if I could get the balance right. I first heard about it when I was at Princeton. I heard that Alan Guth was giving a lecture, and I didn't go to it for some reason. I can't remember why I didn't go, but people described the idea and I thought, "Gosh, that is a horrible idea." My initial reaction was I didn't like it at all, whereas many people thought it was a wonderful, beautiful idea.

Why didn't you like it?

Partly I suspect one has got one's idea of what the universe is like, and this was disruptive to my ideas. I don't know to what extent my reaction was due to that. I suppose it's partly because I didn't think it was really solving the problems of physics. Partly, it was because it was very much dependent on this current idea of symmetry breaking. So much of modern physics is dependent upon symmetry breaking. I never liked the idea right from the start. We know that symmetry breaking takes place in ferromagnetism and so on, and that's part of physics. But whether it's part of physics at this fundamental level . . . I'd always felt that this is an ugly idea because one is looking at a big symmetry group. To me, a big symmetry group isn't a beautiful thing because you have a big group, and a big group is a complicated thing, like a big number.

You would rather it be a smaller one?

I would rather the symmetry group be smaller. The Lorentz group is so beautiful because it's a very simple one, and it even comes from lack of symmetry. Yet these other ideas are imposing symmetry from the start. I don't see the point of that. I've held my own type of view from way back, much before people were worrying about symmetry breaking. So when the symmetry breaking idea came along, it didn't strike me as a beautiful idea. It struck me as an ugly idea. This has colored my thinking so much right through. Guth's idea for inflation was just using symmetry breaking again.

Let me go back to the sociological question. Why do you think that so many other people like the inflationary model?

I think other people haven't felt this way about symmetry. They think symmetry is a beautiful thing. I like symmetry as a mathematical thing that you use. The question is whether symmetry should be there as a fundamental postulate of physics. In GUT models you postulate a big symmetry, not the simplest. This goes back to when I said I like the field of two elements, 0 and 1. But a group of 792 elements, why is that beautiful?

I can see that other people's liking symmetry would have prevented them from rejecting the inflationary universe, but why do you think they were so attracted to it?

I think they thought it was solving these problems. They also saw it solving what I call the self-made problems. In a certain sense, it does solve those problems—the monopole problem and so on—which are produced by some of the grand unified theories. That is legitimate. Then there is also a sociological factor that I believe is important, which I find disturbing. It's a question of which kinds of people are working on what subjects. See, I've spent my life working with general relativity, but a lot of other people have spent their lives doing other things, like particle physics. Then people come in from outside, not being experts on general relativity or cosmology particularly, but knowing about particle physics, symmetry breaking ideas and so on, and bringing this expertise into the subject. I think there are very many more of those people than relativists. Locusts would perhaps be the wrong analogy, but there are huge numbers of people and they see an opening into this subject, and they come in and almost take it over. I felt this a bit with supersymmetry. In general relativity, I felt this again with a lot of the ways people tried to quantize it. This is a bit personal, and I'm not sure I'm being fair, but certainly people often use types of arguments that are very foreign to my way of thinking—very ungeometrical. Bringing ideas in from other subjects is fine. The union of particle physics and general relativity is a magnificent thing, and what one can learn about the early universe from both ways is fascinating. That's fine as long as the particle physicists appreciate the problems of general relativity. I think often they don't. People come in without being aware of the very fundamental problems we have argued over endlessly among general relativists. There are very fundamental difficulties that one has in trying to quantize, and these people just try to sweep them away. I think here in cosmology one has something similar, people coming in from another area, from particle physics, thinking they can sort of take over. There is a lot of input they can produce, but that's different.

Let me ask you about another problem with the big bang model. Do you remember when you first heard about the flatness problem?

Again, it's a thing where I felt I never worried enough about it. I remember Jim Hartle trying to persuade me this was a much bigger problem than he thought I thought it was.

When was that?

This was not all that long ago, it was probably at one of our quantum gravity gatherings. I suspect it might have been about 10 years ago.

Do you remember why you first thought it was not a serious problem?

I've always lumped all these problems into the same thing; they're all part of quantum gravity. You have to understand what made the big bang singularity what it was, and that is something that we just have very little

conception of. We have little conception of why the singularity was produced as extremely uniform. There are two problems here: one is the
uniformity, and the other is the flatness. Jim Hartle correctly persuaded
me. He certainly worried me a bit more than I had been. I thought you can
get the uniformity out of some quantum gravity—the state must be
simple, maybe it comes out of twistor space. So I felt that was something
one could come to terms with. But the flatness problem is why a certain
parameter is very close to o [why omega is so close to 1]. That is something
one has to face up to. But, again, it's the large-number problem. Why is
the time that the universe will take to collapse, whether it is real or
imaginary, so many Planck times? It's a large-number problem. I don't
worry about it quite in the way that other people do. I think of the problem
as the uniformity problem together with the large-number problem. But
the large-number problem is a problem. We see it all over the place. We
see it in the gravitational constant. Why is the Planck mass so much larger
than the electron mass?

> *And you imagine that might be resolved by a fundamental theory of the*
> *singularity?*

The flatness problem, yes. What underlies that is a problem in physics.
The way I look at it, even in physics today—don't worry about the big
bang—we have the large numbers. Dirac, as you know, pointed this out.[8]
We've got these big pure numbers. Are they something that one should
expect to get out of a fundamental theory? I believe that. I think the
numbers should come out of a fundamental theory, rather than Dirac's
view, which is that they are evolving numbers.

> *There is also Dicke's view, the anthropic principle.[9] What do you think*
> *about that argument?*

That's a tricky one. It's a possibility, but I prefer it wasn't right.

> *Why do you prefer it wasn't right?*

The anthropic principle doesn't do as much for you as you'd like. It
doesn't explain the second law of thermodynamics—why the universe was
created in a state of such low entropy. On the whole, one finds that the
anthropic principle is something you bring in when you haven't got a good
theory. People say, "We've got to fix these constants, and the anthropic
principle does it for us." It's a way of stopping and not worrying any
further.

Moving to an observational development, are you familiar with the recent results of Geller, Huchra, and de Lapparent on the bubbles, the large-scale structures, the finding that galaxies seem to be distributed on the surfaces of shells?

I know what you mean, yes.

Does that observation have any effect on your thinking about some of the assumptions of the big bang model, like the assumption of homogeneity?

Yes, it's intriguing. I wouldn't like to put my money one way or the other. Since it's an observational question, to some extent I don't know. I would like to wait and see how impressive those observations are. I have sympathies with both sides. I suppose that if something lies on shells, maybe some explosion or something did it, some wave that's been able to compress everything. We know that galactic centers explode. And there might be some rather intriguing general relativity going on that is responsible for this. On the other hand, if you think it's a false vacuum or some phase change, then that I find disturbing. I would be very troubled if that turned out to be the explanation. For the same reason, cosmic strings would disturb me. Although studying these things is fun, I don't think I believe in them.

Have any of these observations changed your opinion about the basic assumptions of the big bang model?

I don't think they have, no. Obviously one needs some kind of irregularities in the distribution of matter. We can see irregularities. We're pushed back into physics we just don't know. I think the symmetry breaking and bubbles and all of that stuff is premature. We even have ideas in current particle physics that I personally wouldn't want to survive. That's just an opinion. My guess is that broken symmetries, cosmic strings, etc., are going to turn out not to be the right explanations of the observed inhomogeneities. But on the other hand, there might be something quite extraordinary happening in the early stages of the universe, which we've seen with quasars, explosions, and so on, and which might involve physics that we simply don't understand. I don't think that our picture of particle physics is satisfactory at the moment. I believe that there will be some fundamental changes. If you are looking at huge energies, there may be some quite different things from what we understand at the moment. I think we should be prepared to accept big surprises. So if one sees these spherical shells, that doesn't repel me at all. I just have to wait to see what experiments and observations show in the future.

> *Could you tell me briefly what you think are the major problems in cosmology right now?*

Well, clearly, the actual structure of the big bang, and the physics that must have been going on in the early stages. That perhaps isn't a cosmology problem. It's physics, but cosmology depends on it. Galaxy formation, obviously, is a big problem. I just don't know what other people say on that problem now. I haven't studied it myself. It may be that the problem of galaxy formation will be resolved in terms of this other problem. Another problem is what is the missing mass. And there *is* missing mass. We'd like to know whether it's enough to close the universe, which I hope it doesn't, or leave the universe substantially open, which I hope it does. Obviously, that's a big problem. What is that stuff? Again, that depends on current views in particle physics. It could be something that we just don't know. It could be that the missing mass is neutrinos, massive neutrinos. I am no expert in this area.

> *Let me end with a couple of philosophical questions and ask you to put your scientific caution aside. If you could design the universe any way that you wanted to, how would you do it?*

Oh, I'd take the one we've got [laughs]. I can't deny that. There are so many things that make it incredible. I couldn't design a universe that could compete with the one we see. Let's put it like this: The sort of things I hope would underly the actual universe, as I have said before, would be complex numbers. Complex numbers constitute just one aspect of this unity with mathematics. I believe in a deep unity between mathematics and physics. Whether they are, in a sense, the same thing is an intriguing question.

> *You are designing the universe, so you could make them the same.*

My universe would have to be mathematical. Yes. When you put a question that way, I turn the question around myself. This is perhaps more the way Einstein did when he's talking about God and he says, "How would God have done it?" So he tries to put himself in God's place. It certainly would be a mathematical universe. It's the sort of mathematics that is profound as mathematics. Fruitful mathematics tends to be fruitful to physics also. You mentioned the anthropic principle, and there is also the many-worlds point of view towards quantum mechanics. I feel uncomfortable with both of those ideas, particularly the many-worlds interpretation of quantum mechanics. I mention them because they relate to the question of determinism. If we think that there is just one universe, and if it is a determined, mathematical universe, then that universe is not just deterministic in the ordinary sense, but it is completely fixed, with no freedom of initial conditions. I'm not sure what Einstein meant when he asked whether God had any choice in the creation of the world, but perhaps it was this.

Let me ask you one final question. There is a place in Steve Weinberg's book The First Three Minutes *where he says, "The more the universe seems comprehensible, the more it also seems pointless." Have you ever thought about that issue?*

I don't agree with that sentiment at all. Is that his viewpoint? I remember the statement. Of course, he may be sowing seeds. I don't know whether he believes that.

What is your feeling about that?

I suppose my reaction is the opposite of the sentiment that seems to be expressed there, namely, that our comprehension does give the universe a point. It's part of how I look at mathematics. The understanding of something in terms of mathematics doesn't eliminate a problem, it gives it a deeper character. Suppose you have something in nature that you are trying to understand, and finally you can understand its mathematical implications and appreciate it. Yet there is always some deeper significance there. I don't know how to explain it. I don't think our understanding removes the point. In a sense, understanding nature is making it more mathematical. That's what we are doing all the time. Mathematics is logical structure, a disembodied logical structure, and you might think that when you put your physical problem into that disembodied mathematical structure, you have removed its point. Maybe that is the sort of thing Weinberg was saying. Many people might think that. But my view is, once you have put more and more of your physical world into a mathematical structure, you realize how profound and mysterious this mathematical structure is. How you can get all these things out of it is very mysterious and, in a sense, gives the universe more of a point.

For you, the question of a point is intertwined with this mystery somehow?

I think that's true. I suppose the point has to do with one's own existence. When it comes down to it, the question has to do with conscious perception of one's own existence in the world. A world that has no people in it is pointless. A universe that is just chugging away by itself with nobody in it is, in a sense, pointless.

DAVID SCHRAMM

David Schramm was born in St. Louis, Missouri, on October 25, 1945. He received his B.S. from the Massachusetts Institute of Technology in 1967 and his Ph.D. in physics from the California Institute of Technology in 1971. After working at Caltech and the University of Texas at Austin, he moved to the University of Chicago, where he has been a professor since 1974. He is a member of the National Academy of Sciences. Schramm won the Warner prize of the American Astronomical Society in 1978.

Schramm's research interests include theoretical studies of astrophysics, the origin of the elements, cosmic rays, stellar evolution and supernovae, neutrino astrophysics, nucleochronology, the early solar system, black holes, and particle physics. Among Schramm's principal contributions to cosmology is his theoretical work on nuclear physics and the production of elements in the early universe. Schramm was the author, along with Jim Gunn and Gary Steigman, of an influential theoretical calculation in 1977 limiting the possible types of subatomic particles by consideration of nuclear reactions in the infant universe and the observed cosmic helium abundance. Schramm has been one of the most vocal advocates of applying subatomic particle physics to cosmology. Full of enthusiasm and husky in build, Schramm is known for his wrestling ability as well as his physics.

Can you recall any influences in childhood, either from your parents or from other people or books, that may have gotten you interested in science?

I can't think of any specific thing that got me interested in science other than that it was just a natural interest. My father was in real estate. My mother was a librarian. For some reason or another, I found technical things rather fascinating. I liked model airplanes and things like that. Initially, at about the age of 10, aeronautical engineering seemed interesting. Later, pure science began to catch my attention. I used to read the encyclopedia. That was my amusement when I was in about third or fourth grade. I remember the science sections were the ones I found most fascinating—atomic energy, nuclear physics, things like that. I would do all my work in class very, very quickly and then had nothing else to do except for reading the encyclopedia.

Did you ever talk to your parents about science, or did they ever talk to you about science?

Not directly. They were very supportive, but they had no specific science interests themselves. During high school, although I was always interested in science, my main function was athletics. I played football. I won the state wrestling championship. I ran track, and put the shot. Athletics were my main activity in high school. I never studied. It seemed as if everybody I went to high school with went off to local state schools on athletic scholarships. I, without studying, got 800s on the SATs, so I went to MIT. But during high school, I was a standard midwestern boy going to public high school, playing football and wrestling and all those kinds of things.

You said your father was in real estate and your mother was a librarian. Do you remember any hobbies that they had when you were growing up?

My father used to fly a small plane. Then, when he was having some health problems, he switched to flying model airplanes. I did that with him for a while, up until junior high school, at which time I was a juvenile delinquent for a while and then went into athletics.

Do you remember having any interest in astronomy at this age?

I always was interested in great questions: cosmological questions, theology, and the theology-science interface. In fact, not so much on a scientific side but on the religious side, my family was very mixed up. My father was Jewish. My mother was Christian Science. Then my father converted from being Jewish to a branch of Christianity, a fundamental

Christian group. My mother was Christian Science only because her mother was Christian Science. Then my grandmother died, and my mother shifted into a more deistic theology. So the religious debates in the family were fun. I was always somewhat of a skeptic. Having seen all the different religions all at one time, I became skeptical of organized religion. I was looking to the more philosophical sides of things, including scientific things, to try to establish my own theology in some sense, a cosmology.

When you mentioned great questions, can you give an example or two?

Things like everybody thinks of at one time or the other. I probably just dwelled on them longer. Questions like: "Why are we here?" "Where did we come from?" "Where are we going?" "Where did matter come from?" "Where did space come from?" The same questions I'm still asking [laughs].

Let me ask you a few questions about your education. At MIT, did you start off as a physics major?

When I started off, I wasn't sure whether I was going to go into math or physics. I was very excited in high school about math, just because it was very easy for me. As I say, I got a perfect score on the SAT. I would go throughout the year without ever missing a problem on any math assignment, without ever studying. Math just came very easily for me. So I was thinking that maybe that's the direction to go, but then I always found the philosophical side of physics appealing. The articles that appealed to me in the encyclopedia were frequently the things about physics, although history was also fun. So I wasn't sure whether to go into physics or math. I really knew nothing about what it was to be a physicist or a mathematician. In the area I grew up in, a standard middle-class suburban neighborhood, there were no scientists. The closest to scientists were the people who were engineers at McDonnell Aircraft, and they were very applied. That was my only contact with technical people. The most educated people in my family were two uncles who graduated from law school. In fact, it was due to their influence that I ended up at MIT. because I knew nothing about any of these schools. The guidance counselors in my high school were not very sophisticated. It was my uncles, who both mentioned, "If you're interested in science you should go to MIT." In fact, in hindsight, I realize that I would have done just as well, if not better, had I gone to a place like Harvard or Stanford or Chicago, where there's an entire university rather than just one discipline. The science at Harvard is as strong as the science at MIT. But at that time, mainly because of what my uncles had said, I went to MIT.

When you were at MIT, do you remember any particularly influential teachers that you had?

Yes. MIT was a tremendous experience for me. I was coming from an environment where academics really were not what we did in high school. It was athletics and chasing girls. MIT was very different for me. All students with an 800 on the math SAT were put into one class. It must have been a class of over 50 students. So suddenly, instead of being unique, I was one of 50, and a lot of these people had had much more math and science than I had. It really showed in the math. There are two reasons that I didn't go into math. One was just the simple point that it looked as if the competition was hard. They were all ahead of me. It didn't seem like the other students were as far ahead in the physics area. The other reason was the realization that math was more of a game. It was intellectual masturbation rather than playing with the real universe. Physics appealed to me because I was doing math, but applying it to something real. So that was my reason for leaning towards physics. I guess one other thing that may be of relevance is that between high school and going to MIT, I got married, which was not unusual for the area I came from. But it was very unusual for MIT. So I was married as an undergraduate, and that would have made me a little more serious and focused during that period.

Did you know anything about cosmology at that time?

I remember that when I was asked by fellow classmates about what areas in physics I was interested in, my standard response was cosmology.

Had you read any popular books or magazine articles about cosmology?

Oh, yes, bits and pieces here and there. Nothing that stood out. What really got me going in that direction was actually a course by Phil Morrison at MIT. It was a course on galaxies, and Phil is a fantastic teacher, with such a depth of knowledge. I remember getting very excited about this stimulating subject. What I mainly did at MIT, sciencewise, was work in the nuclear physics lab. From my sophomore year on, I spent most of my time there in the nuclear physics lab. Either there or on the wrestling mat. Those were the two main interests that I had. It's so different than what you learn in high school or what you read in the encyclopedia. You discover what scientists really are, what they do, that they are real people and have a good time doing science. It was an experimental lab, and we were doing real experiments. I published two papers on nuclear physics before I graduated.[1] As a result, I knew a lot of nuclear physics by the time I graduated. But also, with Phil Morrison's course I got very excited about astronomy, and that's how I ended up going to Caltech for graduate school. I wanted to do nuclear physics, and I wanted to do something that was astrophysical because of Phil's course. When I mentioned that combination to people, they said, "Oh, you have to go work with Fowler." So that's how I ended up at Caltech.

When you went to Caltech, did you take a course in cosmology or in general relativity?

Well, I was in the physics department. When I came to Caltech, I was initially looking into the space physics program also, and I was taking some courses in applied math. But then I settled in to working with Willie [Fowler]. I took Willie's course, and I took quantum electrodynamics from Richard Feynman, which was great. He'd come in and play bongos. I didn't take any astronomy. I sat in on one course on stellar evolution. There was no formal cosmology course during the time I was there. I took general relativity with Kip Thorne.

He must have talked about cosmology.

Yes, cosmology was certainly in that course. I took that and it was very good. When I took it, the book by Misner, Thorne, and Wheeler was still in note form. That was fun and I really enjoyed it. I remember that Kip had us do a paper, and I did my paper on big bang nucleosynthesis. In fact, it was shortly after the work by Robert Wagoner, William Fowler, and Fred Hoyle.[2] The first time I really looked into that seriously was for Kip's course.

When you had that cosmology section in his course, do you remember having a preference for any particular kind of universe?

I was a student in the late 1960s, and it was shortly after the background radiation was discovered, so there was already a leaning towards the big bang model. Certainly the big bang was what I preferred. I guess my prejudice at that time was that closed models looked prettier. The idea of oscillating universes was an initial prejudice. Kip's course also leaned in that direction. John Wheeler leans in that direction. It was my initial prejudice, which I don't have anymore.

Do you remember why you liked it?

Oh, Willie was, and still is to some extent, a steady state lover. The idea here was that at least you could avoid the problem of an origin to time if you would have many oscillations.

Was this your own feeling, or was it something you had read about?

Oh, more just my own . . . It just appealed to me on some sort of aesthetic ground. Then, as I said, in Kip's course I did the paper on big bang nucleosynthesis, and it was a way of coupling some interesting cosmology with my nuclear physics. My thesis was actually on nuclear chronology and dating the universe. Willie liked to have all of his students do an experiment, and so I did an experiment with Gerry Wasserburg, who's in geophysics. In fact, I ended up as a graduate student spending far more time with Gerry than with Willie, because experimental work takes longer than theoretical work. I also did some theoretical work with Gerry on nuclear chronology,[3] as well as measurements with a mass spectrometer and really getting into laboratory techniques and learning how hard it is to get real numbers. It made theory much more appealing.

Let me skip ahead a few years and ask you a few questions about the merger of particle physics and cosmology. Do you remember the motivating factors for your work on limiting the types of neutrinos from the helium abundance?[4]

Yes. I think in many ways that work was the birth of this whole connection between particle physics and cosmology, because it was the first time when we were able to take something from cosmology and make a statement relevant to particle physics. In order to explain how I got into that and was even aware of the problem, I need to go back a little bit. In the early 1970s there were two main areas I was working on. One was big bang nucleosynthesis and establishing the relevance of deuterium, showing that deuterium couldn't be made in any other way but cosmologically.[5] The other was gravitational collapse and supernova models.[6] And that latter area is the only place in astrophysics where neutrinos and particle physics came in. I was following what was going on in particle physics, because that affected our input into these collapse calculations. In particular, when neutral currents were discovered in the lab, I was one of the first people to be involved in putting them in the stellar evolution codes and realizing that you produce all kinds of neutrinos in stellar collapse, rather than just one kind. But that got me acquainted with the particle community on a first-hand basis.

So those two ingredients came together.

Yes. How the calculation actually ended up being done, though, was the result of Jim Gunn's visiting Chicago. We had Jim as a visiting professor for a few months. He asked me what I was doing, and we started talking. I was talking about both the things I was doing, and in the course of the discussions with Jim, we outlined this argument. It happened just by talking to Jim, and his asking the kind of clever questions that he asks. So we sketched out this thing on Jim's blackboard when he was visiting Chicago. Then I started to write it up. I went to Aspen that summer, and Gary Steigman was there. Gary and I had shared an office when we were both postdocs at Caltech. We were both postdocs with Willie. We shared an office, but we never did any science together. When I got to Aspen, I started talking to Gary about this thing that Jim and I were doing. Gary had evidently done almost the same thing. So instead of writing two separate papers, we merged it together, and it became one paper. That was the summer of 1976, and it got published in 1977. So that's the origin of that paper. I think it was my interactions in the particle community on the supernova problem and the way we worded things that caused a break in how cosmology dealt with particle questions.

If you look earlier, for example in James Peebles's book or at a paper by [V. F.] Shvartsman much earlier, you see that they were talking in terms of energy densities and were commenting that you could limit the energy densities of various kinds of particles.[7] But they were never getting down to the real questions that a particle person asks, such as mass number, number of particle types, and so on. Up until that time, most of the people who were doing theoretical cosmology were relativists. Whereas this approach of coming from the neutrino side involved asking the questions that a particle physicist is concerned with. He's concerned with how many flavors [particle types] there are. That number involves formulating and parametrizing things in a different way. It was that then that really made the connection between particle physics and cosmology.

At the time of our work, the particle community was thinking that every time you go to higher energy you get another new flavor. In fact, this was right after the November revolution, when charm had been discovered,[8] and when the upsilon was found.[9] So it looked like whenever you go up to higher energy, you find new flavors. At that time, we were the first to say, "Hold it. Cosmology wouldn't work that way." Our first limit, which is not as strong as it is now, was around seven or so types of neutrinos. Gradually, we found it would be down to two to four. At the time, even saying that it was a small finite number as opposed to thousands was considered almost blasphemous. Furthermore, somebody was making an argument from an astronomical thing about fundamental physics. That had never been done. The closest to it was probably when Fred Hoyle argued for an excited state of carbon.[10] Then experimentalists went and measured and found it. But that was an excited level, and there are many, many levels. It's not something truly fundamental. Whereas, here was something that was of central importance to a real physics question. We got a lot of flak from a lot of people, saying, "Oh, it's crazy to say something like that. You are talking about numbers of a few and astrophysicists are usually only right to a few orders of magnitude."

Around that same time, other things happened as well, joining particle physics and cosmology. Do you think that for some reason the time was right?

Yes. This particular thing was one of the first to start more communication between the fields. There were a few other things that started to put constraints on the particle types. Lee and Weinberg put limits on neutrino types.[11] Jim Gunn was involved with that as well. Rocky Kolb was a graduate student at Texas, and he did some work at the same time, starting to constrain particle properties from cosmology.[12] A little earlier, actually, Ram Cowsik had done some things on limits on neutrino masses, but at that time, it didn't have much of an impact.[13] Gary had also done some work on the antimatter–matter business.[14] So there had been already some interaction between particle physics and cosmology. There was also the work on the maximum temperature by R. Hagedorn, John Bahcall, Steve Frautschi.[15] None of that work really had a big impact, but by the time the 1970s came, the number of such studies was increasing. Also, the kinds of statements that were being made were more specific to the questions that particle physicists were interested in, rather than just vague numbers that really didn't affect people that much. Then, also, the big bang was really getting established by then. Big bang nucleosynthesis had come of age. It not only fit the helium abundance, which is something people were doing earlier, but we were getting deuterium. That was powerful. Calculations were really fitting the rare isotopes, not just the 25% helium. Big bang nucleosynthesis was working very well. To establish those other isotopes took a lot of the work that we did in the 1970s.[16] In fact, back in the 1960s, when the big bang nucleosynthesis work of Tayler and Hoyle first came out,[17] and then that of Wagoner, Fowler, and Hoyle and others, that work was ignored by the community (and the authors) because Willie had papers saying that you made all these other elements and isotopes in stars.[18] It was only our later work that showed that you couldn't make them in stars and that the big bang was relevant.

At the same time, in the particle community, came grand unified theories. Howard Georgi and Sheldon Glashow were talking about SU(5), which is important at energies of 10^{15} GeV, *way* beyond what you could get in an accelerator.[19] Where else do you have any experimental or observational consequences of GUTs? Well, it's the big bang.

I think the things that really established the connection and made the growth go explosively were three key things all within about a year of each other. One was that GUTs enabled you to understand a long-standing problem in cosmology, the photon-to-baryon ratio. Again, Andrei Sakharov had done it earlier, in 1967, but nobody paid any attention to it.[20] When it was redone by a half dozen different people at different locations, semi-independently and within the context of GUTs, you had a real model.[21] Suddenly, there was success in answering a problem that had not been answered before—a problem in cosmology with a solution using particle physics. So that provided the other half of the relationship, to make it really symbiotic. You really had information going both directions, rather than in the earlier decades, when astronomy was a parasite on physics, just using physics without giving anything back. I think that really cemented it.

Then shortly thereafter were two other critical things. One was that with baryosynthesis, you could have inflation. Alan Guth recognized that the Higgs field from GUTs could drive inflation. But that had to occur after we understood baryosynthesis, because if you inflate a universe and you don't know how to make baryons, you've made it empty. Other people before that time, including my former student Demosthenes Kazanas, had talked about inflationary-type models.[22] But it was the ability to make the baryons at the end that made inflation a viable model—as well as Guth's answering several problems simultaneously.

The next thing that occurred around the same time was curious. Two wrong experiments claimed to have measured a neutrino mass. There was a Soviet experiment, and Reines's experiment.[23] Those experiments motivated a lot of people to follow through on nonzero neutrino masses as a possibility. That really established nonbaryonic dark matter as a legitimate area to work on.

I wanted to ask how you've reacted to some discoveries in the last 10 years, both theoretical and observational. Do you remember when you first heard about the horizon problem?

During graduate school, when I was at Caltech.

How did you react to it? Did it bother you a lot, or did you think it would be solved by some fairly straightforward means?

I can remember being bothered by it. I can remember thinking that it required some rather special initial conditions. It wasn't enough of a bother for me to say that I didn't believe the big bang as a model. But I certainly recognized that there was a problem, and it required setting the initial conditions to be very smooth. I had no idea how to do that.

Did you think that might be a solution to the problem, having very smooth initial conditions?

That's really all inflation does. It makes the initial conditions very smooth. When I first heard about the horizon problem, that was the standard way of talking about the solution to it—the universe started out very, very smooth.

And how did you feel about that explanation?

I accepted that. It probably didn't bother me as much as it should have. But I recognized that you needed to fine-tune things. My reaction was that sometime I'll have to think about that.

Did the horizon problem take on any different significance for you after the inflationary universe model came out?

Yes. Then there was a mechanism to solve it, a mechanism to lay out the initial conditions. Then, certainly, I put more focus on it. I would always mention it in talks, as opposed to just occasionally mentioning it, because there was a solution. You don't always show your warts, but you can always show your medals. So I'd mention the horizon problem with great frequency at that point. In fact, the same thing could be said of the flatness problem.

When you first heard about the inflationary universe model, how did you react to it?

Actually, even before Guth inflated, I had talked to Henry Tye, who had been working with Alan on phase transitions in the early universe, trying to solve the monopole problem.[24] In fact, Jim Fry and I had done some things on the monopole problem.[25] So I was quite up on what they were doing. Then I heard about Alan's inflation idea, and suddenly everything just fit together. It immediately hit me that he had found it. There was the solution, and it all fit together. I was an immediate convert, I guess. There are still a lot of nagging problems with the fluctuations, but I remember right away, I was enthusiastic about it. The inflationary universe model was able to solve a number of things that had been in the back of my mind as problems with the big bang. All of a sudden, in one fell swoop, they went away. I was very pleased with that.

Why do you think the inflationary universe model has been so widely and so quickly accepted by the community?

I think it's *pretty*. With one idea you solve several problems at once. It really gave a boost to this interface of cosmology and particle physics, capturing some of the most exciting aspects of both. It set the initial conditions for the big bang, so it enabled all of the good stuff about the big bang to go there. You don't have to throw away anything. Inflation solves some of the problems, some of the things that used to be nagging embarrassments that you put under the rug. Now you didn't have to sweep them under the rug. So it enabled you as a big bang lover to be even more of a big bang lover. It also enabled you as a particle physicist to say, "Oh, particle physics can do something that's useful to somebody else. It can solve one of their problems." That was not occurring before. In fact, particle physics frequently had the problem of justifying its existence to the general public. Why does anybody care about how many quarks there are or something like that? Here is particle physics able to solve somebody else's problem.

And inflation didn't solve the problems completely. So there was still work to be done. I think that was also an important aspect. If somebody solved something and there's nothing else to be done, you're not going to create an industry. But inflation did—and still does—have some problems, some loose ends. It's not all fixed. There is still room for people to do work. So lots of graduate students, lots of postdocs, and faculty people could then work on this problem and make contributions. It wasn't just a one-fell-swoop and it's solved. Not like Fermat's theorem.

You sound like you're pretty convinced that the inflationary model has a good chance of being right.

Well, the way I would word it is I think the inflation *paradigm* is right—the idea that we went through some de Sitter phase very early. It could very well be that the whole mechanism and the details that we're currently talking about are wrong. Maybe as we learn about the "theory of everything" or some supergravity or string theory, we'll find that inflation occurs in some different way. Andrei Linde has shown that *any* scalar field will inflate.[26] You don't need any sort of special thing. It's very easy to get inflation. So we've gone from where we didn't know about inflation to now, where all you have to do is propose a scalar field at early times and you get inflation. So it might not be Guth's mechanism in detail. In fact, what I *can* say is I'm pretty sure that everything we're working with right now is *not* the way it happens, that we're leaving out some important ingredient. Right now, it's a little too contrived. You have to fine-tune the parameters too much to avoid the fluctuation problem.

Let's return to the flatness problem. When did you hear about it for the first time?

I guess it was a Dicke and Peebles argument. I'm not sure whether it was in a talk that I heard Bob Dicke give or somebody referring to a talk that Dicke gave.

Do you know approximately when that was?

Before Guth, but after graduate school, sometime in the 1970s. In fact, it was back in 1974 when Richard Gott, James Gunn, Beatrice Tinsley, and I did a paper on the open universe.[27] We talked about everything fitting remarkably well at an omega of 0.1. I can remember somebody—it might have been Jim Gunn—going through what is now known as the flatness argument, saying, "Well you realize this requires some very special fine-tuning."

Do you remember how you reacted to the flatness problem when you heard about it, whether you considered it to be a serious problem or not?

Again it was a problem that I assumed just required some very special fine-tuning and very special initial conditions.

Was that an acceptable solution to you?

It was a problem like the horizon problem. And you could solve it that way. You could say, "In the beginning, we had this condition." But it was again always something nagging. It would be nice if there was some better solution. But, at that time, I certainly was not bothered about having an open universe. In fact, I had switched from liking a closed universe to thinking an open universe is fine, and that the data all points to it.

When you found that the data pointed towards an open universe, did that bother you at all, given your earlier philosophical prejudices?

Yes, it took me a while. I didn't like it, but it just seemed to all fit together so well. All the different arguments fit together well for an omega of 0.1.

The thing I'd been working on, the deuterium, and also the age argument were two of the three lines of the triangle. Then Jim Gunn and Richard Gott were doing the bird's-eye view of the universe with the dynamics of halos, and so on, and came up with the density from dynamics being about the same as the density from the deuterium abundance.[28] And then Beatrice Tinsley was the glue, in that she put it all together in a nice way. I can remember when that happened. That was just one of those things when everything fits together, and I realized that was much more powerful than my prejudices. So that convinced me for a while. Now I've gone back the other way [laughs]. Not completely, though.

Do you remember how you reacted to the results of Valerie de Lapparent, Margaret Geller, and John Huchra on the large-scale structure?

Yes. There were a couple different reactions. One was that I had already thought that there were voids out there, from work like what Robert Kirshner had done.[29] My first response when I saw the Geller, Huchra, de Lapparent work was that the voids are real, and maybe we should lean more back toward hot dark matter. We had more reasons to try to save hot dark matter, which had run into difficulties because of the galaxy formation problem. Plus, at the same time, at Chicago, Rich Kron with David Koo, in their pencil beam approach, had seen shells or indications again that there were large structures.[30] I had seen their data even though they hadn't published it then. They had also seen much deeper than the Harvard survey. So I didn't view the Harvard thing as unique, but just one more piece of evidence. They seemed to have better press agents than everybody else, and they've gotten more publicity on it, but it just seemed to me it was all part of the overall package. It all fit together very well.

The other amazing thing—and I mention this when I give talks—was that the de Lapparent, Geller, Huchra work could in principle have been done earlier. The capability was really built up when Marc Davis dedicated a telescope at Harvard and the Smithsonian to do redshift surveys. The thing that was so amazing was why hadn't the astronomical community recognized this structure many years earlier? The reason involves the sociology of the astronomical community: the way telescope time was awarded, and the way people do projects looking for the weird object of the month. If you have only a couple nights a year on a telescope, you're going to get this couple nights a year by promising that you're going to publish some earth-shaking paper, not that you're just going to have all the positions of all the objects. So the net result is that the whole astronomical community had really missed the forest for the trees. They just ignored the background. They just looked at weird objects. They looked at the clusters. They sure knew where Virgo was and how far away it is. But they just hadn't filled in all the other clusters and galaxies. And yet, from a scientific point of view, that's the obvious thing to do. You should know what your background is before you go looking at the weird things. Everybody wants to have an object named after them, so they try to find the strangest object, rather than trying to find what the average is.

How do you think theory and observation have worked together in cosmology in the last decade or so?

I think what's caused the explosion in cosmology has been two-fold. One is the cosmology–particle physics interface, which we've talked about. It's brought in a whole fresh supply of people, smart people who hadn't previously looked at some of the cosmological problems and weren't brought up with the prejudices that others had. So they can come in with new ideas. The other factor has been the explosion of new kinds of observational information. That's ranged from the 3-degree radiation and the constraints on the anisotropy, which has forced us to go to more exotic models than we might have otherwise done. The things we were talking about—like the large-scale structures of Geller, Huchra, and de Lapparent; Koo and Kron; Kirshner, Oemler, and Schechter—have changed our way of thinking or supported it. I had some prejudices already in thinking that way, but it's nice to have your prejudices supported with these things. Also, there is the work on the number of neutrinos. If that had just been some number that was going to sit up in archives, it would be irrelevant. That's so important because people are doing experiments to check it in the laboratory and really see whether it's right.

In fact, I think one of the biggest problems with inflation is it's very hard to prove whether it's right, to really do an experimental check. It's a pretty idea, but you don't have that kind of nice experimental interface. The thing that has been driving a lot of the dark matter stuff lately is that people are going out and trying to search for it. Again, it's no longer just intellectual masturbation. This recent work has changed cosmology from a subject that, up the last decade or so, was just redoing Hubble's observation over and over again—trying to find two numbers, H_0 [Hubble's constant] and q_0 [the deceleration parameter] and not making much progress because of all the systematic errors. Now there's all sorts of different kinds of measurements that you can do.

What do you think are the outstanding problems in cosmology today?

Certainly the whole question of galaxy formation, dark matter, large-scale structure—which I think are all the same problem. It's just the blind man looking at the elephant. Which parts do you focus on? But it's really all one problem. It's the major one, because it's the confrontation of the traditional astronomical observations with all this early universe work. There are ingredients coming in from particle physics, from traditional astrophysical theory, from particle cosmology, and from the observational ends. And the observational ends are in all wavelengths.

I think the number one theoretical problem is: What is the vacuum? It's worded in different ways by different people, but when we talk about the Higgs field, when we talk about the scalar fields that cause inflation, what you're really asking is, "What is a vacuum?" The cosmological constant— why is it zero or why is it so small? Which is again, "What is a vacuum?"

> *Let me finish by asking you something much more speculative, maybe asking you to put some of your natural scientific caution aside. If you could design the universe any way that you wanted to, how would you do it?*

We actually have a pretty good one. The one that we've come up with, the single one-shot big bang, omega-equals-1 universe is a very pretty one. In fact, people used to ask me that question, and the answer I used to give was that the prettiest universe was the steady state universe. This may have been a prejudice of working with Willie Fowler and Fred Hoyle. I used to think that was a very pretty one, because of infinite time. Everything's always the same. Everything always is, always was, and always will be. That's a philosophical way of viewing it. But I guess recently I've come to think that maybe this nice one-shot big bang—but a flat one, right at the boundary—is a very pretty one. Everything is nice and clean. I like the idea that you start out with one grand unified force, a theory of everything, and the rest of the universe is a consequence of that. It takes only one ad hoc assumption—your initial physics—and then everything else follows. I like to minimize the number of ad hoc assumptions. And, in fact, if you can somehow argue that *that* law, which some people are searching for, is the only law that's consistent, then the ultimate statement you're making is that mathematical consistency is the only law. So you just have to say: in the beginning, there was mathematical consistency. Everything else follows, including us [laughs].

> *There's a place in Weinberg's* The First Three Minutes *where Weinberg says that the more that the universe becomes comprehensible, the more it also seems pointless.*

Yes, I've read that. I take some issue with it. If we take what I was just saying about mathematical consistency, that seems to me very beautiful. That's the underlying cause of it all. Now you might ask to what extent then does that affect your daily life, and conclude that it is pointless. That's probably what Steve was saying, that all these other things are just irrelevant consequences of this initial law, or whatever it was. I like to think that it's not that they are irrelevant, but they are a beautiful demonstration of it, that all the intricacies of life develop from *one* simple thing. It's a wonderful demonstration of the beauty of nature.

STEVEN WEINBERG

Steven Weinberg was born May 3, 1933, in New York City. He received his B.A. from Cornell University in 1954 and his Ph.D. in physics from Princeton University in 1957. His professional career has included work at Columbia University (1957–1959), the University of California at Berkeley (1959–1969), MIT (1969–1973), and Harvard (1973–1983). Since 1982 he had been the Josey Regental Professor of Science at the University of Texas, Austin. Among other honors, Weinberg was awarded the Nobel Prize in Physics in 1979 for his work on the electroweak theory in particle physics. He is a member of the National Academy of Sciences and the Royal Society of London.

Principally known for his contributions to elementary particle physics, Weinberg was one of the first scientists to apply new theories of elementary particles to cosmology. In 1979 he and others pointed out that physical processes associated with grand unified theories, operating in the infant universe, might determine the relative proportions of two types of elementary particles, photons and baryons. Previously, the ratio of photons to baryons had been considered a given property of the universe whose value had to be accepted without understanding. As a result of this and related work, Weinberg inspired a generation of particle physicists, including Alan Guth, to take cosmology seriously and to enter the field. Weinberg authored two influential books in the subject: a textbook, *Gravitation and Cosmol-*

ogy (1972), and a popular book, *The First Three Minutes* (1977), which introduced both the general public and many scientists to cosmology. Unintimidated by long and difficult calculations and insistent on understanding things in his own terms, Weinberg has made quantitative a number of topics in cosmology that were discussed only vaguely before.

Can you tell me when you first heard or read about cosmology?

In high school I was very much aware that there was a program underway, called the Sandage program, to observe the large-scale expansion of the universe and determine whether the expansion was slowing down and how fast it was slowing down. I didn't understand it in detail, but I also knew at the time I was in high school that the big telescope at Mt. Palomar was going to start operating soon. If I remember correctly, it started operating in 1952, which is two years after I left high school. We all knew that was coming, and I thought that as soon as they went from a 100-inch to 200-inch telescope, then suddenly all the problems would be solved, and that would be really exciting. We would know whether the universe expands forever or collapses. Of course, it didn't work out that way. We're still wondering about those things. But that was exciting.

So you already knew about the different possible fates for the universe?

I guess so. I had read popular books by George Gamow. I certainly hadn't studied it in a serious way, but it was generally known.

At that time, do you remember whether you had any personal preferences or expectations for what type of universe we might live in?

Yes. I don't remember when I first heard about it, but I do remember quite early, certainly by the time I was in graduate school, I felt that the steady state model of [Hermann] Bondi and [Thomas] Gold, and also [Fred] Hoyle, was the most attractive. There were many things about it that were attractive. I didn't formulate for myself then what I liked about it, but now looking back, I think I can say what I liked about it, and that is that it offered the biggest possibility of making definite predictions. In any other cosmological model, there's a question about the initial conditions. If you believe in the currently popular big bang models, there are all kinds of questions about why the entropy of the universe is what it is and so on. Some questions, there have been efforts to explain, but they generally just push the question back to some earlier time, whereas in the steady state model, there's a self-consistency condition. If you imagine that there is some mechanism for producing matter in intergalactic space, then that would naturally produce matter at a certain rate. Then you could imagine the universe settling down to an equilibrium situation where it was expanding just fast enough so that the matter was being created at a rate that would just fill up the gaps in the expanding system of galaxies. That seemed very attractive to me: a system where there was no gap at all between the things that are physics and the things that are history, where history is determined just as an equilibrium solution of the equations of physics. I thought that would be really beautiful, and, of course, from a practical point of view it makes by far the most precise predictions, the most definite predictions of any cosmological theory. The modern Friedmann–Robertson–Walker big bang models leave some parameters undetermined, whereas the steady state model determines everything. The deceleration parameter is -1, and you don't have to worry about evolution [of the average properties of the universe] when you compare theory with observation, because there's no evolution on the average.

I would say by the time I was in graduate school, I was already strongly prejudiced in favor of the steady state model. And as it happens, it did have the merits that I expected because it was not only falsifiable, it was falsified. It was one of the first things that came out of the Hubble program or the Sandage program, whatever you want to call it, of measuring redshift as a function of distance. This work made it very hard to believe in the steady state model because the deceleration parameter did not look like it was -1. Now, of course, it might *be* -1, and the effects of evolution may be obscuring that fact, but in the steady state model you're committed not to believe in evolution. I don't think that was the thing that actually convinced most people. There were number counts of radio sources that were very important, and probably more important than the Hubble program. And, of course, the discovery of the microwave radiation background. Although, as I pointed out in my book, if you're going to believe in baryon production as the universe expands, you could also believe in the production of low-energy photons.[1]

You were saying that the steady state was the first cosmological theory that was seriously ruled out.

Yes, it showed that we're really doing something serious. In fact, I've quoted it occasionally as an example of something to be proud of in making cosmology a respectable science—that we're not just all making up our favorite cosmological models and then hanging on to them. Some people outside of the field may feel it is that way, but in fact it is *real* science, and observation can kill off even those theories that are favorite models.

Could you talk a little bit about why it is that you started working in cosmology, around 1970?

It really started before that. I think I can be fairly definite. I certainly had some interest in cosmology, as I guess any citizen does who wonders where the universe came from and all that, but I think my scientific interest started by reading an old book by Hermann Bondi, which is just called *Cosmology.* It's short and very readable. It's the sort of book you can read in the bathtub. I read it when I was living in San Francisco, and I lived in San Francisco from 1959 to 1961. Today we would say Bondi's book is terribly out of date. As far as I recall, it has nothing in it about nucleosynthesis, and certainly the microwave radiation background wasn't known about when he wrote it. It didn't have any review of the work of George Gamow and his collaborators on the early universe. But his book showed me for the first time that cosmology was not merely a subject for the observers, but that there was enough mathematical depth to it so that there was something the theorists could learn and do. I hadn't known about the Robertson–Walker metric. The fact that there was a theorem that said that all homogeneous, isotropic universes have this particular metric, irrespective of the dynamics, I thought was great. Bondi did not explain how that theorem was proved. I've never been able to understand the articles by Robertson and Walker, and I finally worked out a proof for myself, which is published in my textbook. But I thought that that was really neat, that you could make such a definite conclusion from something as plausible as homogeneity and isotropy. And then at applying the Einstein equations to that, you could then go farther and get a definite equation of motion for the scale factor. It made it as worth learning as any of the other things that I had learned as a student, in quantum mechanics and atomic physics and so on. It was an interesting body of mathematical formalism that said something useful about the universe. And I hadn't realized that that existed before I read Bondi's book.

Everything you just mentioned was known well before 1960. Why do you suppose that the subject was not taken seriously at that point by more people?

Well, there was that early work of Robertson, Walker, Friedmann, Lemaître, de Sitter, Einstein. It posed the problem of measuring the scale factor. It wasn't clear there was anything else that you could do. It didn't occur to me that if the scale factor was increasing, it was once much smaller and there must have been a time when physics was very different and it would be interesting to study that physics. I mean it's obvious that that would be a physical problem, but it was very far from my mind that there was any useful confrontation between theory and experiment. As far as I knew, the only thing you could do astronomically that would be relevant to cosmological questions was measure distances and redshifts of a lot of things. So the only part of cosmology that seemed worthwhile was that little bit that I'd read about in Bondi's book. I thought that was great, though. I really was enthusiastic about that. I remember very clearly back in that period 1959 to 1961, after reading Bondi's book. I said, "I really want to learn more about cosmology and be a little bit more at home in the field—not just to have read a book." So I gave myself a homework problem. Bondi's book makes a big deal out of the Olbers' paradox. And it occurred to me that the way Bondi described the solution of the Olbers' paradox in terms of the redshift might not apply to neutrinos because neutrinos carry a quantum number, a lepton number, so the number of them is significant quite apart from their energy. So I set myself the problem of learning where do neutrinos go. At first, I was just thinking of the neutrinos that are produced in stars. I thought that would be a good problem to work on, because I'd also have to learn something about stellar nucleosynthesis, which I knew almost nothing about. I had read shortly before that a book by Schwarzschild, *The Structure and Evolution of the Stars*, a beautiful book. That was all I knew about stellar nucleosynthesis, and I thought it would be a good way of learning more about that and also learning more about cosmology to answer this question, do stars produce neutrinos? Why aren't there an infinite number, where do they go, and in different cosmological models. And I wrote a couple of papers about them.[2] I think the most interesting thing I did in those papers was to raise the possibility that there was a sea of degenerate neutrinos up to some definite Fermi level, and that this sea had gotten filled over many cycles of an oscillating universe. The oscillating universe is, of course, in a way like the steady state but with . . .

You don't have to specify initial conditions.

Yes, that's right. It goes on forever. Although it isn't steady, it oscillates, but it's the next best thing to the steady state.

Did it appeal to you for some of the same reasons?

Yes, yes. It still does, in fact, for that reason. But all my work was based on the assumption that somehow or other, even when the universe collapses, at the end of each cycle you essentially have a cold universe. And that's silly. I mean today, of course, we're used to thinking of a very hot early universe. I didn't think that through at all. Nobody knew about the microwave background. But there was a sense of unreality about it. I didn't take that work very seriously. Of course, if a Fermi energy level, if the degenerate sea of low-energy neutrinos was discovered, I would have been in seventh heaven. But I didn't really take it seriously. I didn't expect it to be discovered because, as I have written in *The First Three Minutes*, there was a sense of unreality about any speculations about the universe. The wonderful thing about the microwave background was that it was the kind of thing that I was only dreaming of with the neutrinos. It was really direct observation of something from the very early universe.

> *Going forward in time, was the textbook that you wrote,* Gravitation and Cosmology, *in 1972 part of your same program of further educating yourself about cosmology? Why did you write that textbook?*

I would say the program really took shape in teaching the course. I don't remember the details, but once or twice at Berkeley, and I think twice at MIT, I taught the graduate course in general relativity. I volunteered to do it because I wanted to learn general relativity, and I wanted to learn cosmology. My lecture notes began looking more and more like a book, so eventually it just turned into a book. In a way, I'm sorry I wrote the damn thing because the actual work of writing it was in 1969, '70, '71, and those are the years when I should have dropped everything I was doing and worked on proving that the spontaneously broken gauge theories were renormalizable. I had made that suggestion in 1967,[3] as [Abdus] Salam did in 1968,[4] and I had worked on it in a desultory way, and this book took me away from that problem. I really shouldn't have gone away from that. I should have just worked on that problem, looking back on it. But what can you do?

> *For your own career?*

It's not a question of career. It's wonderful to write a book that has some impact, but it's even more wonderful to make discoveries that have an impact, and between the two of them I'd rather make discoveries than write books. I'm very proud of that book, but I would have been even prouder of proving myself that these theories were renormalizable, rather than just conjecturing it.

The injection of particle physics into cosmology in the 1970s seems to have been a major revolution in the field, very good for astronomers and cosmologists. A number of independent groups, including yourself, started doing particle physics calculations to explain cosmological things such as the photon to baryon ratio.[5] Why do you think that all of this activity happened at approximately the same time?

I don't really know. It's a good question. One thing you can say is that, of course, the mid-1970s was the time when the theoretical work in constructing our standard view of elementary particle physics was essentially completed. The discovery of the instantons by [Gerard] 't Hooft really solved the last outstanding theoretical puzzle. Then, by 1978, the experiments had all fallen in line. So we had a theory of elementary particles that was worth applying and that you could have some confidence in.

There were some questions that arose in particle physics like the number of neutrino species, which suddenly for the first time were taken seriously because a new generation of leptons had just been discovered, and we wondered how many more there were. I don't remember who was the first person to realize that that was a question for which cosmological evidence gave sensitive limits. There were a lot of events like that. For instance, there are cosmological bounds on the mass of the neutrino. I don't know why it all happened at about the same time, but it was just the nature of the problems that particle physicists were considering that they suddenly realized that cosmology could be of help to them. And of course the biggest unsolved question was what is physics like at really short scales of distance, really high energies characterized by the Planck scale, and that seemed to have something to do with the origin of the universe. There, unfortunately, I don't think the interaction between physics and cosmology has been so fruitful. One other thing was the realization of the importance of phase transitions.

Do you think in general that physicists at this time were taking cosmology more seriously than they had, say, 10 years earlier?

The discovery of the microwave radiation background changed everything because it made cosmological speculation back to the first few minutes respectable, normal science. So if someone, for instance, calculated the effect of having extra neutrinos on nucleosynthesis, it was regarded as a respectable thing to do. You were working in a scientific framework that we all understood and that had had some successes. It might not be *true*, but it was at least something worth doing. In that sense, nothing like that has happened for the cosmology of the very early universe [before about one second after the big bang]. There are lots of very good scientists who do highly respected work on the very early universe, but there's still a sense of unreality about all of it, because we have no experimental handle on the very early universe—nothing like the helium abundance or the microwave background.

I think progress has been made, though. I think the general idea of inflation, in all the different versions, from Guth through other people, Linde and so on, is very interesting and the sort of thing I would have been very proud to do myself.[6] But, unfortunately, it leaves you with a very frustrated feeling. There's no way of testing it. It's not fair to say inflation hasn't made any new predictions. Very often theories are tested by using them to explain already known facts. After all, when Newton calculated the length of the month in terms of the acceleration of gravity at the earth's surface and the distance of the moon, that was already known, but it was such a good numerical calculation that worked, it was clearly convincing.

Of course, inflation predicts that omega is equal to 1.

Well, right, but that's like a null test. That means the departure from flatness is very small. There are lots of reasons why the departure from flatness might be very small, including the fact that the universe is absolutely flat.

Would you be surprised if the universe were absolutely flat?

Not at all. In fact, that seems to me the most plausible thing. I think inflation has had other successes which are much more important, such as explaining why there aren't magnetic monopoles around, which really was a puzzle. I would say the real successes of the inflationary picture are solving the monopole problem and also solving the isotropy [horizon] problem—explaining how distant parts of the universe could have gotten into thermal communication with each other even though they're very far apart in the sky, so that you can explain why the microwave background is so isotropic. On the other hand, the fact that the universe contains a huge number of baryons within the horizon, or a huge number of photons, or a huge entropy, however you want to say it, and the very related thing that the omega is so close to 1, I think never seemed to me that puzzling because they're perfectly well understood just in terms of the assumption that the universe is flat, as $k = 0$ [space curvature equal to zero].

You don't consider the flat universe an improbable case, then, when all possible kinds of universe are considered?

Yes, but if you have to decide which is the most aesthetically attractive possibility, I would think that's the most aesthetically attractive possibility. Long before I ever heard of inflation, it always seemed to me that since omega was so close to 1 now, and since the universe is pretty old in terms of any fundamental time scale like the Planck scale, that probably meant that omega was exactly 1.

And you were willing to accept that omega equals 1 just as some initial condition or whatever?

That k is zero. That makes it sound better. When you talk about omega, you're talking about the ratio of matter density to the Hubble constant squared, and you might say, "How could that be so finely tuned?" But if you simply say that of the three possible geometries for an isotropic, homogeneous cosmology—positively curved, negatively curved, and flat—we happen to live in the flat one, then that's clearly something which doesn't require fine-tuning. It just is true.

But it seems that you're making an implicit assumption there that all three of those possibilities are somehow equally likely and that one is more aesthetically pleasing.

Well, yes, it becomes an argument that's hard to settle, but it didn't seem to me anything unnatural in saying that for reasons that we cannot now understand, the universe happens to have $k = 0$. So I don't regard that as a triumph of inflation. I never have. But the other two are good enough.

Getting back to the isotropy problem, which you might call the horizon problem, when did you first hear about that? You wrote about it in your book in 1972, but it was around before then.

Probably a few years before that. It isn't something that has haunted me for years and years. Obviously, the book came out in 1972, and the microwave background wasn't discovered until 1965, so it must have been somewhere in those seven years. I don't know exactly when.

Were you bothered by that problem? How did you think it would be resolved?

I suppose I took a Mr. Micawberish attitude: something will turn up. You know, Mr. Micawber is always saying something will turn up when he's in trouble [laughs]. You can only worry about so many things, and I have never devoted myself professionally to worrying about that. I took due notice in my book that there was such a problem. I must have been worrying about something else, something in elementary particle physics at the time. Usually for me, and I suspect for everyone else, a large part of the answer of why I didn't do something at a particular time is because I was doing something else.

I gather that you did consider it to be a serious problem.

Yes, and I referred to it as such in my book. I certainly didn't have the idea of inflation. When I heard that explanation, I thought it was really very charming. That was what really sold me that something like inflation was probably true. But, of course, the inflationary models have had all sorts of problems, and even Linde now says that the "new inflationary model" due to him and Albrecht and Steinhardt is dead.

Yes, Hawking said the same thing in his new book.[7]

Oh yes, maybe I'm thinking of Hawking, in fact. I think [Andrei] Linde has said something like that also, but I'm not sure.

Linde can afford to say it because now he's got a new model, chaotic inflation, which is in pretty good shape, I guess.[8]

On the other hand, there's a lot of hand-waving in it. From the start, from Guth's work on, one is haunted by the fact that in the end you're just trying to explain some very qualitative features of the universe without any numerical predictions of the kind that really convince you you've understood what's going on. I don't regard omega equal to 1 as a numerical prediction, not in the sense that the Lamb shift is a numerical prediction.

When did you first hear about the flatness problem? Do you remember when you first started thinking about that?

No. Sorry, no recollection.

It's interesting that some of these problems impress some people more than others.

Yes, I know. I've been in that position sometimes. There have been problems in elementary particle physics that I was worried about that nobody else seemed to worry about. I certainly didn't realize there was a monopole problem until John Preskill, who was my student, along with someone in Russia, first pointed it out.[9]

You said that you didn't place the flatness problem in the same category as the horizon problem because you were willing to accept that we could just have a flat universe. Did your view about the flatness problem change any after the inflationary models?

Well, yes, in the sense that it suddenly became plausible, as it hadn't been before, that maybe in fact k is 1 or -1, because if k is 1 . . .

It'll still appear almost zero from observations.

Yes, that's right. Whereas it would be very implausible without [the effects of inflation]. So, in a way, it opened up these other possibilities.

Let's move to some recent observations in cosmology. The work that has been done in the last 5 years or so on the large-scale structure by Huchra, de Lapparent, and Geller and Haynes and Giovanelli—has that work changed your thinking at all about cosmology?

Yes.

Or did you already anticipate it?

No, I certainly didn't anticipate it. I have at various times—for instance, I think in *The First Three Minutes*—referred to the fact that on large scales the universe appears smooth. And it's disturbing to see how unsmooth it does appear, with only the microwave background left to convince you that on *really* large scales, it's smooth. In other words, there's not a hell of a lot of evidence left for a Robertson–Walker universe outside of the microwave background, which is isotropic around us, and the Copernican principle that if the universe is isotropic around you, it's probably isotropic around everyone because there's nothing special about you. Maybe we're misunderstanding something about the microwave background and, in fact, the universe isn't Robertson–Walker at all. That bothers me. There is a mathematical question that I've been curious about for years and years, and I still don't know the answer to. It goes as follows. It's a simple theorem, little more than common sense, but you can also prove it mathematically, as I do in my book. If the universe is isotropic around every point, then it's homogeneous. However, there is no theorem that says that if the universe is *approximately* isotropic around every point, then it's *approximately* homogeneous. It's quite possible that the universe is so constructed that every observer looking around him or her sees the universe as being approximately isotropic, and yet the small departures from isotropy build up as you go to great distances, so that you can go to great distances and find conditions totally different, like, for instance, maybe a different baryon density on the average, or something like that. I go back to this problem every few years, and I never get anywhere. But as time passes, the universe looks less and less isotropic, except for the microwave background. I'm very bothered by it.

Do you think we could be on completely the wrong track?

I don't have any *idea* of how to explain the isotropy of the microwave background if we are on the wrong track. But it's something I worry about every once in a while. I don't have anything substantial to say about it. It's not something I would go into print with, but it bothers me that the whole Robertson–Walker picture may be wrong on very large scales. But that's, by the way, no reason not to continue using it. I think good scientific strategy is to use simple models if they have any chance of being right, and use them to make as many predictions as you can, and then see if they work. And then you may find they don't work. I get very annoyed by a certain kind of heretic cosmologist. I don't mean that it isn't all right to doubt the present model, as I've just been doing with you. But I think there is an attitude that some distinguished astronomers or physicists have had that people who *work* in the context of the big bang, the standard Robertson–Walker big bang model, are wasting their time and being very foolish, because we really don't know that that's true. I've seen that attitude in particle physics. I remember there were people like Landau, who was very skeptical about anyone who would work at all in quantum electrodynamics, because Landau had doubts about the ultimate validity of quantum electrodynamics. I think that's very silly. I think scientific practice is not really in danger from a false consensus. It *needs* a consensus, in order for us to have something to talk about with each other, in order that our work adds up. Without a consensus, you don't have any way of knowing even if the consensus is wrong. You not only can't do anything positive, you can't do anything negative without a consensus. It was in that sense that I called the Robertson–Walker–Friedmann big bang model the standard model in my book. We need standard models. And sometimes we need them because then we can prove they're wrong. So I think it's really terribly important for cosmologists to continue to take the Robertson–Walker–Friedmann model very seriously. If they have ideas on how to go beyond it, fine. But that's no reason for not always going back to the standard model as your reference point.

Until we know for sure that it's wrong.

Absolutely. And, of course, these things never are *wrong*. They always turn out to be right in some domain and wrong in another. Just like the standard model of particle physics may very well require modification if new gauge bosons are discovered, but there will be a certain approximation in which the standard model is right. I don't think there's any doubt of that. And there is some sense in which the standard cosmology is right, but it may be a more limited sense than we think now. If it weren't for the microwave background, I would begin to think real heresies might be worth pursuing right now.

What do you think are the outstanding problems in cosmology today?

I guess there are two different kinds of problems. One is the large-scale problem that the Hubble expansion is still not really pinned down. We still don't know the Hubble constant to better than a factor of 2. We don't know the deceleration parameter, other than to say it's not enormous. I think the space telescope will be a huge help with the first problem. I don't know what will help with the second problem. And of course we may discover that when we understand how to measure distances more accurately, that even at very large distances there are large departures from the Hubble flow. That would be revolutionary, but it's a possibility. So that's the big problem. Then another problem is the origin of structure. People are still arguing whether small things accrete to make big things or big things break up to make small things. There, there's lots of data. We have all these galaxies and they have a certain distribution of masses, a certain distribution of angular momenta. Within the galaxies there are structures like globular clusters and open clusters, and then there are stars and they have themselves a distribution of masses. The whole thing is not understood except in the most qualitative way. That you would think is a doable problem—given an expanding universe, to make a variety of different assumptions about initial perturbations, follow them, and see what happens.

A lot of people have been doing that.

Yes, of course. They have been doing it for 20 or 30 years, and I don't think there's been any real success in explaining the hierarchy of structure in the universe and the distribution of masses at each level in the hierarchy. I think that's the biggest failure of astrophysical theory, because it seems to me that's really a failure of theory. Possibly, it'll be resolved by new observations, but I don't know what observations are needed. It seems to me that's a well-posed theoretical problem. I don't know how to account for what's turned out to be the extreme difficulty. But some problems are very difficult.

Maybe there's too large a range of possible initial conditions, and they just can't all be explored easily.

Yes, but of course it also involves nonlinear hydrodynamics, and we still don't know how to predict the distribution of wave numbers in turbulent flow. We don't know how to calculate convective energy transport in stars because it's again nonlinear hydrodynamics. There are an awful lot of problems where we know the underlying equations, and we don't know how to make progress with them. So maybe this is just another one. But it doesn't seem to be really getting much better. And of course there's always the haunting feeling that the data is just terribly misleading, that all we're seeing is the tip of the iceberg, that the *real* distribution is a distribution of weakly interacting massive particles [WIMPS, for short, one of the candidates for dark matter]. I guess that might be the great breakthrough. If some of these laboratory experiments or maybe progress in elementary particle physics were to confirm that these WIMPs were real, then I think the astrophysicists could focus on that and begin to use that with a sense of confidence. It's hard to cover all possibilities if you have no way of knowing which one is right. I mean, how can you just spend your whole life on the nonlinear hydrodynamics of galaxy formation when you're not sure the particles you're dealing with are the important ones at all. It's kind of discouraging.

Do you think that your views about what are the major problems have changed any in the last 10 years?

No. I think that the deeper problems having to do with inflation, what the universe is like on the Planck scale, and quantum cosmology are so far from any confrontation with experiment that I've suspended thinking about them. It's like Charles Lamb said about space and time, "Nothing puzzles me more. But then nothing puzzles me less, because I never think of it at all" [laughs]. Obviously, if I could see something to do about it, I'd be glad to think about it, but I don't. And I don't regard it as the next step to solve those things because even if one got a completely believable, absolutely debugged new, new, new, new inflationary model, which solved all the problems and was completely natural, I would say, "Good. It's probably true, but what can you do with it?" Of course, it would be nice to have that. I think these other two questions, the question of the Hubble flow and the question of the galaxy formation, are really right up there at the top of the agenda for practical cosmology, if there is such a thing.

Practical cosmology, that's a funny combination of words. Let me ask you a couple of more speculative questions. Maybe putting your scientific caution aside slightly . . .

I thought I had been doing that [laughs].

If you could design the universe any way that you wanted to, how would you do it?

Well, I'd go back to what I said before—that some sort of steady state seems to me tremendously attractive. I don't feel the way my old friend John Wheeler does, that you want a closed universe to avoid boundary conditions. I would be very happy in an infinite universe with infinite time as well as infinite space, and have the whole universe be a self-consistent solution of fundamental physical principles, including all the parameters such as how fast it's expanding and how much mass density there is. Maybe steady state only in some average sense, maybe an oscillating universe. I know one isn't supposed to worry about what happened before the big bang. I've lectured about cosmology to an awful lot of audiences of general public. The question always comes up, "What was there before the big bang?" And I spend so many minutes waffling about that, saying, "Well, maybe there was no time . . ." But I share their perplexity. I find it more comfortable to think about infinite time as well as infinite space, no boundary in time any more than a boundary in space, and with *everything* being explained by physics without any arbitrary historical elements entering it.

No initial conditions.

Yes. No initial conditions at all. That would be appealing, especially if you could show that for a certain set of laws of physics, any set of initial conditions would always evolve into the steady state, so that you wouldn't even have to ask, "Well, why are we in the steady state?"

But if the universe had existed for infinite time, you wouldn't really have to talk about it having initially been in one state and having evolved towards this situation.

No, but you could say to anyone who asked, "Why is the universe in this particular steady state?"—"Well, what else could it be?" And if he or she suggested something else, you could say, "Well, that would evolve into this."

Yes, I see.

And clearly it's comfortable to imagine that although our own sun will get cold and die, there will always be new suns and new life and that things will go on forever. But I don't have any set of equations of physics that would have that property. I have no ideas along those lines, but you asked me what I would like, and that's what I would like.

Yes, I didn't ask you how you would do it.

Thank you [laughs].

One final question. In your book The First Three Minutes *you make a statement that the more the universe seems comprehensible, the more it also seems pointless. Can you elaborate just slightly on that?*

I've gotten more negative comments about that sentence than about anything else I've ever written. Many people who said they liked every other part of the book hated that sentence. Remember, after that sentence, then I go on and I have a cheerful ending and I say, "But one thing that does seem to help, one of the things that makes life worthwhile is doing scientific research." Someone wrote me a letter saying, "You had such a good remark about the universe being pointless, why did you spoil it with that optimism?" [laughs]. I don't think I expressed myself exactly the way I should have, but I think I'm going to leave a more detailed explanation for something else I write. I've thought a little bit about how I would have said that if I had known the reactions to it, and I think I would have said it a little differently. I certainly meant *roughly* what I said, but it didn't come out *exactly* as I wanted it. If you say things are pointless, you have to ask, "Well, what point were you looking for?" And that's what's needed, I think, to be explained. What kind of point would have been there that might have made it *not* pointless. That's what I really would have to explain.

ALAN GUTH

Alan Guth was born on February 27, 1947, in New Brunswick, New Jersey. He received his B.S. and M.S. degrees at the Massachusetts Institute of Technology in 1969 and continued there to receive a Ph.D. in physics in 1972. He was an instructor of physics at Princeton University and then a research associate at Columbia, Cornell, and Stanford. In 1980 he joined the faculty of MIT, where he is now a professor of physics. Guth is a member of the National Academy of Sciences.

Guth was trained as a theorist in elementary particle physics and began his career in that field. In 1978, partly inspired by Steven Weinberg, Guth began working in cosmology, calculating the effects of the new grand unified theories in the early universe. In 1980 Guth proposed a seminal modification of the big bang model, called the inflationary universe model, which made use of new ideas in particle physics and also solved a number of outstanding problems with the standard big bang model. At the time of his pencil-and-paper discovery, in December of 1979, Guth was a postdoctoral fellow at the Stanford Linear Accelerator and was worried about finding a job for the coming year. Guth can often be seen riding his bicycle to and from work and is always willing to sit down and talk, especially with students.

Let me start with a couple of questions on your childhood. Could you tell me about your parents?

Sure. Neither of my parents graduated from college. My father went to two years of junior college. My mother never went to college. All the time I was growing up, my father was working in small businesses. In my younger years, he owned a small grocery store in New Brunswick, New Jersey. From when I was 10 years old or so, he owned a dry-cleaners business, also in New Brunswick.

Do you remember any experiences that you had with your parents that might have been influential in your becoming a scientist?

That's a good question. It's kind of hard to tell from my family background what it was that drove me towards science. Neither of my parents knew much about science, nor were they particularly interested in science. I'm not sure what it was that pointed me towards science.

Do you have any memories of how you thought about cosmology before you became a scientist?

When I was in high school, I read at least one book on cosmology. I don't remember what it was. I remember learning about the possibilities of open and closed universes. Also, whatever book I read talked about an oscillating universe as one possibility, and I think that's the model I liked the best. It seemed natural. It had the appeal that you seemingly didn't have to explain the beginning. And if you can't explain the beginning, it's nice not to *have* to explain the beginning.

Because you could imagine that the oscillations occurred for an infinite period of time?

Yes, that is certainly the way I thought about it. I didn't know about the increase of entropy at that time, and all the problems that brings.

So one attraction about an oscillating universe was that you didn't have to think about the beginning. Did you think about the beginning?

I was aware that if you didn't have an oscillating universe, you'd either need a beginning or a steady state cosmology. I don't remember what I learned at that time about the steady state cosmology, but I think it was already somewhat on the way out.

At that time, did you ever try to imagine how far space extends, or whether the universe was finite or infinite, or any of those kinds of questions?

Yes. That also was a question that came up in some of my popular reading. I did learn about closed universes. I think I had a reasonable, vague understanding of what a closed universe was. I knew about the usual analogy of the 2-dimensional surface of a balloon, increasing it by 1 dimension. It's hard to know now how much sense I was really able to make out of that idea. But I bought it. I thought it made sense. I doubt if I could have calculated the volume of a closed universe.

Tell me a little bit about your undergraduate education at MIT. I'd like to know about any particularly influential experiences that you had. Had you decided from the beginning that you wanted to go into physics?

I had decided from the beginning that I wanted to go into physics, and somewhere, pretty early on, I even went so far as to decide that I wanted to go into particle theory. I think what appealed to me about particle theory was the idea that you're grappling with trying to discover what the funda-mental laws of nature are themselves, and that idea fascinated me—just the idea that there were fundamental laws of nature that could be dis-covered. As far as the things that happened while I was an undergraduate, one of the more influential things for me was a project I got involved in my junior year and then continued in my senior year and the summer follow-ing my senior year. This was not a theoretical physics project. It was a project I did with Aaron Bernstein, who is an experimental nuclear physicist. I had met Aaron during my first year at MIT. He was a recitation instructor for one of the terms of freshman physics. Some time around my junior year I started working with him. I think it was initially a project that we had to do for junior lab and later it became my senior thesis and then later became my master's thesis. It didn't have much to do in any direct way with the work I did later, but it was the first time I really felt like a physicist, and it was a very good feeling. Until then I really just felt like a student.

Because this was original research?

It was original research of sorts. Even though he is an experimentalist and it was an experimentally-oriented thesis, I actually never really got my hands dirty. It was a theoretical analysis of an experiment to be done [laughs]. But it was original in the sense that I felt like what I was doing was important, and had never been done before, and he did a good job of making me feel that what I was doing was important. Another experience I had as an undergraduate was that I spent one summer—it was the summer after my junior year—working at Bell Labs in Murray Hill. That was also a lot of fun, and, at least in terms of my feeling toward physics, I got a lot out of it. I was working with lasers.

When you started thinking about cosmology in 1979, do you remember how you thought about, let's say, the beginning of the universe?

One thing I certainly felt was that the universe was a frighteningly open problem, and it was very different from what I had worked on previously in particle physics. Particle physics is a much cleaner subject than astrophysics and cosmology to begin with. And even within particle physics, I had always sought out the cleanest, most mathematically well-formed problems to work on. So there really was a big change for me to start working on something as messy as cosmology. I really got my arm twisted into it by a friend of mine at Cornell, Henry Tye. I certainly wasn't involved previously. Another influence on me at the time was Steve Weinberg, who was a person I respected very much from particle physics.

He was already working on cosmology.

He was already working on cosmology. He was working on exactly this sort of thing, grand unified theories and cosmology. So if he could work on it, it must make sense.

You felt he gave a certain legitimacy to the subject?

That's right. And for me that was a problem.

Do you mean that in practice cosmology was too messy to pin down, but in principle it might have had answers? Or that even in principle it was hard to pin down?

I doubt that I would have ever said it's impossible in principle to pin down cosmology. I think it's a question of what I thought we were ready to understand at the time.

Tell me about some of the motivating factors or influences in your development of the inflationary universe theory.

The story as it relates to me probably starts in the fall of 1978. It was sort of an odd coincidence. I went to a lecture given by Bob Dicke from Princeton. It was a lecture that I might have gone to and might not have, but I decided to go, and it turned out to be very important in terms of what I learned, because one of the things he talked about was the flatness problem. I don't remember for sure if he used that word, but it's definitely what he was speaking about. As he phrased it, the problem was that if you looked at the universe one second after the big bang, the expansion rate had to be exactly what it was to an accuracy of about one part in 10^{14}, or else the universe would have either flown apart without ever forming galaxies or quickly recollapsed. Now at the time I was not working on cosmology. It was just an odd fact for me, but one that struck me, and I tucked it away in the back of my mind. At the time, I didn't even understand how to derive that fact, but I believed it and was startled by it. Also during the fall of 1978—probably later, I'm not sure if this is the exact chronology—Henry Tye came to me, and this was very important in my getting involved and working in cosmology. He had gotten himself interested in grand unified theories, which were all new to me, and he came to me with a question. He asked me because of some work I'd done earlier on magnetic monopoles. He asked me whether or not grand unified theories would give rise to magnetic monopoles. And at the time I didn't really have any idea what grand unified theories were, so he had to teach that to me, which he did.

You were a postdoc at Cornell at this time, is that right?

That's right. I was a postdoc at Cornell, and so was Henry. So Henry explained to me what a grand unified theory was, and I don't remember how long it took me to figure it out, but I understood at that time enough about magnetic monopoles that it was pretty obvious once I learned about grand unified theories that they would give rise to magnetic monopoles. I came back to Henry and told him that, yes, there would be magnetic monopoles. But I told him that they wouldn't be things that you could expect to find, because they'd weigh something like 10^{16} GeV. So it's just another unobservable consequence of these theories. He immediately came back and said, "Why don't we try to figure out how many of those would have been produced in the big bang." At that time, it sounded to me to be an almost totally crazy thing to think about. I had never worked on cosmology. All the problems that I had worked on up until that time had been very clean, well-defined, mathematically posed problems, and this was the opposite. When you talk about the early universe, it's always hard to know what the relevant physics is and whether or not you've got it right or whether you've missed some other fact that's very important. It's a much more open-ended, less well-defined problem. So I resisted working on it, and it really wasn't until some time in the spring that Henry finally convinced me to begin to work seriously on the problem.

In between, there was one other significant influence on me that was important. I was influenced in part by a visit during the spring of 1979 to

Cornell by Steve Weinberg, who was working on baryogenesis in the context of grand unified theories.[1] He gave several lectures about that at Cornell, and I'd always been very impressed by Steve. It was really his work that convinced me that the early universe was a well-enough-defined question that it made sense to work on it, and that grand unified theories were interesting enough so it made sense to apply them to the early universe. So at that point I started working with Henry, and we fairly quickly came to the tentative conclusion that it looked like far too many of these magnetic monopoles would be produced. We were also in touch—I forget whether it was in the spring or early summer—with John Preskill, who at that time was a graduate student of Steve Weinberg's and who was also working on exactly this question of magnetic monopole production and the big bang. Preskill was the first to publish on this subject, and I think he did have a significantly cleaner argument than Henry and I had, although we were reaching similar conclusions.[2]

In the fall of 1979 I moved to SLAC [Stanford Linear Accelerator Center]. I was on leave from Cornell, intending to spend one year at SLAC, which is also how it turned out. But Henry and I kept working. We communicated by telephone. At this point, our work shifted a little bit because the Preskill paper had been published, so it wasn't enough just to say that too many monopoles would be produced. We had to say something in addition.

You had to solve the problem.

I don't know about solving the problem, but we did gear our work towards trying to figure out whether there was any way around it, whether there was anything that you could change in your assumptions to make grand unified theories compatible with cosmology and the fact that the universe is not swimming with magnetic monopoles. We were led to the idea of supercooling. The magnetic monopoles are produced in the phase transition of the grand unified theories, and the argument that both Preskill and we were using was really just based on timing. Supercooling clearly gets you around that argument—it gives you extra time. So Henry and I wrote a paper about this which we finished in December of 1979. We wrote it as a *Physical Review Letter.*[3] In that paper, we totally ignored any effect that the phase transition might have on the evolution of the universe. We assumed that the scale factor just evolved exactly as it would if the phase transition were not taking place. Somewhere while we were doing that, I'm quite sure that—I never really checked back with him—it was Henry who suggested that we should look at that assumption. It was the beginning of

December 1979 that I sat down and looked at it. And when I did, in one very exciting night, I realized that it would lead to exponential expansion, and I also by that time understood the Dicke problem, the flatness problem, well enough to realize that that would solve the flatness problem. That was the first time I sat down and really went through the arithmetic of the Dicke problem, but I understood enough about how it worked to realize that exponential expansion would avoid the problem. And that was the beginning of inflation.

Meanwhile, Henry and I were still mainly concerned about finishing off this paper. He was about to go on a rather long trip to China, so we wanted to be sure to get the paper tied up before he left. Then I continued working on this exponential expansion idea. There was one other freak coincidence. I'm pretty sure it was sometime in January of 1980 that I learned about the horizon problem for the first time, which was also just an odd piece of information that just fell in my lap one day.

> *I was going to ask you about the horizon problem, but since you learned about that after inflation, let me start with the flatness problem. When you first heard about the flatness problem from Dicke, did you think that it was a problem that demanded a physical explanation, as opposed to a special set of initial conditions? How did you imagine that the problem might be resolved?*

Good question. I don't know if I'm completely sure. I think I initially interpreted it as an indication that there was something missing in the picture of the big bang and that some little ingredient would be invented to explain this fine-tuning.

> *Did you distinguish between initial conditions and physical mechanisms?*

I'm not sure that I would have distinguished between initial conditions and physical mechanisms, but I would have probably said that you would need some new element of physics. I think I've always viewed the initial conditions as things that probably are determined by physics—physics that we may not understand yet. It's hard for me to tell when these ideas emerged in my head, but at least from my present point of view, the ideas of quantum cosmology sound very attractive to me, even though I think we're still only at the primitive stages of trying to understand exactly how that works.

> *Did you know about quantum cosmology at the time that you heard the Dicke talk?*

Probably not. So I'm not sure of what I would have said then, but I'm not sure I would have drawn a sharp distinction between mechanisms and initial conditions. I think I might have said just "some physics" that happens in the early stages that would have to control the mass density of the universe and create this fine-tuned situation.

So you were not willing to accept it then as an accident?

No, I would certainly not be willing to accept it as an accident. I guess Dicke himself at the time—and I don't think he said this in his lecture, but it certainly is in the written version of the Einstein Centennial Volume—was pushing the point of view that [the flatness problem] indicated an oscillating universe, that the entropy builds up from cycle to cycle. I don't know if that idea would have appealed to me or not. I guess I would have considered it a possibility. From my present point of view, the short-coming of that idea is the fact that there's just no physics behind the bounces of the oscillating model. There's no physical theory.

When you heard about the horizon problem, after inflation, did you consider it to be a legitimate problem in the same way that you considered the flatness problem to be legitimate, to indicate something missing in the standard big bang model? By that time, you already had a solution to the horizon problem, so you were in a special position.

Actually, it really did take a while before I regarded the horizon problem as being as significant as the flatness problem. Now I regard the two as being equally significant.

Why did you regard it initially to be less significant than the flatness problem?

I think the reason I would have given at the time was that it's a problem whose statement is less quantitative. With the flatness problem, you have this number, the expansion rate at, say, one second, which has to be tuned to 14 decimal places. If you want to phrase it at grand unified theory times, it's even more—49 decimal places. The horizon problem is always just qualitative—you don't understand why the universe looks the same here as it does there—and for that reason it impressed me somewhat less.

You can certainly state quantitatively the largest angular size over which things were in the horizon at a z of 1,000, when the microwave radiation was produced.

Yes, that's right. But I guess I was less impressed because there's no colossal number that you have to explain. The horizon is not that small. The causal horizon is maybe a factor of 100 smaller than what you'd need to cover the entire observable universe, but not a factor of 10^{15} or 10^{29}. I suspected that there was enough ambiguity so you could find some minor way of changing the physics of the early universe, without something dramatic, that might very well get around the horizon problem.

But you say that since you've been thinking about it more, that you have come to elevate its importance to that of the flatness problem?

Yes, I think so. Why? Well, the strongest influence is probably just the psychological importance of time. In the end, of course, it's not really a scientific question to ask which of these problems is more important. It doesn't affect anything that you've concluded about what's true or false. Emotionally, I regard the two as on equal footing. In terms of the role they play, I think the role of the horizon problem has enlarged somewhat. It enlarges when you consider a question that came up after inflation was proposed. The people who were strongly convinced that omega was really 0.2 rather than 1 advocated the idea of limited inflation—inflation that would cut off at just the right point so that omega doesn't get driven to 1. I think the strongest argument against that is basically the horizon problem of sorts, that is, the uniformity of the cosmic microwave background. If you have inflation that cuts off before it drives omega to 1, it means that you're still leaving significant influences from the initial conditions, before inflation.

Why do you feel the inflationary universe model has caught on so well?

It's a good question, and I think it has caught on better than I would have expected. To some extent it was just an overdue idea. It really is a rather short step beyond what had been done previously in particle physics, and it answered a lot of questions that people had already realized. Also, I guess it is really the only theory that's around right now that makes real predictions about what the early universe should look like. Most ideas that people have make no predictions.

Don't you think people could have used the standard big bang model without the inflationary epoch to make predictions? I guess the initial conditions would have played a larger role.

The initial conditions would have played a much larger role. And for anything that goes beyond the basic issues, like the density fluctuations, the standard model really makes no predictions. You just have to put those in by hand. So people could have just been resistant to the idea. Actually, I am a little surprised that inflation has caught on so well. I also had some lucky breaks as far as public relations, as far as the speed at which [the idea] traveled. When I first came up with the idea, I was at SLAC. That year, Sidney Coleman was at SLAC and Lenny Susskind was at Stanford, spending a good deal of time at SLAC. When I gave the seminar, when I first talked about inflation, Coleman and Susskind were in the audience. Both of them got very excited about it and felt right away that it was a good idea. At least initially, as far as the spread of information, I think they were both instrumental. Both of them went around and talked about it a lot. I was just a lowly postdoc. If I had gone around talking about [the inflationary universe model], no one would have listened for quite some time.

> *Let me change gears. When you sit down to solve some equations for the expansion of the universe, do you have a visual picture in your head? Sometimes I think of a little ball expanding.*

I guess sometimes I do. I agree that the ball is about the only picture you can use for the early universe, even though it's not all that accurate. It doesn't have the same boundaries as the universe. I was not terribly aware of the actual numerical values in cosmology when I started to work on it. I remember a few nights before I gave my first talk [at SLAC], I had to review some material so I would have some idea about what I was talking about. I remember reading Steve Weinberg's popular book on cosmology.

> *Did it bother you any to be making theories about the beginning of the universe?*

You mean in a kind of a religious way?

> *In any kind of way.*

It bothered me in the sense that it was so remote. You didn't know how to test it. To some extent that's still true. Even if inflation is right, it's certainly hard to prove it's right. But once I started working in it, I did gradually come to the view that, even though there are a lot of things we don't know, it is still possible to make reasonable calculations. What impressed me when I started working on cosmology was the way in which order of magnitude estimates work—the fact that if something is wrong, it's typically not wrong by a factor of 2 but by a factor of 10^{20}. So there were things you could learn from it, and it made sense to study cosmology at this time, even though it's not nearly as precise as other physics.

Did you begin taking more seriously a theory for the beginning of the universe when you started working in cosmology?

When I started working, I accepted the standard big bang model quite seriously, which is what everybody else was doing at the time. The view that I might have had, although it's hard to tell over what time period this view emerged, was that it was practically impossible to take the standard big bang model of the universe seriously back to time zero. Certainly, one reason is that you just know that when the temperature gets to be of the order of the Planck scale, there's got to be physics going on that you know nothing about.

How do you picture the creation of the universe? Do you have any ideas of how the universe came into being?

When I started working on cosmology, and before I discovered inflation, I think I would not have asked that question. You only ask questions that you can get some idea of how to answer.

As a scientist that's true, but what about just as a person? What were you thinking about, not necessarily for posing to other scientists?

I was *not* thinking about how the universe was created. I do to some extent try to focus my thoughts on questions that I have some tools for approaching. It's not just that I'm a scientist. How long can you spend thinking about something for which you have no idea what to think? At that time, I was not at all aware that it was possible to make a universe from nothing. I would have thought at that time that conservation of energy would make that impossible. Now, I do think that it's a question that science is beginning to ask. And I think it's very plausible that the universe is a quantum fluctuation, starting from absolutely nothing. Those calculations still involve questions that need answering. But all that is very plausible.

If the universe began as a quantum fluctuation, what was there before the fluctuation? How do you think about that?

I can tell you the image I use for that, although I'm not sure it will hold up in time. What I think now is that in the hypothetical theory of quantum gravity, which doesn't quite exist, there will be a Hilbert space of all the possible states of the universe. The different states of the universe will clearly have different possible geometries for the universe, and among all of those states and those possible geometries I assume that one of the possibilities is a universe with zero points—[in other words], a closed universe with its radius equal to zero or something like that. That's what I take as the definition of "nothing," and that's what I take as the starting point, a universe with no points.

In this quantum era, how do you picture these different possible universes?

The view of quantum mechanics that has always made the most sense to me—although I admit that there are some features that don't make sense to me—is the Everett–Wheeler view, the many-worlds interpretation.[4] You think of all the components of the wave function as representing reality, and our reality is just a branch that we happen to be on. But it's hard to know what words to use. One could at least imagine that there are people on other branches of the wave function—who could be sitting somewhere having a conversation just like the one we're having now.

So all of these possible universes in some way exist.

Yes.

Then what meaning do you give to the words "the universe came into being"? Does that just mean that a particular universe changed from having zero points, from having zero radius, to having a finite radius?

Actually it probably has no precise meaning. I know Hawking tries to avoid talking about [the beginning] for that reason. I think it has meaning to the extent that this classical interpretation of quantum mechanics gives a sense that time flows. But my understanding of quantum gravity, which is certainly not perfect, seems to indicate that there's a level of quantum gravity [in which time cannot] be described. If you follow the universe backwards in time toward the origin of the universe, you don't need quantum gravity while time is evolving, but once you get down toward the beginning of the universe, where you really need quantum gravity, my guess is that Hawking is right, that you really should not talk about the universe beginning at a time. The boundary conditions determine the wave function of the universe.

But in that picture, I take it that you would also imagine that there are other universes simultaneously existing.

Yes, that's right. In my view, there *are* other universes that are simultaneously existing. And there are two levels of that. If inflation is right, even in our component of the wave function the universe is so huge that there are parts that are totally disconnected from us.

What do you think about the anthropic principle?

Emotionally, the anthropic principle kind of rubs me the wrong way. I'm even resistant to listening to it. Obviously, there are some anthropic statements you can make that are true. If we weren't here then we wouldn't be here. As far as the anthropic principle as a way of approaching things, I find it hard to believe that anybody would ever use the anthropic principle if he had a better explanation for something. I've yet, for example, to hear an anthropic principle of world history. Historians don't talk in those terms. They have more concrete things to say. I tend to feel that the physical constants are determined by physical laws that we can't understand now, and once we understand those physical laws we can make predictions that are a lot more precise.

So you would argue that intelligence, or life, doesn't have any special role in the universe?

That's right. I don't think that intelligence or life has any special role. The fact that intelligence or life has evolved may tell us something [about the universe]. But I don't think the laws were contrived in order to allow life to exist. I also think that we don't understand life that well. So it is a rather poor way to try to determine the laws using the fact that life exists. The anthropic principle is something that people do if they can't think of anything better to do.

One of the ideas associated with the anthropic principle is whether the universe is accidental or not. In your view of the quantum mechanical origins of the universe, would you say that these other universes might have different dimensions of space or different values of the fundamental constants?

Yes. Certainly in my grand wave function, there are many different things going on, so any one branch might be accidental, although they all have the same physical laws.

So our particular branch is just one that we happen to be on; there are other creatures possibly living on other branches.

Right. And one of the questions people worry about is what it is that determines the laws of physics. Of course, there are a few coincidences. If the neutron lifetime weren't what it is, the universe would be either all helium or else it would contain almost no helium. Things like that. My emotional feeling is that the general laws are really very restrictive, including how the neutron lifetime comes out, and my guess is that there really is only one consistent theory of nature, which has no free parameters at all.

One of the overtones of the anthropic principle is that the universe has some kind of a purpose.

Yes. Right.

If you thought that the laws of physics were so restrictive that they could not have been accidental, then would that suggest a purpose to the universe?

I don't know. I would say that another way. If there were zillions of possible universes, and our universe happened to have a neutron lifetime in the range to allow life to form—that I think would suggest a purpose. The alternative point of view is that there is only one consistent way the universe could have evolved.

Let me ask you about your reaction to some recent observational work in cosmology, in particular the work on the large-scale structure by Geller, Huchra, and de Lapparent and by Haynes and Giovanelli. Do you remember when you first heard about that work? How did you react to it?

I first heard about that work pretty early on, since I'm here [at the Harvard-Smithsonian Center for Astrophysics] and people were talking about it. My initial reaction was not too different from my feeling about it now. I think it's clearly important work; it has everybody trying to understand the large-scale structure of the universe. It was certainly advertised by some people as being definitive evidence that we didn't understand at all what was going on and that we were back to ground zero in terms of building theories. I don't think that's true. I think we still have the possibility of understanding this. It's a very complicated problem. After the initial observations, the simulations were pursued a little bit more using ideas like cold dark matter, and they began to look a little bit more like the data. I don't think that's entirely artificial. People started to look at circumstances that more closely resemble the circumstances of the observations themselves.

The other idea was biasing. I forget when it was first proposed, but it certainly didn't really catch on until after these observations. Once you assume that biasing is at least a possibility, then it becomes much harder to compare theory with observation because you don't really understand what you mean by biasing. It gives it a lot more uncertainty on the theoretical side. So I think what these observations are showing is that there are certainly things about the large-scale structure of the universe that we don't understand yet. I don't think they mean that inflation or even inflation plus something as specific as cold dark matter is necessarily in trouble.

Did you have any views about the homogeneity or nonhomogeneity of the universe prior to learning about the bubble work?

I always assumed that the universe was homogeneous on the large scale, and I still do. I think that the evidence from the microwave background is still very compelling, no matter what we see at the scales where we actually observe galaxies, which are shorter scales. Unless you believe that the microwave background is something totally different from what we think it is, I think you're still forced to believe that on the largest scales the universe still appears homogeneous. Obviously, if inflation is true, there are scales much, much larger in which the universe might be very inhomogeneous.

What do you consider to be the major problems in cosmology today?

On the observational side, it would be tremendously important if we could get good measurements of the anisotropies in the cosmic microwave background radiation. Certainly at levels that are just a *little* bit beyond what we've observed, there had better be anisotropies in the cosmic background radiation, or else all of our possible theories of galaxy formation fall apart. The other thing that is also important is more and better data of the same type as Huchra, Geller, de Lapparent, and the Seven Samurai.[5] It would help a lot if we had a better map of what the universe in our vicinity looks like. Apparently there's a lot that can be done there. It's just a question of availability of funds and astronomers. Actually, astronomers are available. I think it's really a question of availability of funds more than anything else. But a really good redshift map, many times the amount of data that's currently available, would be a big help. It would not answer the questions by itself, but would help people who are trying to build models of galaxy formation.

What about on the theoretical side?

On the theoretical side, it seems to me that the primary question is the formation of structure. I think there's very little that we're in a position now to learn about the very early universe, except possibly by clues from the structure of the universe. I think on the theoretical side the key question is the question of biasing. It's really the whole question of how galaxies actually formed. The simulations that have been done today, for the most part, simply trace the evolution of mass under the force of gravity. Obviously, in the final stages in which galaxies actually form, there's a lot more physics going on than just gravity, and somehow all of that physics has to be pulled together so that we can get at least some crude understanding of what it is that causes galaxies to form. Hopefully, in doing that, we'd understand how galaxies of different morphologies form, and how to relate the map of mass density produced in these simulations to a map of the light density.

Or maybe getting a handle on the dark matter?

That's right. The dark matter looks crucial to the whole thing. Certainly there's a lot of work to be done before we know what it is, by just trying different things and seeing what the consequences would be. In fact, that may be, in the end, how we'll determine what the dark matter is—by understanding more about the physics that's involved in galaxy formation and then just seeing how it would work for different assumptions about the dark matter. But certainly the dark matter's absolutely crucial in this, and we won't fully understand it until we know what the dark matter is.

A final question. This is much more speculative than the ones that I've asked you so far, and I might ask you to put your ordinary scientific caution aside. If you could design the universe any way that you wanted to, how would you do it?

[Laughs] I don't think that's a fair question.

You don't have to answer it if you don't want to, but . . .

You're giving me the opportunity [laughs]. I really don't think I have any answer.

ANDREI LINDE

Andrei Linde was born in Moscow on March 2, 1948. He received his first diploma in physics from Moscow State University in 1971 and his Ph.D. in physics from the Lebedev Physical Institute in 1974. Since 1983 he has been professor of physics at the Lebedev Physical Institute. In 1978 Linde was awarded the Lomonosov Prize of the Academy of Sciences of the Soviet Union. Linde is married to the physicist Renata Kallosh.

Training in the theory of elementary particle physics, Linde was one of the principal architects of the inflationary universe model, which is an important modification of the standard big bang model. Linde has continued to develop this model and has proposed a variation called chaotic inflation, in which the universe is constantly branching off into regions that have no effect on each other and that may have completely different physical properties—in effect, parallel universes. In 1987 and in 1989 Linde visited several research centers in the United States, lecturing on both the physics of his new cosmological ideas and the broader philosophical implications. Although Linde's work is very mathematical, he describes himself as more intuitive than technical.

How did you get into cosmology? You started as a particle physicist, is that right?

It was a long story. My teacher was [D. A.] Kirzhnits at the Lebedev Physical Institute. First, I came to him as a student. It was at the end of the 1960s, and during this time I studied a theory of weak interactions with him. When I finished my studies as a student, I started postgraduate work with him, and he told me, "Forget about everything. This is a new age, the Weinberg–Salam model is renormalizable, and you should study it. The most interesting thing is that it is very similar to the theory of superconductivity." Kirzhnits understood that symmetry breaking in the Weinberg–Salam model should disappear at high temperatures, similar to the disappearance of superconductivity at high temperatures. They are similar types of theories. He had written the first paper about it.[1] Then he explained it to me, and we started working together, in 1972.

First I did not understand much, being very fresh in this field. Then I started working, and we started publishing papers on this problem.[2] Later, Weinberg, Dolan and Jackiw, and Kirzhnits and I wrote some other papers about symmetry breaking.[3] Symmetry restoration occurs only at extremely large temperatures, and such temperatures can occur only in the early universe. That led me to the investigation of the early universe, and these studies I did by myself. Kirzhnits was interested in elementary particles and superconductivity and other things, and I was very interested in applications of this theory to cosmology. I continued to study the cosmological implications of these things. This was the beginning.

I gather that before 1972 you really hadn't thought very much about cosmology. Is that right?

I think that this was the beginning of my thinking about cosmology seriously. Of course, I was previously interested in such things as cosmology, but when I was just finishing at the Institute, I told my mother, "I am already old. It is too late for me to study gravity" [laughs].

Because you thought there were people already working in it?

Yes, I had studied elementary particle theory, and gravity was something quite different, and so I was afraid. Imagine that you are going to be a postgraduate, and you are choosing your own field of research. At that moment, to start studying new subjects such as gravitational theory after working in weak interaction theory would be too complicated. So I didn't believe that I could do it. But after that, I went into cosmology, and somewhat to my surprise I studied gravity.

When you talk about how it was too late to study gravity, were you thinking of quantum gravity?

I was thinking about everything, and that includes general relativity theory, since actually that theory is rather complicated. It has many branches, and there was a lot of material that had been worked out for many years. People had studied it, and quantum gravity is extremely complicated. I was just lucky that beautiful things were at the surface so I could see them. You see, my mind is not very technical. I work best of all in those places where I can use my intuition.

How do you use images in your work? Do you find images useful or harmful?

Typically, I just use them. Of course, I use mathematics, certainly. But first I usually have a rough idea of how it could work and why, and what is the purpose. Without understanding the purpose of what you are doing, you may try many different ways, and you just solve equations without understanding why it is necessary. For example, you can see how it was that I discovered this new inflation scenario.[4] I was discussing things with many people. The full story is rather complicated. About three years before the paper by [Alan] Guth, I had studied similar problems with one of my colleagues at the Lebedev Institute, and we understood that the vacuum can decay and reheat the universe. We understood that the universe could exponentially expand, and bubbles would collide, and we saw that it would lead to great inhomogeneities in the universe. As a result, we thought these ideas were bad [didn't agree with observations] so there was no reason to publish such garbage. But at that moment we did not realize the main advantages of this exponential expansion, that it could possibly solve the flatness and horizon problems.

I believe that the flatness problem wasn't even stated at that point. My understanding is that the first statement of the flatness problem was by James Peebles and Robert Dicke in 1979. Is that wrong?

Probably so. We did this research in 1978. My colleague, Gennady Chibisov, knew that this problem existed and said to me that we should try to explain why the total entropy of the universe is so large as a result of this phase transition.[5] Why is the total number of photons so large? I did not previously know of the problem. Chibisov lived outside of Moscow, and he was not very rapid, and after all that discussion we did not write anything. But the reason was that we just did not understand the problem correctly. Actually, I wrote about this in my review article in 1979, but our main purpose at that time was not to solve the flatness problem but just to obtain the ratio of photons to protons [baryons].[6] We had observed that it is possible to get this large ratio of 10^8 even if you start from a cold universe filled only with baryons, without photons. So you have a phase transition, you have these bubble collisions, you have reheating, and you can obtain a large ratio. But we saw that these bubble collisions lead to large inhomogeneities. I wrote about two pages on it in my review article, and we never wrote a large article. We just forgot about it. In 1981, when Guth published his paper, some people in Moscow also had written a similar paper. They submitted their paper for publication in *Physics Letters*. But they were a bit too late, and the editors told them that a paper by Guth was already published. Among these people was [V. A.] Rubakov, who is now well-known.

So I easily understood *what* Guth was trying to do. But I did not understand *how* it could be done, since we had seen that the inhomogeneities were large [contrary to actual observations]. I discussed this with people, and I worried about it myself. I really was physically ill at that time. I don't know the reason—maybe this was one of them, since it was very difficult to abandon this simple explanation of many different cosmological problems. I just had the feeling that it was impossible for God not to use such a good possibility to simplify His work, the creation of the universe. I had previously studied a theory of tunneling with bubble formation, and the idea [leading to the "new inflationary model"] came to me from the solution of the corresponding equation by computer. So it was not pure imagination. After that, I could imagine how these bubbles would look when they appear. Then it became clear to me that these bubbles, in some theories, are *not* of that type that had been considered by Guth, not with thin walls. It is not required that inside the bubbles you are already at the minimum of the potential.

> *But you recognized it through these pictures that you formed from the computer calculations?*

There were very many lines of progress. I was simultaneously discussing similar matters with Rubakov, since he had also analyzed this question in the theories I had studied. But what he did not know at that time was that the universe inside the bubble would be exponentially expanding. I discussed this with him by the telephone. I was sitting in my bathroom, since all my children and my wife were already sleeping at the time. I was asking him some technical questions about what he knew about these things, and then I asked him, "But do you know how it works?" I talked to him about these thin bubbles, etc.—all in the bathroom in order not to wake up my children. After that the whole picture had crystallized. I was very excited. I went to my wife, and I woke her up, and I said: "It seems that I know how the universe originated." It was, of course, fantastic.

When was this?

It was in the summer, maybe even in the late spring of 1981. But at that time I did not believe that anybody would easily believe me, since the picture was so strange and the results were so fantastic. Maybe something was incorrect. So I was suffering. I tried to clean the picture up. I tried to find errors. I discussed it with different people, and not everyone believed me. I had serious doubts myself, and I tried to improve the picture and to prove it to myself, and each day I became more and more confident. But still I understood that probably nobody would easily believe me. It was too fantastic.

At this time, what did you think stopped the expansion?

Because of the curvature of the potential, the ball [representing the stored energy in the quantum field] rolls rapidly, then it oscillates and heats the universe, and then inflation stops.

So you knew that the exponential expansion would stop inside the bubble?

Yes, I knew that, and the whole picture had appeared at once. It was not necessary for me to change further any essential features. Though, about half a year or maybe a year later, I improved something connected with the initial stages of the evolution of the bubble. There were some problems connected with the curvature of the universe. So the first picture was rather naive. I understood that it was naive, and I was worried whether the main results were correct. Then I tried to write two large papers in order to prepare people to believe that what I was saying was based on some real calculations. But I had no time to publish them. Then I wrote this short paper and went on vacation and gave it to secretaries to type, since I was absolutely confident that nobody would have the same idea. It was a strange idea at that time. I returned to Moscow and the paper was still not quite ready. In the beginning of October, there was a seminar, a quantum gravity seminar in Moscow, to which [Stephen] Hawking came. At this time, my paper was ready, and I was going to send it to *Physics Letters.* But when I gave a talk at the seminar, Steve raised some objections, so I tried to polish this paper more and more, and actually I sent it for publication only at the end of the month. That is why it appeared in January 1982, not in 1981, which of course would be better [laughs].[7]

You mentioned that when you first read Guth's paper, you thought that there were many good ideas in it.

At first, I did not read Guth's paper; I heard about Guth's paper. I didn't have it before me, but since I had discussed the idea with Rubakov and his collaborators, who had suggested a similar idea, everything was quite clear for me. Lev Okun' from ITEP called me and asked if I had heard anything about Guth's paper explaining the flatness of the universe. I told him, "No, I haven't heard about it, but I know what it's about." And I told him how it works without seeing it.

You said before that it was impossible that God could have missed such a good possibility to simplify his creation. Do you think that the universe was set up by some intelligence? Last night you talked about the possibility of many different universes, all of which would have to obey some laws of physics.[8]

This is a much more complicated question than those which I can confidently . . .

Of course. I understand.

As for myself, I cannot say that I am religious. But I also am not a straightforward materialist who believes that everything is just matter and nothing else. It is also very dangerous to write such things in the press, but in my opinion we have overlooked something very important.

Last night you were talking about the possible role of consciousness.

But it is very complicated. About such things it is probably better to be silent or to say a lot. There are a lot of crazy people who are saying very similar things but without any responsibility. So it is dangerous to say a few words, since, in that case, no one will be sure whether I am speaking the same thing as some crazy person. I am just rather confident that we have lost something important in the way of development of Western science. And what is lost is connected with our behavioristic approach to science, to everything. You see, we have a black box. We have some input and we have some output. What we are interested in is what is the reaction of the black box to the input, what will be the output. Nobody is much interested in what occurs in the black box.

Is our mind the black box?

Yes. I mean, for example, that people typically use the same word for our consciousness, for our feelings, and for the processes that occur in our brains. There are some reactions in our brain, and they say, "This is my anger." You see, they mix reaction and feeling. My red, my blue, my yellow, my bad, and my good—they are my feelings, and they are real to me. They are more real for me than these boxes and this chalk and any things made from metal. I know for sure that I exist, that this is my red, etc. But after further investigation and the development of science, my understanding of my own existence, from which I have started, begins looking like a secondary point. I investigate relations between different objects, which in some sense influence myself. I can study these relations scientifically. I can investigate them and, as a result, very soon all my study, all my science becomes the science connected with matter and with motion of matter and the reactions. How will the computer react when I send such and such information to it, and how will a man react if I send such and such information to him. It becomes unimportant for me how this man *feels* at the moment, if I know all his *reactions*. We often use the word "feels" just in the sense that it helps us describe his reactions. This, in my opinion, goes the wrong way, although it is very productive.

This is the way of science.

Yes. But this way omits something important. I think that in the future development of people's knowledge—I hope that it will be science—this point will not be omitted. But it is a real problem, and this problem has not been seriously considered by science.

I think the scientific community is not ready yet for this.

Absolutely. I was greatly afraid to speak about this yesterday, since any such things expressed before a large auditorium would be taken to mean that you are a little crazy. You are not doing physics, and that is the worst thing you can do.

Do you think that the field of cosmology, more than other fields of physics, has opened up the possibility of the black box playing a role?

I mean that we are already in this black box. We cannot say what will happen to the universe when some input goes to the universe and output goes out. The universe is already in the black box, and we are living in the black box, and our mind is in this separate black box that reacts to the universe. So my interest in cosmology, to a great extent, is connected with my interest in consciousness. I consider cosmology as a model that can be to some extent studied exactly, without making some absolutely metaphysical and unverifiable conjectures.

There are several ways to study such things. One way is to go and do yoga exercises. I don't know. I haven't tried this way, although I understand that maybe it would be necessary sometimes. I am not in a hurry at present. Another possibility is to forget about this problem, and this is the way of most scientists. It just doesn't bother them. This situation is actually very similar to what occurs, for example, if you are not a physicist. If you are just a technician, you can forget about general relativity, and you do not think about space and time very carefully. You just make computers and technology, etc. You should not be bothered with such abstract notions. So many people live and do not know anything about the fact that our space and time is actually space-time, and that it is curved. Nobody is bothered. The same is with physics. Physics is a productive way to do good things, so I cannot say that everything is wrong. However, what is dangerous is to forget some initial points, and that is why philosophy is so important for me. If we forget the starting points, then we are forgetting our roots. And then we will, sometimes, crash.

What philosophers have you read that have been influential for you?

I would say that the main influence on what I am thinking—it may be strange, maybe not—comes from Indian philosophy.

Do you think that your ideas for chaotic inflation and reproducing universes have a root in the Indian archetype of a universe that lives and then dies, expanding and contracting?[9]

It is more complicated. The model of the universe that could be directly inspired by the Indian philosophy is just the model of an oscillating universe, which was known for maybe 50 years, maybe more. The idea that the universe oscillates has a very long history. But, for me, this idea of an oscillating universe became more real after it was combined with my own thoughts about the possibility that the memory of the universe can be washed out by quantum fluctuations of the metric. And these thoughts were also expressed in the Soviet Union by Academician [M. A.] Markov. He is an old, very interesting person who studies gravity, who is the head of the nuclear department of our Academy of Science. He suggested a model of an eternal, oscillating universe, which at each cycle of its evolution forgets what occurred before.

What about the entropy problem?

Yes, entropy. When I heard Markov's ideas for the first time, I thought that maybe he was just an old man, with strange ideas. He did some important work in the past, so he is very respected. Let's forgive him his crazy ideas at present. But my teacher Kirzhnits came and told me, "Be very careful. Markov sometimes says things which sound crazy, and later people understand that they are correct." So I tried to think about it once again. First, it seemed to me that it is impossible to make any sense out of this idea. Then I tried to think about it in a more detailed way, and I suggested a possible realization of this idea under some hypothesis. All these combinations of ideas were together, and they were somewhere at the boundary between philosophy and physics. Thus, the idea of washing out the memory of the universe came to me from Indian philosophy and from Academician Markov. A similar idea is contained in Wheeler's papers, which were written about 15 years ago.[10]

Did you study philosophy in college?

No. To go backwards in time, my interests in physics were deeply related to my interests in philosophy. When I was a schoolboy in Moscow, some woman who was a professor of philosophy came to us to talk. I was in a rather good school, with very good students. I had a lot of discussions with them. Almost all of the boys from our class later went to different Institutes. So it was a rather special class. This woman professor of philosophy came to us, and we tried to ask her questions connected with the origin of knowledge, and so forth. "Why do you believe that matter is primary to our perceptions?" She said so many strange things, and it was quite clear that she just could not answer. It was also quite clear that nobody would listen to what we were thinking. You see, people, when they are growing up, believe the same beliefs they learn when they are young. Then I spoke about these things to a very, very clever man, a professor of physics, who was also very skeptical. Discussions were possible only up to some level. After this level, logic would not help at all. It was something like a wall that nobody can cross. The man surrounds himself by a wall, and he lives inside this safe place, and he does not want to go out. It was so strange an experience for me. I believed that clever people could explain simple things to other clever people.

But you were not asking simple questions.

But actually these questions are not very complicated if you think about them a little. What is complicated is to take consistent steps if you understand something, and not to stop if public opinion says you that you are going in the wrong direction. Public opinion has grown from the experience of our fathers, who were grown on the experience of their fathers, who were peasants. They had no need to think deeply about some problems. What is the first: mind, matter, or something else? Also there was another thing. When I was studying and thinking about philosophy, I created my own reason for what life is, and I suggested some ideas about consciousness and even suggested a very curious theory of telepathic contact. But then it became clear to me that my simple theory of telepathic contact violates Lorentz invariance.

Because the signal's going faster than the speed of light?

Something like that. Therefore, two things became clear to me. First, that my professors of physics—at least those whom I knew—could not at present teach me real philosophy. I must educate myself in some way. Another thing is that even if I educate myself in philosophy, I could still not be prevented from making simple mistakes connected with my absence of understanding of physics. So if you would like to study philosophy, if you would like to study consciousness and mind, you first should guarantee that you will not make simple errors and simple mistakes. That was one of the reasons why I started studying physics. I did not want to make simple mistakes. I wanted to have some basis for studying more fundamental things and not to become crazy.

Did you have any exposure to Indian philosophy in school?

No. At that time, I was mainly thinking about positivists such as Wittgenstein, although I cannot say that I know him very well. The ideas he has expressed are rather dry, but they show you the limited way in which you can be sure of your words, in using your words. Later my own relation to the theory of knowledge and to physics changed somewhat. When I was a student, I tried *ab initio* to investigate most abstract concepts. One of my teachers in mathematics told somebody, "Oh, Linde is very good. He will be a very good physicist. He is a real formalist." This was a compliment in his eyes, that I am a real formalist. If you would say this now to any of my colleagues, they would laugh, since in our department I am probably the least formal man. It proves that when you start formalizing everything without trying to understand in which direction you should go, you sometimes can waste all your life formalizing unnecessary, and maybe just incorrect or at least fruitless, concepts.

We had in the development of physics such a time, when axiomatic field theory first came, and it was believed that everything could be axioma-tized. It was very good mathematics, and it was very well developed, but it was a dry field. A lot of good results still are useful from this area, but most of them are related to theories that are not realized in our world. Everyone believed that the field should satisfy several obvious axioms. Now, all the fields we study do not satisfy these "obvious" axioms. So gradually it became clear—maybe after the influence of my teacher Kirzhnits—that we should not be too formal. It is necessary to use formalism, but if you are just starting from formalism and ending with formalism, then you are in a dangerous position. Only a few people can combine both the intuition and ability to work in a formal way. A good example here is Ed Witten in the United States. He is very bright, of course. He can do very formal things and have a very good understanding as well.

Let me ask you this. Do you think that when we use words like "universe," we understand what those words mean?

If you ask me, do we understand precisely, no, of course we do not. Different people use this word differently. For anyone who tries to under-stand it, at least, the universe is everything which exists. However, this may not be the last word in the interpretation of this subject, since, first of all, the question arises whether we should include consciousness in the definition of everything that exists.

Another question is that it is possible to suggest models in which there are two different times and spaces not intersecting with each other. In some sense, each of them can be described as a universe. So, it will be just as two universes. However, the two universes will affect each other globally. So there are different levels in which one can use this word. That is why I have introduced a concept of a mini-universe. It is more safe and more definite.

Let me just ask one more question. Modern physics has made our intui-tion less and less applicable in some ways. Do you find it equally difficult to think about relativity and quantum theory, or do you find one more familiar to you than the other?

Intuition in general relativity is obviously complicated, and it will remain complicated for many people, since it is another geometry. We are not accustomed to use non-Euclidean geometry, and we must each time think anew. With quantum mechanics, the situation is somewhat different. When we know some basic principles, we can forget for a while about problems of interpretation. We just work and compare our experiments with accelerator data, and it is just an ordinary technical job. However, if we go back to the origins of quantum mechanics, and if we think about the "many-worlds interpretation of quantum mechanics," there are some points that are not clear for me even now. For example, many people just do not want to consider seriously the point of view that our world is split. It is a complicated thing, and also it leads us to a boundary of our knowledge. For myself, I would like to study these questions more deeply. But, for example, when Bryce DeWitt published in *Physics Today* a paper about the many worlds interpretation of quantum mechanics, in 1970 or in 1972, there were about 20 or so letters to *Physics Today*.[11] Some of them were published in *Physics Today,* and each author, from his own point of view, discussed the problem of the interpretation of quantum mechanics. All of them agree that this thing about many universes is just crazy. Maybe it is crazy; maybe it is not crazy enough. On the other hand, the Copenhagen interpretation [of quantum mechanics], which is accepted by most other physicists, seems to suffer from internal inconsistencies. And life becomes even more difficult if one wishes to combine general relativity and quantum mechanics. This is what we are trying to do at present.

Dates and Places of Interviews

Note on Citations and References

Notes

General Reading in Cosmology

References

Glossary

Illustration Credits

DATES AND PLACES OF INTERVIEWS

Marc Davis (University of California, Berkeley): October 14, 1988, Berkeley, California

Gérard De Vaucouleurs (University of Texas, Austin): November 7, 1988, Austin, Texas

Robert Dicke (Princeton University): January 19, 1988, Princeton, New Jersey

Sandra Faber (University of California at Santa Cruz): October 15, 1988, Monte Sereno, California

Margaret Geller (Harvard-Smithsonian Center for Astrophysics): November 23, 1987, Cambridge, Massachusetts

James Gunn (Princeton University): January 18, 1988, Princeton, New Jersey

Alan Guth (Massachusetts Institute of Technology): September 21, 1987, and March 21, 1988, Cambridge, Massachusetts

Stephen Hawking (Cambridge University): January 25, 1989, Cambridge, England

Fred Hoyle (Bournemouth, England): August 15, 1989, Bournemouth, England

John Huchra (Harvard-Smithsonian Center for Astrophysics): December 7, 1987, Cambridge, Massachusetts

Andrei Linde (Lebedev Physical Institute, Moscow): October 22, 1987, Cambridge, Massachusetts

Charles Misner (University of Maryland): April 3, 1989, College Park, Maryland

Jeremiah Ostriker (Princeton University): January 19, 1988, Princeton, New Jersey

Don Page (Pennsylvania State University): May 18, 1988, Andover, Massachusetts

James Peebles (Princeton University): January 19, 1988, Princeton, New Jersey

Roger Penrose (Oxford University): January 24, 1989, Oxford, England

Martin Rees (Cambridge University): March 30, 1988, Tokyo, Japan

Vera Rubin (Carnegie Institute of Washington): April 3, 1989, Washington, D.C.

Allan Sandage (Mt. Wilson Observatory, Pasadena, California): January 11, 1989, Boston, Massachusetts

Wallace Sargent (California Institute of Technology): January 20, 1989, Cambridge, Massachusetts

Maarten Schmidt (California Institute of Technology): March 28, 1988, Tokyo, Japan

David Schramm (University of Chicago): March 29, 1988, Tokyo, Japan

Dennis Sciama (Oxford University): January 25, 1989, Cambridge, England

Joseph Silk (University of California, Berkeley): October 14, 1988, Berkeley, California

Edwin Turner (Princeton University): February 15, 1988, Princeton, New Jersey

Robert Wagoner (Stanford University): October 14, 1988, Stanford, California

Steven Weinberg (University of Texas, Austin): May 6 and May 10, 1988, Cambridge, Massachusetts

NOTE ON CITATIONS
AND REFERENCES

Published works referred to in the Introduction and interviews are briefly cited in the Notes and fully cited in the References at the end of the book. However, books whose title and author are apparent from the interview are cited in the References only. A few works referred to repeatedly in the interviews are also not noted but appear in the References. These are:

Robert Dicke's first printed statement of the flatness problem: Dicke (1969), p. 62.

Robert Dicke and James Peebles' statement of the flatness problem: Dicke and Peebles (1979).

Alan Guth's original paper on the inflationary universe model, sometimes called the "old inflationary model": Guth (1981).

Andrei Linde's "new inflationary model": (Linde 1982).

Andreas Albrecht and Paul Steinhardt's "new inflationary model": Albrecht and Steinhardt (1982).

Valerie de Lapparent, Margaret Geller, and John Huchra's work on the bubble-like distribution of galaxies: de Lapparent, Geller, and Huchra (1986).

H. P. Haynes and R. Giovanelli's work on the filament-like distribution of galaxies: Haynes and Giovanelli (1985).

Steven Weinberg's comment on the pointlessness of the universe: Weinberg (1977), p. 154.

NOTES

Introduction

1. Einstein (1917).
2. Einstein (1915).
3. The astronomer Wilhelm de Sitter's concern about extrapolating from our local observations is evident, for example, in the first paragraphs of de Sitter (1917).
4. Friedmann (1922).
5. Lemaître (1927).
6. Einstein (1922; 1923). In a hand-written draft of this second paper is a crossed-out sentence fragment saying that, to Friedmann's time-dependent solution of the cosmological equations, "a physical significance can hardly be ascribed." (*The Collected Papers of Albert Einstein*, unpublished document 1-026; quoted with permission of the Hebrew University of Jerusalem.)
7. Leavitt (1912).
8. Shapley (1918).
9. Hubble (1924; 1925).
10. Hubble (1929).
11. Slipher, referred to in Eddington (1924), p. 162.
12. The age of the universe cannot be exactly determined by the current rate of expansion of the universe because that rate has not been constant in time. However, the current rate gives a good estimate. To determine the age exactly, both the expansion rate and the change in the expansion rate must be known. The latter is determined by a quantity called omega, to be discussed.
13. For example, Gott, Gunn, Schramm, and Tinsley (1974).
14. For example, the methods of Faber and Jackson (1976) or of Tully and Fisher (1977).
15. Hubble and Humason (1931).
16. For example, Humason, Mayall, and Sandage (1956).
17. If the universe were not expanding, then the size of the observable universe, or the horizon, would be exactly the distance light has traveled since the big bang. However, the universe is expanding. Thus, the *current* distance of an object is somewhat larger than the distance light travels to get from there to here because the object has moved farther away during the voyage of the light ray. This effect is not large. Taking it into account, the horizon is between 1 and 2 times the distance light has traveled since the big bang. If the universe is 10 billion years old, for example, 10 billion light years is still a good estimate for the size of the observable universe.
18. All quantitative estimates in the Introduction assume that the universe is flat and has an expansion rate of 1 per 10 billion years. The estimates would not be substantially changed by other assumptions within the bounds of experimental evidence.
19. Hoyle and Tayler (1964); Zel'dovich (1963; 1965); Peebles (1966); Wagoner Fowler, and Hoyle (1967).

20. Alpher and Herman (1948a; 1948b); Gamow (1948); Alpher, Herman, and Gamow (1948); Dicke, Peebles, Roll, and Wilkinson (1965).
21. Penzias and Wilson (1965).
22. Milne (1933; 1934).
23. A good discussion of this question can be found in Weinberg (1977), chap. 6.
24. Alpher and Herman (1988, p. 26).
25. Tolman (1934, p. 444).
26. Dicke, Peebles, Roll, and Wilkinson (1965, p. 415).
27. Bondi and Gold (1948); Hoyle (1948).
28. Bondi and Gold (1948, p. 254).
29. Schmidt (1963b).
30. Sciama and Rees (1965).
31. The horizon is not precisely equal to the age of the universe multiplied by the speed of light. See note 17.
32. Misner (1969).
33. Dicke (1969, p. 62).
34. Collins and Hawking (1973); Carter (1974). The British cosmologists had discussed the flatness problem before these publications, in the context of the anthropic principle.
35. Dicke and Peebles (1979).
36. Guth (1981).
37. For example, Penrose (1974, 1979a).
38. Charlier (1908, 1922).
39. Shapley (1933).
40. Zwicky (1938).
41. de Vaucouleurs (1953b).
42. de Vaucouleurs (1970).
43. Chincarini and Rood (1975).
44. Tifft and Gregory (1978); Joeveer and Einasto (1978).
45. Gregory and Thompson (1978).
46. Kirshner, Oemler, Schechter, and Shectman (1981).
47. Gregory, Thompson, and Tifft (1981).
48. de Lapparent, Geller, and Huchra (1986).
49. Geller and Huchra (1989).
50. Kirshner, Oemler, Schechter, and Shectman, in progress.
51. Peebles (1980, p. 10).
52. Webster (1976); Peebles (1978).
53. Wolfe (1970); Fabian (1972).
54. Dressler, Faber, Burstein, Davies, Lynden-Bell, Terlevich, and Wegner (1987).
55. Rubin (1951); Rubin, Ford, and Rubin (1973); Rubin, Thonnard, Ford, and Roberts (1976).
56. Zwicky (1933).
57. Ostriker and Peebles (1973).
58. Ostriker, Peebles, and Yahil (1974).
59. Einasto, Kaasik, and Saar (1974).
60. Rubin, Ford, and Thonnard (1978); Bosma (1978).
61. Lemaître (1933).

62. Peebles (1965).

63. Silk (1967).

64. Zel'dovich (1970); Doroshkevich, Ryaberki, and Shandarin (1973); Doroshkevich, Sunyaev, and Zel'dovich (1974).

65. Park (1990). Also, unpublished calculations by Edmund Bertschinger and James Gelb at MIT.

66. Ostriker and Cowie (1981).

67. Doroshkevich, Zel'dovich, and Novikov (1967).

68. Hawking (1982a); Hartle and Hawking (1983); Hawking (1984).

69. Penrose (1965); Hawking and Penrose (1969).

70. Hoyle and Tayler (1964); Wagoner (1967).

71. Shvartsman (1969).

72. Steigman, Schramm, and Gunn (1977).

73. Aarnio et al. (1989); Adeva et al. (1989); Akrawy et al. (1989); Decamp et al. (1989).

74. Abrams et al. (1989).

75. One of the first and simplest grand unified theories, called SU(5), was that of Georgi and Glashow (1974).

76. Dimopoulos and Susskind (1978); Dolgov (1979); Ellis, Gaillard, and Nanopoulos (1979); Ignatiev, Krashikov, Kuzmin, and Tavkhelidze (1978); Toussaint, Trieman, Wilczek, and Zee (1979); Weinberg (1979); Yoshimura (1978).

77. Guth (1981).

78. Brout, Englert, and Spindel (1979); Kazanas (1980); Einhorn and Sato (1981); Starobinsky (1979; 1980).

79. Linde (1983; see also 1987 and references therein).

80. Albrecht and Steinhardt (1982); Linde (1982).

81. For example, Guth and Pi (1982); Hawking (1982b).

82. Dicke (1961).

83. Dirac (1938). Dirac's special combination of fundamental constants is $h^2/(cGm_p^3)$, where h is Planck's constant of quantum physics, c is the speed of light, G is Newton's constant of gravity, and m_p is the mass of the proton. This is numerically equal to about 4×10^{15} seconds, or 10^8 years, which is considered very close to 10^{10} years by cosmological standards.

84. Dirac (1961).

85. Carter, unpublished Cambridge University preprint (1968); first published in Carter (1974).

86. For example, Collins and Hawking (1973); Carter (1974); Carr and Rees (1979); Weinberg (1987).

87. Collins and Hawking (1973).

Fred Hoyle

Extensive revisions were made by the interviewee. Also, several questions were submitted and answered in writing after the interview. These are indicated in the text and below.

1. Robertson (1933).

2. Dirac (1931).

3. Bondi, Gold, and Hoyle (1955). To be exact, this paper proposes giant molecular

clouds to explain recently observed infrared emission and notes that if such clouds were powered by conversion of hydrogen to helium, the resulting helium would roughly agree with observed cosmic ratios of hydrogen to helium. The paper does not explicitly state that the cosmic helium abundance cannot be made in ordinary stars.

4. Hoyle and Tayler (1964); Wagoner, Fowler, and Hoyle (1967).
5. Wagoner, Fowler, and Hoyle (1967).
6. Hoyle and Tayler (1964).
7. The reference to the "temperature" means the temperature of the cosmic background radiation.
8. This question, and Hoyle's answer, were submitted in writing after the interview.
9. Hoyle and Narlikar (1962; 1964a; 1964b; 1964c).
10. See note 8.
11. Hoyle (1989).

Allan Sandage

Extensive revisions throughout the original transcript were made by the interviewee.

1. Grondal (1937).
2. Maxwell (1931).
3. Baade (1952).
4. Arp, Baum, and Sandage (1953); Sandage (1953).
5. Sandage (1961a).
6. Mattig (1958); Mattig (1959).
7. Hubble (1937).
8. Tryon (1973); these ideas were picked up again by Vilenkin (1982).
9. Eggen, Lynden-Bell, and Sandage (1962).
10. ESO-CERN Symposium, Geneva, Switzerland, November 21–25, 1983; proceedings published in Setti and Van Hoire (1984).
11. International School of Astro-Particle Physics and a Unified View of the Macro and Micro Cosmos, organized by A. de Rujula, P. A. Shaver, and D. V. Nanopoulos, January 1987.
12. Zwicky (1933).
13. Meeting of the American Astronomical Society, Boston, January 9–12, 1989.
14. Chincarini and Rood (1975).
15. Gregory and Thompson (1978).
16. In addition to note 14, Tifft and Gregory (1978); Gregory, Thompson, and Tifft (1981).
17. Hubble (1934).
18. Sandage (1970).

Gérard de Vaucouleurs

1. Berget (1925).
2. Lemaître (1931); Lemaître (1950).
3. de Vaucouleurs (1948a; 1948b).
4. de Vaucouleurs (1953a).

5. Shapley (1933).
6. Zwicky (1938).
7. de Vaucouleurs (1953b).
8. Rubin (1951).
9. In addition to the quotes from Walter Baade and Fritz Zwicky denying the reality of large-scale inhomogeneities, there are also quotes on de Vaucouleurs' blackboard on the Hubble constant and one from the French physicist Augustin Fresnel: "Nature does not care for our analytical difficulties" (1826).
10. de Vaucouleurs (1970).
11. de Vaucouleurs (1960).
12. Carpenter (1931; 1938).
13. Charlier (1908).
14. Kiang (1967).
15. Kiang and Saslaw (1969).
16. Kalinkov et al. (1978) and references therein.
17. Bahcall and Burgett (1986).
18. Karachentsev (1967); Ozernoy (1969; 1972; 1975).
19. Fliche and Souriau (1979).
20. Segal (1972; 1976).
21. de Vaucouleurs' response to the last question was augmented after the interview.

Maarten Schmidt

1. Schmidt (1956).
2. Schmidt (1957).
3. Schmidt (1959; 1963a).
4. Minkowski (1960).
5. Schmidt (1965).
6. C. W. Misner described the horizon problem, and tried to solve it, in Misner (1969).
7. Kirshner, Oemler, Schechter, and Shectman (1981).
8. de Vaucouleurs (1970).
9. For example, Dressler, Faber, Burstein, Davies, Lynden-Bell, Terlevich, and Wegner (1987).
10. Dressler (1987).
11. Rubin, Ford, and Rubin (1973).

Wallace Sargent

1. Kahn (1954).
2. Hazlehurst and Sargent (1959).
3. For example, Sargent and Searle (1962).
4. Seyfert (1943).
5. For example, Sargent (1970).
6. Zwicky (1971).
7. Sargent (1968).
8. Readhead, Lawrence, Meyers, Sargent, Hardebeck, and Moffett (1989).

9. Gregory and Thompson (1978).
10. Sargent, Young, Boksenbery, and Tytter (1980).

Dennis Sciama

1. Sciama (1953; 1957).
2. Bondi and Gold (1948).
3. Sciama and Rees (1965).
4. Sciama (1966).
5. Schmidt (1968).
6. Carter (1970; 1973).
7. Carter (1966).
8. Penrose (1965).
9. Hawking (1966); Hawking and Ellis (1968).
10. Friedmann (1922).
11. Hoyle and Tayler (1964); Shvartsman (1969); Steigman, Schramm, and Gunn (1977).
12. Brout, Englert, and Spindel (1979); Starobinsky (1979; 1980); Kazanas (1980); Einhorn and Sato (1981).
13. Sciama may be confusing his earliest recollections of the horizon and flatness problems here. Dicke, in fact, discussed both problems in Dicke (1969) and in Dicke and Peebles (1979).
14. Penrose (1964; 1979a); Hawking (1982a); Hartle and Hawking (1983); Hawking (1984).

Martin Rees

The original transcript of this interview was edited substantially by the interviewee.
1. Penrose (1965).
2. Rees and Sciama (1965a; 1965b); Rees (1966).
3. Sciama and Rees (1965).
4. Rees (1969).
5. Rees (1972).
6. Misner (1968; 1969).
7. Ellis and MacCallum (1969); Ellis and King (1973).
8. Khalatnikov and Lifshitz (1970); Lifshitz and Khalatnikov (1970).
9. Starobinsky (1979; 1980).
10. Brout, Englert, and Spindel (1979).
11. Einhorn and Sato (1981).
12. Longair and Einasto (1978).
13. For example, Doroshkevich, Sunyaev, and Zel'dovich (1975) and references therein.
14. Tinsley (1976).

Robert Wagoner

1. Wagoner (1965; 1966).
2. Kerr (1963).

3. Sciama (1953; 1957).
4. Wagoner, Fowler, and Hoyle (1967).
5. Alpher and Herman (1948a; 1948b); Gamow (1948); Alpher, Herman, and Gamow (1948).
6. Fermi and Turkevich, unpublished, in Alpher and Herman (1950).
7. Alpher, Follin, and Herman (1953).
8. Burbidge, Burbidge, Fowler, and Hoyle (1957).
9. Wagoner, Fowler, and Hoyle (1967).
10. Hoyle and Tayler (1964).
11. Shvartsman (1969).
12. Wagoner (1967).
13. Wagoner (1973).
14. Reeves, Fowler, and Hoyle (1970).
15. Yang, Turner, Steigman, Schramm (1984).
16. Spite and Spite (1981; 1982).
17. Wagoner (1977; 1980).
18. Schneider and Wagoner (1987).

Joseph Silk

1. Silk (1966).
2. For example, Layzer (1968).
3. Silk (1967; 1968a; 1968b).
4. Silk (1967).
5. One of the first statements of the horizon problem was by Misner (1969).
6. For example, Harrison (1967; 1968; 1970); Misner (1968; 1969).
7. For example, Joeveer and Einasto (1978).
8. Matsumoto, Hayakawa, Matsuo, Murakani, Sato, Lange, and Richards (1988).

Robert Dicke

1. Magie (1935).
2. Roll, Krotkov, and Dicke (1964).
3. Brans and Dicke (1961).
4. Dicke, Beringer, Kyhl, and Vane (1946).
5. Dicke, Peebles, Roll, and Wilkinson (1965).
6. Lifshiftz and Khalatnikov (1963a; 1963b).
7. Dicke (1969), p. 62.
8. Dicke (1961).
9. Carter (1968; 1974).

James Peebles

1. Einstein (1917).
2. De Sitter's concern about extrapolating from our local observations is evident in, for example, the first paragraphs of de Sitter (1917).
3. Dicke and Peebles (1965).
4. Dicke, Peebles, Roll, and Wilkinson (1965).

5. Dicke (1961).
6. Bondi and Gold (1948); Hoyle (1948).
7. Lemaître (1927).

Charles Misner

1. Oppenheimer and Synder (1939).
2. Peebles (1967).
3. Misner (1967a; 1967b; 1968; 1969).
4. The following section in brackets was added later by the interviewee in written form.
5. Misner (1969).
6. Misner (1968). The 1968 paper does, however, state that the extreme uniformity of the cosmic background radiation requires a physical explanation and not simply acceptance by *assumption* of homogeneity.
7. Belinsky and Khalatnikov (1969).
8. Doroshkevich and Novikov (1970); Doroshkevich, Lukash, and Novikov (1971).
9. Misner (1969).
10. Belinsky, Khalatnikov, and Lifshitz (1970); see also note 7.
11. Lifshitz and Khalatnikov (1963a; 1963b).
12. Penrose (1965); Hawking and Penrose (1969).
13. Dyson (1979).
14. Hawking (1982a; 1984); Hartle and Hawking (1983).
15. Misner (1977; 1981).
16. Misner (1974).
17. Misner (1984; see also 1977, pp. 97–99; 1981, pp. 66–67).

Jeremiah Ostriker

1. Linde (1983; 1987).
2. Tinsley (1976).
3. Ostriker and Tremaine (1975).
4. Ostriker, Peebles, and Yahil (1974). Ostriker also wrote a paper on massive halos of individual galaxies with Peebles: Ostriker and Peebles (1973).
5. Ostriker and Peebles (1973).
6. Ostriker, Peebles, and Yahil (1974).
7. Zwicky (1933).
8. Schwarzschild (1954).
9. Rubin, Ford, and Thonnard (1978).
10. International Astronomical Union Symposium #100, Besancon, France, August 1982.
11. Ostriker and Cowie (1981).
12. Linde (1983; 1987).

Vera Rubin

1. Alpher (1948); Alpher and Herman (1948a; 1948b); Alpher, Bethe, and Gamow (1948).

2. Rossiter (1982).
3. Rubin (1951a).
4. Humason, Mayall, and Sandage (1956).
5. Rubin (1954).
6. Rubin and Ford (1970; 1971).
7. Burbidge and Burbidge (1975).
8. For example, Rubin, Ford, and Rubin (1973); Rubin, Ford, Thonnard, Roberts, and Graham (1976); Rubin, Ford, Thonnard, and Roberts (1976).
9. Rubin, Ford, and Thonnard (1978; 1980).
10. Ostriker, Peebles, and Yahil (1974); Ostriker and Peebles (1973).
11. For example, Dressler, Faber, Burstein, Davies, Lynden-Bell, Terlevich, and Wegner (1987).
12. Strauss and Davis (1989); Yahil (1989).
13. Dressler, Lynden-Bell, Faber, Burstein, Davies, Wegner, and Terlevich (1987).
14. Rubin, Ford, and Thonnard (1978).

Edwin Turner

The material inside brackets was added later by the interviewee.
1. Sciama (1971).
2. Turner (1976a; 1976b).
3. Page (1960; 1962).
4. Ostriker and Peebles (1973); Ostriker, Peebles, and Yahil (1974).
5. Gott, Gunn, Schramm, and Tinsley (1974).
6. There were several papers in this series. The first was Aarseth, Gott, and Turner (1979).
7. Referred to in Peebles and Groth (1976).
8. Ostriker and Cowie (1981).

Sandra Faber

1. Faber and Jackson (1976).
2. Faber (1973a; 1973b).
3. A series of papers have been published on this project, beginning with Dressler, Lynden-Bell, Faber, Burstein, Davies, Wegner, and Terlevich (1987).
4. Aaronson, Huchra, Mould, Schechter, and Tully (1982).
5. Dressler, Faber, Burstein, Davies, Lynden-Bell, Terlevich, and Wegner (1987).
6. White and Rees (1978).

Marc Davis

1. Peebles (1971).
2. Partridge and Peebles (1967; 1968).
3. Davis and Wilkinson (1974).
4. See, for example, Partridge and Wilkinson (1967).
5. Sandage (1961a; 1961b; 1962).
6. Ostriker and Tremaine (1975); Tinsley (1976).

7. Davis and Peebles (1977).
8. Davis, Geller, and Huchra (1978).
9. Tonry and Davis (1984); Davis, Huchra, Latham, and Tonry (1982); Davis and Huchra (1982); Huchra, Davis, Latham, and Tonry (1983).
10. See Linde (1987) and references therein.
11. Gott, Gunn, Schramm, and Tinsley (1974).
12. da Costa, Pellegrini, Sargent, Tonry, Davis, Meiksin, Latham, Menzies, and Coulson (1988).
13. Dressler, Faber, Burstein, Davies, Lynden-Bell, Terlevich, and Wegner (1987).
14. Aaronson, Huchra, Mould, Schechter, and Tully (1982).
15. Strauss and Davis (1987).

Margaret Geller

1. For example, Misner, Thorne, and Wheeler (1973), chapter 44; see also Wheeler (1977).
2. For example, de Vaucouleurs (1970).
3. Kirshner, Oemler, Schechter, and Shectman (1981).
4. For example, Dressler, Faber, Burstein, Davies, Lynden-Bell, Terlevich, and Wegner (1987).
5. The final question and reply in this interview have been taken from a second interview conducted by R. B. on July 29, 1988.

John Huchra

1. Iben and Huchra (1971).
2. Huchra (1972).
3. Kowal, Sargent, and Huchra (1975); Kowal, Huchra, and Sargent (1976).
4. Huchra (1973).
5. Bond, Green, and Huchra (1974).
6. For example, Turner and Gott (1975).
7. Huchra and Thuan (1977); Huchra and Geller (1982).
8. Tully and Fisher (1977).
9. Aaronson, Huchra, and Mould (1979). There were several more papers in this series; see Aaronson, Mould, Huchra, Sullivan, Schommer, and Bothun (1980).
10. Misner (1969).
11. Huchra, Gorenstein, Kent, Shapiro, Smith, Horine, and Perley (1985).

Stephen Hawking

1. Hawking (1966).
2. Hawking (1982a); Hartle and Hawking (1983); Hawking (1984).
3. For example, Hawking and Penrose (1969).
4. See note 2.
5. Collins and Hawking (1973).
6. Misner recalls that he learned of the flatness problem from Robert Dicke, and he believes this happened about 1969.

7. Carter (1968). This unpublished paper on the anthropic principle did not explicitly raise the flatness problem.
8. *Newsweek*, June 13, 1988.
9. Hawking (1988), p. 132.
10. Hawking (1982b).
11. Linde (1983).

Don Page

1. Fowler (1960).
2. Press and Teukolsky (1972; 1973).
3. Zel'dovich (1971; 1972).
4. Hawking (1974; 1975).
5. Page and Hawking (1976).
6. Page 1976a; 1976b; 1977).
7. Hawking (1982a).
8. Hartle and Hawking (1983).
9. Davies (1983; 1984); Page (1983; 1984).
10. Penrose (1979a).
11. Page (1987); Hawking and Page (1988).

Roger Penrose

1. Penrose (1974; 1977a; 1977b; 1978; 1979a).
2. Penrose (1965); Hawking and Penrose (1969).
3. Hawking and Ellis (1968); Hawking and Penrose (1969).
4. Penrose (1968); Penrose and MacCallum (1973); Penrose and Rindler (1986); Huggett and Tod (1985).
5. Penrose (1972).
6. Penrose (1964).
7. Penrose (1979b); Grunbaum and Shephard (1986); Gardner (1988).
8. Dirac (1938).
9. Dicke (1961).

David Schramm

1. Cosman, Schramm, Enge, Sperduto, and Paris (1967); Cosman, Schramm, and Enge (1968).
2. Wagoner, Fowler, and Hoyle (1967).
3. Schramm and Wasserburg (1970).
4. Steigman, Schramm, and Gunn (1977).
5. Reeves, Audoze, Fowler, and Schramm (1973); Gott, Gunn, Schramm, and Tinsley (1974); Epstein, Arnett, and Schramm (1974); Epstein, Lattimer, and Schramm (1974).
6. Schramm and Arnett (1975a; 1975b).
7. Shvartsman (1969); Peebles (1971).
8. Aubert et al. (1974); Augustin et al. (1974).

9. Herb et al. (1977).
10. Hoyle, referred to as private communication in Dunbar, Wenzel, and Whaling (1953); Hoyle (1954).
11. Lee and Weinberg (1977).
12. Dicus, Kolb, and Teplitz (1977).
13. Cowsik and McClelland (1972).
14. Steigman (1974).
15. Hagedorn (1968); Frautschi (1971); Frautschi, Steigman, and Bahcall (1972).
16. See note 5.
17. Hoyle and Tayler (1964).
18. For example, Fowler, Greenstein, and Hoyle (1961).
19. Georgi and Glashow (1974).
20. Sakharov (1967).
21. Dimopoulos and Susskind (1978); Dolgov (1979); Ellis, Gaillard, and Nanopoulos (1979); Ignatiev, Krashikov, Kuzmin, and Tavkhelidze (1978); Toussaint, Trieman, Wilczek, and Zee (1979); Weinberg (1979); Yoshimura (1978).
22. Kazanas (1980).
23. Kosik, Lyubimov, Novikov, Novik, and Tretyakov (1980); Reines, Sobel, and Pasierb (1980).
24. Guth and Tye (1980).
25. Fry and Schramm (1980).
26. Linde (1983; see also 1987 and references therein).
27. Gott, Gunn, Schramm, and Tinsley (1974).
28. Gott and Gunn (1971).
29. Kirshner, Oemler, Schechter, and Shectman (1981).
30. Koo and Kron (1987).

Steven Weinberg

1. Weinberg (1972).
2. Weinberg (1962a; 1962b).
3. Weinberg (1967). This is Weinberg's work on the electroweak theory, for which he was awarded the Nobel Prize.
4. Salam (1968).
5. Dimopoulos and Susskind (1978); Dolgov (1979); Ellis, Gaillard, and Nanopoulos (1979); Ignatiev, Krashikov, Kuzmin, and Tavkhelidze (1978); Toussaint, Trieman, Wilczek, and Zee (1979); Weinberg (1979); Yoshimura (1978).
6. Guth (1981); Linde (1982); Albrecht and Steinhardt (1982); Linde (1983).
7. Hawking (1988).
8. Linde (1983; see also 1987 and references therein).
9. Preskill (1979); Zel'dovich and Khlopov (1978).

Alan Guth

Alan Guth was interviewed on two different occasions (see interview dates). The text here is based on both interviews, with the order of some of the questions rearranged for a smooth blend.

1. Weinberg (1979).
2. Preskill (1979).
3. Guth and Tye (1980).
4. Everett (1957); Wheeler (1957).
5. The "Seven Samurai" refer to the astronomers D. Burstein, R. L. Davies, A. Dressler, S. M. Faber, D. Lynden-Bell, R. J. Terlevich, and G. Wegner. See for example Lynden-Bell, Faber, Burstein, Davies, Dressler, Terlevich, and Wegner (1988).

Andrei Linde

Linde's interview was conducted before our standard interview format was designed and therefore does not follow that format.

1. Kirzhnits (1972).
2. Kirzhnits and Linde (1972).
3. Weinberg (1974); Dolan and Jackiw (1974); Kirzhnits and Linde (1974).
4. Linde (1982).
5. The large value of the total entropy in the observable universe is closely related to the flatness problem.
6. Linde (1979).
7. Linde (1982).
8. A. D. Linde, "Philosophical Implications of New Cosmology," Loeb Lecture, Harvard University, October 21, 1987.
9. Linde (1983; 1987).
10. Wheeler (1964; 1968); Misner, Thorne, and Wheeler (1973, chap. 44); Wheeler (1977).
11. DeWitt (1970).

GENERAL READING IN COSMOLOGY

Bartusiak, Marcia. 1986. *Thursday's Universe: A Report from the Frontier on the Origin, Nature, and Destiny of the Universe*. New York: Times Books. A good summary of the recent history, ideas, and personalities of cosmology, including the inflationary universe.

Cornell, James, ed. 1988. *Bumps, Voids, and Bubbles in Time: The New Cosmology*. Cambridge: Cambridge University Press. An up-to-date collection of articles on modern cosmology, written by contemporary cosmologists for a popular audience, with articles by Margaret Geller, James Gunn, Alan Guth, Robert Kirshner, Alan Lightman, and Vera Rubin.

Ferris, Timothy. 1979. *The Red Limit: The Search for the Edge of the Universe*. New York: Bantam. A literate account of the history and personalities of modern cosmology, up to the mid-1970s.

Hawking, Stephen. 1988. *A Brief History of Time*. New York: Bantam. A discussion of recent modifications to the big bang model, including the inflationary universe and quantum cosmology.

Munitz, M. K., ed. 1957. *Theories of the Universe*. New York: Free Press. A unique collection of historical papers in cosmology, including selections from Aristotle, Copernicus, Newton, Kant, Einstein, Hubble, and others.

North, J. D. 1965. *The Measure of the Universe*. Oxford: Clarendon Press. An excellent review of the modern history of cosmology, going up to beginning of the twentieth century.

Silk, Joseph. 1988. *The Big Bang*. San Francisco: W. H. Freeman. The big bang model plus a good discussion of many of the astrophysical issues of cosmology, such as dark matter and galaxy formation, with many references to current observational work.

Tucker, Wallace, and Karen Tucker. 1988. *The Dark Matter: Contemporary Science's Quest for the Mass Hidden in Our Universe*. New York: Morrow. A thorough discussion of the observational evidence for dark matter.

Weinberg, Steven. 1977. *The First Three Minutes*. New York: Basic. A classic introduction to the standard big bang model, with emphasis on the early universe, but without treatment of the most recent ideas from particle physics.

REFERENCES

Aarnio, P., et al. 1989. Measurement of the Mass and Width of the $Z°$ Particle from Multihadronic Final States Produced in e^+e^- Annihilations. *Physics Letters B* 231:539.

Aaronson, M., J. Huchra, and J. Mould. 1979. The Infrared Luminosity/HI Velocity Width Relation and Its Application to the Distance Scale. *Astrophysical Journal* 229:1.

Aaronson, M., J. Huchra, J. Mould, P. J. Schechter, and R. B. Tully. 1982. The Velocity Field in the Local Supercluster. *Astrophysical Journal* 258:64.

Aaronson, M., J. Mould, J. Huchra, W. Sullivan, R. Schommer, and G. Bothun. 1980. The Distance Scale from the Infrared Magnitude/HI Velocity-Width Relation. III. The Expansion Rate Outside the Local Supercluster. *Astrophysical Journal* 239:12.

Aarseth, S. J., J. R. Gott, and E. L. Turner. 1979. N-Body Simulations of Galaxy Clustering. I. Initial Conditions and Galaxy Collapse Times. *Astrophysical Journal* 228:664.

Abrams, G. S., et al. 1989. Searches for New Quarks and Leptons Produced in Z-Boson Decay. *Physical Review Letters* 63:2447.

Adeva, B., et al. 1989. A Determination of the Properties of the Neutral Intermediate Vector Boson $Z°$. *Physics Letters B* 231:509.

Akrawy, M. Z., et al. 1989. Measurement of the $Z°$ Mass and Width with the Opal Detector at LEP. *Physics Letters B* 231:530.

Albrecht, A., and P. J. Steinhardt. 1982. Cosmology for Grand Unified Theories with Radiatively Induced Symmetry Breaking. *Physical Review Letters* 48:1220.

Alpher, R. A. 1948. A Neutron-Capture Theory of the Formation and Relative Abundance of the Elements. *Physical Review* 74:1577.

Alpher, R. A., H. A. Bethe, and G. Gamow. 1948. The Origin of Chemical Elements. *Physical Review* 73:803.

Alpher, R. A., J. W. Follin, Jr., and R. C. Herman. 1953. Physical Conditions in the Initial Stages of the Expanding Universe. *Physical Review D* 92:1347.

Alpher, R., and R. C. Herman. 1948a. Evolution of the Universe. *Nature* 162:774.

——— 1948b. On the Relative Abundance of the Elements. *Physical Review D* 74:1737.

——— 1950. Theory of the Origin and Relative Abundance Distribution of the Elements. *Reviews of Modern Physics* 22:153.

——— 1988. Reflections of Early Work on Big Bang Cosmology. *Physics Today* no. 8, 41:26.

Alpher, R., R. C. Herman, G. A. Gamow. 1948. Thermonuclear Reactions in the Expanding Universe. *Physical Review D* 74:1198.

Arp, H. C., W. A. Baum, and A. R. Sandage. 1953. The Color–Magnitude Diagram of the Globular Cluster M 92. *Astronomical Journal* 58:4.

Aubert, J. J., et al. 1974. Experimental Observation of a Heavy Particle J. *Physical Review Letters* 33:1404.

Augustin, J. E., et al. 1974. Discovery of a Narrow Resonance in e^+e^- Annihilation. *Physical Review Letters* 33:1406.

Baade, W. A. 1952. Commission on Extragalactic Nebulae. *Transactions of the International Astronomical Union* 8:397.

Bahcall, N. A., and W. S. Burgett. 1986. Are Superclusters Correlated on a Very Large Scale? *Astrophysical Journal Letters* 300:L35.

Baker, R. H. 1964. *Astronomy*. Princeton: Van Nostrand. 8th edition.

Belinsky, V. A., and I. M. Khalatnikov. 1969. On the Nature of the Singularities in the General Solution of the Gravitational Equations. *Zh. Eksp. Teor. Fiz.* 56:1701. English trans. *Soviet Physics-JETP* 29:911.

Belinsky, V. A., I. M. Khalatnikov, and E. M. Lifshitz. 1970. Oscillatory Approach to a Singular Point in the Relativistic Cosmology. *Usp. Fiz. Nauk* 102:463. English trans. *Advances in Physics* 19:525.

Berget, P. 1925. *Le Ciel*. Paris: Larousse.

Bond, H. E., R. F. Green, and J. P. Huchra. 1974. AU Leonis: An Extragalactic Object in the General Catalogue of Variable Stars. *Publications of the Astronomical Society of the Pacific* 86:688.

Bondi, H. 1952. *Cosmology*. Cambridge: Cambridge University Press.

Bondi, H., and T. Gold. 1948. The Steady-State Theory of the Expanding Universe. *Monthly Notices of the Royal Astronomical Society* 108:252.

Bondi, H., T. Gold, and F. Hoyle. 1955. Black Giant Stars. *The Observatory* 75:80.

Bosma, A. 1978. The Distribution and Kinematics of Neutral Hydrogen in Spiral Galaxis of Various Morphological Types. Ph.D. thesis, University of Groningen.

Brans, C., and R. H. Dicke. 1961. Mach's Principle and a Relativistic Theory of Gravitation. *Physical Review* 124: 925.

Brout, R., F. Englert, and P. Spindel. 1979. Cosmological Origin of the Grand-Unification Mass Scale. *Physical Review Letters* 43:417.

Burbidge, E. M., and G. R. Burbidge. 1975. In *Galaxies and the Universe*, ed. A. Sandage, M. Sandage, and J. Kristian. Chicago: University of Chicago Press.

Burbidge, E. M., G. R. Burbidge, W. A. Fowler, and F. Hoyle. 1957. Synthesis of Elements in Stars. *Reviews of Modern Physics* 29:547.

Carpenter, E. F. 1931. A Cluster of Extra-Galactic Nebulae in Cancer. *Publications of the Astronomical Society of the Pacific* 43:247.

―――― 1938. Some Characteristics of Associated Galaxies. I. A Density Restriction in the Metagalaxy. *Astrophysical Journal* 88:344.

Carr, B. J., and M. J. Rees. 1979. The Anthropic Principle and the Structure of the Physical World. *Nature* 278:605.

Carter, B. 1966. Complete Analytic Extension of the Symmetry Axis of Kerr's Solution of Einstein's Equations. *Physical Review D* 141:1242.

―――― 1968. Large Numbers in Astrophysics and Cosmology. Unpublished preprint. Institute of Theoretical Astronomy, Cambridge University.

―――― 1970. An Axisymmetric Black Hole has Only Two Degrees of Freedom. *Physical Review Letters* 26:331.

―――― 1973. Elastic Perturbation Theory in General Relativity and a Variational Principle for a Rotating Star. *Communications of Mathematical Physics* 30:261.

―――― 1974. Large Number Coincidences and the Anthropic Principle in Cosmology. In *Confrontation of Cosmological Theories with Observational Data*, IAU Symposium 63, ed. M. S. Longair. Dordrecht: Reidel.

Chandrasekhar, S. 1939. *An Introduction to Stellar Structure*. Chicago: University of Chicago.

Charlier, C. V. L. 1908. The Planetary Rotation Problem. *Ark. Mathematics, Astronomy, and Physics* 4:1.

——— 1923. System of an Infinite Universe. *Ark. Mathematics, Astronomy, and Physics* 16:1.

Chincarini, G., and H. J. Rood. 1975. Empirical Properties of the Mass Discrepancy in Groups and Clusters of Galaxies. IV. Double Compact Galaxies. *Nature* 257:294.

Collins, C. B., and S. W. Hawking. 1973. Why Is the Universe Isotropic? *Astrophysical Journal* 180:317.

Cornell, J., ed. 1989. *Bumps, Voids, and Bubbles in Time: The New Cosmology*. Cambridge: Cambridge University Press.

Cosman, E. R., D. N. Schramm, and H. A. Enge. 1968. A Study of ^{61}Ni. *Nuclear Physics* A109:305.

Cosman, E. R., D. N. Schramm, H. A. Enge, A. Sperduto, and C. H. Paris. 1967. Nuclear-Reaction Studies in the Nickel Isotopes. *Physical Review* 163:1134.

Couderc, P. 1930. *L'Architecture de l'Univers*. Paris: Gauthier Villars.

——— 1952. *The Expansion of the Universe*. New York: Macmillan.

Cowsik, R., and J. McClelland. 1972. An Upper Limit to the Neutrino Rest Mass. *Physical Review Letters* 29:669.

da Costa, L. N., P. S. Pellegrini, W. L. W. Sargent, J. Tonry, M. Davis, A. Meiksin, D. W. Latham, J. W. Menzies, and I. A. Coulson. 1988. The Southern Sky Redshift Survey. *Astrophysical Journal* 327:544.

Davies, P. C. W. 1983. Inflation and Time Asymmetry in the Universe. *Nature* 301:398.

——— 1984. Inflation in the Universe and Time Asymmetry. *Nature* 312:524.

Davis, M., M. Geller, and J. Huchra. 1978. The Local Mean Mass Density of the Universe: New Methods for Studying Galaxy Clustering. *Astrophysical Journal* 221:1.

Davis, M., and J. Huchra. 1982. A Survey of Galaxy Redshifts. III. The Density Field and Induced Gravity Field. *Astrophysical Journal* 254:437.

Davis, M., J. Huchra, D. W. Latham, and J. Tonry. 1982. A Survey of Galaxy Redshifts. II. The Large Scale Space Distribution. *Astrophysical Journal* 250:423.

Davis, M., and P. J. E. Peebles. 1977. On the Integration of the BBGKY Equations for the Development of Strongly Non-Linear Clustering in an Expanding Universe. *Astrophysical Journal Supplement* 34:425.

Davis, M., and D. T. Wilkinson. 1974. Search for Primeval Galaxies. *Astrophysical Journal* 192:251.

Decamp, D., et al. 1989. Determination of the Number of Light Neutrino Species. *Physics Letters B* 231:519.

de Lapparent, V., M. J. Geller, and J. P. Huchra. 1986. A Slice of the Universe. *Astrophysical Journal Letters* 302:L1.

de Sitter, W. 1917. On Einstein's Theory of Gravitation and Its Astronomical Consequences. *Monthly Notices of the Royal Astronomical Society* 78:3.

de Vaucouleurs, G. 1946. *Eléments Théoriques et Pratiques de Photographie Scientifique.* Paris: Revue d'Optique.

———— 1947. *La Science de la Photographie.* Paris: Elzevir.

———— 1948a. Sur la Loi de Distribution de la Luminosité dans les Nébuleuses elliptiques et leur Structure. *Comptes-Rendus de l'Académie des Sciences, Paris,* 227:586.

———— 1948b. Recherches sur les Nébuleuses extragalactiques. *Annales d'Astrophysique* 11:247.

———— 1953a. A Revision of the Harvard Survey of Bright Galaxies. Australian National University Mimeogram, Canberra.

———— 1953b. Evidence for a Local Supergalaxy. *Astronomical Journal* 58:30.

———— 1958. *La Photographie Astronomique du Daguerreotype au Télescope Electronique.* Paris: Albin Michel.

———— 1960. The Apparent Density of Matter in Groups and Clusters of Galaxies. *Astrophysical Journal* 131:585.

———— 1970. The Case for Hierarchical Clustering. *Science* 167:1203.

de Vaucouleurs, G., J. Dragesco, P. Selme, H. Faraggi, and H. Tellez-Plasencia. 1956. *Manuel de Photographie Scientifique.* Paris: Revue d'Optique.

de Vaucouleurs, G., and J. Texereau. 1954. *L'Astrophotographie d'Amateur.* Paris: Revue d'Optique.

DeWitt, B. S. 1970. Quantum Mechanics and Reality. *Physics Today* no. 9, 23:30.

Dicke, R. H. 1961. Dirac's Cosmology and Mach's Principle. *Nature* 192:440.

———— 1969. *Gravitation and the Universe: The Jayne Lectures for 1969.* American Philosophical Society.

Dicke, R. H., R. Beringer, R. L. Kyhl, and A. B. Vane. 1946. Atmospheric Absorption Measurements with a Microwave Radiometer. *Physical Review* 70:340.

Dicke, R. H., and P. J. E. Peebles. 1965. Gravitation and Space Science. *Space Science Reviews* 4:419.

———— 1979. The Big Bang Cosmology—Enigmas and Nostrums. In *General Relativity: An Einstein Centenary Survey,* ed. S. W. Hawking and W. Israel. Cambridge: Cambridge University Press.

Dicke, R. H., P. J. E. Peebles, P. G. Roll, and D. T. Wilkinson. 1965. Cosmic Blackbody Radiation. *Astrophysical Journal* 142:414.

Dicus, D. A., E. W. Kolb, and V. L. Teplitz. 1977. Cosmological Upper Bound on Heavy Neutrino Lifetimes. *Physical Review Letters* 39:168.

Dimopoulos, S., and L. Susskind. 1978. Baryon Number of the Universe. *Physical Review D* 18:4500.

Dirac, P. A. M. 1931. Quantized Singularities in the Electromagnetic Field. *Proceedings of the Royal Society* A133:60.

———— 1935. *The Principles of Quantum Mechanics.* Oxford: Clarendon Press.

———— 1938. New Basis for Cosmology. *Proceedings of the Royal Astronomical Society of London* A165:199.

———— 1961. Reply to Dicke. *Nature* 192:441.

Dolan, L., and R. Jackiw. 1974. Symmetry Behavior at Finite Temperature. *Physical Review D* 9:3320.

Dolgov, A. D. 1979. Baryon Asymmetry of the Universe and Thermodynamical Equilibrium Disturbance. *Zh. Eksp. Teor. Pis'ma Red.* 29:254.

Doroshkevich, A. G., V. N. Lukash, and I. D. Novikov. 1971. Impossibility of Mixing in the Bianchi Type IX Cosmological Model. *Zh. Eksp. Theor. Fiz.* 60:1201. English trans. in *Soviet Physics-JETP* 33:649.

Doroshkevich, A. G., and I. D. Novikov. 1970. Mixmaster Universes and the Cosmological Problem. *Astron. Zh.* 47:948. English trans. 1971. *Soviet Astronomy-AJ* 14:763.

Doroshkevich, A. G., V. S. Ryabenki, and S. F. Shandarin. 1973. Nonlinear Theory of Development of Potential Perturbations. *Astrofizika* 9:257.

Doroshkevich, A. G., R. A. Sunyaev, and Ya. B. Zel'dovich. 1974. The Formation of Galaxies in Friedmannian Universes. In *Confrontation of Cosmological Theories with Observational Data*, IAU Symposium 63, ed. M. S. Longair. Dordrecht: Reidel.

Doroshkevich, A. G., Ya. B. Zel'dovich, and I. D. Novikov. 1967. The Origin of Galaxies in an Expanding Universe. *Soviet Astronomy-AJ* 11:233.

Dressler, A. 1987. The Large-Scale Streaming of Galaxies. *Scientific American* 257:38.

Dressler, A., S. M. Faber, D. Burstein, R. L. Davies, D. Lynden-Bell, R. J. Terlevich, and G. Wegner. 1987. Spectroscopy and Photometry of Elliptical Galaxies: A Large-Scale Streaming Motion in the Local Universe. *Astrophysical Journal Letters* 313:L37.

Dressler, A., D. Lynden-Bell, S. Faber, D. Burstein, R. Davies, G. Wegner, and R. Terlevich. 1987. Spectroscopy and Photometry of Elliptical Galaxies. I. New Distance Estimator. *Astrophysical Journal* 313:42.

Dunbar, R. E. P., W. A. Wenzel, and W. Whaling. 1953. The 7.68 MeV State of C^{12}. *Physical Review D* 92:649.

Dyson, F. J. 1979. Time without End: Physics and Biology in an Open Universe. *Reviews of Modern Physics* 51:447.

Eddington, A. S. 1924. *The Mathematical Theory of Relativity.* Cambridge: Cambridge University Press.

—— 1926. *The Internal Constitution of the Stars.* Cambridge: Cambridge University Press.

—— 1928. *Nature of the Physical World.* New York: Macmillan.

—— 1933. *The Expanding Universe.* New York: MacMillan.

Eggen, O. J., D. Lynden-Bell, and A. R. Sandage. 1962. Evidence from the Motions of Old Stars that the Galaxy Collapsed. *Astrophysical Journal* 136:748.

Einasto, J., A. Kaasik, and E. Saar. 1974. Dynamic Evidence on Massive Coronas in Galaxies. *Nature* 250:309.

Einhorn, M. B., and K. Sato. 1981. Monopole Production in the Very Early Universe in a First-Order Phase Transition. *Nuclear Physics* B180:385.

Einstein, A. 1915. The General Theory of Relativity. *Sitzungsberichte der Preussischen Akad. d. Wissenschaften* 778.

—— 1917. Cosmological Considerations on the General Theory of Relativity. *Sitzungsberichte der Preussischen Akad. d. Wissenschaften* 142. English trans. 1952. *The Principle of Relativity*, ed. W. Perrett and G. B. Jeffery. New York: Dover.

—— 1922. Remark on the Paper by A. Friedmann, 'On the Curvature of the World.' *Zeitschrift für Physik* 11:326.

——— 1923. Note on the Paper by A. Friedmann, 'On the Curvature of the World.' *Zeitschrift für Physik* 16:228.

Ellis, G. F. R., and A. R. King. 1973. Tilted Homogeneous Cosmological Models. *Communications in Mathematical Physics* 31:209.

Ellis, G. F. R., and M. A. H. MacCallum. 1969. A Class of Homogeneous Cosmological Models. *Communications in Mathematical Physics* 12:108.

Ellis, J., M. K. Gaillard, and D. V. Nanopoulous. 1979. Baryon Number Generation in Grand Unified Theories. *Physics Letters B* 80:360.

Epstein, R. I., W. D. Arnett, and D. N. Schramm. 1974. Can Supernovae Produce Deuterium? *Astrophysical Journal Letters* 190:L13.

Epstein, R. I., J. M. Lattimer, and D. N. Schramm. 1974. The Origin of Deuterium. *Nature* 263:198.

Everett, H., III. 1957. 'Relative State' Formulation of Quantum Mechanics. *Reviews of Modern Physics* 29:454.

Faber, S. M. 1973a. Variations in Spectral-Energy Distributions and Absorption-Line Strengths among Elliptical Galaxies. *Astrophysical Journal* 179:731.

——— 1973b. Ten-Color Intermediate-Band Photometry of Stars. *Astronomy and Astrophysics Supplement* 10:201.

Faber, S. M., and R. E. Jackson. 1976. Velocity Dispersions and Mass-to-Light Ratios for Elliptical Galaxies. *Astrophysical Journal* 204:668.

Fabian, A. C. 1972. Analysis of X-ray Background Fluctuations. *Nature* 237:19.

Ferris, T. 1979. *The Red Limit: The Search for the Edge of the Universe.* New York: Bantam.

Fliche, H. H., and J. M. Souriau. 1979. Quasars and Cosmology. *Astronomy and Astrophysics* 78:87.

Fowler, W. A. 1960. Origin of the Elements. *Saturday Evening Post* 232:40.

Fowler, W. A., J. Greenstein, and F. Hoyle. 1961. Deuteronomy: Synthesis of Deuterons and the Light Nuclei during the Early History of the Solar System. *American Journal of Physics* 29:393.

Frautschi, S. 1971. A Statistical Bootstrap Model for Hadrons. *Physical Review D* 3:2821.

Frautschi, S., G. Steigman, and J. Bahcall. 1972. Quarks as a Thermometer for Cosmology. *Astrophysical Journal* 175:307.

Friedmann, A. 1922. On the Curvature of the World. *Zeitschrift für Physik* 10:377.

Fry, J.N., and D. N. Schramm. 1980. Unification, Monopoles, and Cosmology. *Physical Review Letters* 44:1361.

Gamow, G. 1940a. *Birth and Death of the Sun.* New York: Viking.

——— 1940b. *Mr. Tompkins in Wonderland.* Cambridge: Cambridge University Press.

——— 1944. *Mr. Tompkins Explores the Atom.* Cambridge: Cambridge University Press.

——— 1947. *One, Two, Three . . . Infinity.* New York: Viking.

——— 1948. The Evolution of the Universe. *Nature* 162:680.

——— 1952. *The Creation of the Universe.* New York: Viking.

Gardner, M. 1988. *Penrose Tiles to Trapdoor Ciphers.* New York: W. H. Freeman.

Geller, M. J., and J. P. Huchra. 1989. Mapping the Universe. *Science.* 246:897.

Georgi, H., and S. L. Glashow. 1974. Unity of All Elementary Particle Forces. *Physical Review Letters* 32:438.

Gott, J. R., and J. E. Gunn. 1971. The Coma Cluster as an X-ray Source: Some Cosmological Implications. *Astrophysical Journal* 169:113.

Gott, J. R., III, J. E. Gunn, D. N. Schramm, and B. M. Tinsley. 1974. An Unbound Universe? *Astrophysical Journal* 194:543.

Gregory, S. A., and L. A. Thompson. 1978. The Coma/A 1367 Supercluster and Its Environs. *Astrophysical Journal* 222:784.

Gregory, S. A., L. A. Thompson, and W. G. Tifft. 1981. The Perseus Supercluster. *Astrophysical Journal* 243:411.

Grondal, F. A. 1937. *The Romance of Astronomy.* New York: Macmillan.

Grunbaum, B., and G. C. Shephard. 1986. *Tilings and Patterns.* New York: W. H. Freeman.

Guth, A. 1981. Inflationary Universe: A Possible Solution to the Horizon and Flatness Problems. *Physical Review D* 23:347.

Guth, A. H., and S-Y Pi. 1982. Fluctuations in the New Inflationary Universe. *Physical Review Letters* 49:1110.

Guth, A., and S.-H. H. Tye. 1980. Phase Transitions and Monopole Production in the Very Early Universe. *Physical Review Letters* 44:631.

Hagedorn, R. 1968. Hadronic Matter Near the Boiling Point. *Nuovo Cimento* 56A:1027.

Hardy, G. H. 1938. *A Mathematician's Apology.* Cambridge: Cambridge University Press.

Harrison, E. R. 1967. Matter, Antimatter, and the Origin of Galaxies. *Physical Review Letters* 18:1011.

———— 1968. Baryon Inhomogeneities in the Early Universe. *Physical Review D* 167:1170.

———— 1970. Galaxy Formation in the Early Universe. *Monthly Notices of the Royal Astronomical Society* 148:119.

Hartle, J. B., and S. W. Hawking. 1983. Wave Function for the Universe. *Physical Review D* 28:2960.

Hawking, S. W. 1966. The Occurrence of Singularities in Cosmology. *Proceedings of the Royal Society of London A* 294:511.

———— 1974. Black Hole Explosions? *Nature* 248:30.

———— 1975. Particle Creation by Black Holes. *Communications of Mathematical Physics* 43:199.

———— 1982a. The Boundary Conditions of the Universe. *Pontificae Academiae Scientarium Scripta Varia* 48:563.

———— 1982b. The Development of Singularities in a Single Bubble Inflationary Universe. *Physics Letters* 115B:295.

———— 1984. The Quantum State of the Universe. *Nuclear Physics B* 239:257.

———— 1988. *A Brief History of Time.* New York: Bantam.

Hawking, S. W., and G. R. F. Ellis. 1968. The Cosmic Blackbody Radiation and the Existence of Singularities in Our Universe. *Astrophysical Journal* 152:25.

Hawking, S. W., and D. N. Page. 1988. How Probable Is Inflation? *Nuclear Physics B* 298:789.

Hawking, S. W., and R. Penrose. 1969. The Singularities of Gravitational Collapse and Cosmology. *Proceedings of the Royal Society of London* A314:529.

Haynes, H. P., and R. Giovanelli. 1985. A 21 Centimeter Survey of the Perseus-Pisces

Supercluster. I. The Declination Zone +27.5 to 33.5 degrees. *Astronomical Journal* 90:2445.

Hazlehurst, J., and W. L. W. Sargent. 1959. Hydrodynamics in a Radiation Field—A Covariant Treatment. *Astrophysical Journal* 130:276.

Heckmann, O. 1942. *Theorien der Kosmologie.* Berlin.

Herb, S. W. et al. 1977. Observation of a Dimuon Resonance at 9.5 GeV in 400 GeV Proton-Nucleus Collisions. *Physical Review Letters* 39:252.

Hoyle, F. 1948. A New Model for the Expanding Universe. *Monthly Notices of the Royal Astronomical Society* 108:372.

——— 1950. *The Nature of the Universe.* New York: Harper.

——— 1955. *Frontiers in Astronomy.* London: Heinemann.

——— 1962. *Astronomy.* Garden City, N.J.: Doubleday.

——— 1965. *Galaxies, Nuclei, and Quasars.* New York: Harper.

——— 1966. *Man in the Universe.* New York: Columbia University Press.

——— 1989. The Steady-State Theory Revived? *Comments on Astrophysics* 13:81.

Hoyle, F., and J. V. Narlikar. 1962. Mach's Principle and the Creation of Matter. *Proceedings of the Royal Society* A270:334.

——— 1964a. Time Symmetric Electrodynamics and the Arrow of Time in Cosmology. *Proceedings of the Royal Society* A277:1.

——— 1964b. The C-Field as a Direct Particle Field. *Proceedings of the Royal Society* A282:178.

——— 1964c. On the Gravitational Influence of Direct Particle Fields. *Proceedings of the Royal Society* A282:184.

Hoyle, F., and R. J. Tayler. 1964. The Mystery of the Cosmic Helium Abundance. *Nature* 203:1108.

Hubble, E. P. 1924. *Annual Report of the Mount Wilson Observatory.*

——— 1925. N.G.C. 6822, a Remote Stellar System. *Astrophysical Journal* 62:409.

——— 1929. A Relation between Distance and Radial Velocity among Extra-Galactic Nebulae. *Proceedings of the National Academy of Science* 15:168.

——— 1934. The Distribution of Extra-Galactic Nebulae. *Astrophysical Journal* 79:8.

——— 1935. *The Realm of the Nebulae.* Oxford: Oxford University Press.

——— 1937. *The Observational Approach to Cosmology.* Oxford: Oxford University Press.

Hubble, E. P., and M. L. Humason. 1931. The Velocity–Distance Relation among Extra Galactic Nebulae. *Astrophysical Journal* 74:43.

Huchra, J. 1972. An Analysis of the Magnetic Field of 53 Camelopardais and Its Implications for the Decentered Dipole Rotator Model. *Astrophysical Journal* 174:435.

——— 1973. Comet Huchra 1973h" International Astronomical Union Circular #2533.

Huchra, J., M. Davis, D. Latham, and J. Tonry. 1983. A Survey of Galaxy Redshifts. IV. The Data. *Astrophysical Journal Supplement* 52:89.

Huchra, J., and M. Geller. 1982. Groups of Galaxies. I. Nearby Groups. *Astrophysical Journal* 257:423.

Huchra, J., and T. X. Thuan. 1977. Isolated Galaxies. *Astrophysical Journal* 216:694.

Huchra, J., M. Gorenstein, S. Kent, I. Shapiro, G. Smith, E. Horine, and R. Perley.

1985. 2237+0305: A New and Unusual Gravitational Lens. *Astronomical Journal* 90:691.

Huggett, S. A., and K. P. Tod. 1985. *An Introduction to Twistor Theory*. Cambridge: Cambridge University Press.

Humason, M. L., N. U. Mayall, and A. R. Sandage. 1956. Redshifts and Magnitudes of Extra-Galactic Nebulae. *Astronomical Journal* 61:97.

Iben, I., and J. Huchra. 1971. Comments on the Instability Strip for Halo Population Variables. *Astronomy and Astrophysics* 14:293.

Ignatiev, A. Yu, N. V. Krashikov, V. A. Kuzmin, and A. N. Tavkhelidze. 1978. Universal CP-Noncovariant Superweak Interaction and Baryon Asymmetry of the Universe. *Physics Letters B* 76:436.

Jeans, J. 1928. *Astronomy and Cosmogony*. Cambridge: Cambridge University Press.

—— 1929. *The Universe around Us*. New York: Macmillan.

—— 1954. *The Stars and Their Courses*. Cambridge: Cambridge University Press.

—— 1932. *The Mysterious Universe*. Cambridge: Cambridge University Press.

Joeveer, M., and J. Einasto. 1978. Has the Universe a Cell Structure? In *The Large Scale Structure of the Universe*, IAU Symposium 79, ed. M. S. Longair and J. Einasto. Dordrecht: Reidel.

Kac, M. 1985. *Enigmas of Chance: An Autobiography*. New York: Harper and Row.

Kahn, F. D. 1954. The Acceleration of Interstellar Clouds. *Bulletin of the Astronomical Institute of the Netherlands* 12:189.

Kalinkov et al. 1978. Superclustering of Galaxies. *IAU Symposium* no. 79:276. Dordrecht: Reidel.

Karachentsev, I. D. 1967. *Communications from the Byurakan Observatory* 39:95.

Kazanas, D. 1980. Dynamics of the Universe and Spontaneous Symmetry Breaking. *Astrophysical Journal Letters* 241:L95.

Kerr, R. P. 1963. Gravitational Field of a Spinning Mass as an Example of Algebraically Special Metrics. *Physical Review Letters* 11:237.

Khalatnikov, I. M., and E. M. Lifshitz. 1970. General Cosmological Solution of the Gravitational Equations with a Singularity in Time. *Physical Review Letters* 24:76.

Kiang, T. 1967. On the Clustering of Rich Clusters of Galaxies. *Monthly Notices of the Royal Astronomical Society* 135:1.

Kiang, T., and W. C. Saslaw. 1969. The Distribution in Space of Clusters of Galaxies. *Monthly Notices of the Royal Astronomical Society* 143:129.

Kirshner, R. P., A. Oemler, Jr., P. L. Schechter, and S. A. Shectman. 1981. A Million Cubic Megaparsec Void in Bootes? *Astrophysical Journal Letters* 248:L57.

Kirzhnits, D. A. 1972. Weinberg Model and the 'Hot' Universe. *Soviet Physics-JETP Letters* 15:519.

Kirzhnits, D. A., and A. D. Linde. 1972. Macroscopic Consequences of the Weinberg Model. *Physics Letters* B42:471.

—— 1974. Relativistic Phase Transitions. *Zh. Eksp. Teor. Fiz.* 67:1263. English trans. 1975. *Soviet Physics-JETP* 40:628.

Koo, D. C., and R. G. Kron. 1987. Evolution of Very Faint Field Galaxies and Quasars. In *Observational Cosmology*, IAU Symposium 124, ed. A. Hewitt, G. Burbidge, and L. Z. Fang. Dordrecht: Reidel.

Korzybski, A. 1958. *Science and Sanity: An Introduction to Non-Aristotelian Systems and*

General Semantics. Lakeville, Conn.: International Non-Aristotelian Library Pub. Co.

Kosik, V. S., V. A. Lyubimov, E. G. Novikov, V. S. Novik, and E. F. Tretyakov. 1981. An Estimate of the ν_e Mass from the β Spectrum of Tritium in the Valine Molecule. *Yad. Fiz.* 32:301. English trans. *Physics Letters B* 94:266.

Kowal, C. T., J. Huchra, and W. L. W. Sargent. 1976. The 1975 Palomar Supernovae Search. *Publications of the Astronomical Society of the Pacific* 88:521.

Kowal, C. T., W. L. W. Sargent, and J. Huchra. 1975. The 1974 Palomar Supernovae Search. *Publications of the Astronomical Society of the Pacific* 87:401.

Layzer, D. 1968. Black-Body Radiation in a Cold Universe. *Astrophysics Letters* 1:99.

Leavitt, H. S. 1912. Harvard Circular No. 173. Reprinted in 1966 *Source Book in Astronomy,* ed. H. Shapley. Cambridge: Harvard University Press.

Lee, B. W., and S. Weinberg. 1977. Cosmological Lower Bound on Heavy Neutrino Masses. *Physical Review Letters* 39:165.

Lemaître, G. 1927. A Homogeneous Universe of Constant Mass and Increasing Radius Accounting for the Radial Velocity of Extra-Galactic Nebulae. *Annals of the Scientific Society of Brussels* 47A:49. English trans. 1931. *Monthly Notices of the Royal Astronomical Society* 91:483.

——— 1931. *Nature (Supplement)* 128:704.

——— 1933. Spherical Condensations in the Expanding Universe. *Comptes Rendus de l'Academie des Sciences, Paris* 196:903.

——— 1950. *The Primeval Atom.* New York: D. Van Nostrand.

Ley, W. 1949. *The Conquest of Space.* New York: Viking.

——— 1951. *Rockets, Missiles, and Space Travel.* New York: Viking.

Lifshitz, E. M., and I. M. Khalatnikov. 1963a. Investigations of Relativistic Cosmology. *Advances in Physics* 12:185.

——— 1963b. Problems of Relativistic Cosmology. *Usp. Fiz. Nauk* 80:391.

——— 1970. Oscillatory Approach to a Singular Point in the Open Cosmological Model. *Soviet Physics-JETP Letters* 11:123.

Linde, A. D. 1979. Phase Transitions in Gauge Theories and Cosmology. *Reports on the Progress of Physics* 42:389.

——— 1982. A New Inflationary Universe Scenario: A Possible Solution of the Horizon, Flatness, Homogeneity, Isotropy, and Primordial Monople Problems. *Physics Letters B* 108:389.

——— 1983. Chaotic Inflation. *Physics Letters B* 129:177.

——— 1987. Particle Physics and Inflationary Cosmology. *Physics Today* no. 9 40:61.

Longair, M. S., and J. Einasto, eds. *The Large Scale Structure of the Universe,* IAU Symposium 79. Reidel: Dordrecht.

Lynden-Bell, D., S. M. Faber, D. Burstein, R. L. Davies, A. Dressler, R. J. Terlevich, and G. Wegner. 1988. Spectroscopy and photometry of Elliptical Galaxies. V. Galaxy Streaming toward the New Supergalactic Center. *Astrophysical Journal* 326:19.

Lyttleton, R. A. 1956. *The Modern Universe.* New York: Harper Brothers.

Magie, W. F. 1935. *A Source Book in Physics.* New York: McGraw Hill.

Matsumoto, P., S. Hayakawa, M. Matsuo, H. Murakani, S. Sato, A. E. Lange, and P. L. Richards. 1988. Submillimeter Spectrum of the Cosmic Background Radiation. *Astrophysical Journal* 329:567.

Mattig, W. 1958. On the Connection between Redshift and Apparent Brightness. *Astronomische Nachrichten* 284:109.

—— 1959. On the Connection between Number of Extragalactic Objects and Their Apparent Brightness. *Astronomische Nachrichten* 285:1.

Maxwell, R. W. 1931. *Stars for Sam*, ed. C. E. St. John. New York: Harcourt, Brace.

Millikan, R. A. 1935. *Electrons (+ and −), Protons, Photons, Neutrons, and Cosmic Rays*. Chicago: University of Chicago Press.

Milne, E. A. 1933. World-Structure and the Expansion of the Universe. *Zeitschrift fur Astrophysik* 6:1.

—— 1934. A Newtonian Expanding Universe. *Quarterly Journal of Mathematics (Oxford)* 5:64.

Minkowski, R. 1960. A New Distant Cluster of Galaxies. *Astrophysical Journal* 132:908.

Misner, C. W. 1967a. Neutrino Viscosity and the Isotropy of Primordial Blackbody Radiation. *Physical Review Letters* 19:533.

—— 1967b. Transport Processes in the Primordial Fireball. *Nature* 214:40.

—— 1968. The Isotropy of the Universe. *Astrophysical Journal* 151:431.

—— 1969. The Mixmaster Universe. *Physical Review Letters* 22:1071.

—— 1974. Some Topics for Philosophical Inquiry Concerning the Theories of Mathematical Geometrodynamics and of Physical Geometrodynamics. *Proceedings of the 1972 Biennial Meeting of the Philosophy of Science Association*. Dordrecht: Reidel.

—— 1977. Cosmology and Theology. In *Cosmology, History, and Theology*, eds. W. Yourgrau and A. D. Breck. New York: Plenum.

—— 1981. Infinity in Physics and Cosmology. In *Infinity*, ed. D. O. Dahlstrom, D. T. Ozar, and L. Sweeney. Washington: American Catholic Philosophical Association.

—— 1984. The Only Reliable Theories Are Fallen Theories. Conference on Foundation Problems of Physics, University of Texas, Austin.

Misner, C. W., K. S. Thorne, and J. A. Wheeler. 1973. *Gravitation*. San Francisco: W. H. Freeman.

Munitz, M. K., ed. 1957. *Theories of the Universe*. New York: Free Press.

North, J. D. 1965. *The Measure of the Universe*. Oxford: Clarendon.

Oppenheimer, J. R., and H. Synder. 1939. On Continued Gravitational Collapse. *Physical Review D* 56:455.

Ostriker, J. P., and L. L. Cowie. 1981. Galaxy Formation in an Intergalactic Medium Dominated by Explosions. *Astrophysical Journal Letters* 243:L127.

Ostriker, J. P., and P. J. E. Peebles. 1973. A Numerical Study of the Stability of Flattened Galaxies: Or, Can Cold Galaxies Survive? *Astrophysical Journal* 186:467.

Ostriker, J. P., P. J. E. Peebles, and A. Yahil. 1974. The Size and Mass of Galaxies and the Mass of the Universe. *Astrophysical Journal Letters* 193:L1.

Ostriker, J. P., and S. D. Tremaine. 1975. Another Evolutionary Correction of the Luminosity of Giant Galaxies. *Astrophysical Journal Letters* 202:L113.

Ozernoy, L. 1969. Traces of Photon Eddies. *Soviet Physics-JETP Letters* 10:251.

—— 1972. Dynamical Parameters of Galaxy Clusters as a Consequence of Cosmological Turbulence. *Soviet Astronomy* 15:923.

———— 1975. Regularities in Systems of Galaxies and the Relation to the Problem of Hidden Mass. *Soviet Physics Uspekhi* 18:260.

Page, D. N. 1976a. Particle Emission Rates from a Black Hole: Massless Particles from an Uncharged, Nonrotating Hole. *Physical Review D* 13:198.

————. 1976b. Particle Emission Rates from a Black Hole. II. Massive Particles from a Rotating Hole. *Physical Review D* 14:3260.

———— 1977. Particle Emission Rates from a Black Hole: Charged Leptons from a Nonrotating Hole. *Physical Review D* 16:2402.

———— 1983. Inflation Does Not Explain Time Asymmetry. *Nature* 304:5921.

————. 1984. Can Inflation Explain the Second Law of Thermodynamics? *International Journal of Theoretical Physics* 23:725.

———— 1987. Probability of R^2 Inflation. *Physical Review D* 36:1607.

Page, D. N., and S. W. Hawking. 1976. Gamma Rays from Primordial Black Holes. *Astrophysical Journal* 206:1.

Page, T. 1960. Average Masses and Mass–Luminosity Ratios of the Double Galaxies. *Astrophysical Journal* 132:9110.

———— 1962. M/L for Double Galaxies, A Correction. *Astrophysical Journal* 136:685.

Park, C. 1990. Large N-Body Simulations of a Universe Dominated by Cold Dark Matter. *Monthly Notices of the Royal Astronomical Society* 242:59p.

Partridge, R. B., and P. J. E. Peebles. 1967. Are Young Galaxies Visible? *Astrophysical Journal* 147:868.

———— 1968. Are Young Galaxies Visible II? *Astrophysical Journal* 148:377.

Partridge, R. B., and D. T. Wilkinson. 1967. Isotropy and Homogeneity of the Universe from Measurements of the Cosmic Microwave Background. *Physical Review Letters* 18:557.

Peebles, P. J. E. 1965. The Black-Body Radiation Content of the Universe and the Formation of Galaxies. *Astrophysical Journal* 142:1317.

———— 1966. Primordial He Abundance and Fireball II. *Astrophysical Journal* 146:542.

———— 1967. Microwave Radiation from the Big Bang. In *Relativity Theory and Astrophysics. I. Relativity and Cosmology*, ed. J. Ehlers. American Mathematics Society.

———— 1971. *Physical Cosmology*. Princeton: Princeton University Press.

———— 1978. Stability of a Hierarchical Clustering Pattern in the Distribution of Galaxies. *Astronomy and Astrophysics* 68:345.

———— 1980. *The Large Scale Structure of the Universe*. Princeton: Princeton University Press.

Peebles, P. J. E., and E. J. Groth. 1976. An Integral Constraint on the Evolution of the Galaxy Two-Point Correlation Function. *Astronomy and Astrophysics* 53:131.

Penrose, R. 1964. Conformal Treatments of Infinity. In *Relativity, Groups, and Topology*, ed. C. DeWitt and B. S. DeWitt. New York: Gordon and Breach.

———— 1965. Gravitational Collapse and Space-Time Singularities. *Physical Review Letters* 14:57.

———— 1968. Twistor Quantization and the Curvature of Spacetime. *International Journal of Theoretical Physics* 1:61.

———— 1972. On the Nature of Quantum Gravity. In *Magic without Magic: John Archibald Wheeler*, ed. J. Klauder. San Francisco: W. H. Freeman.

———— 1974. Singularities in Cosmology. In *Confrontation of Cosmological Theories with Observational Data.* IAU Symposium 63, ed. M. S. Longair. Dordrecht: Reidel.

———— 1977a. Spacetime Singularities. In *Proceedings of the First Marcel Grossman Meeting on General Relativity, TCTP Trieste,* ed. R. Ruffini. Amsterdam: North Holland.

———— 1977b. Aspects of General Relativity. In *Physics and Contemporary Needs,* ed. Riazuddin. New York: Plenum.

———— 1978. Singularities of Spacetime. In *Theoretical Principles in Astrophysics and Relativity,* ed. N. R. Lebovitz, W. H. Reid, and P. O. Vandervoort. Chicago: University of Chicago.

———— 1979a. Singularities and Time-Asymmetry. In *General Relativity: An Einstein Centenary Survey,* ed. S. W. Hawking and W. Israel. Cambridge: Cambridge University Press.

———— 1979b. Pentaplexity: A Class of Nonperiodic Tilings of the Plane. *Mathematical Intelligencer* 2:32.

Penrose, R., and M. A. H. MacCallum. 1973. Twistor Theory: An Approach to the Quantization of Fields and Spacetime. *Physics Reports* 6C 4:2421.

Penrose, R., and W. Rindler. 1986. *Spinors and Space-Time,* vol. 2. Cambridge: Cambridge University Press.

Penzias, A. A., and R. W. Wilson. 1965. A Measurement of Excess Antenna Temperature at 4080 Mc/s. *Astrophysical Journal* 142:419.

Preskill, J. P. 1979. Cosmological Production of Superheavy Magnetic Monopoles. *Physical Review Letters* 43:1365.

Press, W. H., and S. A. Teukolsky. 1972. Floating Orbits, Superradiant Scattering, and the Black-Hole Bomb. *Nature* 238:211.

———— 1973. Perturbations of a Rotating Black Hole. II. Dynamical Stability of the Kerr Metric. *Astrophysical Journal* 185:649.

Readhead, A. C. S., C. R. Lawrence, S. T. Meyers, W. L. W. Sargent, H. E. Hardebeck, and A. T. Moffet. 1989. A Limit on the Anisotropy of the Microwave Background Radiation on Arc Minute Scales. *Astrophysical Journal* 346:566.

Rees, M. J. 1966. The Appearance of Relativistically Expanding Radio Sources. *Nature* 211:468.

———— 1969. The Collapse of the Universe: An Eschatological Study. *The Observatory* 89:193.

———— 1972. Origin of the Cosmic Microwave Background Radiation in a Chaotic Universe. *Physical Review Letters* 28:1669.

Rees, M. J., and D. W. Sciama. 1965a. The Kinetic Temperature and Ionization Level of Intergalactic Hydrogen in a Steady State Universe. *Astrophysical Journal* 145:6.

———— 1965b. Structure of the Quasi-Stellar Radio Source 3C 273 B. *Nature* 208:371.

Reeves, H., J. Audoze, W. A. Fowler, and D. N. Schramm. 1973. On the Origin of Light Elements. *Astrophysical Journal* 179:909.

Reeves, H., W. A. Fowler, and F. Hoyle. 1970. Galactic Cosmic Ray Origin of Li, Be, B in Stars. *Nature* 226:727.

Reines, F., H. W. Sobel, and E. Pasierb. 1980. Evidence for Neutrino Instability. *Physical Review Letters* 45:1307.

Robertson, H. P. 1933. Relativistic Cosmology. *Reviews of Modern Physics* 5:62.

Roll, P. G., R. Krotkov, and R. H. Dicke. 1964. The Equivalence of Inertial and Passive Gravitational Mass. *Annals of Physics* 26:442.

Rossiter, M. W. 1982. *Women Scientists in America: Struggles and Strategies to 1940.* Baltimore: The Johns Hopkins University Press.

Rubin, V. C. 1951. Differential Rotation of the Inner Metagalaxy. *Astronomical Journal* 56:47.

———— 1954. Fluctuations in the Space Distribution of the Galaxies. *Proceedings of the National Academy of Sciences* 40:541.

Rubin, V. C., and W. K. Ford, Jr. 1970. Rotation of the Andromeda Nebula from a Spectroscopic Survey of Emission Regions. *Astrophysical Journal* 159:379.

———— 1971. Radial Velocities and Line Strengths of Emissions Lines across the Nuclear Disk of M31. *Astrophysical Journal* 170:25.

Rubin, V. C., W. K. Ford, Jr., and J. S. Rubin. 1973. A Curious Distribution of Radial Velocities of ScI Galaxies with $14.0 \leq m \leq 15.0$. *Astrophysical Journal Letters* 183:L111.

Rubin, V. C., W. K. Ford, Jr., and N. Thonnard. 1978. Extended Rotation Curves of High-Luminosity Spiral Galaxies. IV. Systematic Dynamical Properties. *Astrophysical Journal Letters* 225:L107.

———— 1980. Rotational Properties of 21 Sc Galaxies with a Large Range of Luminosities and Radii, from NGC 4605 (R = 4 kpc) to UGC 2885 (R = 122 kpc). *Astrophysical Journal* 238:471.

Rubin, V. C., W. K. Ford, Jr., N. Thonnard, M. S. Roberts, and J. A. Graham. 1976. Motion of the Galaxy and the Local Group Determined from the Velocity Anisotropy of Distant ScI Galaxies. I. The Data. *Astronomical Journal* 81:687.

Rubin, V. C., N. Thonnard, W. K. Ford, Jr., and M. S. Roberts. 1976. Motion of the Galaxy and the Local Group Determined from the Velocity Anisotropy of Distant Sc I Galaxies. II. The Analysis for the Motion. *Astronomical Journal* 81:719.

Russell, B. 1958. *The ABC of Relativity,* rev. Felix Pirani. 1925, rpt. London: G. Allen & Unwin.

Sagan, C. 1985. *Contact.* New York: Simon and Schuster.

Sakharov, A. D. 1967. Violation of CP Invariance, C Asymmetry, and Baryon Asymmetry of the Universe. *Zh. Eksp. Teor. Fiz. Pis'ma Red.* 5:32. English trans. in *Soviet Physics-JETP Letters* 5:24.

Salam, A. 1968. *Elementary Particles,* ed. N. Svartholm. Stockholm: Almquist and Wiksels.

Sandage, A. R. 1953. The Color–Magnitude Diagram for the Globular Cluster M3. *Astronomical Journal* 58:61.

———— 1961a. Ability of the 200-inch Telescope to Discriminate between World Models. *Astrophysical Journal* 133:355.

———— 1961b. The Light Travel Times and the Evolutionary Correction to Magnitudes of Distant Galaxies. *Astrophysical Journal* 134:916.

———— 1962. The Change of Redshift and Apparent Luminosity of Galaxies due to the Deceleration of Selected Expanding Universes. Appendix by G. C. McVittie. *Astrophysical Journal* 136:319.

———— 1970. Cosmology: A Search for Two Numbers. *Physics Today* 23:34.

Sargent, W. L. W. 1968. New Observations of Compact Galaxies. *Astronomical Journal* 73:893.

———— 1970. The Spectroscopic Survey of Compact and Peculiar Galaxies. *Astrophysical Journal* 160:405.

Sargent, W. L. W., and L. Searle. 1962. Studies of the Peculiar A Stars. I. The Oxygen-Abundance Anomaly. *Astrophysical Journal* 136:408.

Sargent, W. L. W., P. J. Young, A. Boksenbery, and D. Tytter. 1980. The Distribution of Lyman-Alpha Absorption Lines in the Spectra of Six QSO's: Evidence for an Intergalactic Origin. *Astrophysical Journal Supplement* 42:41.

Schmidt, M. 1956. A Model of the Distribution of Mass in the Galactic System. *Bulletin of the Astronomical Institutes of the Netherlands* 13:15.

———— 1957. The Distribution of Mass in M31. *Bulletin of the Astronomical Institutes of the Netherlands* 14:17.

———— 1959. The Rate of Star Formation. *Astrophysical Journal* 129:243.

———— 1963a. The Rate of Star Formation. II. The Rate of Formation of Stars of Different Mass. *Astrophysical Journal* 137:758.

———— 1963b. 3C 273: A Star-like Object with Large Redshift. *Nature* 197:1040.

———— 1965. Optical Spectra and Redshifts of 31 Radio Galaxies. *Astrophysical Journal* 141:1.

———— 1968. Space Distribution and Luminosity Functions of Quasi Stellar Radio Sources. *Astrophysical Journal* 151:393.

Schneider, P., and R. V. Wagoner. 1987. Amplification and Polarization of Supernovae by Gravitational Lensing. *Astrophysical Journal* 314:154.

Schramm, D. N., and W. D. Arnett. 1975a. Neutral Currents and Supernovae. *Physical Review Letters* 34:113.

———— 1975b. The Weak Interaction and Gravitational Collapse. *Astrophysical Journal* 198:629.

Schramm, D. N., and G. J. Wasserburg. 1970. Nucleochronologies and the Mean Age of the Elements. *Astrophysical Journal* 162:57.

Schwarzschild, M. 1954. Mass Distribution and Mass–Luminosity Ratio in Galaxies. *Astronomical Journal* 59:273.

———— 1958. *The Structure and Evolution of the Stars.* Princeton: Princeton University Press.

Schwarzschild, M., B. Schwarzschild, L. Searle, and A. Meltzer. 1957. A Spectroscopic Comparison between High and Low Velocity K Giants. *Astrophysical Journal* 125:123.

Sciama, D. W. 1953. On the Origin of Inertia. *Monthly Notices of the Royal Astronomical Society* 113:34.

———— 1957. Inertia. *Scientific American* no. 2, 196:99.

———— 1959. *The Unity of the Universe.* Garden City: Doubleday.

———— 1966. On the Origin of the Microwave Background. *Nature* 211:277.

———— 1969. *The Physical Foundations of General Relativity.* Garden City: Doubleday.

———— *Modern Cosmology.* Cambridge: Cambridge University Press.

Sciama, D. W., and M. J. Rees. 1965. Cosmological Significance of the Relation between Redshift and Flux Density for Quasars. *Nature* 211:1283.

Segal, I. E. 1972. Covariant Chronogeometry and Extreme Distances. *Astronomy and Astrophysics* 18:143.

———— 1976. *Mathematical Cosmology and Extragalactic Astronomy.* New York: Academic Press.

Setti, G., and L. Van Hove, eds. 1984. *First ESO-CERN Symposium: Large Scale Structure of the Universe, Cosmology, and Fundamental Physics.* Geneva: ESO-CERN.

Seyfert, C. K. 1943. Nuclear Emission in Spiral Nebulae. *Astrophysical Journal* 97:28.

Shapley, H. 1918. Studies Based on the Colors and Magnitudes in Stellar Clusters. VI. On the Determination of the Distances of Globular Clusters. *Astrophysical Journal* 48:89.

——— 1933. Note on the Distribution of Remote Galaxies and Faint Stars. *Harvard Bulletin* 890:1.

——— 1943. *Galaxies.* Philadelphia: Blakiston.

Shvartsman, V. F. 1969. Density of Relic Particles with Zero Rest Mass in the Universe. *Soviet Physics-JETP Letters* 9:184.

Silk, J. 1966. Local Irregularities in a Gödel Universe. *Astrophysical Journal* 143:689.

——— 1967. Fluctuations in the Primordial Fireball. *Nature* 215:115.

——— 1968a. Cosmic Blackbody Radiation and Galaxy Formation. *Astrophysical Journal* 151:459.

——— 1968b. When Were Galaxies and Galaxy Clusters Formed? *Nature* 218:453.

——— 1988. *The Big Bang.* San Francisco: W. H. Freeman.

Spite, M., and F. Spite. 1981. Lithium Abundance in Two Halo Stars. *C.R. Seances Academy of Science*, Serial II (France) 293:299.

——— 1982. Lithium Abundance at the Formation of the Galaxy. *Nature* 297:483.

Starobinsky, A. 1979. Spectrum of Relict Gravitational Radiation and the Early State of the Universe. *Soviet Physics-JETP Letters* 30:682.

——— 1980. A New Type of Isotropic Cosmological Models without Singularity. *Physics Letters* 91B:99.

Steigman, G. 1974. Confrontation of Antimatter Cosmologies with Observational Data. In *Confrontation of Cosmological Theories with Observational Data*, IAU Symposium 63, ed. M. S. Longair. Dordrecht: Reidel.

Steigman, G., D. N. Schramm, and J. E. Gunn. 1977. Cosmological Limits to the Number of Massive Leptons. *Physics Letters* 66B:202.

Strauss, M. A., and M. Davis. 1989. The Peculiar Velocity Field Predicted by the Distribution of IRAS Galaxies. In *Large Scale Motions in the Universe*, ed. V. Rubin and Coyne. Princeton: Princeton University Press.

Tifft, W. G., and S. A. Gregory. 1978. Observations of the Large-Scale Distribution of Galaxies. In *The Large Scale Structure of the Universe*, IAU Symposium 79, ed. M. S. Longair and J. Einasto. Dordrecht: Reidel.

Tinsley, B. 1976. Effects of Main Sequence Brightening on the Luminosity Evolution of Elliptical Galaxies. *Astrophysical Journal* 203:63.

Tolman, R. C. 1934. *Relativity, Thermodynamics, and Cosmology.* Oxford: Clarendon Press.

Tonry, J., and M. Davis. 1984. A Survey of Galaxy Redshifts. I. Data Reduction Techniques. *Astronomical Journal* 84:1511.

Toussaint, D., S. B. Trieman, F. Wilczek, and A. Zee. 1979. Matter–Antimatter Accounting, Thermodynamics, and Black-Hole Radiation. *Physical Review D* 19:1036.

Tryon, E. P. 1973. Is the Universe a Vacuum Fluctuation? *Nature* 246:396.

Tully, R. B., and J. P. Fisher. 1977. A New Method of Determining Distances to Galaxies. *Astronomy and Astrophysics* 54:661.

Turner, E. L. 1976a. Binary Galaxies. I. A Well Defined Statistical Sample. *Astrophysical Journal* 208:20.

——— 1976b. Binary Galaxies. II. Dynamics and Mass-to-Light Ratio. *Astrophysical Journal* 208:304.

Turner, E. L., and J. R. Gott. 1975. Evidence for a Spatially Homogeneous Component of the Universe: Single Galaxies. *Astrophysical Journal Letters* 197:L87.

Vilenkin, A. 1982. Creation of Universes from Nothing. *Physics Letters B* 117:25.

Wagoner, R. V. 1965. Rotation and Gravitational Collapse. *Physical Review D* 138:B1583.

——— 1966. Possible Mechanism for Producing Strong Radio Sources. *Physical Review Letters* 16:249.

——— 1967. Cosmological Element Production. *Science* 155:1369.

——— 1973. Big Bang Nucleosynthesis Revisited. *Astrophysical Journal* 179:343.

——— 1977. Determining q_o from Supernovae. *Astrophysical Journal Letters* 214:L5.

——— 1980. *Physical Cosmology*, ed. R. Balian, J. Audouze, and D. N. Schramm. Amsterdam: North-Holland.

Wagoner, R. V., W. A. Fowler, and F. Hoyle. 1967. On the Synthesis of Elements at Very High Temperatures. *Astrophysical Journal* 148:3.

Watson, J. D. 1968. *The Double Helix*. New York: Atheneum.

Webster, A. 1976. The Clustering of Radio Sources. II. The 4C, GB, and MC1 Surveys. *Monthly Notices of the Royal Astronomical Society* 175:71.

Weinberg, S. 1962a. The Neutrino Problem in Cosmology. *Nuovo Cimento*, series X, 25:15.

——— 1962b. Universal Neutrino Degeneracy. *Physical Review* 128:1457.

——— 1967. A Model of Leptons. *Physical Review Letters* 19:1264.

——— 1972. *Gravitation and Cosmology*. New York: John Wiley.

——— 1974. Gauge and Global Symmetries at High Temperature. *Physical Review D* 9:3357.

——— 1977. *The First Three Minutes*. New York: Basic Books.

——— 1979. Cosmological Production of Baryons. *Physical Review Letters* 42:850.

——— 1987. Anthropic Bounds on the Cosmological Constant. *Physical Review Letters* 59:2607.

Wheeler, J. A. 1957. Assessment of Everett's 'Relative State' Formulation of Quantum Theory. *Reviews of Modern Physics* 29:463.

——— 1964. Geometrodynamics and the Issue of the Final State. In *Relativity, Groups, and Topology*, ed. C. DeWitt and B. S. DeWitt. New York: Gordon and Breach.

——— 1968. Superspace and the Nature of Quantum Geometrodynamics. In *Battelle Rencontres: 1967 Lectures in Mathematics and Physics*, ed. C. DeWitt and J. A. Wheeler. New York: Benjamin.

——— 1977. Genesis and Observership. In *Foundational Problems in the Special Sciences*, ed. R. Butts and J. Hintikka. Dordrecht: Reidel.

White, S. D. M., and M. J. Rees. 1978. Core Condensation in Heavy Haloes: A Two-Stage Theory for Galaxy Formation and Clustering. *Monthly Notices of the Royal Astronomical Society* 183:341.

Wolfe, A. M. 1970. New Limits on the Shear and Rotation of the Universe from the X-ray Background. *Astrophysical Journal Letters* 159:L61.

Yahil, A. 1989. The Structure of the Universe to 10,000 km s^{-1} as Determined by IRAS Galaxies. In *Large Scale Motions in the Universe,* ed. V. Rubin and Coyne. Princeton: Princeton University Press.

Yang, J., M. S. Turner, G. Steigman, D. N. Schramm, and K. Olive. 1984. Primordial Nucleosynthesis: A Critical Comparison of Theory and Observation. *Astrophysical Journal* 281:493.

Yoshimura, M. 1978. Unified Gauge Theories and the Baryon Number of the Universe. *Physical Review Letters* 41:281.

Zel'dovich, Ya. B. 1963. The Theory of the Expanding Universe as Originated by A. Friedmann. *Usp. Fig. Nauk* 80:357. English trans. 1964. *Soviety Physics-Uspekhi* 6:475.

———— B. 1965. Survey of Modern Cosmology. *Advances in Astronomy and Astrophysics,* vol. 3. New York: Academic Press.

———— 1970. Gravitational Instability: An Approximate Theory for Large Density Perturbations. *Astronomy and Astrophysics* 5:84.

———— 1971. Generation of Waves by a Rotating Body. *Zh. Eksp. Teor. Fiz. Pis. Red.* 14:270. English trans. *Soviet Physics-JETP Letters* 14:180.

———— 1972. Amplification of Cylindrical Electromagnetic Waves Reflected from a Rotating Body. *Zh. Eksp. Teor. Fiz.* 62:2076. English trans. *Soviet Physics-JETP* 35:1085.

Zel'dovich, Ya. B., and M. Yu. Khlopov. 1978. On the Concentration of Relic Magnetic Monopoles in the Universe. *Physics Letters B* 79:239.

Zwicky, F. 1933. Spectral Displacement of Extra-Galactic Nebulae. *Helvetian Physics Acta* 6:110.

———— 1938. On the Clustering of Nebulae. *Publications of the Astronomical Society of the Pacific* 50:218.

———— 1971. *Catalogue of Selected Compact Galaxies and of Post-Eruptive Galaxies.*

GLOSSARY

absolute magnitude   A unit expressing the intrinsic luminosity, or energy per second, emitted by an astronomical object.

abundances   The relative amounts of chemical elements. For example, hydrogen makes up about 75% of the mass of the universe, so its "cosmic abundance" is 75%.

acausal initial conditions   Initial conditions that could not have been caused by any prior physical process.

accretion disk   A disk of orbiting gas around a central mass. The gas in the disk, attracted by the gravity of the central mass (for example, a star), spirals toward it.

adiabatic perturbations   A slight bunching up of the nearly smooth, primal cosmic fluid, such that the photons and baryons are bunched up the same way. (See baryons; photon.)

alpha particle   An atomic nucleus consisting of two protons and two neutrons, held together by the strong nuclear force. (See neutron; proton.)

angular momentum   The product of mass, speed, and distance from a central axis. When a spinning body has no forces acting on it, its angular momentum remains constant. Thus, it can spin faster, but it must shrink accordingly, so that the product of size and rotational speed remain constant.

anisotropy   The condition in which the universe appears different in different directions.

anisotropic models   See mixmaster model.

anthropic principle   The weak form of the anthropic principle states that life can exist only during a brief period of the history of our universe. The strong form of the principle states that out of all possible values for the fundamental constants of nature and the initial conditions of the universe, only a small fraction could allow life to form at all, at any time. (See boundary conditions; fundamental constants of nature.)

arcsecond   A unit of angle in astronomy. One arcsecond is $1/3600$ of a degree.

axisymmetric collapse   Collapse of mass in such a way that the mass maintains the symmetry of a cylinder.

background radiation   See cosmic background radiation.

balloon metaphor   An analogy in which the expansion of the universe is likened to the surface of an expanding balloon with dots painted on it. Each dot represents a galaxy. From the point of view of any one dot, all the other dots are moving away from it, and this view is the same from any dot. The analogy applies only to the surface of the balloon, not to its interior or exterior. Since the surface is 2-dimensional and actual space is 3-dimensional, the geometry of the analogy is one dimension lower than that of the actual universe. Furthermore, the analogy applies only to closed universes, since the surface of a balloon has a finite size. (See closed universe.)

baryogenesis   The production of baryons, a type of subatomic particle. It is believed that baryons were produced in the early universe. (See baryons.)

baryons   Subatomic particles that interact via the strong nuclear force. The proton and the neutron are examples of baryons. Until recently, it was believed that baryons were conserved, that is, the net number of baryons before and after any physical process was unchanged. Grand unified theories (GUTS) of physics, first proposed in the mid-1970s, suggest that baryons may not be conserved.

baryon-to-photon ratio   (See photon-to-baryon ratio.)

beta decay   The process in which a neutron disintegrates into a proton, an electron, and an antineutrino. The escaping electron is sometimes called a beta ray. (See neutrino; neutron; proton.)

biasing   A hypothesized feature of the condensation of galaxies out of a background medium of gas. In any theory of galaxy formation, galaxies form in the places where gas is bunched up and concentrated, like ripples on a pond. According to the assumption of biasing, galaxies form only where the mass concentrations are very large and do not form at all where the concentrations are moderate.

big bang model   An evolutionary model of cosmology in which the universe began about 10 billion years ago, in a state of extremely high density and temperature. According to this model, the universe has been expanding, thinning out, and cooling since its beginning. It is an observational fact that distant galaxies are all moving away from our own galaxy, as predicted by the big bang model. (See closed universe; flat universe; Friedmann models; open universe.)

binary galaxies   Two galaxies orbiting each other owing to their mutual gravitational attraction.

blackbody radiation   A unique type of radiation whose spectrum and other properties are completely characterized in terms of a single quantity, temperature. Blackbody radiation is produced after a group of particles and photons have come into thermal equilibrium with each other, with every reaction between the particles balanced by the reverse reaction, so that the system as a whole has stopped changing. In this situation, all parts of the system, including the radiation, have come to the same temperature. Blackbody radiation would be produced, for example, inside an oven that is maintained at a constant temperature and in which the door has been left closed for a long time. (See photon; spectrum; thermal equilibrium.)

black hole   A mass that is sufficiently compact that not even light can escape its intense gravity. Thus it appears black from the outside. If the sun were compressed to a sphere about four miles in diameter, it would become a black hole. It is believed that some massive stars, after exhausting their nuclear fuel, collapse under their own weight to form black holes.

boundary conditions   Conditions needed to determine the evolution of a physical system, given the laws of nature. For example, the swing of a pendulum is determined both by the laws of mechanics and gravity and by the initial height at which the pendulum is released. This latter is called a boundary condition, or an initial condition.

Brans–Dicke theory   A theory of gravity formulated by Robert Dicke and Carl Brans in 1961 and differing from Einstein's theory of gravity, general relativity. In Einstein's theory, Newton's "gravitational constant," which measures the strength of the gravitational force, is a constant. In the Brans–Dicke theory, the magnitude of Newton's gravitational constant depends on the distribution of distant masses in the universe. (See general relativity.)

broken symmetry   (See symmetry breaking.)

bubbles (in the inflationary universe model)   Regions of empty space surrounded by a thin wall of energy. According to an early version of the inflationary universe model, bubbles condensed out of the surrounding medium of energy in the manner that chunks of ice might condense out of a liquid carefully cooled to below the freezing point.

bubbles (in the large-scale distribution of galaxies)   The name for the structures formed by the observed distribution of galaxies in space. Some surveys of nearby galaxies show that galaxies are located on roughly spherical shells, called bubbles, of about a hundred million light years in diameter (about a thousand times the diameter of a single galaxy). Few galaxies reside in the interior of a "bubble."

causality puzzle   See horizon problem.

CCDs   The acronym for charge-coupled devices, which are highly sensitive photoelectric devices that can electronically record the intensity and point of arrival of tiny amounts of light. CCDs are placed at the receiving end of telescopes, to "take pictures" of very faint astronomical objects; they have almost completely replaced photographic plates.

Cepheid   A type of star that oscillates in brightness, growing dim, then bright, then dim again in a regular way, with the cycle time closely correlated with its intrinsic luminosity. Thus, by measuring the light cycle time of a Cepheid star, one knows its intrinsic luminosity. A comparison of a star's intrinsic luminosity to its *apparent* brightness then gives the distance to the star. Cepheid stars are among the few astronomical objects whose distances can be reliably determined. (See intrinsic luminosity.)

Chandrasekhar mass limit   The maximum mass attainable by burned-out stars, beyond which the star's inward gravitational force exceeds its outward pressure force and causes the star to collapse. Quantitatively, the Chandrasekhar mass limit is about 1.4 times the mass of our sun. A star that has burned up all its nuclear fuel and that has a mass exceeding the Chandrasekhar mass will collapse, possibly forming a black hole.

chaotic inflation   A variation of the inflationary universe model in which random quantum fluctuations are continually forming new universes. (See inflationary universe model; new inflation; old inflation; quantum fluctuations; quantum mechanics.)

charm   The property that distinguishes one of the types of quarks. At present, there are six types of quarks known, one of which is the "charmed" quark. (See quark.)

chemical enrichment   The process in which a star manufactures chemical elements, such as carbon and oxygen, in the nuclear reactions in its interior and then ejects these elements into space. The gas in the surrounding space is said to be chemically enriched.

Christoffel symbols   Mathematical quantities used in the mathematical formalism of general relativity, Einstein's theory of gravity.

closed universe   A universe that has a finite size. Closed universes expand for a finite time, reach a maximum size, and then collapse. In closed universes, the inward pull of gravity dominates and eventually reverses the outward flying apart of matter; that is, gravitational energy dominates the kinetic energy of expansion. The value of omega is greater than 1 for a closed universe. If a universe begins closed, it remains closed; if it begins open, it remains open; if it begins flat, it remains flat. In the big bang model of the universe, the question of whether the universe is closed, open, or flat is determined by the initial conditions, just as the fate of a rocket launched from earth is determined by its initial upward velocity relative to the strength of earth's gravitational pull. If the initial rate of expansion of the universe was lower than a critical value, determined by the mass density, the universe will expand only for a certain period of time and then collapse, just as a rocket launched with a velocity below a critical value, dependent on the strength of earth's gravity, will reach a maximum height and then fall back to earth. This is the behavior of a closed universe. If the initial rate of expansion of the universe was larger than a critical value, the universe is open and will keep expanding forever. If the initial rate of expansion was precisely the critical value, the universe is flat and will expand forever, but with a rate of expansion that approaches zero. (See flat universe; omega; open universe.)

cluster   In cosmology, a group of galaxies closer together than could be expected if galaxies were randomly scattered through space. The size of a typical cluster is about 15 million light years. Clusters of galaxies can further bunch together to form a "supercluster" of galaxies, the typical size of which is about 150 million light years.

clustering   In cosmology, the observed tendency of galaxies to bunch together, rather than to distribute themselves uniformly and independently of each other. (See galaxy.)

cold dark matter   See dark matter.

Coma cluster   The nearest massive cluster of galaxies. The Coma cluster, about 300 million light years from us, contains about 1,000 galaxies in a region about 10 million light years across. (See cluster; galaxy.)

compactification   The process in which a space of many dimensions effectively reduces its dimensions. Some new theories of particle physics—the superstring theories—claim that the universe actually has 10 spatial dimensions but that 7 of these dimensions have become "compactified" down to subatomic size and thus are unobservable. (See superstring theory.)

complex analytic A particular property of mathematical representations of physical or mathematical systems. To have complex analytic structure, a system must be able to be represented by complex numbers, among other things. The representation must also have the property that its value at one point suffices to define its value at all other points. (See complex number.)

complex number A type of number that is the sum of two parts, the first of which produces a positive number when multiplied by itself (like ordinary numbers), and the second of which produces a negative number when multiplied by itself (unlike ordinary numbers). Complex numbers were discovered mathematically in the nineteenth century and have been found to play a major role in physics.

Copenhagen interpretation of quantum mechanics The view of quantum mechanics holding that a physical system exists in one and only one of its possible states after a measurement is made. Prior to the measurement, the system has no physical existence and is describable only in terms of the probability of each possible result of a measurement. (See many-worlds; quantum mechanics.)

correlation function A quantitative measure of the clustering of objects, often galaxies in cosmology. The correlation function measures the probability of objects lying within a certain distance of each other. If the objects are uniformly and randomly distributed, the correlation function is zero. If the objects tend to cluster together, the correlation function is big.

cosmic background radiation Often called simply background radiation, or cosmic microwave radiation, a pervasive bath of radio waves coming from all directions of space. According to the big bang theory, this radiation was produced by the collisions of particles when the universe was much younger and hotter, and it uniformly filled up space. The collisions between the radiation and matter stopped when the universe was about a million years old, and the cosmic background radiation has been traveling freely through space ever since. The cosmic background radiation is now in the form of radio waves.

cosmic rays High-energy subatomic particles from space. The origin of cosmic rays is not known with any certainty.

cosmogony The study of the origin of the universe. Cosmogony is, therefore, one part of cosmology.

cosmological constant A contribution to gravity that results from the effective mass density, or energy density, in the vacuum. A positive cosmological constant acts as if it were negative gravity—it makes two masses repel each other instead of attract each other. Einstein's first cosmological model contained a cosmological constant, which appeared as an additional term in the equations of general relativity. (See false vacuum; vacuum.)

cosmological principle The statement that the universe is homogeneous and isotropic on the large scale, that is, it appears the same at all places and, from any one place, looks the same in all directions.

covariance function Essentially the same as the correlation function. (See correlation function.)

CP violation   A reaction between subatomic particles is said to be a "CP violating" reaction if the reaction produces a different result when the electrical charges of the particles are changed to their opposites and the mirror image of the particle trajectories is used.

CPT invariant   A theory is "CPT invariant" if for every possible reaction between subatomic particles, a reaction can also occur in which the electrical charges of the particles changed to their opposites, the mirror image of the particle trajectories is used, and the directions of motion are reversed. Assuming general notions of modern physics, all conceivable theories of nature are CPT invariant.

critical mass density   The value of average cosmic mass density above which the universe is closed. The average mass density of the universe is obtained by measuring the mass in a very large volume of space, including many galaxies, and dividing by the size of the volume. The critical mass density is determined by the current rate of expansion of the universe. According to estimates of the current rate of expansion, the current critical mass density is about 10^{-29} grams per cubic centimeter. According to the best measurements, the average mass density of our universe appears to be about one tenth the critical mass density. (See closed universe; omega; open universe.)

cross section   A measure of the likelihood of interaction between two particles.

curvature   The departure of the geometry of the universe from Euclidean (flat) geometry. Qualitatively, the curvature is indicated by the curvature parameter, denoted by k. The values $k = 0, 1, -1$ refer to flat (uncurved) geometry, closed geometry, and open geometry, respectively. In a flat geometry, for example, the circumference of a circle is twice pi times its radius. In a closed geometry, the circumference is smaller than twice pi times the radius; in an open geometry, it is larger. (See closed universe; flat universe; open universe.)

dark matter   Matter in the universe that we detect by its gravitational influences, yet do not see. Dark matter that has small random speed and is easily concentrated by gravity is called cold dark matter. Dark matter that has large random speed and is thus able to resist gravitational clumping is called hot dark matter. Recent models to explain the observed pattern of galaxy clustering can be characterized, in part, as to whether they invoke hot dark matter or cold dark matter. However, since we do not know what the dark matter is, we do not have any direct evidence of whether it is cold or hot.

deceleration parameter   A parameter that measures the rate of slowing down of the expansion of the universe. Gravity causes the slowing down. The deceleration parameter equals omega (another cosmological parameter) when the universe is dominated by radiation, approximately the first 100,000 years after the big bang, and $1/2$ omega when the universe is dominated by matter. Since the deceleration parameter is equivalent to omega (assuming a cosmological constant of zero, as often done), it determines the ultimate fate and spatial geometry of the universe. The deceleration parameter is often denoted by the symbol q_o. (See omega.)

density fluctuations   Random inhomogeneities in an otherwise smooth distribution of matter.

density gradient   A spatial variation in the density of matter.

density parameter   Another name for omega. (See omega.)

de Sitter model   A particular solution to Einstein's cosmological equations, found by Wilhelm de Sitter in 1917, in which space expands at a rapid, exponential rate. This solution was very different from the solutions of Friedmann and of Lemaître, in which the universe expands at a much slower rate (a rate with the distance between any two points increasing as something between the square root of time and linearly with time). The Friedmann and Lemaître type solutions became incorporated in the standard big bang model. Recent modifications of the big bang model, such as the inflationary universe model, propose that the universe went through a period of exponential growth, or a de Sitter phase, early in its evolution.

deuterium   An atomic nucleus consisting of a proton and a neutron. It is believed that deuterium was the first compound nucleus formed in the infant universe.

differential equation   An equation that describes the evolution of a system over time, given boundary conditions for the system. Almost all of the laws of physics are expressed in the mathematics of differential equations. (See boundary conditions.)

differential forms   Mathematical quantities used in advanced geometrical concepts.

dipole anisotropy   A variation in the intensity of the cosmic background radiation caused by the motion of the Milky Way. Because the Milky Way, our galaxy, is not moving precisely with the uniform expansion of the universe as a whole, the cosmic background radiation appears slightly more intense in the direction of our peculiar motion and slightly less intense in the opposite direction. This effect, which can actually be used to measure the speed and direction of our peculiar motion, is called the dipole anisotropy.

Dirac equation   An equation that describes the behavior of subatomic particles called electrons. The Dirac equation incorporates both the laws of special relativity and the laws of quantum theory. (See quantum mechanics; special relativity.)

Dirac large number hypothesis   The current age of the universe, divided by the time it takes light to cross the radius of a proton, is about 10^{40}. This number is also approximately equal to the ratio of the strengths of the electromagnetic and gravitational forces. Dirac felt that the approximate equality of these two large numbers was too unlikely to be accidental and that some physical process must be at work to maintain the equality. Since the first number clearly changes in time (because the age of the universe is increasing), Dirac proposed that the "fundamental constants of nature" entering the second number should also change in time, to maintain the equality.

double galaxies   Two galaxies in orbit about each other; same as "binary galaxies."

dynamics   The physics that explains how particles and systems move under the influence of forces. The dynamical laws of a theory give a quantitative statement of the response of a particle to an applied force.

Einstein–de Sitter theory   A particular solution to Einstein's cosmological equations in which the universe is flat. (See flat universe.)

Einstein equations   The equations of Einstein's theory of gravity, called general relativity. The Einstein equations quantitatively specify the gravity produced by matter and energy. Since gravity is believed to be the principal force acting over very large distances, the Einstein equations are used in modern theories of cosmology.

electromagnetic force   One of the four fundamental forces of nature. Electricity and magnetism arise from the electromagnetic force. The other three fundamental forces are the gravitational force, the weak nuclear force, and the strong nuclear force. (See gravitational force; nuclear forces.)

electron volt   A unit of energy or of mass. The electron weighs about 10^{-27} grams, which is equivalent to about 500 thousand electron volts of energy. Thus, the electron volt is tiny by ordinary standards. The energy released by dropping a penny (3 grams) to the floor is about 4×10^{17} electron volts.

electroweak theory   The theory that unifies the electromagnetic force and the weak nuclear force into a single force. This theory was developed in the 1960s by Sheldon Glashow, Steven Weinberg, and Abdus Salam and has been subsequently confirmed in the laboratory. One of the mathematical properties of this theory is called the electroweak symmetry.

ellipticity (of a galaxy)   A quantitative measure of the shape of a galaxy. A completely spherical galaxy has zero ellipticity. A galaxy shaped like a cigar has a very high ellipticity.

ensemble (of universes)   A hypothetical group of many universes of varying properties. Some physicists attempt to estimate how "probable" are the properties of our universe by imagining it as a sample from an ensemble of universes.

entropy   A quantitative measure of the degree of disorder of a physical system. Highly disordered systems have a large entropy; highly ordered systems have low entropy. One of the laws of physics, the second law of thermodynamics, says that the entropy of any isolated physical system can only increase in time.

Eötvös experiment   An experiment carried out by Baron Eötvös in the nineteenth century confirming that different objects experience identical accelerations in a gravitational field. (In the absence of air resistance, a cannon ball and a feather, released from the same height, hit the floor at the same time.)

equation of state   An equation that describes a relationship between the energy, pressure, and density of a quantity of mass. The equation of state of a substance depends on the material it is made of and on the forces operating within it.

equivalence principle   The statement that a gravitational force is completely equivalent in all of its physical effects to an overall acceleration in the opposite direction. For example, a person in an elevator in space accelerating upward at 32 feet per second per second would feel the floor pushing upward against her feet in exactly the same way as if the elevator were at rest on earth, where gravity pulls downward with an acceleration of 32 feet per second per second. The "weak equivalence principle," which is not as strong as the equivalence principle, states that all objects, independent of their mass or composition, fall with the same acceleration in the presence of gravity. The Eötvös experiment, and later refinements of this experiment, have proven the weak equivalence principle.

Euclidean geometry   The geometry developed by the Greek Euclid about 300 BC. Euclidean geometry, like all geometries, deduces certain results from a set of starting assumptions. One of the critical assumptions of Euclidean geometry is that given any straight line and a point not on that line, there is exactly one line that can be drawn through that point parallel to the first line. One of the results of Euclidean geometry is that the interior angles of any triangle sum to 180 degrees. Euclidean geometry is the geometry we learn in high school.

explosive galaxy formation   A theory of galaxy formation wherein the explosion of a large number of stars creates a giant shock wave that travels outward and compresses the surrounding gas. Galaxies form in the regions of high-density gas.

exponential expansion   Extremely rapid expansion. "Exponential" is a mathematical term that precisely defines the rate of expansion. For example, a balloon that doubles its size every second is expanding exponentially. By contrast, a balloon whose radius is one inch after one second, two inches after two seconds, three inches after three seconds, and so on, is expanding linearly with time, rather than exponentially. According to the inflationary universe model, the early universe went through a brief period of exponential expansion, during which its size increased enormously.

extragalactic distance scale   The set of distances to astronomical objects outside our galaxy. It is difficult to obtain distances to objects further than about 10 million light years with accuracies better than about 25%.

Faber–Jackson relation   An empirically observed correlation between the speeds of stars in the center of a galaxy and the intrinsic luminosity of the galaxy—the higher the random speeds, the more luminous the galaxy. Since the speeds of stars can be directly measured by the Doppler shift in their colors, the Faber–Jackson relation permits an estimate of the intrinsic luminosity of a galaxy. By comparison of this with the *observed* brightness, the distance to the galaxy may be inferred.

Fabry–Perot interferometer   A device that measures distances and changes of distance very accurately, using the pattern of overlap of waves of light.

false vacuum   A region of space that appears to be empty (a vacuum), but actually contains stored energy. When this stored energy is released, the false vacuum is said to decay. (See vacuum.)

Fermi level   The maximum energy of any particle in a group of low-temperature subatomic particles called fermions. Fermions, such as electrons, cannot occupy the same space at the same energy. Thus, if many fermions are placed close together, their energies must all be different. The energy of that particle with the largest energy is the Fermi energy of the system.

field theory   A theory in which forces are communicated between two particles by a "field," which fills up the space between the two particles. In a field theory, any particle, such as an electron, is surrounded by a field. The field continuously creates and destroys intermediary particles, which transmit the force of the electron to other particles. In fact, particles themselves are considered to be concentrations of energy in the field. An axiomatic field theory is a field theory that begins with certain assumptions or axioms, as in Euclid's geometry, and deduces the necessary consequences of these assumptions.

fine structure constant   A parameter that measures the strength of the electromagnetic force. The fine structure constant is a combination of other fundamental constants of nature—the electrical charge of the electron, the speed of light, and Planck's constant of quantum mechanics.

fine-tuning   A phrase meaning a highly constrained and implausible adjustment of the parameters of a theory.

flavor   In particle physics, another word for "type." For example, there are 6 flavors of quarks, meaning that there are 6 different types of quarks.

flat universe   A universe that is at the boundary between an open and closed universe. In a flat universe, the average mass density always has precisely the critical value. A flat universe has zero total energy and an infinite size. Flat universes have the geometry of an infinite, flat surface, that is, Euclidean geometry. The value of omega is 1 for a flat universe. (See closed universe; critical mass density; Euclidean geometry; omega; open universe.)

flatness problem   The puzzle of why the universe today is so close to the boundary between open and closed, that is, why it is almost flat. Equivalently, why should the average mass density today be so close to the critical mass density, but not exactly equal to it? If omega begins bigger than 1, it should get bigger and bigger as time goes on; if it begins smaller than 1, it should get smaller and smaller. For omega to be near 0.1 today, about 10 billion years after the big bang, it had to be extraordinarily close to 1 when the universe was a second old. Some people consider such a fine balance to have been highly unlikely according to the standard big bang model, and thus are puzzled as to why the universe today is almost flat. (See closed universe; critical mass density; flat universe; open universe.)

fluctuations   Deviations from uniform conditions. For example, a mass of gas that bunches up to a higher density than the surrounding gas would be referred to as a fluctuation. The amount of bunching at each mass scale is called the fluctuation spectrum. Most cosmologists attempt to explain the observed structures in the universe (such as groups of galaxies) by the gravitational condensation and growth of small fluctuations of mass in the past.

Fourier component   A measure of the fluctuations of some physical quantity on a particular length scale. The density of sand on a beach, for example, would have a small Fourier component on the scale of a few feet, where the beach appears smooth, but a large Fourier component on the scale of a hundredth of an inch, where the beach appears grainy owing to individual sand particles.

Four-shooter   An astronomical instrument comprised of four highly sensitive photoelectric cells (CCDs). The four-shooter is placed at the end of a telescope and used to electronically record incoming light.

fractal A pattern that repeats itself or nearly repeats itself on many different scales of magnification. For example, suppose that some ink on a piece of paper appears to form a star. If you look at the piece of paper with a magnifying glass, you see that the dark areas are not solid black, but are formed of tiny stars themselves. If you look at one of these small stars with a microscope, you see that the dark areas of each of the tiny stars is formed from an arrangement of even tinier stars. Such a repeating pattern of stars would be called a fractal.

Friedmann equation An equation for the evolution of the universe. The Friedmann equation can be derived from Einstein's theory of gravity, plus the assumptions that the universe is homogeneous (looks the same at every point) and isotropic (looks the same in every direction). The solution of the Friedmann equation tells, among other things, how the distance between galaxies changes with time. (See homogeneity; isotropy.)

Friedmann models A general class of cosmological models that assume the universe is homogeneous and isotropic on large scales and that allow the universe to evolve in time. Most calculations in the standard big bang model assume a Friedmann cosmology. (See Friedmann equation; homogeneity; isotropy.) A cosmological model that has the same properties as a Friedmann model under some conditions is said to have a Friedmann limit.

fundamental constants of nature Physical quantities, like the speed of light or the mass of an electron, that enter into the laws of physics in a basic way and are believed to be the same at all times and everywhere in the universe. Most physicists take the fundamental constants of nature as given properties of the universe.

galaxy An isolated aggregation of stars and gas, held together by their mutual gravity. A typical galaxy has about 100 billion stars, has a total mass equal to about a trillion times the mass of the sun, is about 100,000 light years in diameter, and is separated from the nearest galaxy by a distance of about 100 times its own diameter. Thus, galaxies are islands of stars in space. Our galaxy is called the Milky Way. Galaxies come in two major shapes: flattened disks with a central bulge, called spirals, and amorphous, semispherical blobs, called ellipticals. If galaxies are found bunched up next to each other, they are said to lie in groups or clusters. Clusters with a particularly large number of galaxies in them are called rich clusters. Galaxies that do not lie in such groups but rather seem to be scattered uniformly and randomly through space are called field galaxies. Some galaxies are characterized by the dominant type of radiation they emit. For example, radio galaxies are unusually strong emitters of radio waves.

galaxy counts A quantitative measure of how many galaxies there are in each range of luminosity and at each range of distance from earth.

galactic halo The unseen mass that is believed to surround each galaxy and whose gravitational effects are believed to hold the galaxy together.

gauge field An energy field that permits a gauge symmetry. (See gauge symmetry.)

gauge symmetry A property of modern theories of physics, formulated in the

1960s and confirmed by experiment. A symmetry, in general, is a property that allows a system to behave in the same way even though it has undergone some change. For instance, a snowflake has a 6-sided "rotational symmetry"—a snowflake appears identical after every 60° rotation. A gauge symmetry is something like a rotation, in which the amount of rotation can vary randomly from one point of space to the next. (See field theory.)

Gaussian A random distribution of initial conditions is often referred to as a Gaussian distribution. Also, a certain kind of bell-shaped curve is called a Gaussian. (See boundary conditions.)

Gaussian noise Random fluctuations in an otherwise smooth distribution of something.

general relativity Einstein's theory of gravity, formulated in 1915. The theory of general relativity prescribes the gravity produced by any arrangement of matter and energy.

geometrodynamics The study of how gravity alters the geometrical properties of space, distorting those properties from the familiar properties of Euclidean geometry. Einstein's theory of gravity, general relativity, is a theory of geometrodynamics.

GeV A unit of energy, or of mass. A proton weighs about 10^{-24} grams, which is equivalent to about 1 GeV. One GeV equals 1 billion electron volts. (See electron volt.)

globular cluster A spherical congregation of stars within a galaxy that orbit each other because of their mutual gravity. A typical globular cluster has about a million stars; thus globular clusters are much smaller than galaxies. There are about 100 globular clusters in the Milky Way.

Gödel's theorem A theorem discovered and proved by the mathematician Kurt Gödel in 1931. Gödel's theorem says that there are certain propositions in each branch of mathematics (such as arithmetic or geometry) whose truth or falsity cannot be proven using methods and results only from that branch of mathematics.

grand unified theories (GUTs) Theories in physics that attempt to explain the forces of nature as manifestations of a single underlying force.

gravitational constant A fundamental constant of nature that measures the strength of the gravitational force. The gravitational constant is also called Newton's gravitational constant and is denoted by G.

gravitational force The weakest of the four fundamental forces of nature, the gravitational force between any two masses is proportional to the product of the masses and varies inversely as the square of the distance between them. The other three fundamental forces are the electromagnetic force and two kinds of nuclear forces. (See electromagnetic force; nuclear forces.)

gravitational lens A galaxy that intervenes between us and a distant astronomical object and that gravitationally deflects the light from that distant object. (Light, like matter, is attracted by gravity.) Gravitational lenses can focus, distort, and split light beams in the same way that ordinary glass lenses do.

gravity (gravitational) waves   Traveling waves of energy that transmit a gravitational force whose strength is changing in time. Gravitational waves are analogous to the electromagnetic waves that transmit the electromagnetic force.

Great Attractor   A hypothesized large mass, some hundred million light years from earth, that seems to be affecting the motions of many nearby galaxies by virtue of its gravity.

GUTs   See grand unified theories.

HI region   A region of space containing neutral hydrogen gas, that is, hydrogen atoms each consisting of the atomic nucleus and the orbiting electron.

HII region   A region of space containing ionized hydrogen gas, that is, hydrogen atoms without their electrons.

H–R diagram   A diagram of stars arranged according to their luminosity (measured on the y axis) and temperature (on the x axis). In the early part of the twentieth century, the Danish astronomer E. Hertzsprung and the American astronomer H. Russell plotted known stars in such a diagram and found a definite correlation between luminosity and temperature. (See main sequence.)

Hamiltonian theory   A theory for calculating the trajectory of a particle under an applied force. Hamiltonian theory, developed in the nineteenth century, is equivalent to Newton's laws of mechanics but is reformulated in a mathematically elegant way to allow easier solutions to some problems.

Hartle–Hawking proposal   See no-boundary proposal.

Higgs field   An energy field predicted by certain new theories of elementary particles and forces, particularly the grand unified theories. The energy stored in a false vacuum is the energy of a Higgs field. (See false vacuum; field theory; grand unified theories.)

hierarchical clustering model   A model of galaxy clustering in which different patterns appear at different scales of distance and in which the "average" density of matter depends on the size of the volume over which the average is performed. In a homogeneous model, on the other hand, the average density is independent of the size of the volume over which the average is performed. (See pancake model.)

Hilbert space   A mathematical tool used in the formalism of quantum mechanics. The dimensions of a Hilbert space consist of wave functions, instead of length, width, and breadth. (See quantum mechanics; wave function.)

holomorphic   Same as complex analytic.

homogeneity   In cosmology, the property that any large volume of the universe looks the same as any other large volume. Most cosmological models assume homogeneity.

homogeneity problem   Same as horizon problem.

horizon   The maximum distance that an observer can see. In cosmology, our horizon is the distance from us that light has traveled since the beginning of the universe. Objects more distant than our horizon are invisible to us because there hasn't been enough time for light to have traveled from there to here.

horizon problem The puzzle that widely separated regions of the universe are observed to share the same physical properties, such as temperature, even though these regions were too far apart when they emitted their radiation to have exchanged heat and homogenized during the time since the beginning of the universe. In particular, we detect the same intensity of cosmic radio waves (cosmic background radiation) from all directions of space, suggesting that the regions that emitted that radiation had the same temperature at the time of emission. However, at the time of emission, when the universe was about 1 million years old, those regions were separated by roughly 100 million light years, much exceeding the distance light or heat could have traveled since the big bang. The horizon problem is also called the causality puzzle. (See horizon.)

Hubble constant The rate of expansion of the universe. The Hubble constant actually changes in time, even though it is called a constant, because gravity is slowing down the rate of expansion of the universe. The Hubble constant is equal to the recessional speed of a distant galaxy, divided by its distance from us. Assuming a homogeneous and isotropic universe, the recessional speed of a distant galaxy is proportional to its distance; thus the Hubble constant as determined by any receding galaxy should be the same, yielding a universal rate of expansion of the universe. According to estimates, the current value of the Hubble constant is approximately 1 per 10 billion years, meaning that the distance between any two distant galaxies will double in about 10 billion years at the current rate of expansion. Astronomers measure the Hubble constant in units of kilometers per second per megaparsec. For example, a Hubble constant of 100 kilometers per second per megaparsec—which astronomers would refer to simply as a Hubble constant of 100—corresponds to 1 per 10 billion years. The Hubble constant is denoted by the symbol H_0.

Hubble flow See Hubble law.

Hubble law The law that recessional speed is proportional to distance for a homogeneous and isotropic universe. Galaxies moving away from us with a speed precisely following this law are said to follow the Hubble flow. Because the actual universe is not precisely homogeneous, with lumpiness arising from clustering of galaxies and voids of empty space, the motions of actual galaxies deviate somewhat from the Hubble flow.

Hubble program The research program carried out in the 1920s and 1930s by Edwin Hubble to measure the recessional speeds and distances of a large number of galaxies and to attempt to measure the deceleration parameter. This last parameter can in principle be determined by measuring the apparent brightness and redshift of a large number of objects of identical intrinsic luminosity. (See deceleration parameter; Sandage program; standard candle.)

Hubble time The reciprocal of the Hubble constant. The Hubble time gives an estimate for the age of the universe. To obtain a precise value for the age of the universe, the deceleration parameter must also be known, since the expansion rate has changed in time. (See deceleration parameter; Hubble constant.)

hydrodynamics The study of how gases and fluids flow under applied forces.

image dissector scanner   A specialized television camera used as a light detector (instead of a photographic plate) in the 1970s.

image tubes   Electronic devices that amplify incoming light while preserving its direction.

inflationary universe model   A recent modification of the standard big bang model in which the infant universe went through a brief period of extremely rapid (exponential) expansion, after which it settled back into the more leisurely rate of expansion of the standard model. The period of rapid expansion began and ended when the universe was still much less than a second old, yet it provides a physical explanation for the flatness and horizon puzzles. The inflationary universe model also suggests that the universe is vastly larger than the portion of it that is visible to us. (See exponential expansion.)

infrared photometry   The measurement of light intensities using infrared light instead of optical (visible to the human eye) light. Infrared light has longer wavelengths than optical light. (See photometry.)

initial conditions   See boundary conditions.

interstellar medium   The medium of gas and dust that fills the space between the stars.

intrinsic luminosity   The energy per second emitted by an astronomical object, analogous to the wattage of a light bulb.

IRAS samples   Astronomical objects detected in infrared radiation by the Infrared Astronomical Satellite launched in 1983. Infrared light has longer wavelengths than visible light.

isotope   An atomic nucleus having the same number of protons as a more commonly found atomic nucleus but a different number of neutrons. For example, the hydrogen nucleus has a single proton; deuterium has one proton and one neutron and would be called an isotope of hydrogen. (see neutron; proton.)

isotropy   In cosmology, the property that the universe appears the same in all directions. The uniformity of the cosmic background radiation, coming from all directions of space, suggests that on the large scale the universe is isotropic about our position. If we then assume that our position is not unique, we conclude that the universe appears isotropic about all points. This last result requires that the universe be homogeneous. (See cosmic background radiation; homogeneity.)

Jodrell Bank   A telescope in England designed to detect radio waves emitted by astronomical objects.

k-correction   A correction to the intensities of astronomical objects recorded on photographic plates. Since the colors of distant astronomical objects are shifted to the red, their intensities may not be accurately recorded by photographic plates, which are sensitive only in a limited band of colors. Correcting for the mismatch is the k-correction. (See redshift.)

Keplerian rotation curve   A rotation curve is a plot of speed versus distance from the center of an astronomical system. If most of the mass of the system is concentrated at the center, as in a solar system, then the speed of any orbiting body, such as a planet, is inversely proportional to the square root of its distance from the center. This last situation produces a Keplerian rotation curve. (See rotation curve.)

Kerr solution   A solution to Einstein's equations describing the gravity produced by a spinning point of mass.

kiloparsec   A measure of distance equal to 1,000 parsecs, or about 3,000 light years.

Kitt Peak   A mountain in Arizona where several large telescopes are located.

Lagrangian   A mathematical starting point of most theories in particle physics. The Lagrangian of a theory contains the assumptions of the theory; from the Lagrangian, the laws and consequences of the theory may be derived.

Lamb shift   A tiny change in the energy of an electron orbiting the nucleus of an atom, caused by the interaction of the electron with other particles that appear and quickly disappear in "quantum fluctuations." (See quantum fluctuations.)

lambda term   Same as cosmological constant.

large-scale structure   The distribution of galaxies and other forms of mass on large distance scales, covering hundreds of millions of light years and larger. A perfectly homogeneous and isotropic universe would have no large-scale structure; a universe with all the galaxies lined up in single file would have enormous large-scale structure.

large-scale motions   Bulk motions of distant galaxies deviating from the Hubble flow. (See Hubble law.)

leptons   Fundamental particles of nature, which may interact via all of the fundamental forces except the strong nuclear force. The electron is an example of a lepton.

light cone   Any event, such as an explosion, affects other regions of the universe by sending out light and heat and other forms of energy in all directions. None of these signals can travel faster than the speed of light. Therefore, at each moment after the event, the most distant region of space that can have been affected by the event lies at the distance light has traveled since the event. This most distant affected region, at each moment of time, is called the "light cone" of the event. The light cone spreads out in time, just like ripples spreading out from a stone thrown in a pond.

Lorentz invariance   The assumption that the laws of physics are identical for all observers moving at constant velocity relative to each other. Lorentz invariance is a basic tenet of special relativity and, consequently, of all of modern physics. (See special relativity.)

luminosity function   A quantitative description of the variation of luminosities among galaxies. The luminosity function gives the number of galaxies shining at each level of luminosity in each volume of space.

Lyman alpha clouds   Gas lying between us and quasars that absorbs some of the radiation from those quasars. (See quasars.)

M31   The nearest large galaxy to our galaxy, the Milky Way. Also called the Andromeda galaxy, M31 is about 2 million light years away.

Mach's principle The hypothesis that the inertia of bodies—that is, their resistance to acceleration by applied forces—is determined not by any absolute properties of space but by the effects of distant matter in the universe. Equivalently, Mach's principle proposes that the distinction between accelerated and nonaccelerated frames of reference is determined by the effects of distant matter.

magnetic monopole A hypothesized particle that would have either a magnetic north pole or a magnetic south pole but not both. All magnetic particles and magnets ever observed have both poles. Magnetic monopoles are predicted by grand unified theories of physics. That grand unified theories predict the existence of large numbers of magnetic monopoles, when none have been discovered, is called the monopole problem. (See grand unified theories.)

magnitude In astronomy, a measure of the brightness of an astronomical object.

main sequence The pattern of stars in a luminosity–temperature diagram, or H–R diagram. There is a correlation between intrinsic luminosity and temperature for stars powered by hydrogen-burning nuclear reactions, and such stars are said to be main sequence stars, or to lie along the main sequence of the H–R diagram. (See H–R diagram.)

many-worlds (Everett–Wheeler) interpretation of quantum mechanics The view of quantum mechanics holding that a physical system simultaneously exists in all of its possible states prior to and after a measurement of the system. (Compare with the Copenhagen interpretation of quantum mechanics.) In the many-worlds interpretation, each of these simultaneous existences is part of a separate universe. Every time we make a measurement of a physical system and find it to be in a particular one of its possible states, our universe branches off to one of the universes in which the system is in that particular state at that moment. The system, however, continues to exist in its other possible states, in parallel universes. (See Copenhagen interpretation of quantum mechanics; quantum mechanics.)

mass models Models that attempt to infer the distribution of mass in an astronomical system by comparing the observed properties of the system (such as the distribution of light) with those properties predicted by various theoretical distributions of mass.

mass-to-light ratio (M/L) The ratio of total mass in a physical system to the amount of light produced by that system. Often, mass that does not produce light or radiation of any kind can nevertheless be detected by its gravitational effects. Systems with a large amount of such dark matter would have a high mass-to-light ratio.

massive halos Spherical distributions of dark matter surrounding galaxies. (See dark matter.)

matter-to-antimatter ratio The ratio of mass in particles to mass in antiparticles. For every type of particle, there is an antiparticle counterpart. The positron, for example, is the antiparticle of the electron and is identical to the electron except for having opposite electrical charge. The abundances of particles and antiparticles do not have to be equal. It appears that our universe is made up almost entirely of particles, rather than antiparticles, although there is no fundamental difference between the two kinds of matter.

Maxwell equations The equations of electromagnetism, formulated by the Scottish physicist James Maxwell in the nineteenth century.

megaparsec A measure of distance equal to 1 million parsecs, or about 3 million light years.

Messier catalogue One of the earliest catalogues of nebulous-appearing astronomical objects, compiled in 1781 by the French astronomer Charles Messier. Messier's catalogue included many objects that were later realized to be galaxies.

metals In astronomy, all elements heavier than hydrogen and helium, the two lightest elements, are called metals.

metric A mathematical description of the geometrical properties of the universe.

micron A unit of distance, equal to one ten-thousandth of a centimeter.

microwave radiation See cosmic background radiation.

Minkowski space Space that is empty of matter or energy.

minute of arc A unit of angle equal to 1/60 of a degree.

missing mass The cosmic mass that some scientists hypothesize so that the universe will have the critical density of matter, with an exact balance between gravitational energy and kinetic energy of expansion. Such mass is called missing because it represents about 10 times as much mass as has actually been detected. (See closed universe; critical mass density; dark matter.)

mixmaster model A non-Friedmannian cosmological model that begins with a highly anisotropic infant universe and shows how anisotropies are reduced in time. (See Friedmann models.)

monopole See magnetic monopole.

Mt. Wilson The location, in California, of the 100-inch diameter telescope used by Edwin Hubble and others.

nanosecond A billionth of a second.

N-body simulations Computer simulations of the behavior of a large number of bodies under their mutual interactions. In cosmological N-body simulations, the bodies are usually galaxies and the interactions are gravitational. Thus, the computer simulates how a group of galaxies should behave under their mutual gravitational attraction. The law of gravity and the initial positions and velocities of the hypothetical galaxies and other masses are fed into the computer. The computer then calculates the evolution of the system.

neutral current A type of reaction between particles subject to the weak nuclear force. In this reaction, all intermediary particles that transmit the forces are electrically neutral, which is the origin of the word "neutral."

neutrino A subatomic particle that has no electrical charge, has little if any mass, and interacts with other particles only through the weak nuclear force and the gravitational force. (See nuclear forces.)

neutron A type of subatomic particle that, together with the proton, makes up the atomic nucleus. The neutron has no electrical charge and is composed of three quarks. (See proton; quark.)

neutron lifetime   The time it takes an isolated neutron at rest to disintegrate into other elementary particles, about equal to 15 minutes.

neutron star   An extremely dense star, made up almost entirely of neutrons. A typical neutron star has a mass about equal to our sun and a diameter of about 10 miles.

new inflation   A 1982 modification of the original inflationary universe model. While the old inflationary universe model solved a number of cosmological problems, it led to the result that the universe was very inhomogeneous during the inflationary epoch and contained bubbles of empty space surrounded by a medium filled with energy. In new inflation, no such bubbles appear, although the universe still undergoes a brief epoch of extremely rapid expansion. (See inflationary universe model.)

no-boundary proposal   An initial (boundary) condition for the universe proposed by James Hartle and Stephen Hawking. In this proposal, the mathematics of general relativity is reformulated so that time is replaced by a space-like coordinate, in effect representing the universe as having 4 space dimensions instead of 3 space dimensions and a time dimension. (In such a formulation, "time" does not have its usual meaning.) Hawking and Hartle suggest that the geometry of this representation of the universe should be analogous to the geometry of the surface of a sphere, that is, a shape with no edges—hence the name no-boundary proposal. When translated back into ordinary time and space, this suggested boundary condition takes the form of a specific initial condition for the universe. The no-boundary proposal is formulated within a quantum mechanical calculation of the behavior of the early universe.

no-hair theorem   The notion that all detailed effects of the initial conditions of the universe have been erased by subsequent physical processes. According to this view, a wide variety of initial conditions would have led to the same universe today; the initial conditions were unimportant in determining the future state of the universe.

non-Euclidean geometry   Geometry that does not follow the postulates and results of Euclidean geometry. For example, in a non-Euclidean geometry, the sum of the interior angles of a triangle differs from 180 degrees. According to Einstein's general relativity theory, gravity distorts space into a non-Euclidean geometry.

nuclear chronology   A method of dating an object by measuring how many atomic nuclei have disintegrated and changed into other nuclei. Uranium dating of the earth is an example of nuclear chronology.

nuclear forces   There are two kinds of nuclear forces: the strong nuclear force and the weak nuclear force. These two forces, plus the gravitational and electromagnetic forces, comprise the four fundamental forces of nature. The strong nuclear force, which is the strongest of all four forces, is the force that holds protons and neutrons together in the atomic nucleus. The weak nuclear force is responsible for certain kinds of radioactivity—for example, the disintegration of a neutron into a proton, electron, and antineutrino.

nucleosynthesis   The production of heavy nuclei from the fusion of lighter ones. According to the big bang theory, the infant universe consisted of only hydrogen, the lightest of all atomic nuclei, because any heavier nuclei would have come apart in the intense heat. All other elements would have to be formed later, in nucleosynthesis processes. It is believed that most of the helium, the next lightest element after hydrogen, was formed when the universe was a few minutes old, and that most of the other elements were made much later, in nuclear reactions at the centers of stars.

Occam's razor   The notion that the simplest explanation of a problem is the preferred explanation, unless it is known to be wrong.

Olbers's paradox   The puzzle of why the sky is dark at night. If the universe extends infinitely in space, as it might, then the accumulated light from an infinite number of distant stars should seemingly cause the sky to be bright at all times, whether our sun is visible or not. This paradox, first posed in the eighteenth century, has been resolved by the big bang theory. In a universe with a beginning, we can receive light only from that part of the universe close enough so that light has had time to travel from there to here since the big bang (about 10 billion years ago). Thus, even if space extends infinitely far, only a limited region, and a limited number of stars, are visible to us. And the accumulated light from this limited number of stars is not sufficient to spoil the darkness of the night sky.

old inflation   The original (1981) inflationary universe model. (See inflationary universe model; new inflation.)

omega   The ratio of the average density of mass in the universe to the critical mass density, the latter being the density of mass needed to eventually halt the outward expansion of the universe. In an open universe, omega is always less than 1; in a closed universe, it is always greater than 1; in a flat universe it is always exactly equal to 1. Unless omega is exactly equal to 1, it changes in time, constantly decreasing in an open universe and constantly increasing in a closed universe. Omega has been measured to be about 0.1, although such measurements are difficult and uncertain. (See critical mass density; closed universe; flat universe; open universe.)

opacity   A measure of the resistance of a medium to the transmission of visible light or other forms of radiation. A material that is opaque is said to have a high opacity; one that is transparent has low opacity.

open cluster   A loose aggregate of stars, much smaller than a globular cluster.

open universe   A universe fated to expand forever. In an open universe, the kinetic energy of expansion is always greater than the gravitational energy, and the value of omega is always less than 1. Open universes have the geometry of an infinite curved surface with the same amount of curvature at every point. (See closed universe; omega.)

order-of-magnitude estimate   An approximate estimate of the magnitude of something, accurate to within a range of 10 times too big to 10 times too small.

For example, given that the population of the United States is 250 million, any estimate of the population lying between 25 million and 2,500 million would be an acceptable order-of-magnitude estimate. Astronomers are accustomed to order-of-magnitude estimates.

oscillating universe model   A universe that expands, then contracts, then expands, then contracts, and so on through many cycles.

Palomar   The mountain in California upon which sits the largest telescope in the United States, 200 inches in diameter. The telescope itself is sometimes referred to as the Mt. Palomar telescope.

pancake model   A model of galaxy formation in which the first structures to condense out of the smooth background of primordial gas were very large in size. These large masses then collapsed into thin sheets (pancakes) and fragmented into many smaller pieces the size of galaxies. A competing theory, sometimes called the hierarchical clustering model, proposes that the first structures to form were the size of galaxies. As galaxies clustered together, due to gravity, larger and larger structures were formed. (See hierarchical clustering model.)

parallax   The change in apparent position of a star or other astronomical object caused by the movement of the earth around the sun. A shift in our location causes a shift in viewing direction when we sight a star, just as a tree in the front yard appears to shift relative to the neighbor's house across the street as we sight it from different positions.

particle physics   That branch of physics that attempts to understand the fundamental particles and forces of nature.

particle-to-antiparticle ratio   Same as matter-to-antimatter ratio.

Perseus–Pisces region   A region of space containing a huge congregation of galaxies called a supercluster. The galaxies in this supercluster appear to be distributed in a long chain.

peculiar A stars   Stars can be classified according to their surface temperatures, which determine, in large part, the spectrum of radiation they emit. A stars have surface temperatures between about 7,500 and 11,000 degrees centigrade. Peculiar A stars are A stars whose emitted radiation spectra have many of the characteristics of A stars but are peculiar in certain ways. (See spectrum.)

peculiar velocity   A deviation in the velocity of a galaxy from that expected on the basis of a uniform expansion of the universe. (See Hubble law.)

phase transition   A sudden transition between one state of matter or energy and another state. For example, when hot water turns to steam or when ice crystallizes out of a liquid that has been cooled to below freezing, a phase transition has occurred. According to the grand unified theories of particle physics, the infant universe may have undergone one or more overall phase transitions. In this case, the energy uniformly filling all space corresponded to the supercooled liquid. (See false vacuum.)

photometry   The measurement of the intensity of light from an astronomical object.

photomultiplier   A device that converts incoming light into an electrical signal.

photon   The subatomic particle that transmits the electromagnetic force. Light consists of a stream of photons.

photon-to-baryon ratio   The ratio of the number of photons to the number of baryons in any typical, large volume of space. (See baryons; photon.)

Planck's constant   A fundamental constant of nature that measures the magnitude of quantum mechanical effects. Visible light, for example, consists of discrete particles of light, or photons, each carrying an amount of energy equal to Planck's constant multiplied by the frequency of visible light. (The energy of one photon of visible light is approximately 10^{-18}, or a billionth of a billionth, the energy of a penny dropped to the floor from waist high.) By combining Planck's constant with two other fundamental constants of nature—Newton's gravitational constant and the speed of light—one obtains other "Planck units" that mark critical densities and times when quantum mechanics and gravity were both extremely important. For example, the Planck density, or Planck scale, is the density of matter above which the structure, and perhaps meaning, of space and time break down due to quantum mechanical effects. Numerically, the Planck density is about 10^{93} grams per cubic centimeter. The infant universe had this enormous density when it was about 10^{-43} seconds old, which is called the Planck time, and when it had a temperature of about 10^{22} Centigrade. At this temperature, the mean energy per particle was equivalent to the Planck mass, about 10^{-5} grams. (See quantum mechanics.)

Planck mass   See Planck's constant.

Planck time   See Planck's constant.

population I, II, and III stars   The youngest observed stars are called population I stars; older observed stars are called population II; and it is postulated that an even older generation of stars, called population III, existed still earlier. Population II stars formed mostly from hydrogen and helium. Population I stars, like our sun, formed from hydrogen, helium, and a large range of heavier elements (like carbon and oxygen) believed to have been created in the interiors of earlier population II and III stars and then blown out into space.

potential   A quantitative measure of how much energy is associated with each possible arrangement of a physical system. An arrangement with a relatively high value of the potential is one with a relatively large amount of energy. For example, the gravitational potential of a pendulum at the top of its swing is large, the potential is small when the pendulum is at the bottom of its swing. Since systems in nature generally evolve toward arrangements of lower energy, as in the tendency of upended books to fall over, the final resting point of a system is in a configuration at the minimum value of the potential. The minimum of the potential for empty space corresponds to the vacuum. (See vacuum.)

primordial black holes   Small black holes hypothesized to have formed during the first 10^{-43} seconds of the universe, when quantum effects were very large.

primordial fireball radiation   Same as cosmic background radiation.

proper motion   The angular motion of a star or object per year. For example, the proper motion of a star might be 0.001 degrees per year.

proton   A type of subatomic particle that, together with the neutron, makes up the atomic nucleus. The proton has a positive electrical charge and is made up of three quarks. (See neutron; quark.)

pulsars   Neutron stars that spin rapidly and have strong magnetic fields, which produce electromagnetic radiation. (See neutron star.)

q_0   The deceleration parameter. (See deceleration parameter.)

quantum cosmology   The subfield of cosmology that deals with the universe during its first 10^{-43} seconds, when quantum mechanical effects and gravity were both extremely important. (See Planck's constant; quantum mechanics.)

quantum electrodynamics   The quantum theory of electromagnetism. (See quantum mechanics.)

quantum field   A distribution of energy that is constantly creating and destroying particles, according to the probabilities of quantum mechanics, and transmitting the forces of nature. (See field theory; quantum mechanics.)

quantum fluctuations   Continuous variations in the properties of a physical system, caused by the probabilistic character of nature as dictated by quantum mechanics. For example, the number of photons in a box with perfectly reflecting walls is constantly varying because of quantum fluctuations. Quantum fluctuations can cause particles to appear and disappear. Some theories hold that the entire universe was created out of nothing, in a quantum fluctuation.

quantum gravity   A theory of gravity that would properly include quantum mechanics. To date, there is no complete and self-consistent theory of quantum gravity, although successful quantum theories have been found for all the forces of nature except gravity. (See quantum mechanics.)

quantum mechanics   The theory that explains the dual wave-like and particle-like behavior of matter and the probabilistic character of nature. According to quantum mechanics, it is impossible to have complete and certain information about the state of a physical system, just as a wave cannot be localized to a single point in space but spreads out over many points. This uncertainty is an intrinsic aspect of the system or particle, not a reflection of our inaccuracy of measurement. Consequently, physical systems must be described in terms of probabilities. For example, in a large collection of uranium atoms, it is possible to accurately predict what *fraction* of the atoms will radioactively disintegrate over the next hour, but it is impossible to predict *which* atoms will do so. As another example, an electron with a well-known speed cannot be localized to a small region of space but behaves as if it occupied many different places at the same time. Any physical system, such as an atom, may be viewed as existing as a combination of its possible states, each of which has a certain probability. Quantum theory has been extremely successful at explaining the behavior of nature at the subatomic level, although many of its results violate our common-sense intuition. (See Copenhagen interpretation of quantum mechanics; many-worlds interpretation of quantum mechanics; uncertainty principle; wave function.)

quark   One of the fundamental, indestructible particles of nature, out of which many other subatomic particles are made. The neutron, for example, is built of three quarks. Five types of quarks have been discovered, and it is believed that a sixth also exists. Quarks interact mainly via the strong nuclear force and the electromagnetic force.

quark–hadron phase transition   A phase transition in the early universe when freely roaming quarks combined to form neutrons, protons, and other strongly interacting particles called hadrons. (See neutron; phase transition; quark.)

quasars   Extremely distant and luminous astronomical objects that are much smaller than a galaxy and much more luminous. Quasars may be the central regions of certain very energetic galaxies at an early stage of their evolution. It is believed that the power of a quasar derives from a massive black hole at its center.

radiative transfer   The process by which radiation travels through a medium.

radiative vicosity   The friction produced by the collisions between matter and radiation.

radio source counts   A radio source is an astronomical object that emits radio waves. A compilation of how many radio sources there are in space at each epoch and in each range of luminosity is called a radio source count.

radiometer   A device that detects radio waves from space and measures their direction.

Rayleigh–Jeans spectrum   The low-frequency portion of a blackbody spectrum. (See blackbody radiation; spectrum.)

reaction rate   The rate at which a chemical or nuclear reaction proceeds. Particles interacting via the strong nuclear force react together in roughly 10^{-23} seconds, while particles interacting via the weak nuclear force, such as the disintegration of a neutron, might take seconds.

reddening   The process by which light from an astronomical object grows red as the light travels through interstellar dust. Dust scatters blue light more than red light, thus leaving predominantly red light.

red giant star   A large, cool star of high luminosity. Many stars become red giant stars after they have exhausted their hydrogen fuel and start burning helium.

redshift   A shift in color toward the red end of the spectrum, caused when a source of light (and color) is moving away from the observer. The magnitude of the redshift is directly related to the magnitude of recessional speed; thus measurement of the redshift of an object measures its recessional speed. In an expanding universe, the colors of galaxies are shifted to the red, and in a uniformly expanding universe, the redshift is directly proportional to an object's distance from earth (except for extremely distant objects). Measurement of an astronomical object's redshift provides the distance to the object.

redshift survey   The methodical tabulation of the redshifts of a large number of galaxies in a particular region of the sky. Redshifts directly measure the recessional speeds of galaxies. If Hubble's law is assumed, this speed can be translated to distance. Under such an assumption, a redshift survey provides the third dimension, depth, for the galaxies in a survey. The other two dimensions for each galaxy are provided by its position on the sky. The redshift of a galaxy is obtained by measuring its spectrum of light; in this way it is possible to see how much its colors are shifted. (See spectrum.)

refractor   A telescope in which the light is focused by a lens at the viewing side of the telescope. By contrast, a reflecting telescope is one in which light is focused by a mirror.

renormalization   A mathematical procedure in quantum mechanics that allows one to make a correspondence between the formal quantities of the theory and the actual quantities observed in the lab. Some of the formal quantities, such as the mass of an electron, have infinite values before renormalization.

Riemannian geometry   A large class of non-Euclidean geometries. The mathematics of general relativity uses Riemannian geometry. (See general relativity.)

relativity   The theory of how motion and gravity affect the properties of time and space. The special theory of relativity establishes, among other things, the nonabsolute nature of time. The amount of time elapsed between two events will not be the same for two observers or clocks in relative motion to each other. The general theory of relativity describes how gravity affects the geometry of space and the rate at which time passes. (See general relativity; special relativity.)

reproducing universes   The process in some inflationary universe models whereby the universe is constantly spawning new universes, causally disconnected from each other and from the parent universe. (See chaotic inflation.)

rich clusters   Clusters with a particularly large number of galaxies. (See galaxy.)

Robertson–Walker metric   A mathematical description of the geometrical properties of a homogeneous and isotropic universe. Friedmann cosmologies all use the Robertson–Walker metric. (See Friedmann models; homogeneity; isotropy; metric.)

rotation curve   A quantitative description of how fast each part of a galaxy is rotating about the center. A rotation velocity is the velocity of a rotating galaxy at a certain distance from the center of the galaxy. A "flat" rotation curve is one in which the rotational velocity of the galaxy remains constant with distance away from the center of the galaxy. This indicates that the mass of the galaxy increases linearly with distance from its center. (See Keplerian rotation curve.)

Sandage program   A series of observations carried out by Allan Sandage to measure how the Hubble constant varies with distance. Since the universe was expanding faster in the past, the Hubble constant should be bigger at greater distances, which convey information from earlier epochs. Measuring how the Hubble constant changes in time gives the rate of deceleration of the universe. This, in turn, is equivalent to a determination of the omega parameter. In practice, such a program attempts to measure the apparent brightnesses of a group of objects with the same intrinsic luminosities but different redshifts and thus different distances. This program is difficult because the intrinsic luminosity of an object does not stay constant in time, so there is actually no group of objects meeting the requirements. The Sandage program was a continuation of the Hubble program. (See Hubble program; omega; redshift.)

scalar field   A field of energy generated by scalar particles. These hypothesized particles have no intrinsic spin. All known elementary particles have some intrinsic spin; thus scalar particles and scalar fields are theoretical to date. (See quantum field.)

scalar-tensor theory   A class of theories of gravity more complex than Einstein's theory, general relativity. The best known scalar-tensor theory is the Brans–Dicke theory. In some scalar-tensor theories, the gravitational constant is not constant, as it is in general relativity. (See Brans–Dicke theory; general relativity.)

scale factor   A measure of changing distances in cosmology. The distance between any two galaxies, for example, is proportional to the scale factor, which is always increasing in an expanding universe. If the scale factor doubles in size, then the distance between any two galaxies doubles.

scale length   A measure of the size of a physical system or region of space.

Schmidt plates   Photographic plates obtained with a Schmidt telescope, which is a type of telescope with a particularly large field of view.

Schwarzschild radius or limit   The "surface" of a black hole, within which the strength of gravity is so strong that light cannot escape. The Schwarzschild radius is proportional to the mass of the black hole and would be about 2 miles for a black hole of the mass of our sun. Black holes were first "theoretically" discovered by Karl Schwarzschild in 1917. (See black hole.)

Schwarzschild singularity   The center of a black hole. According to Einstein's theory of general relativity, the entire mass of a black hole is concentrated at a point at its center, the "singularity." It is believed that quantum mechanical effects, not included in the theory, would cause the mass to spread out over a tiny but nonzero region, thus preventing an infinite density of matter and doing away with the singularity.

second law of thermodynamics   A physical law formulated in the nineteenth century and stating that any isolated system becomes more disordered in time. (See entropy.)

Seyfert galaxy   A type of spiral galaxy first discovered by Karl Seyfert in the 1940s. The central region of a Seyfert galaxy is distinguished by powerful radiation, much of it focused into narrow frequencies.

Shane–Wirtanen catalogue   A catalogue of all galaxies brighter than seventeenth magnitude (a measure of brightness). There are about a million galaxies in the Shane–Wirtanen catalogue.

Shapley–Ames catalog   A catalogue of galaxies brighter than thirteenth magnitude, completed in 1932. There are about 1200 galaxies in this catalogue.

sigma   In astronomy, a quantitative measure of the random speeds of stars in a collection of stars. If the stars were molecules of gas, darting this way and that, then sigma would be directly related to the temperature of the system. A high sigma is analogous to a high temperature. Sigma is also called the velocity dispersion.

simulations   In science, simulations of physical systems with a computer. (See N-body simulations.)

singularity   A place, either in space or in time at which some quantity, such as density, becomes infinite. The laws of physics cannot describe infinite quantities and, in fact, physicists believe that infinities do not exist in nature. All singularities, such as the Schwarzschild singularity, are therefore probably the artifacts of inadequate theories rather than real properties of nature. According to Einstein's theory of general relativity, the universe began in a singularity of infinite density, the big bang. Physicists believe that an improved and yet-to-be discovered modification of general relativity, incorporating quantum mechanics, will show that the universe did not begin as a singularity. (See Schwarzschild singularity.)

singularity theorems   In astronomy and cosmology, mathematical proofs that show the conditions under which a mass will gravitationally collapse to form a singularity. The singularity theorems of cosmology, proved in the 1960s, indicate that the current behavior of the universe, together with the laws of general relativity without quantum mechanical corrections, require that at some definite time in the past the universe was compressed to a state of zero size and infinite density, called a singularity. The laws of physics break down at a singularity and cannot be used to predict anything during or before the singularity occurred. (See singularity.)

space curvature   See curvature.

spallation   The process in which an incoming beam of particles or energy collides with a substance, reacts with it, and knocks off pieces of it.

special relativity   Einstein's theory of time and space, formulated in 1905, which shows how measurements of length and time differ for observers in relative motion.

spectrograph   An instrument that records the amount of light in each range of wavelength, that is, in each range of color. In general, each type of astronomical object, such as a star or a galaxy, will emit a characteristic spectrum of light. (See spectrum.)

spectroscopy   The study of what wavelengths of light an object or substance will emit under various conditions.

spectrum   The amount of light in each range of wavelength, that is, in each range of color. The term spectrum can also be applied more generally to the intensity of something at each length scale. An object that emits radiation in a continuous range of colors is said to have a continuous spectrum. An object that emits radiation only at certain wavelengths is said to have emission lines; objects that *absorb* radiation only at certain wavelengths are said to have absorption lines.

spinor   A mathematical object that reverses sign after a rotation by 360 degrees and returns to itself only after a rotation by 720 degrees. (More familiar mathematical and physical objects return to themselves after a rotation by 360 degrees.) A physical example showing spinor behavior is the following: Paint each face of a cube a different color and connect each of the eight corners of the cube to the corresponding corners of the room with threads. Now rotate the cube by 360 degrees. The threads are hopelessly tangled up, even though the cube has returned to its original position. Rotation of the cube by *another* 360 degrees, however, allows one to untangle the threads. Spinors involve complex numbers. (See complex numbers.)

spontaneous emission   Radiation emitted by an isolated body.

standard candle   In astronomy, any class of objects with the same intrinsic luminosity, or with some property that allows a reliable determination of the intrinsic luminosities. (See intrinsic luminosity.)

statistical distribution   The range of variation of some quantity in a population, obtained by sampling many members of the population. For example, the statistical distribution of the height of American males could be obtained by sampling 10,000 randomly chosen males and counting the number of them within each range of heights. In cosmology, the distance between pairs of galaxies, averaged over a large number of galaxies, would constitute a statistical distribution.

statistical mechanics   The area of physics that analyzes the behavior of a system with very many members, such as a gas with many individual molecules. In such a situation, the behavior of the whole system is obtained by averaging over the behavior of individual members.

steady state model   A model of the universe in which the universe does not change in time. A modern steady state model was proposed in the late 1940s. The big bang model is not a steady state model.

stimulated emission   Radiation emitted by a body, such as an atom, when it is bombarded by radiation. The stimulated radiation has the same wavelength and direction as the bombarding radiation.

strings   The hypothesized, basic constituents of matter, according to new theories of physics. In earlier theories of physics, the basic constituents of matter were point-like particles, such as electrons, which interacted with other particles at a point. According to the string theory, the basic constituents are 1-dimensional structures called strings. There are completely different strings, called cosmic strings, which can form according to some theories and which may extend for great distances in space. Postulated to have formed as a result of processes in the early universe, cosmic strings are 1-dimensional structures of enormous energy, extending for perhaps thousands or millions of light years in space. There is no good observational evidence that either kind of strings exist. (See superstring theory.)

SU(5)   The simplest type of grand unified theory, proposed in the 1970s. (See grand unified theories.)

supercluster   See cluster.

supercooling   The process by which a substance is cooled below the temperature at which a phase transition should occur, such as water that has been cooled to below zero degrees Centigrade but that has not yet formed ice. (See phase transition.)

superconductivity   The phenomenon in which some substances, when cooled to a sufficiently low temperature, lose all resistance to the flow of electricity.

supernova   The violent collapse and explosion of a star, producing an enormous amount of energy in a short amount of time. The stellar debris remaining after a supernova is called a supernova remnant. Astronomers believe that pulsars are formed by supernovae.

super radiance   A process by which energy may be extracted from a rotating black hole. A beam of radiation approaches the black hole and "bounces off" with more energy than it had before, analogously to a marble scattering off a rapidly spinning top. The source of the gained energy is the rotational energy of the black hole or top, which slows down in the process.

supersymmetry   A mathematical property of some theories of physics proposing that every particle of integer spin (intrinsic angular momentum) has a partner of half integer spin. For example, the photon, which is the particle of light, has a spin of 1 unit. Its hypothesized super symmetric partner is called the photino, which would have a spin of 1/2 units.

superstring theory   A new type of theory in physics that unifies all the forces of nature, including the gravitational force, and that may be capable of explaining all of the fundamental laws and particles of nature. In superstring theories, the basic constituent of matter is a 1-dimensional structure, called a string, rather than a point-particle structure. According to superstring theory, space has more than 3 dimensions. (See strings.)

symmetry   The property of being unchanged after some transformation. A square, for example, has a 4-sided rotational symmetry. It appears the same after it is rotated by 90 degrees.

symmetry breaking   The process by which an intrinsic symmetry of a system is disrupted. For example, a compass, in the absence of any outside magnetic field, has rotational symmetry and is equally likely to point in any direction. The magnetic field of the earth breaks the symmetry and causes the compass to point in a particular direction, toward the earth's north magnetic pole. In some cosmological models, the infant universe was much more symmetric than it is today. As the universe aged and cooled, some of these symmetries were permanently broken.

symmetry group   A set of mathematical transformations that represent a symmetry.

symmetry principle   A principle that requires a physical system to have a symmetry. For example, the notion that empty space should be devoid of any preferred directions and should thus be unchanged by rotations is a symmetry principle. A collection of compasses in empty space would be expected to point in all directions, reflecting this hypothesized rotational symmetry.

tensor   A mathematical object. A simple type of tensor is a vector, which may be thought of as an arrow in space. Three quantities are needed to describe a vector, namely its projected lengths in the three spatial dimensions. An example of a vector is the velocity of a particle; its total length represents the speed of the particle and its direction gives the direction of motion. More complex tensors require more than three quantities to specify them. The laws of physics are expressed with tensors.

thermal equilibrium   The condition of a system in which all its parts have exhanged heat and come to the same temperature. An isolated system in thermal equilibrium does not change over time. This is also a state of maximum disorder. (See blackbody radiation; entropy.)

thermodynamics   The study of how bodies change as they exert forces and exchange heat with other bodies.

3C 295   A radio source in a distant cluster of galaxies, about 5 billion light years away.

topology   The global property of a surface that indicates what other surfaces it can be deformed into without changing the number of interior holes. For example, a solid cylinder has the same topology as a sphere because a cylinder may be reshaped into a sphere without adding holes. A donut has a different topology than a sphere because it cannot be reshaped into a sphere without removing the hole in its middle. A donut has the same topology as a coffee cup with one handle.

Tully–Fisher relation   An observed relation between the intrinsic luminosity of a spiral galaxy and the rotational speed of its stars. More luminous galaxies have stars that are moving faster. (See galaxy.)

uncertainty principle   Also called the Heisenberg uncertainty principle, a fundamental result of quantum mechanics stating that the position and speed of a particle cannot be simultaneously known with complete certainty. If one is known with high certainty, the other becomes very uncertain. The product of uncertainty in position and uncertainty in momentum (mass multiplied by speed) is equal to a constant, Planck's constant. Since both initial position and initial speed are required to forecast the future position of a particle, the uncertainty principle prevents completely accurate predictions of the future from the past, even if all the laws of physics are known. (See Planck's constant; quantum mechanics.)

upsilon   A type of heavy lepton. (See leptons.)

vacuum   A state of minimum energy. Empty space is often referred to as the vacuum. Because of the uncertainty principle, even empty space has a minimum energy content.

velocity dispersion   See sigma.

velocity field   The velocities of a group of objects with different velocities at different positions of space.

Virgo infall   The observed gravitational motion of nearby galaxies toward the Virgo cluster of galaxies, about 50 million light years away. The Virgo cluster represents a strong concentration of mass, a strong departure from a uniform distribution of matter, and it therefore causes galaxies in its vicinity to deviate from the Hubble flow. (See Hubble law.)

virial theorem (or method)   In gravitational physics, a quantitative relationship between the amount of gravitational energy and the amount of kinetic (motional) energy of an isolated physical system in equilibrium. Thus, for such a system, only one of the two kinds of energy need be directly measured; the other can be inferred by use of the virial theorem. The universe as a whole is not in a state of equilibrium; its gravitational energy and kinetic energy of expansion are thus not required to obey the virial theorem.

voids   Large regions of space without galaxies.

W and Z particles   Particles that transmit the unified electromagnetic and weak nuclear forces. These particles were predicted by the Weinberg–Salam theory of the 1960s and later discovered in the 1980s. (See electroweak theory; Weinberg–Salam theory.)

wave function   The mathematical description of a physical system according to the laws of quantum mechanics. The wave function tells what possible states the physical system could be in and what is the probability of being in any particular state at any given moment.

weak interactions   Interactions between certain particles and caused by the weak nuclear force between them. (See nuclear force.)

Weinberg–Salam model   A theory in physics, developed by Steven Weinberg, Abdus Salam, and Sheldon Glashow, that unifies two fundamental forces of nature, the electromagnetic force and the weak nuclear force. This is the same as the electroweak theory.

Wheatstone bridge   A device that measures the resistance of an electrical circuit to the flow of electricity.

white dwarf   An extremely dense star that has burned out its nuclear fuel and is supported against collapse by the pressure of its electrons.

WIMP   Acronym for weakly interacting massive particles that might make up the missing mass. (See missing mass.)

wormhole   A bridge to another universe created by a black hole.

z   Notation for redshift. Objects of higher z are further away. A redshift of z near 0 corresponds to objects nearby; a redshift of $z = 1$ corresponds to objects at a distance of about 10 billion light years.

Zel'dovich spectrum   A particular prescription for how much clumping of matter should occur on each length scale. Specifically, the Zel'dovich spectrum proposes that the strength of inhomogeneity, of clumping, should be the same for each length scale at the moment when that length scale is equal to the size of the horizon. (See horizon.)

Zwicky compact galaxy   A type of very dense galaxy thought by Fritz Zwicky to be near the end point of galactic evolution.

ILLUSTRATION CREDITS

Photograph of Albert Einstein by Johan Hagemeyer, Bancroft Library, courtesy of the American Institute of Physics Niels Bohr Library.

Photograph of Wilhelm de Sitter, Yerkes Observatory Photograph.

Photograph of Georges Lemaître courtesy of Owen Gingerich.

Photograph of Henrietta Leavitt courtesy of the Harvard College Observatory.

Photograph of Alexander Friedmann courtesy of the Leningrad Physico-Technical Institute and the American Institute of Physics Niels Bohr Library.

Photograph of Edwin Hubble courtesy of The Observatories of the Carnegie Institution of Washington.

Photograph of the Sombrero galaxy courtesy of the National Optical Astronomy Observatories.

Wedge Diagram of the Center for Astrophysics Redshift Survey courtesy of V. de Lapparent, M. J. Geller, and J. P. Huchra, *Astrophysical Journal Letters* (1986).

Illustration of the rotation speeds of galaxies by Laszlo Meszoly.

Illustration of gravitational condensation by Laszlo Meszoly.

Photograph of the 200-inch Hale Telescope courtesy of Palomar Observatory.

Illustration of the cosmic time line by Laszlo Meszoly.

Illustration of the expansion of the early universe by Laszlo Meszoly.

Photograph of Fred Hoyle by Lotte Meitner-Graf.

Photograph of Allan Sandage by Douglas Cunningham, courtesy of The Observatories of the Carnegie Institution of Washington.

Photograph of Gérard de Vaucouleurs courtesy of the University of Texas.

Photograph of Maarten Schmidt by Robert Paz, courtesy of the California Institute of Technology.

Photograph of Wallace Sargent courtesy of the California Institute of Technology.

Photograph of Dennis Sciama by Godfrey Argent Studio, London, England.

Photograph of Martin Rees courtesy of Martin Rees.

Photograph of Robert Wagoner by Leo Holub.

Photograph of Joseph Silk courtesy of the Lawrence Berkeley Laboratory, University of California.

Photograph of Robert Dicke by Robert P. Matthews, Princeton University.

Photograph of James Peebles by Robert P. Matthews, Princeton University.

Photograph of Charles Misner by Cindy Grim, courtesy of Campus Photo Services, University of Maryland, College Park.

Photograph of James Gunn by Robert P. Matthews, Princeton University.

Photograph of Jeremiah Ostriker courtesy of Jeremiah Ostriker and Princeton University Department of Physics.

Photograph of Vera Rubin by Robert J. Rubin.

Photograph of Edwin Turner by Robert P. Matthews, Princeton University.

Photograph of Sandra Faber by Don Fukuda, courtesy of the University of California, Santa Cruz.

Photograph of Marc Davis courtesy of Marc Davis.

Photograph of Margaret Geller by Steven Seron, courtesy of the Smithsonian Astrophysical Observatory.

Photograph of John Huchra by Steven Seron, courtesy of the Smithsonian Astrophysical Observatory.

Photograph of Stephen Hawking by M. M. P., Cambridge, England.

Photograph of Don Page courtesy of William Jewell College.

Photograph of Roger Penrose courtesy of Roger Penrose.

Photograph of David Schramm © Patricia Evans 1990.

Photograph of Steven Weinberg courtesy of Steven Weinberg.

Photograph of Alan Guth by Alan Lightman.

Photograph of Andrei Linde courtesy of Andrei Linde.

Library of Congress Cataloging-in-Publication Data

Lightman, Alan P., 1948–
    Origins : the lives and worlds of modern cosmologists /
Alan Lightman and Roberta Brawer.
        p.   cm.
    Contains interviews with 27 persons.
    Includes bibliographical references and index.
    ISBN 0-674-64470-0 (alk. paper)
        1. Cosmology—Miscellanea. 2. Astronomers—Biography.
I. Brawer, Roberta. II. Title.
QB981.L54 1990
523.1—dc20

90-4623
CIP